Techniques and
Experiments for
Organic Chemistry

Techniques and Experiments for Organic Chemistry

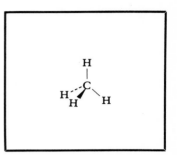

THIRD EDITION

Addison Ault
Cornell College

Allyn and Bacon, Inc.
Boston · London · Sydney · Toronto

Production Editor: Jane Dahl
Interior Designer: Nancy McJennett
Manufacturing Buyer: Karen Mason

Library of Congress Cataloging in Publication Data

Ault, Addison.
  Techniques and experiments for organic chemistry.

  Includes bibliographical references and index.
  1. Chemistry, Organic — Laboratory manuals.
I. Title.
QD261.A94   1979   547′.0028   78-23920
ISBN 0-205-06528-7

Printed in the United States of America.

Second printing . . . November, 1979

Contents

viii

Preface

This book is intended for use in the laboratory part of an introductory course in organic chemistry. The overall organization of this third edition is the same as it was in both the first and second editions. In Part I, I have given general descriptions of the theory and practice of many of the most common laboratory techniques of organic chemistry. I have also suggested a number of exercises to illustrate some of the techniques. For the experiments of Part II, I have included a variety of separations and preparations, and also a number of projects that involve organic synthesis.

The first sections of Part I contain discussions of laboratory safety, cleaning up, the laboratory notebook, the chemical literature, and a new section of the glassware used in the organic chemistry laboratory. Following this are discussions of a number of separation procedures including crystallization, distillation, extraction, and chromatographic methods; physical methods of identification including boiling point, melting point, and infrared and nuclear magnetic resonance spectrometry; and chemical methods of identification and characterization including qualitative tests for the elements, qualitative tests for the functional groups, and procedures for the preparation of derivatives. Finally, there are sections describing the apparatus and techniques for carrying out chemical reactions.

Part II presents first some experiments that involve only separation and purification, such as the isolation of cholesterol from gallstones, lactose from milk, isolation of the two enantiomeric forms of carvone (these enantiomers have different odors) and the resolution of α-phenylethylamine, then a number of one-step transformations, some multi-step syntheses, including the preparation of Δ^4-cholestene-3-one from cholesterol, sulfanilamide from benzene, and 1-bromo-3-chloro-5-iodobenzene from benzene, and finally a collection of synthesis projects. The synthesis projects include the preparation of flavors and fragrances such as coconut aldehyde, some unusual hydrocarbons, two cyclopropene derivatives, some insect pheromones, and a number of steroid transformations that lead ultimately to testosterone and progesterone.

All of the experiments of the second edition have been retained in this third edition. New experiments include the isolation of (R)-(+)-limonene from grapefruit or orange peel, the addition of dichlorocarbene to cyclohexene by phase-transfer catalysis, the preparation of isoamyl acetate (a component of the alarm pheromone of the honey bee), the preparation of cholesteryl benzoate (a substance that can behave as a "liquid crystal"), the preparation of triphenylcarbinol, the preparation of N,N-diethyl-m-toluamide (the insect repellent "Off"), the thiamine-catalyzed formation of benzoin from benzaldehyde, and N-benzyldihydronicotinamide, a model compound for the biochemical reducing agent NADH.

In my experience I have found that students enjoy lab work more when they understand why a procedure works and when they can make choices. I have, therefore, emphasized explanations and have often provided choices among similar experiments and, sometimes, between alternative procedures. I have also tried to make it possible to emphasize independent work such as qualitative organic analysis and the preparation of interesting substances.

During the preparation of this third edition I have enjoyed the assistance that many have given. Chemical Abstracts Services, the Division of Chemical Education of the American Chemical Society, and other publishers have been generous with their permission to reproduce copyrighted material, and I have had the benefit of comments from students and teachers who have used the first two editions. Helpful reviews for the third edition were provided by Professor Janet V. Hamilton, Tarrant County Junior College, Professor Bruce B. Jarvis, University of Maryland, Professor William B. Martin, Lake Forest College, Professor J. W. Timberlake, University of New Orleans, and Professor William N. Washburn, University of California at Berkeley. The editorial and production assistance of Allyn and Bacon, Inc., especially Jane Dahl, and the continuing interest and encouragement of my wife, Janet, and of my family were also very helpful during the preparation of this book.

Addison Ault

LABORATORY OPERATIONS

Preliminary Topics

Separation of Substances; Purification of Substances

Determination of Physical Properties

Determination of Chemical Properties; Qualitative Organic Analysis

Apparatus and Techniques for Chemical Reactions

Many of the practical aspects of the work that a person may do in the chemistry laboratory are discussed in Part I of this book. The first sections in this part include discussions of laboratory safety, organic laboratory glassware, the laboratory notebook, the literature of organic chemistry, and cleaning up. The remaining sections have been organized according to the following topics: (1) methods of separation and purification, (2) the determination of physical properties, (3) the determination of chemical properties, and (4) carrying out a chemical reaction.

The discussion of the theory and techniques of separation or purification includes filtration, recrystallization, distillation, steam distillation, sublimation, extraction, chromatography, and removal of water.

The sections on the determination of physical properties and their interpretation in terms of molecular structure include boiling point, melting point, density, index of refraction, optical rotation, molecular weight, and solubility characteristics, as well as the more recently developed spectrometric methods: infrared, ultraviolet-visible, nuclear magnetic resonance, and mass spectrometry.

The sections concerning the determination of chemical properties include qualitative tests for elements, qualitative tests for functional groups, and reactions for the formation of derivatives.

The sections on apparatus and techniques for chemical reactions include methods for heating and cooling, stirring, addition of reagents, and working up the reaction.

1

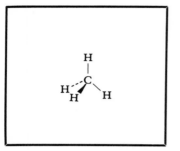

Preliminary Topics

The first six sections provide information and procedures that should be of use for all experiments in the organic lab. In order to avoid accidents, hazards must be recognized and certain precautions must be taken; the first section describes some of these dangers and precautions. The next section describes and illustrates some of the special items of glassware used in the organic laboratory. The third section gives some suggestions about cleaning up and disposal of waste. A fourth section describes the purposes of a laboratory notebook and recommends ways to keep such a notebook. The next section tells how to find certain kinds of information in the chemical literature, and the last section includes a periodic table of the elements with atomic weights, and some information about solutions of common acids and bases.

§1 SAFETY

Chemists must sometimes use hazardous materials, and therefore certain precautions must be regularly observed in order to minimize the probability and consequences of an accident.

1.1 FIRE

Know the location of: the safety shower, the fire blanket, and the fire extinguishers. Never work alone in the laboratory.

The danger of fire is the most obvious hazard in the organic chemistry laboratory. This is an unavoidable result of the fact that most of the liquids commonly used are relatively volatile and flammable (see Table 8.1). The danger and consequences of a fire may be decreased merely by minimizing the amount and size of the containers of flammable solvents stored in the laboratory.

The two most common reasons for a fire are

1. Boiling a flammable solvent with a flame and without a condenser.
2. Using a volatile and flammable solvent, especially during an extraction, without noticing that your neighbor's burner is on (or lighting your burner when your neighbor is using such a solvent).

If possible, then, all heating should be done with a steam bath, hotplate, heating mantle, or electric immersion heater rather than a burner. If a burner is used for heating a flammable solvent, it is absolutely essential that a condenser be used; otherwise, the vapors that escape from the neck of the flask will flow down to the flame and ignite. When working with the most volatile and flammable solvents such as carbon disulfide, ether, petroleum ether, or pentane, all flames on the bench must be extinguished, as the vapors of these solvents can travel over the desk top and be ignited by a distant burner. Solvents should never be distilled into the atmosphere, but should be condensed and collected in a receiver. *Flammable liquids*

If the vapors from a flask do ignite, the fire can often be extinguished by

1. Turning off the burner.
2. Gently placing a notebook or clipboard over the top of the vessel containing the burning solvent.

If solvent spilled on the desk top ignites,

1. Move bottles and flasks of unspilled solvent away, if possible.
2. Use a carbon dioxide fire extinguisher on the fire. The extinguisher is operated as follows: *Fire extinguisher*
   a. Pull out the pin that is held in place by a thin wire and lead seal.
   b. Aim the cone at the base of the fire from a few feet away.
   c. Release carbon dioxide snow by squeezing the hand grip or turning open the valve.

Do not use a fire extinguisher with wild abandon, as it is easy to make a fire worse by causing the blast from the nozzle to knock over and break bottles or flasks containing more flammable materials.

If your clothing is set on fire, move under the nearest safety shower (the location of which you should know well enough that you can get there with your eyes shut) and pull the chain to turn the shower on.

If someone else's clothing has been set on fire, assist him to the nearest safety shower and pull the chain. It is very easy for a person to panic under these conditions; the person must not be allowed to panic and run.

First aid for burns from flames, hot pieces of glass, hot iron rings, or hot-plates consists *only* of the application of cold water. Never apply any kind of dressing. Call the doctor if the burn is more than a very localized injury. *Only water on burns*

1.2 EXPLOSIONS

Always wear eye protection in the laboratory, preferably shatterproof goggles.

An explosion is an exothermic reaction that accelerates in rate until it gets out of control and shatters its container. The severity of the blast depends upon how much material is involved and how fast it all reacts.

Explosive mixtures are usually mixtures of oxidizing and reducing agents, as redox reactions are most likely to be highly exothermic. Gunpowder and many commercial explosives are mixtures of this type.

Explosive substances are (1) compounds that can undergo internal redox reactions, such as the polynitro compounds trinitrotoluene, picric acid (trinitrophenol), and nitroglycerine, or (2) compounds that can decompose to give very much more stable molecules, such as acetylene, nitrogen triiodide, diazonium salts, diazo compounds, peroxides, azides, and fulminates.

Use a safety shield

If you must work with a substance or mixture that is potentially explosive or known to be explosive, it is best to work on as small a scale as possible and behind a safety shield of shatterproof glass (hood fronts should be, but are not always, made of shatterproof glass). Just because a reaction has been run many times without incident by you or someone else does not mean that the potential danger has disappeared and that no explosion will occur the next time. New reactions should be run on a small scale, behind a safety shield, at least until their explosive potential has been estimated.

It is always a good idea to place a safety shield between yourself and the apparatus when carrying out a vacuum distillation. There is always the possibility that the flask will collapse under vacuum because of a crack or flaw in the glass. The safety shield will protect you from flying glass and the hot contents of the flask and oil bath.

Explosions are infrequent, because most people know about potentially explosive systems and try to avoid them. However, spectacular and tragic explosions continue to occur.

The three most common explosion hazards in the laboratory are

1. An exothermic reaction that gets out of control (explosion and fire).
2. Explosion of peroxide residues upon concentration of ethereal solutions to dryness (see Section 1.7).
3. Explosion upon heating, drying, distillation, or shock of unstable compounds (diazonium salts, diazo compounds, peroxides, polynitro compounds).

1.3 POISONING

Know the location of a chart indicating first-aid measures for various types of poisoning.

Almost anything is harmful in large enough doses, but it is easier to get a harmful dose of certain relatively dangerous materials in the chemical laboratory than in most other places.

There is absolutely no reason to be poisoned by mouth, as nothing should be eaten or drunk in the laboratory and pipetting should be done with a suction bulb or pipetter, not by mouth. Everyone, especially the fingernail biter, should wash his hands after working in the laboratory.

Poisoning through intact skin

Certain harmful materials can be absorbed relatively quickly through intact skin. These include dimethyl sulfate, nitrobenzene, aniline, phenol, and phenylhydrazine (see Section 1.7). Fatal doses of cyanide can be acquired through a cut in the skin. Therefore, you should not work with cyanide salts or solutions if you have a cut on the hand.

When harmful or flammable gases are used (and this includes almost all gases but oxygen, nitrogen, and the inert gases), they should be used in the hood. The harmful properties of certain specific gases are described in Section 1.7.

The most suitable first-aid treatment for poisoning depends to a very great extent upon the nature of the poison. Consult a chart, which should be posted in the laboratory or the stockroom. Call the doctor.

1.4 CUTS

Most cuts in the laboratory result from trying to force a thermometer or glass tube into a hole in a rubber or cork stopper. The thermometer or tube breaks and is driven into the palm of the hand or base of the fingers. Instead of forcing, you should ream out the hole with a round file or, in the case of a rubber stopper, lubricate the hole with glycerine. The excess glycerine can be washed off with water. The use of thermometer and tubing adapters with ground-glass-jointed glassware should greatly reduce the frequency of this type of accident.

Any time glass is forced, as in trying to unfreeze a joint, protect your hands by wrapping the glass in several thicknesses of a clean cloth towel.

First aid in the case of a cut consists first of removing any large, obvious pieces of glass and then stopping the bleeding. For venous bleeding, the cut can be pinched together or pressure can be applied with a gauze pad or clean towel. Arterial bleeding is much more dangerous and must be controlled additionally *Control of bleeding* by application of hand or thumb pressure to the appropriate pressure point. For the arms and hands, this is where the pulse can be felt at the wrist or inside the upper arm just below the armpit.

Cuts should be treated by a medical doctor.

1.5 SPILLS

In general, spills on your skin or clothing should be treated by washing with plenty of water, either in the sink or under the safety shower. Section 1.7 describes *Safety shower* treatment for spills involving a number of specific substances. Refer to this section after you wash thoroughly, or have someone look up the substance you have spilled while you wash. You should reread Section 1.7 before you work with any of the substances listed there.

Spills on the desk or floor should be treated as recommended in Section 1.7.

Both acids and bases should be neutralized with sodium bicarbonate. Never use strong acids or bases for this purpose.

1.6 CHEMICALS IN THE EYE

Always wear eye protection, preferably shatterproof goggles. Know the location of the eyewashing fountain.

If your eyes are protected by shatterproof goggles, you would have to be very unlucky indeed to get something in your eyes.

First aid for chemicals in the eyes is to wash them thoroughly (for 15 minutes or so), either by means of a special water fountain, which can direct a large but gentle flow of water into the eye, or with the eyewash dispenser. If these cannot be found immediately, use a gentle stream from a hose attached to a water faucet or a beaker of water. Contact lenses must be removed in order to wash the eyes. *Remove contact lenses* Be alert to the possibility that a person with something in his eyes may not be able to see well enough to get to the eyewash fountain and may have to be led there by someone else.

1.7 A SHORT LIST OF HAZARDOUS MATERIALS AND SOME OF THEIR PROPERTIES

A discussion of the dangerous properties of many substances can be found in *Dangerous Properties of Industrial Materials*, 4th edition, by N. I. Sax (New York: Van Nostrand Reinhold, 1975). If you plan to work with an unfamiliar substance, you should look up its dangerous properties in Chapter 12 of this book.

The purpose of the following list is to alert you to the greatest dangers presented by substances that are often found in the organic laboratory. The laboratory assistants and the instructor in charge should determine the dangerous properties of all substances present in the laboratory.

- *Carcinogens:* Benzidine, α-naphthylamine, β-naphthylamine, certain poly-nuclear hydrocarbons.
- *Chemicals that can be rapidly absorbed in fatal doses through* intact *skin:* Aniline, dimethyl sulfate, nitrobenzene, phenylhydrazine, phenol, 1,1,2,2-tetra-chloroethane.
- *Explosives:* Polynitro compounds such as picric acid, trinitrotoluene, tri-nitrobenzene, 2,4-dinitrophenylhydrazine.
- *Lachrymators and vesicants:* Benzylic halides, allylic halides, α-halocarbonyl compounds (such as the phenacyl halides), dicyclohexylcarbodiimide, iso-cyanates.
- *Volatile and flammable solvents:* Pentane, petroleum ether, diethyl ether, and carbon disulfide present the greatest fire hazard of the common solvents (see Section 1.1). Carbon disulfide vapors can spontaneously ignite at steam bath temperatures.
- *Acetic acid:* First aid for spills: dilute with large amounts of water.
- *Acetic anhydride:* Corrosive; will quickly blister the skin if not washed off. First aid for spills on the skin: wash off with water and finally with dilute ammonia solution.
- *Acetonitrile:* Poisonous on its own, but it can also produce hydrogen cyanide.
- *Acetyl chloride:* Reacts violently with water to produce HCl and acetic acid; see *Acid chlorides.*
- *Acid chlorides (and other acid halides):* Acetyl chloride, benzoyl chloride, benzenesulfonyl chloride, p-toluenesulfonyl chloride: corrosive, lachryma-tors. First aid for spills on the skin: wash with water and finally with dilute ammonia solution.
- *Acids:* Corrosive. First aid for spills: dilute with large amount of water, then wash with sodium bicarbonate solution.
- *Aluminum chloride:* Corrosive; reacts violently with water to produce hydrogen chloride.
- *Ammonia:* First aid for inhalation: fresh air; inhalation of steam. First aid for ammonia in the eyes: irrigation with water for 15 minutes.
- *Aniline:* Fatal doses can be absorbed through intact skin.
- *Benzene:* According to the *Merck Index,* chronic toxicity involves bone marrow depression and aplasia; rarely, leukemia. Under the O.S.H.A. emergency standard, effective May 21, 1977, benzene exposure should be no more than one part per million during an eight-hour day (a time-weighted average; 3.1 mg/cubic meter), and no more than 5 ppm during a fifteen-minute period. Benzene should be dispensed in the hood so that vapors and spills will not contaminate the lab.
- *Benzoyl peroxide:* Explosive; should not be heated for recrystallization or for melting-point determination; see *Peroxides.*
- *Bromine:* Exceedingly corrosive. Should be poured wearing gloves, face shield, and laboratory apron. Should be dispensed by means of a buret with a Teflon stopcock. First aid for spills on skin: wash *instantly* with water, rinse with ethanol, and rub in glycerine. First aid for inhalation: see *Chlorine.*
- *Carbon tetrachloride:* Poisoning by carbon tetrachloride can be accomplished by inhalation, ingestion, or absorption through the skin. O.S.H.A. regula-

tions, effective as of May, 1976, state that exposure to carbon tetrachloride vapors should be limited to 10 ppm (65 mg/cubic meter) during an eight-hour day.

- *Chlorine:* Exceedingly corrosive. Should be handled in the hood. First aid for inhalation: fresh air; inhalation of vapors from a very dilute solution of ammonia. First aid for spills on skin: see *Bromine*.

- *Chloroform:* O.S.H.A. regulations, effective as of May, 1976, state that exposure to chloroform vapors should be limited to 50 ppm (240 mg/cubic meter) during an eight-hour day.

- *Chlorosulfonic acid:* Exceedingly corrosive; reacts with water with explosive violence to form sulfuric acid and hydrogen chloride. First aid for spills on skin or clothing: wash with very large amounts of water (safety shower).

- *Dichloromethane (methylene chloride):* Dichloromethane vapors are said to be narcotic in high concentrations. O.S.H.A. regulations, effective as of May, 1976, state that exposure to methylene chloride vapors should be limited to 500 ppm (1740 mg/cubic meter) during an eight-hour day. Dichloromethane appears to be the least hazardous of the halogenated solvents.

- *Dimethyl sulfate:* Very dangerous; fatal doses can be absorbed quickly through intact skin. First aid for spills on skin: remove by washing with dilute ammonia solution; remove contaminated clothing.

- *Dimethyl sulfoxide:* Not particularly harmful in itself, but is rapidly absorbed through intact skin and can apparently aid the absorption of other materials through the skin.

- *Ethers:* Ethyl ether, isopropyl ether, tetrahydrofuran, dioxane. In addition to being highly flammable, ethers can also absorb and react with oxygen upon storage to form dangerously explosive peroxides. Ether which has not been stored in full, airtight, amber bottles should be routinely discarded within two months. Ether of unknown vintage should be treated with the respect due to an equal amount of an exceedingly unstable high explosive; if crystals can be observed in the ether or if it appears to contain a viscous layer, the bottle should not even be touched but should be disposed of by explosives experts. Large amounts of ether should never be concentrated unless the absence of peroxides has been experimentally verified. Ethereal solutions should never be concentrated to dryness by heating with a flame. Isopropyl ether seems to be especially treacherous with respect to peroxide formation.

  Test for presence of peroxides in ethers or hydrocarbons: Add 0.5–1 ml of the material to be tested to an equal volume of glacial acetic acid to which has been added about 100 mg of sodium or potassium iodide. A yellow color indicates a low concentration of peroxides, and a brown color a high concentration. A blank determination should be run.

  Removal of peroxides from ethers: Stir or shake the ether with portions of a solution prepared by dissolving 60 grams of ferrous sulfate and 6 ml of conc. sulfuric acid in 100 ml of water.

  For information concerning the formation of peroxides, detection and estimation of peroxides, inhibition of peroxide formation, and removal of peroxides, see N. V. Steere, *Handbook of Laboratory Safety*, 2nd edition (Cleveland, Ohio: The Chemical Rubber Company, 1971), pp. 190–194, and H. L. Jackson et al., *J. Chem. Educ.* **47**, A175 (1970).

- *Fuming nitric acid:* 95% nitric acid, containing oxides of nitrogen: extremely corrosive. First aid for spills: wash with large amounts of water and finally with sodium bicarbonate solution.

- *Fuming sulfuric acid:* Oleum; concentrated sulfuric acid containing dissolved sulfur trioxide: extremely corrosive. First aid for spills: wash with large amounts of water and finally with sodium bicarbonate solution.

- *Halogenated solvents:* Carbon tetrachloride, chloroform, etc. 1,1,2,2-Tetra-chloroethane is the most dangerous, as it can be absorbed rapidly through the skin. Avoid breathing the vapors of these solvents.
- *Hydrazine:* Explosive; dangerous in combination with oxidizing agents.
- *Hydrides:* Lithium aluminium hydride, sodium hydride: react instantly and explosively with water, liberating and possibly igniting hydrogen. Calcium hydride is slightly less vigorous in its reaction with water. Contact with water must be scrupulously avoided. Borohydrides react less vigorously with water, but rapidly with acidic solutions, to liberate hydrogen. Further information concerning the properties and procedures recommended for the safe handling of individual hydrides may be found in Reference 4.
- *Hydriodic acid:* First aid for spills: wash with large amounts of water and finally with dilute sodium bicarbonate solution.
- *Hydrobromic acid:* First aid for spills: wash with large amounts of water and finally with dilute sodium bicarbonate solution.
- *Hydrochloric acid:* First aid for spills: wash with large amounts of water and finally with dilute sodium bicarbonate solution.
- *Hydrogen bromide gas:* First aid for inhalation: fresh air; lie down and rest.
- *Hydrogen chloride gas:* First aid for inhalation: fresh air; lie down and rest.
- *Hydrogen peroxide:* Concentrated solutions can explode; powerful oxidizing agent; see *Peroxides.*
- *Hydrogen sulfide gas:* First aid for inhalation: fresh air; artificial respiration if necessary.
- *Lithium metal:* Reacts with water to produce hydrogen gas.
- *Lithium aluminum hydride:* Reacts instantly and explosively with water, liberating and possibly igniting hydrogen gas. Excess lithium aluminum hydride can be decomposed by dropwise addition of ethyl acetate. See *Hydrides.*
- *Methylene chloride:* see *Dichloromethane.*
- *Nitric acid:* Corrosive. First aid for spills: dilute with large amounts of water, then wash with sodium bicarbonate solution.
- *Nitrobenzene:* Fatal doses can be absorbed through intact skin.
- *Oleum:* See *Fuming sulfuric acid.*
- *Peracids:* Peracetic acid, trifluoroperacetic acid. Concentrated solutions can explode; powerful oxidizing agent; see *Peroxides.*
- *Perchloric acid:* Concentrated solutions can explode; powerful oxidizing agent; see *Peroxides.* For dangers in the use of perchloric acid and perchlorates, see N. V. Steere, *Handbook of Laboratory Safety,* 2nd edition (Cleveland, Ohio: The Chemical Rubber Company, 1971), pp. 205–216.
- *Peroxides:* All peroxides are potentially explosive, especially if heated. They are also powerful oxidizing agents, and mixtures with oxidizable materials are also potentially explosive. Ethers and other substances, such as tetralin, decalin, cumene, and certain other hydrocarbons, form peroxides when stored without exclusion of oxygen. Such liquids should not be distilled or concentrated without first determining that peroxides are not present; see *Ethers,* test for presence of peroxides.
- *Phenol:* Corrosive. Should not be handled with bare hands. Fatal doses can be absorbed through intact skin.
- *Phenylhydrazine:* Fatal doses can be absorbed through intact skin.
- *Phosphoric acid:* First aid for spills: wash with large amounts of water and finally with sodium bicarbonate solution.

- *Potassium metal:* Reacts instantly and explosively with water to form and ignite hydrogen gas. On storage without exclusion of oxygen, it forms exceedingly dangerous and explosive peroxides. If such material is possibly present, the sample should be disposed of by explosives experts. See *Sodium metal.*

- *Potassium cyanide:* Source of cyanide ion; fatal in small amounts; fatal doses can be ingested via a cut in the skin. Spills should be carefully cleaned up and disposed of immediately. Acidification of cyanide solutions will release deadly hydrogen cyanide gas.

- *Potassium hydroxide:* Caustic; corrosive solution. First aid for spills: wash with large amounts of water and finally with dilute bicarbonate solution.

- *Sodium metal:* Reacts instantly and explosively with water to form and usually ignite hydrogen gas. Sodium and potassium metals are usually stored under xylene. If oxide-free metal is required, transfer a piece of the metal to a mortar containing xylene and cut off the oxide coating with a knife. To weigh the metal, remove a piece from the xylene, briefly blot it with a piece of filter paper and add it to a tared beaker of xylene. For further details and information concerning the handling of sodium and potassium metals, see Reference 4. Scrap sodium should be disposed of by adding little pieces to a large volume of methanol; scrap potassium may be disposed of similarly, using *tert*-butanol.

- *Sodium amide:* Reacts violently with water. Aged samples have been reported to decompose explosively.

- *Sodium cyanide:* See *Potassium cyanide.*

- *Sodium hydride:* See *Hydrides.*

- *Sodium hydroxide:* Caustic; corrosive solution. First aid for spills: wash with large amounts of water and finally with dilute bicarbonate solution.

- *Sulfuric acid:* Corrosive. First aid for spills: dilute with large amounts of water, then wash with sodium bicarbonate solution.

- *1,1,2,2-Tetrachloroethane:* Fatal doses can be absorbed through intact skin.

- *Thionyl chloride:* Corrosive, volatile. First aid for spills: wash with large amounts of water and finally with dilute ammonia solution.

References

Sources from which one may obtain more information about laboratory safety in general, specific information about the storage and handling of dangerous substances, and hazardous aspects of specific chemicals include:

1. *Handbook of Laboratory Safety*, 2nd edition, N. V. Steere, editor, The Chemical Rubber Co., Cleveland, Ohio, 1971.
2. N. I. Sax, *Dangerous Properties of Industrial Materials*, 4th edition, Van Nostrand Reinhold, New York, 1975.
3. Wall-size chart: "Emergency Procedures for Dangerous Materials," Sargent-Welch Scientific Co., Catalog No. S-18812.

Information about properties and methods of handling of many compounds may be found in:

4. L. F. Fieser and M. Fieser, *Reagents for Organic Synthesis*, Wiley, New York; Volume 1, 1967; Volume 2, 1969; Volume 3, 1972; Volume 4, 1974; Volume 5, 1975; Volume 6, 1977.

Information about the toxicity of many compounds may be found in:

5. *The Merck Index*, 9th edition, Merck and Co., Inc., Rahway, New Jersey, 1976.

6. *The Toxic Substances List 1974 Edition*, U.S. Department of Health, Education, and Welfare, HEW Publication No. (NIOSH) 74-134.

Information about methods of disposal of many compounds may be found in:

7. *Aldrich Catalog Handbook of Fine Chemicals*, Aldrich Chemical Co., Inc., Milwaukee, Wisconsin, 1979–1980.

Questions

1. Draw a plan of your laboratory that shows
   a. the location of the safety shower, the fire blanket, and the fire extinguishers.
   b. the location of the eyewash fountain.
2. Describe where the "Poison Chart" is located.

§2 GLASSWARE USED IN THE ORGANIC CHEMISTRY LABORATORY

Organic chemists routinely use a variety of special items of glassware. Flasks with a round bottom and a narrow neck (*boiling flasks;* Figure 2.1) are used to contain a

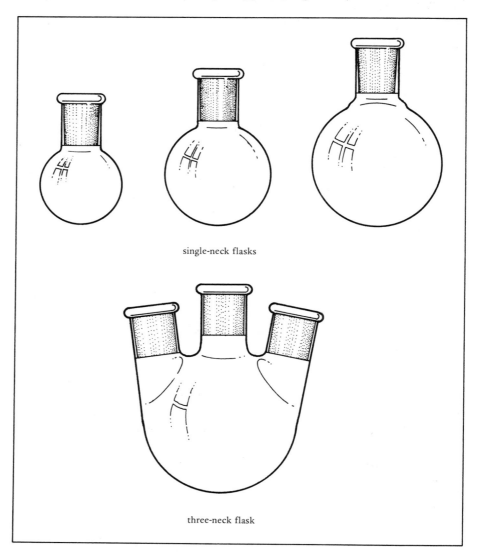

single-neck flasks

three-neck flask

Figure 2.1. Boiling flasks.

reaction mixture that is to be heated or boiled. In order to condense and return the vapors, the flask is usually fitted with a *reflux condenser* (Figure 2.2) as described in Section 30. Boiling flasks are made with multiple necks as well (Figure 2.1).

Flasks with a flat bottom and a narrow neck (*Erlenmeyer flasks;* Figure 2.3) are used for reaction mixtures that will be allowed to stand for a period of time, or to hold a solution that is to be set aside for crystallization (Section 8). The narrow neck retards evaporation and can accept a cork, rubber, or glass stopper.

Beakers (Figure 2.4) are used only as temporary containers or in a situation in which a flask with a narrow neck would be inconvenient.

Condensers and flasks can be connected together and fit with other items of equipment by means of various *adapters* (Figure 2.5). As examples, two common arrangements for distillation are shown in Figures 9.17 and 10.2.

A *separatory funnel* (Figure 2.6) is used to separate mixtures of immiscible liquids (Section 13.3) or it can be used as a dropping funnel to add a liquid to a reaction mixture, as shown in Figure 30.2.

Porcelain filtering funnels, either the Buchner or Hirsch style, are used with a suction filtration flask (Figure 2.7) in order to separate solids from liquids, as described in Section 8.

Other items of glassware commonly used by organic chemists include graduated cylinders for measuring liquids, distillation flasks (round bottom flasks with a long neck and a side-arm), and drying tubes (Figure 2.8).

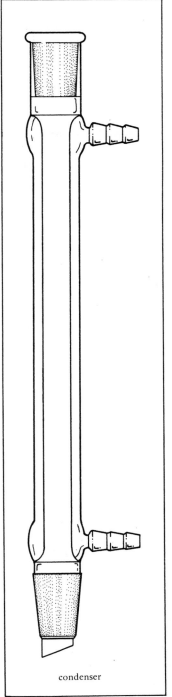

condenser

Figure 2.2. Condenser.

Figure 2.3. Erlenmeyer flasks.

§3 CLEANING UP

In chemistry, as in many things, cleaning up is a part of the job that seems to take a lot more time than it should. Some ways by which time and effort of cleaning may be minimized will be described in this section.

One way to expedite cleanup is to distinguish glassware that is merely wet with a volatile solvent, glassware that can be cleaned simply by rinsing, and glassware that is too dirty to be cleaned by rinsing. Glassware that is just wet with a

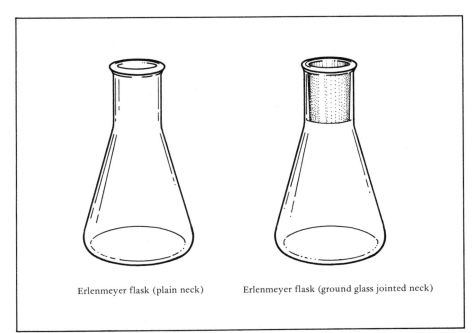

Erlenmeyer flask (plain neck)   Erlenmeyer flask (ground glass jointed neck)

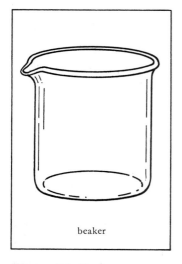

beaker

Figure 2.4. Beaker.

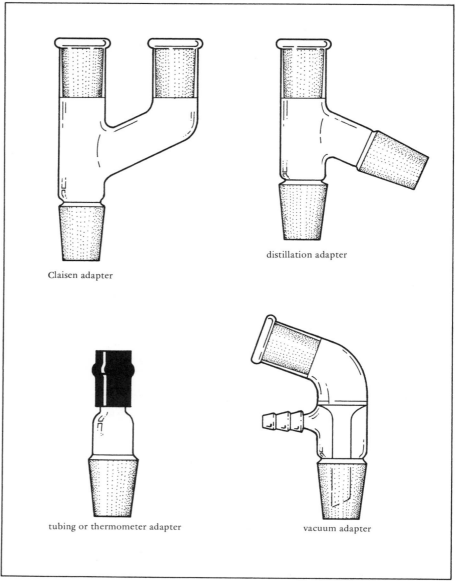

Claisen adapter

distillation adapter

tubing or thermometer adapter

vacuum adapter

Figure 2.5. Adapters.

volatile solvent does not need to be cleaned, but simply set aside on a towel spread out on the desk top and allowed to drain and dry. Glassware that can be cleaned by rinsing should be rinsed as soon as possible either with water if it is wet with an aqueous solution, or with an organic solvent such as technical acetone if it is wet with an organic solution. Once rinsed, this glassware is only wet with a volatile solvent and need merely be set aside to drain and dry. You can save a lot of work if you avoid scrubbing with soap and water any glassware that does not need to be cleaned in this way.

Glassware that will not come clean merely by rinsing—the flasks in which a reaction has been run or from which a distillation has been carried out are examples—can almost always be cleaned by first adding an organic solvent such as technical acetone (or a solvent appropriate to the nature of the material in the flask) and allowing the flask to stand with occasional swirling in order to dissolve the bulk of the material. After discarding the solvent (and repeating this procedure if necessary), any stubborn residue can be removed by scrubbing with technical acetone and a brush, or soapy water and a brush, or both in succession. It is very rarely necessary to use anything stronger than these to clean glassware.

Much cleaning of glassware can be done while waiting for something to filter or cool or warm up or crystallize or distill or react. Using odd moments to rinse, clean, and put away will save a lot of time at the end of the day. Also, the sooner something is cleaned after using, the easier it is to clean, as you should know from your dishwashing experience. This is especially true of items that cannot be cleaned inside with a brush, such as pipets or Buchner or Hirsch funnels. It should not be necessary to point out how short-sighted it is to put glassware away dirty to be cleaned before use!

3.1 CARE OF GROUND-GLASS-JOINTED GLASSWARE

Ground-glass-jointed glassware is very nice and most convenient, but it is expensive (about $3.00 per joint). All glassware can be broken; however, ground-glass-jointed glassware can also become useless if the pieces cannot be gotten apart

Clean up instead of watching

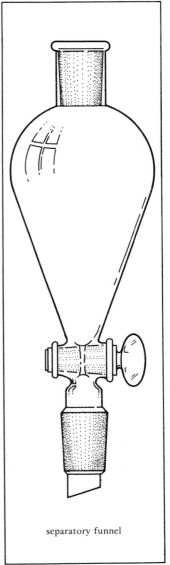

Figure 2.6. Separatory funnel.

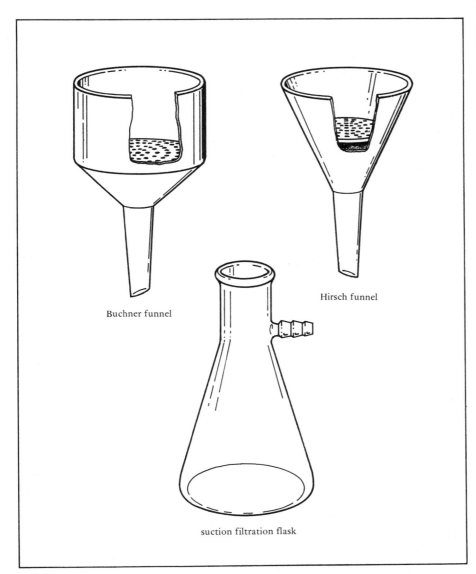

Buchner funnel

Hirsch funnel

suction filtration flask

separatory funnel

Figure 2.7. Porcelain filtering funnels and suction filtration flask.

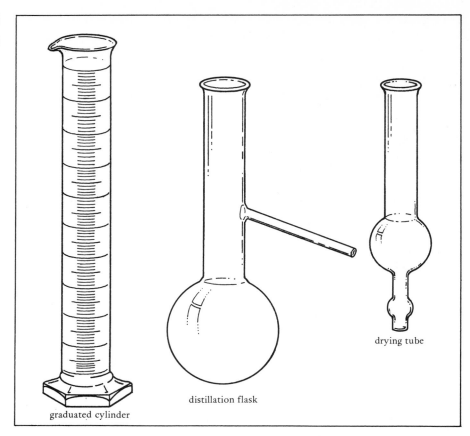

Figure 2.8. Other items of glassware.

graduated cylinder

distillation flask

drying tube

because the joints have "frozen." The best way to prevent freezing is to disassemble the apparatus immediately after use, before the apparatus cools if it is hot, and not by lubricating the joints (with the exceptions described in Section 30). In cases in which a lubricant has been used, the joints should be wiped as clean of grease as possible before washing so that the grease will not be spread over the rest of the piece. Hydrocarbon-based greases can be removed by rinsing with dichloromethane, but silicone greases resist removal almost completely. Dichloromethane or ether followed by soap and water have been recommended. Glassware that has been used with silicone-based greases can easily be recognized because it is not wet by water; the water stands up in beads on the surface.

3.2 SEPARATORY FUNNELS AND GLASSWARE WITH STOPCOCKS

Separatory funnels should be stored with the plug removed so that it cannot become frozen. Exceptions are separatory funnels with Teflon stopcocks, or those whose plugs have been removed, cleaned, and regreased just before storage. When rinsing a separatory funnel with a glass stopcock plug, the plug should be removed and the plug and barrel wiped free of grease unless the apparatus has been used only briefly and the grease has not been leached out. The plug should be regreased (Teflon plugs need no lubricant) by applying a thin stripe of lubricant along the plug on the two sides without holes. After the plug is inserted into the barrel, the grease is spread out evenly by rotating the plug a few times in the barrel. More than a minimum amount of grease will cause the holes in the plug to become partially or completely stopped up. Other items with stopcocks, such as burets and distilling heads, should be treated similarly.

If a stopcock plug becomes frozen, pullers are available, which will either pull out the plug or break the piece in the attempt. While the jack screw of the puller is being tightened, the item should be wrapped in towels to contain the fragments if the piece breaks.

It is sometimes possible to loosen frozen stopcocks or joints by very judicious heating and/or tapping; these techniques are best learned by watching an expert.

3.3 DRYING OF GLASSWARE

Glassware that has been rinsed with either water or acetone will dry upon storage. The outside of all items and the insides of beakers or other open items may be dried with a towel. If it is necessary to quickly dry an item of glassware like a graduate or flask for immediate reuse, a stream of air may be drawn through the inside by inserting to the bottom a piece of glass tubing connected to the hose of the aspirator. Acetone or other relatively volatile solvents will be removed from most items within a minute. A pipet can be connected directly to the aspirator. If the item is wet with water, it should be rinsed with technical acetone to remove the water and then dried as just described. Obviously, something to be used with water or an aqueous solution will not normally need to be dried of water before use! In many cases, it will be fastest and most satisfactory to rinse the item with a little of the solvent that will be used next.

Use it wet

3.4 DISPOSAL OF WASTE

Make use of the different containers provided for waste paper and broken glass.

Chemicals should be disposed of as instructed. Relatively harmless solids may be discarded along with other solid waste, but others may have to be decomposed before discarding, or washed down the sink drain. Organic liquids should not be poured down the drain but put into the containers provided for them. Aqueous solutions or water-soluble liquids may usually be disposed of down the sink drain along with a lot of water.

Radioactive wastes should be placed in the containers reserved for them.

The *Aldrich Catalog Handbook of Fine Chemicals* suggests methods to be used to dispose of various substances.

---

§4 THE LABORATORY NOTEBOOK

The lab notebook serves two purposes. First, it is a place to record and keep information that should be available in the lab while you are doing an experiment. Second, it is a place to write down and preserve both the description of the experiment as it was actually done and the results of the experiment.

The kind of information you need to have in the lab always includes a description of the procedure you plan to follow. When running a reaction you want also to know certain things about the materials you are working with—information such as molecular weight, melting and boiling points, density, and solubility properties. At some time a balanced equation should be written down and the suitability of the molar and gram amounts called for should be verified by reference to the balanced equation and the molecular weights. All of this should be written in the lab notebook so that at some later time it will be possible to discover the source of any mistakes that might have been made.

During the course of the experiment you try to record all relevant information about what was done and what happened. The easiest way to do this is to state how the actual experimental conditions and results differed from the conditions and results that were anticipated. Sometimes, for example, it is difficult or unnecessary to duplicate exactly certain of the conditions called for (conditions such as time and temperature, if indeed they are precisely specified), but the actual conditions should be recorded in the notebook. Similarly, the behavior of the reaction may not be as anticipated, and the variations should be written down. An adequate notebook will enable another person to know exactly what you did and what happened, and it should make it possible for your results to be verified. An adequate notebook is an essential part of good graduate school and industrial research.

A general lab notebook format might be as follows:

1. *Descriptive title and date.* This will serve to identify the experiment that is the subject of the notebook entry. If the notebook is the record of a research project, the different attempts to carry out a particular procedure should be distinguished by a notation such as "Run One," "Run Two" in the title. The record of the first experiment should start on page 3 or page 5 of the notebook so as to leave room to include a table of contents at the beginning of the book.

2. *Balanced equation.* If you are planning to run a reaction, the equation will show at a glance exactly what you are trying to do. It also shows the basis for the stoichiometric calculations.

3. *Molecular weights and molar, gram, and volume amounts of reagents, solvents, and products.* When an experiment is carefully planned, all of this information will be needed. It makes sense to record it in a well-organized and permanent way.

4. *Relevant physical properties of reagents, solvents, and products.* It is often necessary to know the melting point, boiling point, density, and solubility properties of the material with which you are working. In a recrystallization, for example, it is necessary to know which solvents might freeze at the temperature of the ice bath. In a distillation you need to know at what temperature the various components of the mixture may be expected to boil. In an extraction you need to know which solvent should contain each substance, and which solution should be the more or less dense. You must apply a certain amount of judgment as to what is relevant. For example, it should seldom be necessary to know the exact melting point or boiling point of solid sodium chloride or anhydrous magnesium sulfate. Similarly, while it makes sense to measure liquids by volume, solids are usually measured by weight and the corresponding densities and volumes are irrelevant.

Items 3 and 4 may be combined, perhaps into a comprehensive table. The necessary information concerning the physical properties can usually be found in one of the handbooks or dictionaries (see Section 5.2).

5. *A description of the experimental procedure.* If you plan to follow a procedure from the lab manual, it should be sufficient to refer to the page in the book where the procedure is given since you will have the manual in the lab with you. If you are following a procedure from a hand-out or from the chemical literature, you can tape or rubber-cement a copy in your notebook; be sure to include a reference to the source of the procedure. It is usually not permitted to bring a library book into the lab, where it might be damaged.

6. *Departures from a planned procedure.* Here, with reference to the planned procedure given in item 5, you tell what you actually did. Perhaps a slightly different amount was used than was called for. The actual time, temperature, and concentration should be recorded if they were different from those

called for in the procedure or if they were unspecified. Here is where you admit that something was spilled or boiled over. Don't forget that if you have an extraordinary loss the rest of the procedure must be scaled down accordingly.

7. *Results, including percent yield.* The outcome of the experiment should be described. If a substance has been prepared, you should report the amount obtained, both in grams and as a percent of what could theoretically be expected on the basis of the balanced equation. For example, if you prepare methyl benzoate from benzoic acid and methanol (Section 56.3), the balanced equation and the gram and molar amounts of starting materials are as follows:

$$
\text{benzoic acid} \quad + \quad CH_3OH \quad \longrightarrow \quad \text{methyl benzoate} \quad + \quad H_2O
$$

benzoic acid	methanol	methyl benzoate
12.2 g	19.7 g	
m.w. = 122 g/mole	m.w. = 32 g/mole	m.w. = 136 g/mole
0.1 mole	0.62 mole	

It should be clear that the limiting reagent in this case is benzoic acid and that the maximum possible (or theoretical) yield of methyl benzoate is 0.1 mole, which, for a molecular weight of 136 g/mole, would correspond to 13.6 grams. If 10 grams were actually obtained, the actual yield would be 10/13.6, or 74% of theoretical.

If you have determined any physical or chemical properties, such as color, melting point, boiling point, index of refraction, density, or specific rotation, these should be recorded along with theoretical or literature values. The sources of the literature values (lab manual, handbook, etc.) should also be recorded. If you determine the infrared spectrum of your product you may wish to fasten the spectrum into your notebook. If you are working on a research project it may be preferable to file the spectra separately. In this case you should make a notation in your book that a spectrum was obtained.

8. *Comments.* In case you or someone else might wish to repeat the experiment you should point out any difficulties you encountered and write down any suggestions for changes in the procedure.

Since a good lab notebook will *show all original data* (such as the weight of the paper upon which a substance was to be weighed and the weight of the paper plus the sample) and will *indicate all calculations* (such as the difference between the two weights just mentioned) the notebook tends to get messy. Yet the notebook will be the only record of your lab work and you should want it to be intelligible and an example of your best efforts. One way out of this dilemma is to use the left-hand pages for recording data and calculations in a preliminary form, if desired, and the right-hand pages for the final entries. Thus if the product of a reaction were being weighed as just described, the two weights and the subtraction could be recorded on the left-hand page, and the difference (the weight of the sample) would be recopied in the appropriate place on the right-hand page. Figure 4.1 illustrates a sample page from a laboratory notebook.

Cyclohexene from Cyclohexanol   10-27-78

In this experiment we will convert cyclohexanol to cyclohexene by heating the alcohol with 85% H_3PO_4:

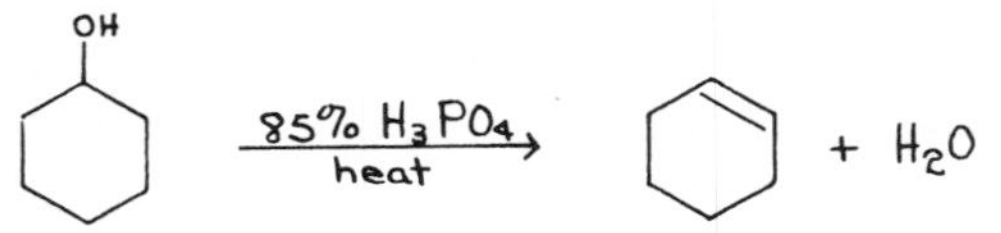

$$\text{cyclohexanol} \xrightarrow[\text{heat}]{85\% \ H_3PO_4} \text{cyclohexene} + H_2O$$

0.20 mole   0.20 mole
20.0 grams   16.4 grams

The product will be isolated by distillation.

| Substance | Amount | | | M.W. | density | m.p. | b.p. | solubility | |
	moles	grams	ml					H_2O	organics
cyclohexanol	0.20	20.0	21	100.2	0.96	25	161	no	yes
85% H_3PO_4			5	98.0		liquid	dec.	yes	no
xylene			20	106.2	mixture of isomers			no	yes
anh. $MgSO_4$		~0.5		120.4		high		yes	no
cyclohexene	0.20	16.4	20.2	82.2	0.81	-131	83	no	yes

Procedure: Ault: Techniques and Experiments for Organic Chemistry, 2nd ed., Holbrook Press, Boston (1976) p. 177

Departures from procedure: washed with 20 ml. of water instead of saturated NaCl solution.

Yield: I obtained 5.2 grams of product boiling between 82° and 84° C. Theoretical yield of cyclohexene: 16.4 g
Percent yield: $(5.2/16.4)(100) = 32\%$

Figure 4.1. Sample page from lab notebook.

Questions

1. Look up in a handbook the molecular weight, boiling point, and density of acetic anhydride.

2. Look up in a handbook the molecular weight and melting point of salicylic acid.

3. Aspirin can be made by treating salicylic acid with acetic anhydride according to the procedure described in Section 56.5. Assume the experiment calls for 10 grams of salicylic acid and 20 ml of acetic anhydride:

   a. Write a balanced equation for the reaction.

   b. Calculate the number of moles of salicylic acid and of acetic anhydride that will be put into the flask.

   c. Which of these substances will determine the maximum amount of product that will be produced? That is, which substance is the limiting reagent?

   d. What is the maximum amount (moles and grams) of aspirin that could be formed in the experiment?

e. If all the salicylic acid were converted to aspirin, how many moles of acetic acid would be formed, and how many moles of acetic anhydride would remain unused?

f. If 11.0 grams of aspirin were actually isolated after the reaction had taken place, what would have been the percent yield of aspirin?

---

§5 THE CHEMICAL LITERATURE

Journals

The results of chemical research are normally recorded in notes, communications, or papers in various chemical journals. If you wanted to look up the reported physical properties of a compound, or methods of analysis for a compound, methods of preparation or purification, you could search this primary literature through the index volumes of the various journals. Since there are several dozen major journals, and thousands of journals altogether, the time and effort required would be enormous.

Chemical Abstracts

To cope with this problem, abstracting journals have been established. Abstracting journals publish a brief summary, or abstract, of a paper and then provide detailed indexes to the abstracts. Today, *Chemical Abstracts* is the most important abstracting journal in the field of chemistry. *Chemical Abstracts* was started in 1907 by the Americal Chemical Society, and in 1976—to take a single year—it published abstracts of over 367,000 papers, reports, and patents. *Chemical Abstracts* currently produces five principal indexes: Author, General Subject, Chemical Substance, Formula, and Numerical Patent. An index entry refers you to the abstract, and if the abstract indicates that the desired information is in the original article or patent, you then look up the original by means of the reference provided with the abstract. A complete search of *Chemical Abstracts* requires that you look in the various cumulative indexes and, for recent volumes, the two semiannual indexes for each year. The use of *Chemical Abstracts* is outlined briefly in Section 5.1.

Secondary sources

While this is an improvement over looking things up in the indexes of individual journals, it is still a lot of work if all you want to know is the melting point of a compound or what would be a good solvent for recrystallization. For this reason, a great many specialized secondary sources have been developed. The most familiar is probably the *Handbook of Chemistry and Physics*, published by the Chemical Rubber Company, or N. A. Lange's *Handbook of Chemistry*. These and other secondary sources make certain kinds of information easy to obtain. For example, there are collections of methods of synthesis; methods of analysis; physical properties such as melting point, boiling point, solubility, vapor pressure, heat of combustion, and infrared, ultraviolet, NMR, and mass spectra; physiological properties; and hazardous properties. The key to success in quickly finding answers to specific questions about a compound is to be familiar with the secondary sources for that kind of information. Sections 5.2 and 5.3 describe the most commonly used secondary sources for information about physical properties and methods of preparation of organic compounds, and Section 5.4 lists several collections of infrared, ultraviolet, NMR, and mass spectra. A few other books that organic chemists find useful are listed in Section 5.5.

The three articles by J. E. H. Hancock provide a more complete introduction to the literature of organic chemistry (1).

5.1 CHEMICAL ABSTRACTS

Chemical Abstracts consists of print or microfilm abstracts of notes, papers, reviews, patents, etc., which have been published to report results of research and scholarship in the field of chemistry and related areas. The abstracts have been classified and indexed in order to make it as easy as possible to locate those abstracts that contain or refer to specific information. Without the indexes the abstracts themselves would be of very limited value. In addition to the indexes referred to above—Author, General Subject, Chemical Substance, Formula, Numerical Patent—there are the Index of Ring Systems, the Patent Concordance, and the Index Guides. An introduction to the Index Guide appears in Volume 76, pages 1I through 140I, and it provides descriptions and instructions for use of the various indexes. Supplements to the Index Guides are published annually.

The first job in searching the literature by means of *Chemical Abstracts* is to locate the relevant entries in the Subject or Chemical Substance Index. If you wish to find information about a specific compound the task is not particularly difficult, but it may take more than a few minutes. The initial step is to determine the name used to index the compound (the *index name*). An approach that usually works is to look up the substance according to its molecular formula in the Formula Index. All references to the compound in which you are interested will be made under the particular name employed as the index name by *CA* at that time. Then you look up the index name in the corresponding Subject or Chemical Substance Index, where the entries will be much more descriptive. The 9th edition of the *Merck Index* also provides the *CA* index name for most of the compounds listed.

Each entry in the index will refer to an abstract by means of two numbers. The first, in bold face, is the *CA* volume number. The second serves to locate the abstract within the volume. For entries from 1907 through 1933 the second number refers to the page, with a smaller superscript number as a suffix indicating the fractional distance down the page on a scale of 1–9, thus: $21:138^5$. For entries from 1934 through 1966 the second number refers to the column, since two columns have been printed per page from 1934 to the present. In addition, for entries from 1947 to 1966 a letter takes the place of the superscript number, and the scale of distance down the column runs from a–i. In 1967 a new system was adopted wherein the abstracts of each volume are consecutively numbered and reference is made to an abstract by volume number and, within the volume, by abstract number. The letter that follows these abstract numbers has no meaning except as a computer check character.

Occasionally the abstract itself contains the desired information, perhaps a melting point, but usually one has to take the final step of looking up the original article by means of the reference cited in the abstract. The reference will include the name of the author, an abbreviated title of the journal, the volume and page numbers, and the year of publication.

A search of all the volumes of *Chemical Abstracts*, from 1907 to the present, will require the use of several cumulative indexes and in addition the individual volume indexes for the most recent years. Table 5.1 presents a partial summary of the indexes available for *Chemical Abstracts*. In 1972, the Subject Index was divided into two parts: the Chemical Substance Index and the General Subject Index. The first contains references to chemical substances by name, and the second contains references to all other subjects.

5.2 SECONDARY SOURCES FOR PHYSICAL PROPERTIES OF ORGANIC COMPOUNDS

For physical properties such as molecular weight, melting point, density, index of refraction, color, and solubility, the following handbooks are very convenient:

· *Handbook of Chemistry and Physics*, 57th edition, R. C. Weast, editor, CRC Press, Cleveland, Ohio, 1976–1977. Contains physical properties for about

Table 5.1. Partial summary of indexes for *Chemical Abstracts*

Name	Years	Volumes	S	GS	CS	A	F	IG
Decennial Index	1907–1916	1–10	x			x		
Decennial Index	1917–1926	11–20	x			x		
Decennial Index	1927–1936	21–30	x			x		
Decennial Index	1937–1946	31–40	x			x		
Collective Formula Index	1920–1946	14–40					x	
Fifth Decennial Index	1947–1956	41–50	x			x	x	
Sixth Collective Index	1957–1961	51–55	x			x	x	
Seventh Collective Index	1962–1966	56–65 [a]	x			x	x	
Eighth Collective Index	1967–1971	66–75 [a]	x			x	x	x
Semiannual	1972	76, 77 [a]		x	x	x	x	x
Semiannual	1973	78, 79 [a]		x	x	x	x	x
Semiannual	1974	80, 81 [a]		x	x	x	x	x
Semiannual	1975	82, 83 [a]		x	x	x	x	x
Semiannual	1976	84, 85 [a]		x	x	x	x	x
Semiannual	1977	86, 87 [a]		x	x	x	x	x

S = Subject Index GS = General Subject Index CS = Chemical Substance Index
A = Author Index F = Formula Index IG = Index Guide

[a] Two volumes per year.

14,000 organic compounds in addition to much other information. A reference to Beilstein is given for almost every compound.

· *Lange's Handbook of Chemistry*, 11th edition, J. A. Dean, editor, McGraw-Hill, New York, 1974. Contains physical properties for about 6500 organic compounds in addition to much other information. A reference to Beilstein is given for each compound. It is much easier to locate the entry for a compound in the Lange's Handbook than in the *Handbook of Chemistry and Physics*.

· *Aldrich Catalog Handbook of Fine Chemicals*, Aldrich Chemical Co., Inc., Milwaukee, Wisconsin, 1979–1980. In addition to molecular weight and selected physical properties it gives methods for disposal, references to Beilstein and *The Merck Index*, and references to infrared and NMR spectra published in The Aldrich Library of Infrared Spectra and The Aldrich Library of NMR Spectra (see Section 5.4).

· *Handbook of Tables for Identification of Organic Compounds*, 3rd edition, The Chemical Rubber Co., Cleveland, Ohio, 1967. Contains melting-point and boiling-point data, and melting points of derivatives for over 8,000 organic compounds. The organization is by functional group and by increasing melting point or boiling point within each functional group.

Three other useful secondary sources are the following:

· *Dictionary of Organic Compounds* (Heilbron), 4th edition, Volumes 1–5, Oxford University Press, New York, 1965. This dictionary is an alphabetical listing of over 25,000 organic compounds. It contains valuable information about solvents used for recrystallization, as well as some reactions, derivatives, and literature references. An annual supplement is published.

· *The Merck Index*, 9th edition, Merck and Co., Inc., Rahway, New Jersey, 1976. This index contains information about solubility, purification, and hazardous properties as well as medicinal uses for about 10,000 compounds. The *Chemical Abstracts* index name is also given in most cases.

· *Beilsteins Handbuch Der Organischen Chemie*, 4th edition, Springer-Verlag, Berlin, 1918–present. This German language work is the most complete   *Beilstein*

How to find it in Beilstein

secondary source for information about the properties, preparation, and reactions of organic compounds. For all information, references are given to the primary literature sources. The fourth edition, in 31 volumes (Bände), was published during 1918–1938. This principal edition (Hauptwerk; H) covers the organic chemical literature from the beginning until 1909. Since 1938, there have been published a First Supplement (Erstes Ergänzungswerk; E I) covering from 1910 to 1919 in an organization parallel to that of the Hauptwerk, and a Second Supplement (Zweites Ergänzungswerk; E II) similarly covering from 1920 to 1929. The Hauptwerk and the first two supplements list every organic compound known through 1929. A Third Supplement (Drittes Ergänzungswerk; E III) covers the literature from 1930 through 1949, with many more recent references. The publication of a Fourth Supplement (Viertes Ergänzungswerk) was begun in 1972. This supplement covers the chemical literature from 1950 through 1959. Starting with Volume 17, published in 1974, the third and fourth supplements are being published together in common volumes which review the literature from 1930 through 1959.

There are four ways to find a compound in Beilstein. Possibly the easiest is to look it up in either *Lange's Handbook of Chemistry* or the *Aldrich Catalog Handbook of Fine Chemicals*. For example, *p*-bromoacetanilide is found in the Lange's Handbook under "Bromoacetanilide (*p*)"; the Beilstein reference is given as XII-642, which means that an entry for this compound is given on page 642 of Volume 12 of the Hauptwerk. In the Aldrich Catalog/Handbook the reference given under "*p*-Bromoacetanilide" is *Beil.* 12, 642, which provides the same information as the Lange's Handbook. The *Handbook of Chemistry and Physics* may also be used. Under the entry "Acetic acid, amide, N(4-bromophenyl)" the handbook gives the reference B12² 348, which means that an entry for *p*-bromoacetanilide can be found in Beilstein in Volume 12 of the Second Supplement on page 348. However, as the example indicates, it is sometimes difficult to find the entry in the *Handbook of Chemistry and Physics*.

A second method for finding a substance in Beilstein is to use the cumulative Formula Index (General-Formelregister) of the Second Supplement (Volume 29). For each entry, the names of all the isomers are given with references to the Hauptwerk, the Erstes Ergänzungswerk, and the Zweites Ergänzungswerk. For example, there are eighteen entries under C_8H_8BrNO. The fifteenth is "4-Brom-acetanilid **12** 642, I 319, II 348," which means that information about *p*-bromoacetanilide may be found in Volume 12 of the Hauptwerk on page 642, in Volume 12 of the Erstes Ergänzungswerk on page 319, and in Volume 12 of the Zweites Ergänzungswerk on page 348. This method requires a little understanding of the German nomenclature in order to recognize the name of the desired isomer.

A third way is to use the corresponding cumulative Subject Index (General Sachregister) of the Second Supplement (Volume 28). Since this requires a working knowledge of German nomenclature, it is relatively unsatisfactory for most Americans.

The fourth way requires some understanding of the organization of Beilstein. It sounds a little complicated, but after some familiarity with it, it is usually the fastest. Beilstein is divided into four major parts:

1. Acyclische Reihe (non-ring compounds) Vols. 1–4

2. Isocyclische Reihe (ring compounds; only carbon atoms in ring) Vols. 5–16

3. Heterocyclische Reihe (ring compounds; atoms other than carbon in ring) Vols. 17–27

4. Natural Products Vols. 30 and 31.

first two major divisions, compounds are listed in the following order of functioning classes:

1. Kohlenwasserstoffe (hydrocarbons)
2. Oxy-Verbindungen (alcohols)
3. Oxo-Verbindungen (aldehydes and ketones)
4. Carbonsäuren (carboxylic acids)
5. Sulfinsäuren (sulfinic acids)
6. Sulfonsäuren (sulfonic acids)
8. Amine
9. Hydroxylamine
10. Hydrazine
11. Azo-Verbindungen (azo compounds)

Numbers 7 and 12–28 are for other more unusual classes of compounds.

Polyfunctional compounds are found under the class that comes latest in the list: hydroxy acids are found under carboxylic acids, while amino acids are under amines (principle of latest position). Within each class, compounds appear in order of increasing unsaturation, and within these groups in order of increasing molecular weight.

Compounds that are not members of functioning classes are organized as follows:

1. Halogen, nitroso, nitro, and azido substitution products are found following the unsubstituted (or parent) compound. For example, the halo-, nitro-, and halo, nitrobenzenes appear after benzene.
2. Compounds that give members of the functioning classes upon hydrolysis are found under the last possible entry (principle of latest position). For example, methyl propionate is found under propionic acid, while N-methyl-propionamide is found under methylamine. Propionic anhydride, propion-amide, and propionitrile are found after propionic acid in that order. Also, methyl benzoate is found under benzoic acid, and phenyl acetate under phenol, as is anisole (methyl phenyl ether). Phenyl acetate is after anisole (principle of latest position again).

Once the entry is found in either the Hauptwerk or any Supplement, the corresponding entry in any other series can be found easily through a system of double numbering of pages. In each volume of each Supplement, in addition to the ordinary page numbers, a cross-reference page number appears at the top center of each page. These numbers refer to the corresponding pages in the Hauptwerk. Thus, having found a compound in the Hauptwerk (or having located the page where it would have been entered had it been known before 1909), one can find and search the corresponding pages of the same volume in each Supplement using these cross-reference page numbers. Similarly, after an entry is located in a volume of the supplement, the cross-reference page numbers can be used to search for corresponding entries in the Hauptwerk or other Supplements. Finally, direct cross-references are often given in the Second and Third Supplements. For example, at the beginning of the entry for *p*-bromoacetanilide in Volume 12, page 348, of the Second Supplement appears the notation (H 642; E I 319), which states that the corresponding entries appear (in Volume 12) on page 642 of the Hauptwerk and on page 319 of the First Supplement.

5.3 SECONDARY SOURCES FOR METHODS OF PREPARATION OF ORGANIC COMPOUNDS

Lab manuals

One of the most available and easily used sources of detailed directions for the preparation of specific compounds is the various laboratory manuals used in undergraduate courses in organic chemistry. Since these procedures are designed for use by students, a moderately competent organic chemist can usually make them work. Because most of the compounds whose preparation is described in a laboratory manual are available commercially, this source of information is of limited value. The most useful book of this type is:

- *Vogel's Textbook of Practical Organic Chemistry, Including Qualitative Organic Analysis*, 4th edition, revised by B. S. Furniss, A. J. Hannaford, V. Rogers, P. W. G. Smith, and A. R. Tatchell, Longman Inc., New York, 1978.

There are two useful collections of specific procedures for the preparation of individual compounds. The first (Shirley) is a single volume, and the second (*Organic Syntheses*) is a continuing series.

- *Preparation of Organic Intermediates*, D. A. Shirley, Wiley, New York, 1951. A collection of more than 500 preparations of compounds not available commercially at the time of publication.

Organic Syntheses
- *Organic Syntheses*, Wiley, New York. A continuing series of annual volumes, initiated in 1921. Each volume presents 30 to 35 detailed and tested preparations of compounds not available commercially at the time of publication. Five Collective Volumes have been published, which include, respectively, annual volumes 1–9, 10–19, . . . , 40–49. In addition to a general index, the Collective Volumes contain indexes for type of compound, type of reaction, molecular formula, apparatus, and author.

There are four reference works that describe general methods for the preparation of different classes of compounds. The first two (Wagner and Zook, and Migrdichian) are one- and two-volume works, respectively, and the last two (*Organic Reactions* and *Newer Methods*) are continuing series.

- *Synthetic Organic Chemistry*, R. B. Wagner and H. D. Zook, Wiley, New York, 1953. Each chapter summarizes known methods for the preparation of a type of organic compound. The tables at the end of each chapter list compounds of the type discussed in the chapter and indicate for each compound the methods used for its preparation, the yield, and references to procedures in the chemical literature.
- *Organic Synthesis*, V. Migrdichian, 2 volumes, Reinhold, New York, 1957. This work is similar to that of Wagner and Zook.

Organic Reactions
- *Organic Reactions*, Wiley, New York. A continuing, approximately biennial, series of volumes, initiated in 1943. A typical volume presents several thorough discussions of a type of reaction or the methods of preparation of a type of compound. Exhaustive tables are included, which summarize the applications of each reaction. The information given includes reaction conditions, yields, and references to the literature.
- *Newer Methods of Preparative Organic Chemistry*, W. Foerst, editor, Academic Press, New York. A series of occasional volumes. Each volume is a collection of review articles which originally appeared in *Angewandte Chemie*. Each article discusses a particular type of reaction and includes some specific preparations and many references to the literature.

A more general reference work that contains information and references concerning the preparation of a great many compounds is the multi-volume treatise edited by E. H. Rodd:

- *Chemistry of Carbon Compounds*, E. H. Rodd, editor, Elsevier, Amsterdam and New York, 1951. A systematic presentation of the preparation and properties of organic compounds. The overall organization is according to structural type. The last volume contains an extensive general index.

Finally, the most comprehensive and most generally useful reference work for the preparation of organic compounds, as well as their properties and reactions, is Beilstein:

- *Beilsteins Handbuch Der Organischen Chemie*, 4th edition, Springer-Verlag, Berlin, 1918–present.

The use of Beilstein has been described in the preceding section.

5.4 COLLECTIONS OF SPECTRA

- *The Aldrich Library of Infrared Spectra*, C. J. Pouchert, editor, Aldrich Chemical Co., Inc., Milwaukee, Wisconsin; first edition (8000 spectra), 1970; second edition (10,000 spectra), 1975.
- *High Resolution NMR Spectra Catalog*, Varian Associates, Palo Alto California; Volume 1, 1962; Volume 2, 1963; 700 spectra.
- *The Aldrich Library of NMR Spectra*, C. J. Pouchert and J. R. Campbell, editors, Aldrich Chemical Co., Inc., Milwaukee, Wisconsin; ten volumes plus index; 6000 spectra; 1974.
- *Sadtler Standard Spectra*, Sadtler Research Laboratories, Inc., Philadelphia, Pennsylvania. The Sadtler standard spectra are large and continuing collections of infrared, ultraviolet, and NMR spectra. They are available both in printed form and on microfilm. The largest collection, that of infrared spectra obtained with prism instruments, contains over 40,000 spectra. In addition there are special collections available (pharmaceuticals, commonly abused drugs, etc.) and commercial special collections (agricultural chemicals, food additives, etc.).

5.5 MISCELLANEOUS

Other useful books include the following:

- L. F. Fieser and M. Fieser, *Reagents for Organic Synthesis*, Wiley, New York; Volume 1, 1967; Volume 2, 1969; Volume 3, 1972; Volume 4, 1974; Volume 5, 1975; Volume 6, 1977. These volumes contain a wealth of interesting information about the availability, preparation, properties, and use of many substances in organic synthesis. The more recent volumes update and supplement the first volume.
- D. D. Perrin, W. L. F. Armarego, and D. R. Perrin, *Purification of Laboratory Chemicals*, Pergamon Press, New York, 1966. This book first offers a brief discussion of various methods of purification and then gives procedures for purification of individual organic, inorganic, and organometallic compounds.
- J. A. Riddick and W. B. Bunger, *Organic Solvents*, Volume II in *Techniques of Chemistry*, 3rd edition, A. Weissberger, editor, Wiley-Interscience, New

York, 1970. This volume contains physical properties and purification methods for 354 solvents.

- N. I. Sax, *Dangerous Properties of Industrial Materials*, 4th edition, Van Nostrand Reinhold, New York, 1975. This book provides discussions of methods of hazard control followed by specific hazard-analysis information for more than 13,000 common industrial and laboratory materials.

References

1. J. E. H. Hancock, "An Introduction to the Literature of Organic Chemistry," *J. Chem. Educ.* **45**, 193, 260, 336 (1968).

2. E. H. Huntress, *A Brief Introduction to the Use of Beilstein's Handbuch der Organischen Chemie*, 2nd edition, Wiley, New York, 1938.

3. O. Weissbach, *The Beilstein Guide*. A Manual for the Use of *Beilsteins Handbuch der Organischen Chemie*, Springer-Verlag, New York, Inc., 1976.

4. *Searching the Chemical Literature*, revised and enlarged edition, Advances in Chemistry Series #30, R. F. Gould, editor, American Chemical Society, Washington, D.C., 1961. A collection of papers on various topics concerning the chemical literature.

5. M. G. Mellon, *Chemical Publications*, 4th edition, McGraw-Hill, New York, 1965. An introduction to the nature and use of the chemical literature.

§6 TABLES

Table 6.1. Solutions of acids

Solution	Density (grams/ml)	Concentration (moles/liter) [a]	To make a liter of solution
95–98% H_2SO_4	1.84	18.1	Concentrated sulfuric acid
6 M H_2SO_4	1.34	6.0	332 ml conc. H_2SO_4 + 729 ml H_2O
3 M H_2SO_4	1.18	3.0	166 ml conc. H_2SO_4 + 875 ml H_2O
10% H_2SO_4	1.07	1.09	60 ml conc. H_2SO_4 + 957 ml H_2O
1 M H_2SO_4	1.06	1.0	55 ml conc. H_2SO_4 + 962 ml H_2O
69–71% HNO_3	1.42	15.7	Concentrated nitric acid
6 M HNO_3	1.19	6.0	382 ml conc. HNO_3 + 648 ml H_2O
3 M HNO_3	1.10	3.0	191 ml conc. HNO_3 + 831 ml H_2O
10% HNO_3	1.06	1.67	106 ml conc. HNO_3 + 905 ml H_2O
1 M HNO_3	1.03	1.0	64 ml conc. HNO_3 + 943 ml H_2O
36.6–38% HCl	1.18	12.0	Concentrated hydrochloric acid
6 M HCl	1.10	6.0	500 ml conc. HCl + 510 ml H_2O
3 M HCl	1.05	3.0	240 ml conc. HCl + 756 ml H_2O
10% HCl	1.05	2.9	242 ml conc. HCl + 763 ml H_2O
1 M HCl	1.02	1.0	83 ml conc. HCl + 920 ml H_2O
99.7% CH_3COOH	1.05	17.5	Glacial acetic acid
6 M CH_3COOH	1.04	6.0	343 ml glacial acetic acid + 682 ml H_2O
3 M CH_3COOH	1.03	3.0	171 ml glacial acetic acid + 845 ml H_2O
10% CH_3COOH	1.01	1.69	97 ml glacial acetic acid + 912 ml H_2O
1 M CH_3COOH	1.01	1.0	57 ml glacial acetic acid + 949 ml H_2O
48% HBr	1.50	8.9	Constant-boiling HBr
57% HI	1.70	7.6	Constant-boiling HI
85% H_3PO_4	1.70	14.7	Syrupy phosphoric acid
70% $HClO_4$	1.67	11.7	Concentrated perchloric acid

[a] Also mmole/ml.

Table 6.2. Molecular weights and molar volumes of acids

Compound	Molecular Weight	Molar Volume; ml
H_2SO_4, 98%	98.1	55
HNO_3, 70%	63.0	64
HCl, 37%	36.5	83
CH_3COOH	60.0	57
HBr, 48%	80.9	51
HI, 57%	127.9	132
H_3PO_4, 85%	98.0	68
$HClO_4$, 70%	100.5	86

Table 6.3. Molecular weights of bases

Compound	Molecular Weight
NaOH	40.0
KOH	56.1
NH_3	17.0
K_2CO_3	138.2
Na_2CO_3	106.0
$NaHCO_3$	84.0
$NaC_2H_3O_2$	82.0
$NaC_2H_3O_2 \cdot 3H_2O$	136.1

Table 6.4. Molecular weights, densities, and molar volumes of selected liquid reagents

Compound	Molecular Weight	Density	Molar Volume, ml
acetic anhydride	102	1.08	94
ammonium hydroxide, 28%	17	0.90	67
aniline	93	1.02	91
bromine	160	3.12	51
hydrazine, 64%	32	1.04	48
phosphorus oxychloride, $POCl_3$	153	1.68	91
phosphorus trichloride, PCl_3	137	1.58	87
pyridine	79	0.98	81
sodium hydroxide, 50%	40	1.53	52
thionyl chloride, $SOCl_2$	119	1.66	72

Table 6.5. Solutions of bases

Solution	Density (grams/ml)	Concentration (moles/liter)[a]	To make a liter of solution
50% NaOH	1.53	19.1	Concentrated NaOH solution
6 M NaOH	1.21	6.0	314 ml 50% NaOH + 734 ml H_2O
3 M NaOH	1.11	3.0	157 ml 50% NaOH + 873 ml H_2O
10% NaOH	1.11	2.77	145 ml 50% NaOH + 889 ml H_2O
1 M NaOH	1.04	1.0	52 ml 50% NaOH + 963 ml H_2O
45% KOH	1.45	11.7	Concentrated KOH solution
6 M KOH	1.26	6.0	512 ml 45% KOH + 512 ml H_2O
3 M KOH	1.14	3.0	256 ml 45% KOH + 764 ml H_2O
10% KOH	1.09	1.95	167 ml 45% KOH + 850 ml H_2O
1 M KOH	1.05	1.0	85 ml 45% KOH + 927 ml H_2O
28–30% NH_3	0.90	15.0	Concentrated ammonium hydroxide
6 M NH_3	0.95	6.0	400 ml conc. NH_4OH + 590 ml H_2O
10% NH_3	0.96	5.63	375 ml conc. NH_4OH + 620 ml H_2O
3 M NH_3	0.98	3.0	200 ml conc. NH_4OH + 797 ml H_2O
1 M NH_3	0.99	1.0	67 ml conc. NH_4OH + 933 ml H_2O
10% K_2CO_3	1.09	0.79	109 g anh. K_2CO_3 + 982 ml H_2O
5% K_2CO_3	1.04	0.38	52.2 g anh. K_2CO_3 + 992 ml H_2O
10% Na_2CO_3	1.10	1.04	52.5 g anh. Na_2CO_3 + 998 ml H_2O
5% Na_2CO_3	1.05	0.50	110.3 g anh. Na_2CO_3 + 993 ml H_2O
Saturated $NaHCO_3$	1.06	1.0	85 g $NaHCO_3$ + 973 ml H_2O
5% $NaHCO_3$	1.04	0.62	52.8 g $NaHCO_3$ + 983 ml H_2O

[a] Also mmole/ml.

Table 6.6. Periodic table of the elements

METALS NONMETALS

TRANSITION METALS

PERIODS	IA	IIA	IIIB	IVB	VB	VIB	VIIB	VIII	VIII	VIII	IB	IIB	IIIA	IVA	VA	VIA	VIIA	O
1	1.0079 H 1																1.0079 H 1	4.00260 He 2
2	6.94 Li 3	9.01218 Be 4											10.81 B 5	12.011 C 6	14.0067 N 7	15.9994 O 8	18.9984 F 9	20.179 Ne 10
3	22.9898 Na 11	24.305 Mg 12											26.9815 Al 13	28.086 Si 14	30.9738 P 15	32.06 S 16	35.453 Cl 17	39.948 Ar 18
4	39.098 K 19	40.08 Ca 20	44.9559 Sc 21	47.90 Ti 22	50.9414 V 23	51.996 Cr 24	54.9380 Mn 25	55.847 Fe 26	58.9332 Co 27	58.71 Ni 28	63.546 Cu 29	65.38 Zn 30	69.72 Ga 31	72.59 Ge 32	74.9216 As 33	78.96 Se 34	79.904 Br 35	83.80 Kr 36
5	85.4678 Rb 37	87.62 Sr 38	88.9059 Y 39	91.22 Zr 40	92.9064 Nb 41	95.94 Mo 42	98.9062 Tc 43	101.07 Ru 44	102.9055 Rh 45	106.4 Pd 46	107.868 Ag 47	112.40 Cd 48	114.82 In 49	118.69 Sn 50	121.75 Sb 51	127.60 Te 52	126.9046 I 53	131.30 Xe 54
6	132.9054 Cs 55	137.34 Ba 56	57–71 *	178.49 Hf 72	180.9479 Ta 73	183.85 W 74	186.2 Re 75	190.2 Os 76	192.22 Ir 77	195.09 Pt 78	196.9665 Au 79	200.59 Hg 80	204.37 Tl 81	207.2 Pb 82	208.9804 Bi 83	(210) Po 84	(210) At 85	(222) Rn 86
7	(223) Fr 87	(226.0254) Ra 88	89–103 †	104	105	106	107	108										

* LANTHANIDE SERIES	138.9055 La 57	140.12 Ce 58	140.9077 Pr 59	144.24 Nd 60	(145) Pm 61	150.4 Sm 62	151.96 Eu 63	157.25 Gd 64	158.9254 Tb 65	162.50 Dy 66	164.9304 Ho 67	167.26 Er 68	168.9342 Tm 69	173.04 Yb 70	174.97 Lu 71
† ACTINIDE SERIES	(227) Ac 89	232.0381 Th 90	231.0359 Pa 91	238.029 U 92	237.0482 Np 93	(242) Pu 94	(243) Am 95	(245) Cm 96	(245) Bk 97	(248) Cf 98	(253) Es 99	(254) Fm 100	(256) Md 101	(253) No 102	(257) Lr 103

Separation of Substances; Purification of Substances

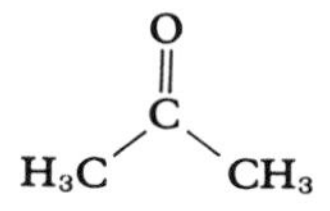

A pure substance contains only one kind of molecule; an impure substance is a mixture of molecules. When the various kinds of molecules of a mixture can be made to behave differently under the conditions of some procedure, the procedure may form the basis for a separation. The theory behind each of the separation procedures described in this section is presented from the point of view of the different behavior that may be expected of different molecules under the experimental conditions.

By the definition given above for a pure substance, purification procedures must be separation procedures. The definition also implies that the ultimate experimental criterion for purity is that a substance has been shown to be inseparable by all known separation procedures. Thus the possibility always remains that a substance currently believed to be pure may someday be shown to be separable into components. This situation is appreciated especially well by biochemists, who have repeatedly had the experience of finding that a new separation procedure shows that a substance considered pure is, in fact, a mixture. Separation of many biochemical materials is very difficult because often the molecules to be separated behave in nearly the same way in almost all separation procedures.

In practice, the question of the purity of a particular substance is usually settled indirectly. That is, the properties of a sample of the substance in question are compared with the properties of a sample that is judged to be pure because it could not be separated into components. If the two samples agree in all properties, they may be judged to be of the same purity—both pure or, sometimes, both impure. Some of the properties, for example boiling point, index of refraction, and infrared spectrum, and the ways in which they may be determined, are discussed in Part I, Determination of Physical Properties, Sections 16–26. If the comparison shows that the samples differ in some property, at least one is not pure.

As an example, suppose you isolate eugenol from cloves (Section 44). After purification of the sample, you determine its boiling point, index of refraction, and infrared spectrum. Comparison of these properties with those determined for a sample of eugenol obtained from a chemical supply house shows agreement within experimental precision. The strongest statement that can be made concerning the purity of your product is that you have not shown the composition of your material to be different from that of the sample obtained from the commercial source. It is possible that if both samples were obtained from the same source and purified using the same methods, both could be grossly contaminated by the same impurities. If each sample were prepared in a different way from different starting materials (so that the product mixtures would contain different impurities) and each were subjected to extensive purification by powerful separation procedures, your confidence in the degree of purity of both samples would be much greater. You would have to be very unlucky to have both samples contaminated to the same degree with the same impurities.

In most cases, however, the comparison with a reference sample is not a direct one like that just described. Usually, you simply compare your experimental values with values given in the chemical literature. To use the example of eugenol from cloves again, you would compare your experimentally determined boiling point and index of refraction with the literature values (which may not be self-consistent) and the infrared spectrum with a photoduplicated reference spectrum, if possible. In this kind of indirect comparison, you are usually more tolerant of discrepancies between your experimental values and the literature values, since it is unlikely that the experimental conditions were exactly the same. For example, two samples whose melting points differed by five degrees when compared directly and simultaneously could not be called identical. However, a five-degree discrepancy between an experimental melting point and a literature value would not usually in itself cause you to conclude that the samples were different. Factors such as different apparatus, different thermometer, different rates of heating, and different experimenters could easily account for the disagreement; the samples might really have been identical.

It should now be apparent that the question "Is this sample pure?" can never be answered with an unqualified Yes. The strongest statement that can be made regarding purity is "The sample was shown to be homogeneous by separation procedures A, B, and C," or "The sample had the same properties as a sample prepared by method X and purified by procedures Y and Z."

§7 FILTRATION

Filtration involves the separation of insoluble solid materials from a liquid. In this operation, the liquid passes through a porous barrier (sintered glass or filter paper) and the solid is retained by the barrier. The liquid may be made to pass through the barrier by gravity alone, in which case the procedure is called a *gravity filtration*. Alternatively, the liquid may be caused to pass through by a combination of gravity and air pressure. Such an operation is called a *vacuum* or *suction filtration*.

7.1 GRAVITY FILTRATION

A piece of filter paper and a conical glass funnel to support it are all that are required for gravity filtration. In order to maximize the rate at which the liquid may flow through the filter paper, the paper should be folded as indicated in Figure 7.1. The folded paper is then dropped into the funnel (see Figure 7.2).

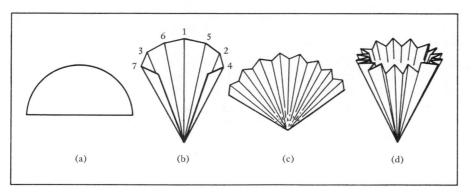

Figure 7.1. Folding of filter
paper for gravity filtration:
(a) Fold the filter paper circle
(11 cm diameter) in half.
(b) Crease the half to divide
it into eight equal pie-shaped
sections; it is easiest to make
the creases in the numerical
order shown.
(c) Turn the piece over and
pleat it into a fan by folding
each pie-shaped section in
half in the direction opposite
to the previous creases.
(d) Pull the two sides apart.

The funnel is best supported in an iron ring, as shown in the figure. The material to be filtered is poured into the filter paper cone, in portions if necessary. This type of operation is used in hot filtration during recrystallization (Section 8.4), in which case a stemless funnel is used, or in removal of a drying agent from a solution (Section 15.2).

7.2 VACUUM OR SUCTION FILTRATION

In vacuum or suction filtration, a partial vacuum is created below the filter, causing the air pressure on the surface of the liquid to increase the rate of flow through the filter. A typical apparatus is illustrated in Figure 7.3.

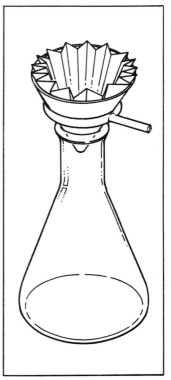

Figure 7.2. Arrangement of filter paper, funnel, and flask for gravity filtration.

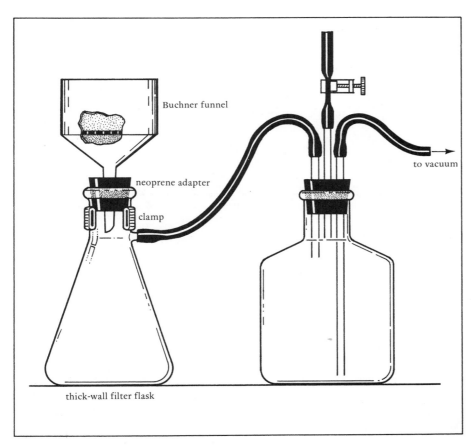

Figure 7.3. Apparatus for vacuum filtration.

Don't make this mistake

A circle of filter paper just large enough to cover the holes in the bottom of the Hirsch or Buchner funnel should be used. A common error is to try to use a piece of filter paper so large that it must be turned up at the edges. If this is done, it is almost impossible to create a vacuum in the suction flask. Not only will the filtration take much longer, but any material that flows over the edge of the filter paper will run down into the suction flask without being filtered.

Filtration is carried out by connecting the side arm of the suction flask to the source of vacuum, which is almost always the water aspirator. When a water aspirator is used, the flask should be connected to the aspirator through a trap, as shown in Figure 7.3. The trap prevents water from the aspirator from being sucked back into the filter flask. Turn the aspirator on just a little at first so as to create a gentle vacuum, wet the filter paper with a small portion of the same solvent used in the solution being filtered while making sure that the paper is being pushed down over the holes, and pour the mixture to be filtered onto the center of the paper. Once the mixture has been added, the vacuum may be increased. When using the water aspirator, be sure to break the vacuum by disconnecting the tubing attached to the side arm of the filter flask before turning off the water. This procedure of suction filtration is used to collect a solid after recrystallization.

---

§8 RECRYSTALLIZATION

Purification of a solid by recrystallization from a solvent depends upon the fact that different substances are soluble to differing extents in various solvents. In the simplest case, all the unwanted materials are much more soluble than the desired compound. In this case, the sample is dissolved in the hot solvent to form a saturated solution, the solution is cooled, and the crystals, which will have separated upon cooling, are collected by suction filtration (Section 7.2). The soluble impurities remain in solution after cooling and pass through the filter paper with the solvent upon suction filtration.

If insoluble impurities are present in the sample they should be removed by filtering the hot solution by gravity (Section 7.1) before it is allowed to cool.

These procedures are summarized in the flow chart on the facing page.

8.1 CHOICE OF SOLVENT

The choice of solvent is crucial in purification by recrystallization, but there is no easy way to know which solvent will work best. If you wish to recrystallize a known compound, the chemical literature (perhaps a lab manual) may report or recommend the use of certain solvents. However, these solvents may not be useful if different impurities are involved. Quite often, several solvents must be tried.

Test the solvent

One essential characteristic of a useful solvent is that the desired compound must be considerably more soluble in the solvent when it is hot than when it is cold. A way of testing whether or not this requirement is met is to add as much of the compound as will cover the tip of a spatula (100 mg) to a small test tube, add a few drops of solvent, and (if the substance does not dissolve in the cold solvent) heat the mixture to boiling on the steam bath or over a low burner flame. If the material does not go into solution at this point, more solvent may be added, a little at a time, with continued heating, until it does. Then the solution is cooled by placing the test tube in a beaker of cold water to see whether the compound will crystallize out.

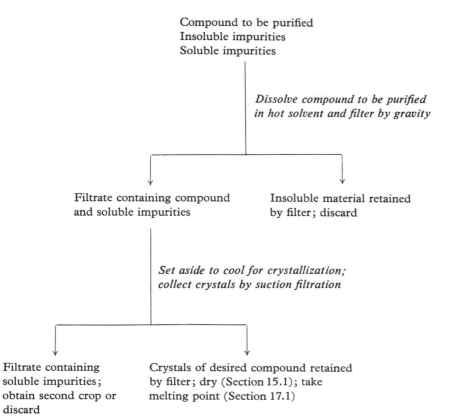

Compound to be purified
Insoluble impurities
Soluble impurities

*Dissolve compound to be purified
in hot solvent and filter by gravity*

Filtrate containing compound
and soluble impurities

Insoluble material retained
by filter; discard

*Set aside to cool for crystallization;
collect crystals by suction filtration*

Filtrate containing
soluble impurities;
obtain second crop or
discard

Crystals of desired compound retained
by filter; dry (Section 15.1); take
melting point (Section 17.1)

There are several points at which you may decide that a certain solvent is not suitable for use in recrystallizing a given compound. If a very small amount of solvent (say 1 ml per gram) serves to dissolve the compound when it is cold, the amount of solvent that can be used will be so small that at the point of suction filtration you will be working with a damp mush rather than a suspension of crystals. When the compound dissolves this readily in the cold solvent, it will not be possible to use fresh cold solvent either to help in transferring the mush to the funnel for suction filtration or to wash the crystals free of the remaining solution containing the impurities, without involving unacceptable losses.

If a very large amount of solvent is required to dissolve the compound (say 100 ml per gram), you may not have large enough flasks and may not be able to afford the amount of solvent required. If only 1 gram needs to be recrystallized, there is no problem, but 100 grams would require 10 liters of solvent!

Finally, of course, the solvent might have to be rejected because the compound is not much more soluble in the hot solvent than in the cold solvent.

The relatively fast, qualitative test just described will be satisfactory in many cases. Sometimes you will want to be more quantitative and to weigh the amounts of compound used and recovered, as well as to measure the amount of solvent used. In order to obtain a valid measurement of the amount of solvent used, you will have to fit the flask or test tube used for the trial recrystallization with a reflux condenser that will condense the vapors of the boiling solvent and return them to the flask.

With experience, you will find that this quick test of the suitability of a solvent for recrystallization often works satisfactorily. But at first, it may give you misleading indications. There are several situations you should watch out for. If the compound dissolves only slowly in the boiling solvent, you may prematurely reject the solvent as incapable of dissolving the compound. Or you may decide that more solvent will be required than is actually necessary. As a result, you will

Don't be fooled

add too much solvent; your estimate of the amount of solvent needed will be too big, and your estimate of the percent of compound that can be recovered on cooling will be too small, since the additional solvent also will dissolve more compound when it is cold. As an extreme case, you may use so much extra solvent that the solution does not become saturated on cooling and nothing is recovered.

Supersaturation

Sometimes a compound starts to separate from a cooled solution only after a considerable length of time. When this occurs, the solution is said to be *supersaturated*. In this case, it is possible to mistakenly decide that the substance is as soluble in the cold solvent as in the hot solvent, or that too much solvent was used to dissolve the sample. Since supersaturation occurs when the process of forming new crystal nuclei is unusually slow, it can be relieved by adding a powdered crystal of the substance (seed crystals) or by scratching the walls of the flask below the surface of the solution with a stirring rod. This is thought to relieve supersaturation by producing burrs and bits of glass that, by chance, may act as points of crystal growth.

Seeding; scratching

Although crystallization is sometimes essentially complete in a few minutes and often within a half hour, it *can* take several hours or even several days. In such cases it is easy to mistakenly decide that only a small percent of the sample will ever separate from solution, simply because you did not wait long enough for crystallization to become complete.

The mistakes that can be made in estimating the suitability of a solvent can be summarized by saying that you mistakenly assume that equilibrium has been established when actually it has not. Either not all that will dissolve has yet dissolved, or not all that will crystallize has yet crystallized.

So far, it has been assumed that the sample of the compound being used to determine the suitability of a solvent for recrystallization is relatively pure. If the sample contains appreciable amounts of "insoluble" impurities, another problem arises. At some point in your attempt to dissolve the sample in the hot solvent, the residue will be the insoluble impurity, not the desired compound. What you want to do is stop at this point, separate the solution from the residue, and allow the solution to cool. The problem is to distinguish this situation from the one in which the residue is composed of both the desired compound and the impurity, or is entirely the desired compound. A careful observer may be able to detect a different appearance of the residue, or a slowing in the process of dissolution since the more soluble material is already in solution and the addition of more solvent does not dissolve as much material now as a similar amount did earlier. If this situation is not recognized, you may succeed only in recrystallizing the relatively insoluble impurity.

*Problem of insoluble
impurities*

Probably the best approach in such a situation is to use sufficient solvent to dissolve the desired compound and to leave as residue what appears to be the insoluble impurity, then to separate the solution from the residue (1) by allowing the latter to settle and decanting the supernatant solution, (2) by removing the solution with a medicine dropper, or (3) by carrying out a gravity filtration on the hot mixture. The properties of the residue (melting point, infrared spectrum) can be compared with those of the original sample.

Even if a solvent is found, which in convenient amounts (say 5–10 ml per gram) quickly dissolves the compound when hot, and from which the compound separates quickly on cooling as nice crystals with 80–90% recovery, the solvent still may not be suitable for the purification of a particular sample. It may be that certain impurities are neither much more soluble nor much less soluble than the desired substance; they neither remain in solution upon cooling nor remain "entirely undissolved" in the hot solvent so that they can be removed by gravity filtration of the hot solution. Instead, there can be intermediate situations in which the relative solubilities of the desired compound and an impurity are such that it is only the impurity that can be purified by recrystallization. When this occurs,

Table 8.1. Properties of common solvents

35

SECTION 8

Solvent	Boiling Point (°C)	Density (g/ml)	Solubility in Water	Flammability
Acetic acid	118[a]	1.05	∞[e]	+
Acetone	57	0.79	∞	+ + +
Acetonitrile	80	0.79	∞	+ +
Benzene	80[b]	0.88	i[f]	+ + +
Carbon disulfide	45	1.26	i	+ + + +
Carbon tetrachloride	77	1.59	i	−
Chloroform	61	1.49	i	−
Cyclohexane	81	0.78	i	+ + +
Dichloromethane	41	1.34	i	−
Dimethylformamide	153	0.94	∞	+
Dimethylsulfoxide	189[c]	1.10	∞	+
Dioxane	101[d]	1.03	∞	+ + +
Ethanol, 95%	78	0.81	∞	+ +
Ether, diethyl	35	0.71	7[g]	+ + + +
Ethyl acetate	77	0.90	8[g]	+ + +
Hexane	68	0.66	i	+ + +
Ligroin	60–90	0.67	i	+ + +
Methanol	65	0.79	∞	+
Pentane	36	0.63	i	+ + + +
Petroleum ether	30–60	0.64	i	+ + + +
Tetrachloroethane	146	1.60	i	−
Tetrahydrofuran	65	0.89	∞	+ + +
Toluene	111	0.87	i	+ +
Water	100	1.00	∞	−

[a] Freezes at 17°C
[b] Freezes at 5°C
[c] Freezes at 18°C
[d] Freezes at 12°C
[e] Miscible in all proportions
[f] "Insoluble"
[g] Grams per 100 ml

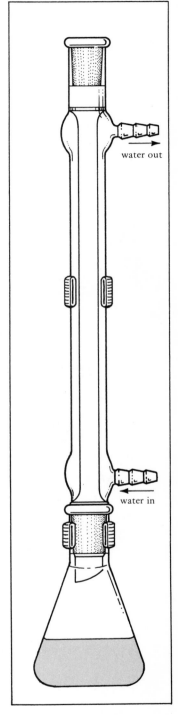

Figure 8.1. Apparatus for dissolving a sample for recrystallization.

another solvent must be used or this particular impurity must be removed in another way.

If you are working with a new compound or if you can find no recommendation for the use of a particular solvent in the recrystallization of a compound, you must simply test various solvents. A rough guiding principle is that solvents of molecular structure similar to that of the compound may be good solvents for that compound.

In some cases, mixtures of solvents may be necessary or useful (Section 8.9). Table 8.1 lists certain properties of solvents commonly used for crystallization.

8.2 DISSOLVING THE SAMPLE

After a solvent has been chosen, either through a recommendation in the literature or by the test procedure just described, a solution of the sample in the hot solvent must be prepared. Figure 8.1 shows a typical apparatus.

Choose an Erlenmeyer (conical) flask of such a size that it will be less than half filled with solution. Place the solid in the flask, add about 75 percent of the amount of solvent thought to be required, fit the flask with a reflux condenser, and bring the mixture to a boil. In the case of solvents boiling below 95°C, use the steam bath; for higher-boiling solvents, a burner or an oil bath is appropriate (Section 32.1). If a little more solvent is needed, add it down the condenser (if a flammable solvent is being heated with a burner, extinguish the burner before adding the solvent). Use of a boiling stone is recommended (Section 9.6).

If the amount of solvent needed is not specified or has been determined only roughly, place only a portion of the sample in the flask (for example, 1.0 gram in a 125-ml Erlenmeyer flask) and add measured amounts of solvent down the condenser. From the amount of solvent required to dissolve this first portion, the amount needed for the whole sample can be calculated. If the total will fill the flask much more than half full, transfer the first solution to a larger flask of appropriate size and prepare the rest of the solution in the larger flask. If no transfer is necessary, add the remainder of the sample (remove the flask from the heat and remove the condenser from the flask) followed by the rest of the solvent, using the solvent to wash the solid down from the neck of the flask as necessary.

If all the sample is added to the flask at once and then it turns out that more solvent is required to dissolve the sample than will fit in the flask, a very, very messy transfer will be necessary in which much material can be lost.

You should keep in mind the possibilities that the compound may dissolve only slowly in the boiling solvent and that "insoluble" impurities may be present in your sample.

8.3 DECOLORIZING THE SOLUTION

The presence of colored impurities in a sample or in a solution of a colorless compound is obvious. Sometimes during recrystallization the colored substances remain in solution upon cooling and are removed with the solvent upon suction filtration. Often the colored substances are adsorbed by the crystals as they are formed, giving an obviously impure product. Since the types of molecules that impart the color to the solution are often the types that are preferentially adsorbed by activated charcoal (decolorizing carbon),* the addition of a small amount of activated charcoal (1–2% by weight of the sample; or 1 mg per milliliter of solution; or enough to cover the tip of a spatula in 50 ml of solution) followed by gravity filtration of the hot solution can serve to remove some if not most of the color. The molecules responsible for the color will be adsorbed by the carbon and will be separated along with it during the filtration.

*Warning: cool below b.p.
before adding carbon*

The carbon should be added only after the solution has cooled a bit below the boiling point; otherwise, the solution will boil over when the addition is made. After the addition, the solution may be reheated to boiling. In general, it is quite all right to use decolorizing carbon in any recrystallization unless it is obvious from the lack of color of the solution that this step is not necessary.

Activated charcoal can also serve to adsorb small amounts of resinous material or very finely divided solid impurities that might not otherwise be removed by filtration.

Adsorption is not as efficient at higher temperatures, and sometimes it may be preferable to prepare an ethereal solution of the substance at room temperature, treat this solution with activated charcoal, filter it by gravity, and remove the ether by evaporation. This process is very similar to that of chromatography (Section 14), and the principle is just the same.

8.4 HOT FILTRATION

When the desired substance is in solution in the hot solvent, insoluble impurities (including dust, pieces of filter paper, glass, and cork) and decolorizing carbon,

---

* Activated carbon prepared from wood is desirable, having a surface area of hundreds of square meters per gram. Animal charcoal (bone black) contains a large proportion of inorganic salts and has less adsorptive power. Excellent decolorizing carbons are sold under the trade names of Norit and Nuchar.

if used, can be separated by filtering the hot solution by gravity. Vacuum filtration cannot be used, because the reduced pressure in the suction flask will cause the filtrate to boil and material in solution will be deposited over a large area.

The main problem in hot filtration is that the hot solution cools a little before it runs through the filter. This means that some crystallization can take place in the filter. You can try to avoid this undesired crystallization by warming the funnel (with steam, quickly wiping it dry with a towel; or with a flame) and pouring only a little of the solution into the filter at a time, keeping the remainder at the boiling point. If a large volume of solution must be filtered, or if the solubility of the substance decreases greatly in the range just below the boiling point, a heated funnel is a necessity. Often it is a good idea to use a small excess (10–25%) of solvent so that the solution will not become saturated until the temperature has fallen somewhat below the boiling point. A stemless funnel is always recommended: with a stemless funnel, you can avoid the problem of crystallization, and subsequent clogging, in the stem.

Use a stemless funnel

Although usually the filtrate should be collected directly in the Erlenmeyer flask in which the cooling for crystallization is to take place, an arrangement such as that illustrated in Figure 8.2 can be very useful in cases where there is trouble with crystallization of the compound in the filter. The filtrate in the beaker can be boiled up around the funnel, keeping it hot. Some care must be used with a flammable solvent if the heating must be done with a flame. After filtration is complete, the filtrate can be transferred to an Erlenmeyer flask for crystallization.

If an excess of solvent has been used to minimize the problem of crystallization in the filter paper, it may be removed at this point by distillation. This will also serve to bring back into solution any crystals that may have separated during filtration.

In all cases, the filtrate should be reheated, if necessary, to dissolve any crystals that may have formed and make a clear solution. Crystals that form when the hot filtrate hits the cold flask are likely to be less pure than those that separate more slowly from solution.

Reheat filtrate

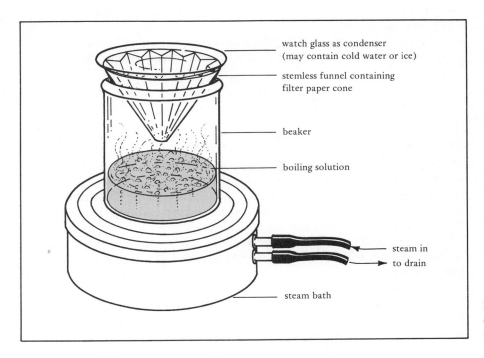

watch glass as condenser (may contain cold water or ice)

stemless funnel containing filter paper cone

beaker

boiling solution

steam in

to drain

steam bath

Figure 8.2. A way to prevent crystallization in the funnel during hot filtration.

Don't cool too fast !

After the filtrate from the hot filtration has been adjusted to the desired volume and all solid has been brought back into solution, the filtrate is allowed to cool.

The rate of cooling determines the size of the crystals. Slow cooling tends to favor fewer and larger crystals; fast cooling tends to favor more and smaller crystals. Very large crystals are to be avoided since they often occlude the solvent and its dissolved impurities. Very small crystals are undesirable because it is difficult to wash them free of the solvent and the soluble impurities, and it takes longer to dry them. Needles between 2 and 10 mm in length are fine, as are prisms 1 to 3 mm in each dimension.

Usually the best compromise of speed, convenience, and quality of crystals is reached by allowing the solution to cool to room temperature on a non-heat-conducting surface such as a cork ring. Sometimes it will be possible to hurry the cooling without getting overly small crystals by swirling the flask in a beaker of water at room temperature, or even by placing the flask directly in an ice bath. The rate of cooling can be slowed greatly by supporting the flask in a beaker of water at the temperature of the solution and allowing both to cool spontaneously to room temperature. The fastest and most satisfactory procedure can only be determined empirically.

Since solubility decreases with decreasing temperature, it is often a good idea to finally cool the mixture from room temperature to 0°C (or to the freezing point of the solvent if it is above 0°C) in a mixture of ice and water. Cooling below 0°C is not often done, since colder baths are not so easily prepared and other problems related to the condensation of water vapor from the air in the form of water or frost require special techniques and apparatus.

If the crystals are collected too soon, some material will be lost, which would have separated from solution on further standing. The minimum acceptable time for crystallization—varying from a few minutes for some substances to days for others, as mentioned in Section 8.1—can be determined only by experiment. In order to tell whether one-half hour is enough, two identical samples must be prepared. One is filtered after standing for half an hour at room temperature, and the other perhaps the next day. A comparison of the amount recovered in the two cases will tell whether anything is to be gained by letting the sample stand for more than half an hour.

*When is crystallization
complete ?*

Similar experiments can be performed to answer the question of whether it is worth cooling the mixture in an ice bath before the filtering operation.

8.6 COLD FILTRATION

When crystallization is complete, the product is collected by suction filtration. The size of the funnel used should be such that it will not be more than half filled with crystals. The suction flask should be large enough so that the solution will not fill it above the tip of the funnel or the side arm.

In some cases it is possible to decant the solvent through the funnel until the mass of crystals remaining in the flask is just covered with solvent, and then to pour the entire mass of crystals into the funnel in one smooth operation. More often, it is necessary to suspend the crystals in the solvent by swirling, and then to quickly pour part of the suspension into the funnel, wait until the liquid level has fallen almost to the level of the crystals in the funnel, and finally repeat the swirling and pouring operation.

Crystals will often remain in the flask after all the solvent has been poured out. These must be scraped into the funnel, rinsed into the funnel with fresh cold solvent (if the compound is relatively insoluble in the cold solvent), or rinsed into the funnel with portions of the filtrate. The last process can easily lead to a mess. It is far better to work hard to pour the crystals out with the liquid in the first place!

During the entire operation of getting the crystals into the funnel, it is best if the level of the liquid in the funnel does not fall below that of the crystals. Air should never be drawn through the crystals until they have been rinsed with fresh solvent (see next section). Sometimes it is necessary to break the vacuum in the flask by disconnecting the hose at the side arm in order to slow the rate of filtration sufficiently.

8.7 WASHING THE CRYSTALS

After the crystals have been transferred to the funnel for suction filtration and the liquid has been drawn off, some fresh solvent should be poured over the crystals in order to wash off the liquid containing the soluble impurities. If this is not done, the soluble impurities will be deposited on the crystals when the solvent evaporates. If the product is relatively soluble in the cold solvent, one washing will usually have to suffice. Two washes are generally appropriate.

If the crystals are relatively soluble, a minimum amount of solvent must be used, and it should be cooled thoroughly in an ice bath. If the crystals are not very soluble, larger amounts of solvent may be used and it need not be chilled. In either case, it is possible to use some of the wash liquid to rinse any remaining crystals out of the flask in which crystallization took place.

Don't use too much solvent

If the crystals are not matted down tight into a solid cake, the vacuum may be released and the wash liquid poured over them evenly and then drawn off by reestablishing the vacuum. If the crystals do form a solid cake, the wash liquid will have to be added to the crystallized matter in the funnel and the product mass carefully pulled apart and suspended evenly by means of a small spatula. Some care is required to suspend all of the product and not tear or dislodge the filter paper. It is best if a very slight vacuum can be maintained during this operation without drawing the wash liquid through too fast. When a procedure says "wash thoroughly," it is probably calling for this rather tedious and delicate operation of carefully breaking up and suspending the filter cake in the wash liquid. After this has been done, the wash liquid is drawn off by suction. An alternative procedure that is sometimes appropriate is to transfer the filter cake to a beaker, add solvent, and break up and suspend the material in the beaker. Following this, the product is again collected by suction filtration.

Sometimes, if a relatively nonvolatile solvent such as acetic acid or nitrobenzene is used for recrystallization, it may be possible to wash this solvent off with a more volatile solvent so as to speed the drying of the crystals. It should hardly be necessary to mention that the crystals should not be soluble in the more volatile solvent.

8.8 DRYING THE CRYSTALS

After the crystals have been collected by suction filtration and washed, they can usually be dried satisfactorily by continuing to draw air over them and then spreading them out on a piece of filter paper.

If the crystals collect as a solid cake, air should be drawn over them in the funnel until the solvent has almost stopped dripping. Failure to suck the filter cake as dry as possible before breaking it up and spreading it out to dry is a very common mistake. The filter cake should be a damp, friable solid, not a paste or mush, when spread out to dry.

Suck crystals dry as possible

More ways of drying solids are described in Section 15.1.

The preceding sections described techniques that will be generally useful in recrystallization. At some point, however, you are likely to encounter various difficulties in an attempted recrystallization. This section describes some of these difficulties, and suggests some possible solutions.

Solvent Pairs

Occasionally you will wish to recrystallize a compound that is too readily soluble in some of the available solvents and not soluble enough in others. In this situation, a mixture of solvents can be useful. To test the suitability of a pair of solvents, prepare a hot solution of a sample of the compound in a small amount of the better solvent, and add slowly, while keeping the mixture hot, some of the poorer solvent. When a cloudiness is produced by slight crystallization, add a little of the better solvent, still keeping the mixture hot, until the cloudiness has been dispelled, and then allow the solution to cool. Although the solubility of a substance in a mixture of solvents usually changes gradually with the proportion of the solvents, sometimes the solubility of a substance can be greatly increased or decreased by the addition of only a small amount of a better or a poorer solvent.

Solvent pairs that are most often used include benzene/ligroin, acetic acid/water, and alcohol/water. Any pair of miscible liquids can be used. Although alcohol/water mixtures are very commonly used, they seem to promote oiling out.

When a recrystallization is carried out using a pair of solvents, it is often a good idea to do the hot filtration before adding the poorer solvent. The addition should be carried out as previously described. A very common error is to add more of the poorer solvent than is needed to just achieve saturation of the hot solution. In extreme cases, this results in the precipitation of everything in solution, impurities as well as the desired compound.

Oiling Out

Sometimes, during cooling for crystallization, the product separates not as crystals but as a liquid (an oil). This may be indicated first by the formation of a cloudiness or opalescence, and then by the formation of visible droplets. It is undesirable to allow the product to separate as an oil because often the oil is an excellent solvent for impurities. When (or if) the oil finally freezes, the impurities that have dissolved in the oil will be in the crystals.

Separation of the product as an oil occurs most often in the recrystallization of low-melting substances, or when mixtures of alcohol and water are used as solvent.

Sometimes oiling out can be prevented by using a little more solvent (or more of the better solvent of a pair) so that the solution will become saturated at a lower temperature. The lower the temperature at which the product separates, the more likely it is to separate as a solid rather than as a liquid.

If the first traces of oil can be caused to solidify by the addition of seed crystals, by vigorous stirring or swirling of the mixture, or by scratching the walls of the flask with a stirring rod, the remainder of the product will usually separate as crystals if the rate of cooling is not too great. If oiling out cannot be prevented—that is, if most of the product separates as an oil before it can be caused to solidify—you can hope that recrystallization of the solidified oil will give a better result, you can try a different solvent, or, probably best, you can purify the product by another method before attempting to recrystallize it.

It has been assumed so far that, at equilibrium, crystals are present rather than oil. In some cases, oil formation cannot be prevented, since oil is present in the equilibrium state. An example of this situation occurs in the recrystallization of *pure* acetanilide from water at acetanilide concentrations of greater than 5.2% by weight. The only possible remedy in these cases is to change the composition of the system by adding solvent or by changing the solvent.

Failure to Crystallize

Occasionally, crystallization will not occur when a solution is cooled, even though it is supersaturated. The most stubborn cases are explained in terms of impurities, or "tar," acting as a protective colloid. If the normal expedients of adding a seed crystal or scratching the flask with a stirring rod fail, crystallization can sometimes be initiated by cooling the mixture in a salt/ice bath (about $-10°C$) or a Dry Ice/acetone bath (about $-70°C$), depending upon the freezing point of the solvent.

Since the rate of crystal growth is lower at low temperatures and in the more viscous solutions that are obtained at low temperatures, a higher temperature is needed for a good rate of crystal growth than for crystal initiation. For this reason, it sometimes works to cool the mixture for a while to $-70°C$ to initiate crystal formation, and then to allow it to warm slowly to room temperature. If crystals have been initiated at the low temperature, they may have an opportunity to grow at an optimum rate in an intermediate temperature range attained during the warming process. Placing the flask on a piece of Dry Ice for a few minutes is sometimes helpful. Alternatively, bits of Dry Ice may be dropped into the solution.

For crystallization to occur in some cases, the solution must be stored in a refrigerator or freezer for long periods of time, even years.

Solubility Differences Caused by Impurities

As pointed out, impurities frequently cause oiling out and can inhibit crystallization. Often, when the level of impurity has been reduced by one recrystallization, the sample will behave appropriately in succeeding crystallizations. Occasionally an impurity will be present that will greatly increase the solubility of the substance. As this material is removed by recrystallization, the solubility of the substance may decrease dramatically, and in some cases the compound will become "insoluble" in the solvent previously used for recrystallization.

Another aspect of this phenomenon is that in one case a certain solvent may serve to recrystallize a compound very well, whereas in another it cannot be used. The reason may be the unsuspected presence of a small amount of some other material.

Wet Samples

Often a solid will be isolated by pouring a reaction mixture into water and collecting the resulting precipitate by suction filtration. When, in this procedure, the product separates as a fine powder, as it frequently does, it is difficult to suck it dry on vacuum filtration. The crude, damp material may easily be more than half water. Recrystallization of such damp material from a water-miscible solvent will often require the use of more solvent than expected, since the water in the crude product will reduce its solubility. If the damp material is recrystallized from a water-immiscible solvent, an extra water phase will be present along with the hot solution. The water should be removed with a medicine dropper or pipet before hot filtration. The solution will also be saturated with water at this point.

Recrystallization of low-melting compounds is a real challenge. Low melting point and high solubility in nonpolar solvents usually go together, as described in Section 21.2. This leaves you to choose between using very small volumes of nonpolar solvents and using solvent pairs including water. If water is involved, there is the possibility of oiling out due to equilibrium existence of two immiscible phases. Neither alternative is very attractive.

In addition, since crystal formation is impossible above the melting point of the compound, all useful cooling of the hot solution must take place from a maximum temperature somewhat below the melting point of the substance. The lower the melting point, the smaller the range of cooling and the smaller the difference in solubility between the hot solution and the cold solvent. This leads to relatively large losses on recrystallization unless the mixture can be cooled below 0°C.

Small Samples

When the amount of material is less than about 50 mg or the volume of the hot solution is less than about 5 ml, the usual techniques of recrystallization give unacceptably large losses in the two filtration steps.

With small amounts, the hot solution should be prepared in a small flask or a short test tube so that it can be withdrawn by a medicine dropper for the hot filtration. With more than a couple of milliliters, you can use a very small funnel and, instead of using filter paper, put a small plug of cotton or glass wool in the stem of the funnel. It is easy to use so much plugging material that the liquid will not run through, and it is very wise to test the funnel with a sample of the pure hot solvent. With less than 1 or 2 ml, you can put a small plug of cotton into the tip of a medicine dropper, draw the solution into the dropper through the cotton, remove the cotton with tweezers while holding the tip of the dropper over the test tube to which the filtrate is to be added, and then release the hot solution into the test tube.

An advantage in cooling the solution for crystallization in a test tube is that the crystals may be collected by centrifugation followed by decanting of the solvent. Washing can be done by suspending the crystals in a little cold solvent, centrifuging, and decanting. Alternatively, the solvent can be removed with a dropper that has had its tip drawn down to a capillary of about 1 mm. The wash liquid can be removed in the same way.

The crystals can be dried in the test tube by laying the tube on its side with the bottom slightly raised or by connecting it to the vacuum as shown in Figure 8.3.

When the crystals are dry, they can be removed from the test tube by inverting it over a piece of filter paper and tapping it with a stirring rod.

Second Crops

The filtrate that is removed when the crystals are collected by suction filtration is saturated with respect to the compound. It is often possible to concentrate the filtrate by distilling off some of the solvent, and then to obtain a further crop of crystals by allowing the concentrated filtrate to cool. The second crops are usually not as pure as the first, and this technique is most useful in those cases in which there is a relatively small difference in solubility of the compound in the hot and cold solvent.

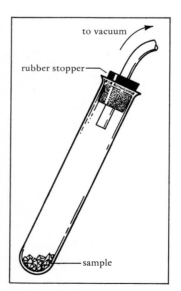

to vacuum

rubber stopper

sample

Figure 8.3. A way to dry crystals in a test tube.

1. Suppose you are recrystallizing a sample of benzoic acid from water. The original sample is contaminated with sand and salt.
   a. Briefly describe the procedure that you would use.
   b. Explain how the two impurities would be separated from the benzoic acid by your procedure.

2. Explain what effect each of the following mistakes would have on the success of a recrystallization:
   a. Too much solvent was used.
   b. Too little solvent was used.
   c. The hot solution was filtered by suction.
   d. The decolorizing carbon was not completely removed by the hot filtration.
   e. The hot solution was immediately placed in an ice bath.
   f. The crystals were washed with warm solvent.
   g. The crystals were not washed at all.
   h. There was inadequate suction during the suction filtration.

Problems

1. What is the expected percent recovery upon recrystallization of 2.00 grams of benzoic acid from 100 ml of water, assuming the solution cools to 18°C for crystallization?

   Solubility of benzoic acid in water:

   2.2 g/100 ml water at 75°C
   0.27 g/100 ml water at 18°C

2. What is the expected percent recovery upon recrystallization of 30 grams of benzoic acid from 100 ml of carbon tetrachloride, assuming that the solution will be cooled to 20°C for crystallization?

   Solubility of benzoic acid in carbon tetrachloride:

   32 g/100 ml CCl₄ at 60°C
   5.6 g/100 ml CCl₄ at 20°C

3. Compare the percent recovery to be expected upon recrystallization of 2.00 grams of benzoic acid from 200 ml of water with that to be expected if only 100 ml of water is used, cooling to 18°C in either case.

4. If the percent recovery to be expected increases when less solvent is used, other things being equal, would it be a good idea to try to recrystallize 2.00 grams of benzoic acid from less than 100 ml of water? Explain.

5. **a.** Calculate the percent recovery to be expected upon recrystallization of 2.00 grams of benzoic acid from 100 ml of water if the solution is allowed to cool to 4°C before suction filtration.

   Solubility of benzoic acid in water at 4°C:
   0.18 g/100 ml

   b. Do the same for the recrystallization of 30 grams of benzoic acid from carbon tetrachloride, assuming that the solution can be cooled to 0°C before suction filtration.

   Solubility of benzoic acid in carbon tetrachloride at 0°C:
   1.5 g/100 ml

6. **a.** Would you choose to recrystallize a 100-milligram sample of benzoic acid from water or from carbon tetrachloride? Present the reasons for your choice.
   b. Would you choose to recrystallize a 100-gram sample of benzoic acid from water or from carbon tetrachloride? Present the reasons for your choice.

7. A mixture contains 95% by weight A and 5% by weight B. Assume that you must obtain pure A by recrystallization of a 100-gram sample of the mixture.

a. What is the minimum amount of solvent necessary for the recrystallization? What percent of A in the sample should crystallize out upon cooling of the hot solution? Assume that the solubilities are:

	Hot	Cold
A	10 g/100 ml	2 g/100 ml
B	10 g/100 ml	2 g/100 ml

b. Answer the same two questions as in **a.**, but assume the solubilities of B are:

  5 g/100 ml (hot)   and   1 g/100 ml (cold)

c. Answer the same two questions as in **a.**, but assume the solubilities of B are:

  2 g/100 ml (hot)   and   0.4 g/100 ml (cold)

d. Answer the same two questions as in **a.**, but assume the solubilities of B are:

  1 g/100 ml (hot)   and   0.2 g/100 ml (cold)

e. Answer the same two questions as in **a.**, but assume the solubilities of B are:

  0.5 g/100 ml (hot)   and   0.1 g/100 ml (cold)

Exercises

Any skill improves with practice. The purpose of an exercise is to provide practice. For this reason, I have included a few suggestions for practice in recrystallization.

endo-5-norbornene-
2,3-dicarboxylic acid

1. Recrystallization of *endo*-5-norbornene-2,3-dicarboxylic acid.

This substance may be recrystallized very nicely from water. Its tendency to remain supersaturated for a little while makes it easier to complete the hot filtration before crystallization begins. If the solution is allowed to cool slowly, the crystals separate in beautiful long spars. Since the corresponding anhydride is much less expensive and is converted to the diacid on boiling with water, the instructions specify that you start with it rather than the acid. The anhydride may be prepared according to the procedure of Section 66.2.

Procedure. Place 4.0 grams of the anhydride in a 125-ml Erlenmeyer flask and add 50 ml of water. Heat the mixture to boiling and continue to heat until the oily liquid goes into solution. Filter the hot solution by gravity and allow it to cool. Collect the resulting crystals by suction filtration and wash them with a little water.

endo-5-norbornene-
2,3-dicarboxylic anhydride

2. Recrystallization of acetanilide. Water is often recommended for the recrystallization of acetanilide. Solubility information for acetanilide may be found in the *Handbook of Chemistry and Physics* (older editions).

3. Recrystallization of *m*-nitroaniline. Both water and 75% aqueous ethanol have been recommended for the recrystallization of this compound. It usually forms lovely yellow needles.

4. Recrystallization of *p*-nitroaniline. Water at 100 ml per gram has been recommended for the recrystallization of this substance.

5. Recrystallization of benzil. If benzil is dissolved in 95% ethyl alcohol at the rate of 6.5 ml per gram in a relatively large flask, and if care is taken to leave no traces of solid in any part of the flask, the solution may be cooled to room temperature without crystallization. When a minute seed crystal is added, a very beautiful phenomenon of crystal growth may be observed. If no seed crystal is added, crystallization may take a long time to occur, but sometimes the entire sample of benzil will separate as a single crystal.

Sources from which more information about recrystallization processes may be obtained include:

1. K. B. Wiberg, *Laboratory Technique in Organic Chemistry*, McGraw-Hill, New York, 1960, p. 98.

2. A. I. Vogel, *Practical Organic Chemistry*, 3rd edition, Wiley, New York, 1956, p. 122.

3. R. S. Tipson, *Technique of Organic Chemistry*, Vol. III, Part I, 2nd edition, A. Weissberger, editor, Interscience, New York, 1956, p. 395.

§9 DISTILLATION

The process of distillation consists of heating a liquid to a temperature at which it is converted to a vapor, and then condensing the vapor back to a liquid in another part of the apparatus. Purification of a mixture by distillation depends upon the fact that substances can differ in the degree to which they can be vaporized under the experimental conditions.

9.1 VAPOR PRESSURE

If a sample of a liquid is placed in an otherwise completely empty space, some of it will vaporize. As this happens, the pressure in the space above the liquid will rise and will finally reach some constant value. The pressure under these conditions is due entirely to the vapor of the liquid, and is called the *equilibrium vapor pressure*.

Equilibrium vapor pressure

The equilibrium vapor pressure increases with temperature according to Equation 9.1-1 (see Figure 9.1), where C is a constant and T is the absolute temperature:

$$P \propto e^{-C/T} \quad \text{or} \quad P \propto \frac{1}{e^{C/T}} \qquad (9.1\text{-}1)$$

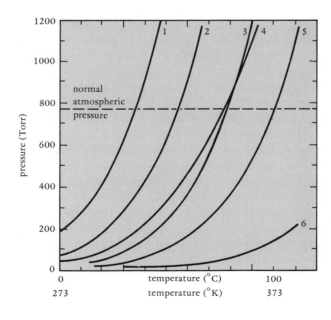

Figure 9.1. Graph of the equilibrium vapor pressure versus temperature for several liquids: (1) diethyl ether, (2) acetone, (3) ethyl alcohol, (4) carbon tetrachloride, (5) water, (6) bromobenzene.

This relationship between vapor pressure and temperature can be rewritten in a logarithmic form:

$$\ln P = \frac{-C}{T} + \text{constant} \qquad (9.1\text{-}2)$$

or, using common logarithms:

$$2.3 \log P = -\frac{C}{T} + \text{constant} \qquad (9.1\text{-}3)$$

$$\log P = -\frac{C}{2.3T} + \frac{\text{constant}}{2.3}$$

In these last equations, "constant" is the natural log of the proportionality constant implied in Equation 9.1-1. The logarithmic form of the equation makes it apparent that a graph of log P versus $1/T$ should give a straight line of slope $-C/2.3$ (see Figure 9.2).

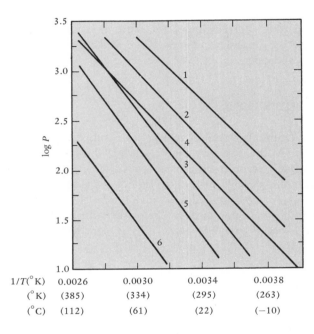

Figure 9.2. Graph of log (equilibrium vapor pressure) versus $1/T$, in degrees K, for several liquids. The liquids are the same as those of Figure 9.1: (1) diethyl ether, (2) acetone, (3) ethyl alcohol, (4) carbon tetrachloride, (5) water, (6) bromobenzene.

Molecular interpretation

The phenomenon of vapor pressure is interpreted in terms of molecules of liquid escaping into the empty space above the liquid. As the number of molecules in the vapor space above the liquid becomes larger, the rate of return of molecules from the vapor space to the liquid increases until the rate of return has risen to equal the constant rate of escape. This is the equilibrium condition, and the corresponding concentration of molecules in the vapor space gives rise to the equilibrium vapor pressure. At higher temperatures, the greater kinetic energy of the molecules in the liquid results in a greater constant rate of escape. Equilibrium is established at higher temperatures, then, with larger numbers of molecules in the vapor phase, and at correspondingly higher pressures.

The equilibrium vapor pressure will be exerted in a closed container whether or not there are other molecules present in the gas phase (air, for example). If there are other molecules present, the equilibrium vapor pressure will not be equal to the total pressure as in the previous example. The total pressure will *Partial pressure* be the sum of the equilibrium vapor pressure of the liquid (its *partial pressure*) plus the pressure due to the other molecules (the sum of their various partial pressures).

Different substances have different vapor pressures at any given temperature, as illustrated by the examples in Figures 9.1 and 9.2. This is equivalent to saying that the values for the constants C and "constant" in Equations 9.1-2 and 9.1-3 are different for different molecules. In Section 16.2, boiling points will be interpreted in terms of the heat of vaporization (ΔH_{vap}; the amount of heat required to convert a mole of the liquid to a vapor at the normal boiling point; always positive) and the entropy of vaporization (ΔS_{vap}; the entropy increase that accompanies the conversion of a mole of liquid to a vapor at the normal boiling point; always positive). Since it can be shown that in Equation 9.1-3

$$C = \frac{\Delta H_{vap}}{R} \quad \text{and} \quad \text{Constant} = \frac{\Delta S_{vap}}{R}$$

where R is the ideal gas constant, then

$$\log P = -\frac{\Delta H_{vap}}{2.3RT} + \frac{\Delta S_{vap}}{2.3R}$$

A high vapor pressure at any given temperature is thus the result of either a small heat of vaporization or a large entropy of vaporization, or both, and to say that a compound generally has a high vapor pressure is equivalent to saying that it has a low boiling point.

9.2 DISTILLATION OF A PURE LIQUID

If a pure liquid is heated in a flask connected to a condenser that is open to the atmosphere at the other end (see Figure 9.3), its vapor pressure will rise, as

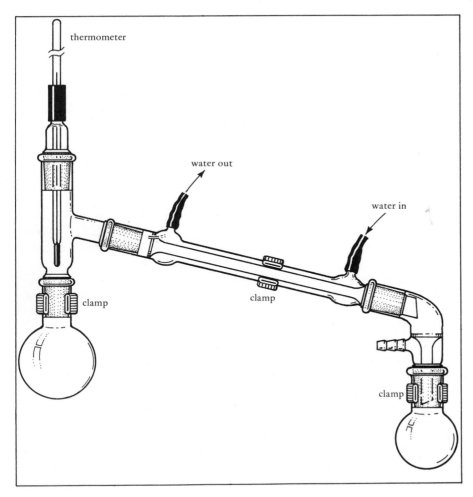

Figure 9.3. A simple apparatus for distillation at atmospheric pressure.

explained in the preceding section. When the temperature of the liquid rises to a certain point, the vapor pressure of the liquid will slightly exceed that of the atmosphere and the vapor will start to expand out of the apparatus through the condenser. It is the function of the condenser to cool the vapors and reconvert them to liquid. In a distillation, the condenser is arranged so that the condensate does not return to the flask, in contrast to its use as a reflux condenser. As long as liquid remains in the flask, the temperature of the distilling vapor will not rise. The continuing input of heat serves only to supply the required heat of vaporization and thus to convert more liquid to vapor.

The temperature at which distillation takes place at a total pressure of 1 atmosphere is called the *normal boiling point* of the liquid. If the pressure in the apparatus, which is usually open to the atmosphere, is not equal to exactly 1 atmosphere, the temperature at which boiling will begin (when equilibrium vapor pressure equals external pressure) will be different from the normal boiling point. It is sometimes desirable to distill a substance at as low a temperature as possible, using an apparatus in which the pressure can be reduced (see Section 10).

The significant features in the distillation of a pure liquid are that (1) the compositions of the liquid, the vapor, and the condensate (or distillate) are identical and constant during the process, and (2) the temperatures of the liquid and the vapor are constant and, ideally, equal throughout the distillation.

9.3 MISCIBLE PAIRS OF LIQUIDS

When two liquids that are completely soluble in one another are mixed, the vapor pressure of each, at a particular temperature, is diminished by the presence of the other. Such mixtures can be characterized in terms of the contribution of each component to the total vapor pressure as a function of the composition of the mixture.

Miscible pairs of liquids are said to behave ideally if the contribution of each component to the total vapor pressure is directly proportional to its mole fraction. That is,

$$P_A = x_A P_A^0 \qquad P_B = x_B P_B^0 \qquad (9.3\text{-}1)$$

$$P_{\text{total}} = P_A + P_B = x_A P_A^0 + x_B P_B^0 \qquad (9.3\text{-}2)$$

where P_A and P_B are the vapor pressures of A and B above a solution of mole fraction x_A and x_B, and P_A^0 and P_B^0 are the vapor pressures of pure A and pure B at that particular temperature. This type of behavior, often referred to as behavior according to *Raoult's law*, may also be represented graphically as in Figure 9.4.

Ideal behavior is approximated by mixtures such as benzene/toluene, *n*-hexane/*n*-heptane, carbon tetrachloride/silicon tetrachloride, and *n*-butyl bromide/*n*-butyl chloride, in which the mixture is composed of molecules of similar size and type of intermolecular interaction. A vapor pressure–composition diagram is given for the system benzene/toluene in Figure 9.5.

For the purpose of discussing the separation of a pair of miscible liquids by distillation, a boiling-point diagram is very helpful. This is a diagram showing the temperature at which mixtures of various composition boil (at a given total external pressure, usually 1 atmosphere) and the compositions of the liquid and vapor that are in equilibrium at this temperature. The way this information is stored in a boiling-point diagram is best shown by considering how one is constructed. Suppose, for example, that you prepare mixtures of benzene and toluene of 0.1, 0.3, 0.5, 0.7, and 0.9 mole fraction benzene and heat each mixture to boiling in an apparatus open to 1 atmosphere of pressure and in which the condensed vapor is returned to the boiler. A thermometer can be used to deter-

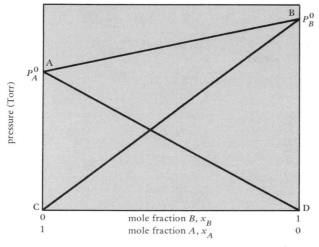

Line AD: partial pressure of component $A = P_A = x_A P_A^0$.

Line BC: partial pressure of component $B = P_B = x_B P_B^0$.

Line AB: total vapor pressure $P_{\text{total}} =$ line AD + line BC.

Figure 9.4. Vapor pressure–composition diagram for the ideal system A/B at a particular temperature.

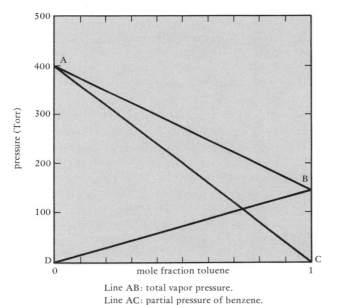

Line AB: total vapor pressure.
Line AC: partial pressure of benzene.
Line DB: partial pressure of toluene.

Figure 9.5. Vapor pressure–composition diagram for the system benzene/toluene at 60°C.

mine the temperature at which the mixture boils, and a sample of the condensate can be removed to determine the composition of the vapor. (The composition of the liquid is known since the mixture was made up of known amounts of the components; it could also be determined experimentally.) Suppose the results are as shown in the table overleaf.

These data can be plotted as shown in Figure 9.6. A smooth line can then be drawn to interpolate all the other possible experimental points, and the result is shown in Figure 9.7. Now, from Figure 9.7, the composition of the vapor in equilibrium with any mixture of benzene and toluene at its boiling point (under 1 atmosphere of pressure) can be determined. For example, if a 1:4 (by moles) mixture of benzene and toluene is heated for distillation, you would expect that

b.p. (°C)	Liquid Composition (mole fraction benzene)	Vapor Composition (mole fraction benzene)
110.6	0.00	0
105.7	0.10	0.21
98.3	0.30	0.51
92.4	0.50	0.71
87.3	0.70	0.86
82.6	0.90	0.96
80.0	1.00	1.00

the mole fraction of benzene in the vapor would be about 0.4 at an initial boiling point of about 102°C.

It is generally true, as in the particular case of benzene and toluene, that the vapor in equilibrium with the liquid will be richer than the liquid in the more volatile component. This seems intuitively reasonable in that the molecules of the component with the higher vapor pressure at any given temperature should tend to escape more frequently and thus be overrepresented in the vapor phase.

As the distillation of a mixture of miscible liquids progresses, the mixture will gradually be depleted of the more volatile component. As this happens, according to the distillation diagram, *the boiling point will gradually rise*, and the distillate, though always richer than the residue in the more volatile component, will contain a continually decreasing proportion of the more volatile component. You can

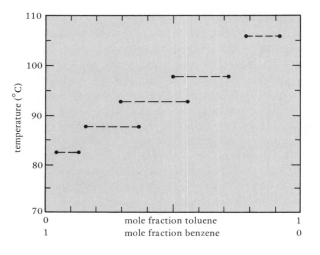

Figure 9.6. Construction of the boiling-point diagram for the system benzene/toluene at 1 atmosphere. Each pair of points indicates the boiling point of a mixture of benzene and toluene and the compositions of the liquid and the vapor that are in equilibrium.

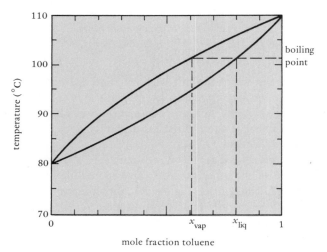

Figure 9.7. Boiling-point diagram for the system benzene/toluene at 1 atmosphere.

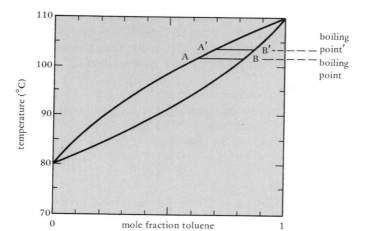

Figure 9.8. Boiling-point diagram for the system benzene/toluene at 1 atmosphere. The horizontal lines AB and A'B' connect points on the two curves that represent the compositions of liquid and vapor that are in equilibrium at two different temperatures.

think of this process as being represented by the gradual movement of line AB in Figure 9.8 upward and to the right to, say, A'B'. Thus, the significant features in the distillation of a miscible pair of liquids that serve to distinguish it from the distillation of a pure liquid are that (1) the compositions of the liquid and vapor (or distillate) are not the same, and (2) the boiling point of the liquid will gradually rise during the distillation. Exceptions to this generalization are discussed in the section on azeotropic mixtures (Section 9.5).

9.4 FRACTIONAL DISTILLATION

From Figure 9.7 and the example cited in the previous section, you can see that if you started to distill a mixture of benzene and toluene that was 0.2 mole fraction in benzene, the vapor initially in equilibrium with the mixture would be about 0.4 mole fraction in benzene. If you started with a large sample, this would be the composition of the first part of the distillate. If the first part of the distillate were then redistilled, the vapor in equilibrium with it would be approximately 0.6 mole fraction in benzene and the first few drops of the distillate would have this composition. You can see that if you started with a large enough sample and repeated this process several times, collecting only the very first part of the distillate each time, you could obtain a very small sample of fairly well purified benzene. A four-step process of this type could be represented by the movement of a point from A to B . . . to I.

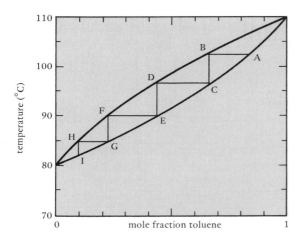

Figure 9.9. Boiling-point diagram for the system benzene/toluene at 1 atmosphere. The series of four horizontal and four vertical lines connecting the points A and I indicate that after four successive steps of vaporization followed by condensation, a large sample of liquid whose boiling point and composition correspond to point A would yield a small sample of a liquid whose boiling point and composition would correspond to point I.

The process just described would be too inefficient to be practical. If 10 percent of the material were distilled each time, then, starting with 1 liter, only 0.1 milliliter would be obtained at the end of the fourth stage. Also, if a finite amount were collected in each step, the final purity would be less than that indicated in Figure 9.9.

A more practical procedure, which was used in the past, is to distill the entire sample, collecting the distillate over successive, arbitrary temperature ranges. The first fraction (the most volatile fraction) is then redistilled in a similar manner until the temperature of the distillation rises to that of the second fraction. The second fraction is then added to the boiler and distilled in a similar manner until the temperature rises to that of the third fraction, and so on. The entire process may be repeated four or five times. In each series of redistillations, the distillate is collected according to the temperature range over which it boils; in the later series, the temperature ranges for the initial and final fractions are made smaller, and those of the intermediate fractions made larger. In a successful fractional redistillation process such as this, most of the distillate in the last stage will be in the first and last fractions.

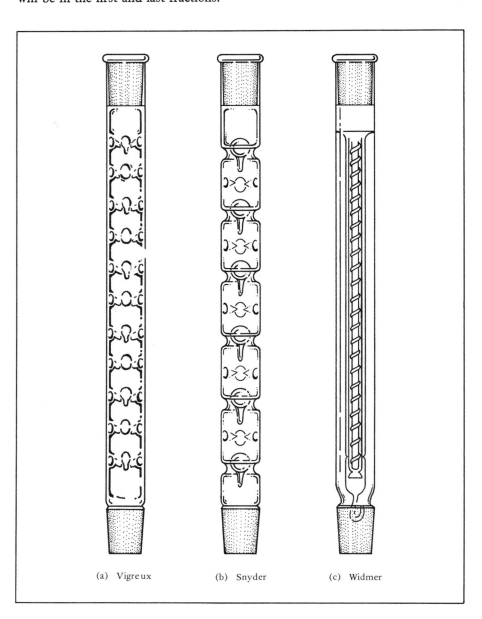

(a) Vigreux   (b) Snyder   (c) Widmer

Figure 9.10. Some different types of fractionating columns.

Fortunately, an apparatus called a *fractionating column* has been developed that can effect a high degree of purification relatively quickly and easily, and with little loss or rejection of material. Some different types of fractionating columns are illustrated in Figure 9.10. The significant feature of a fractionating column is that it provides for efficient exchange of heat and material between the condensate flowing down the column and the vapors flowing up. The material finally coming out of the top of the column as a vapor has been subjected to multiple condensations and evaporations on the way up, each of which has served to enrich the vapor in the more volatile component. A good column can produce a distillate in which the enrichment corresponds to between 25 and 100 steps like the four in Figure 9.9. Since some industrial fractionating columns are built up of units called plates, each of which theoretically provides enrichment corresponding to one step, the efficiency of enrichment of a column is expressed as *theoretical plates*, rather than "steps." Mixtures for which the boiling-point diagram is known may be used to determine the efficiency of a column.

Theoretical plates

While it might seem that the more theoretical plates a fractionating column has, the better, there are other factors that need to be considered in choosing a column for a particular distillation. *Column holdup* is the volume of material that would not flow out the bottom of the column if poured into the top—the volume of liquid required to wet the column. It would not be possible to distill a sample whose volume is less than the volume of the holdup; the loss on distillation will always be at least equal to the holdup. HETP is the height (length of column) equivalent to one theoretical plate. A low HETP is desirable both to keep columns that are highly efficient from being also inconveniently tall and to minimize holdup. *Throughput* is the maximum volume of liquid that can be boiled up through the column per unit of time while still maintaining equilibrium throughout the column. A high throughput is desirable so that a separation may be done quickly. The *reflux ratio*, R, is the ratio of the ratio of the amount of condensate formed at the top of the column and returned to the column to the amount removed as distillate. A reflux ratio of 19 would mean that 5 percent of the condensate at the top of the column is removed as distillate. The higher the reflux ratio, the greater the operating efficiency of any given column, since the column will be operating more nearly at the equilibrium conditions, which exist at total reflux (infinite R)—the conditions used in determining the theoretical plate rating of a column. The ideal column, then, will have a high theoretical plate rating, a low HETP, low holdup, and high throughput, and will not suffer a great loss in operating efficiency with decreasing reflux ratio. All real columns involve compromises among these factors and, of course, cost. Table 9.1 compares some simple columns in terms of these criteria. These numbers are estimates of what might be expected from certain simple fractionating columns with a minimum of insulation. A vacuum jacket would increase the efficiency about 25 percent. A larger column diameter would increase HETP and holdup, but would allow a greater throughput. The HETP was estimated for total reflux, and it will increase with a finite rate of distillation.

Holdup

HETP

Throughput

Reflux ratio

Table 9.1. Comparison of simple fractionating columns

Type	Diameter (mm)	Throughput (ml/min)	HETP (cm)	Holdup (ml/plate)
Glass tube	6	2	~40	~1.3
Vigreux	12	2–5	~7	~2/3
Glass tube with ⅛-in helices	10	3–8	~4	~2
Glass tube with stainless steel sponge	12	2–5	~4	~1.5

Since column height and holdup increase with increasing theoretical plates, it is undesirable to use a more efficient fractionating column than is necessary to effect the desired separation.

9.5 AZEOTROPIC MIXTURES

In many mixtures of pairs of miscible liquids, the vapor pressure of each component cannot be represented by Equation 9.3-1, and a plot of the total vapor pressure of the mixture versus composition at any given temperature will not give a straight line such as the line AB in Figures 9.4 and 9.5. As long as the total vapor pressure of any mixture at a certain temperature is neither greater nor less than the vapor pressure of either pure component at that temperature, mixtures of these substances will approximate the expected behavior of ideal mixtures upon distillation. The system water/methanol is such a system; its vapor pressure–composition diagram is given in Figure 9.11, and its distillation diagram in Figure 9.12.

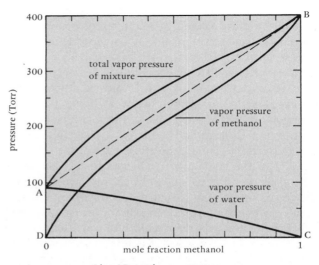

Figure 9.11. Vapor pressure–composition diagram for the system water/methanol at 49.76°C.

Line AB: total vapor pressure.
Line AC: partial pressure of water.
Line DB: partial pressure of methanol.

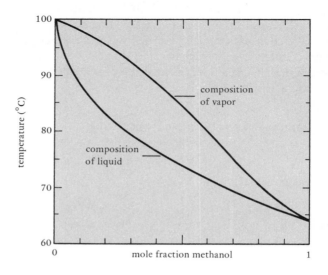

Figure 9.12. Boiling-point diagram for the system water/methanol at 1 atmosphere.

If the total vapor pressures of mixtures go through a maximum at intermediate compositions (positive deviations from Raoult's law), the boiling point of the mixture whose composition corresponds to the vapor-pressure maximum will be lower than that of either pure component (or any other mixture), and the vapor in equilibrium with this mixture will have the same composition as the liquid. Benzene/methanol is such a system; its vapor pressure–composition diagram is given in Figure 9.13, and its distillation diagram in Figure 9.14.

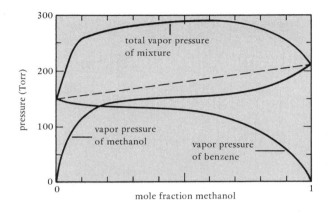

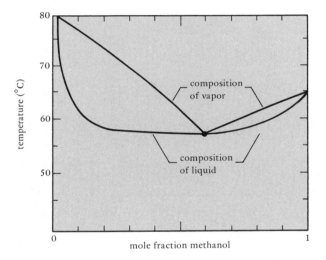

Figure 9.13. Vapor pressure–composition diagram for the system benzene/methanol at 35°C. The vapor-pressure maximum occurs at 0.567 mole fraction methanol.

Figure 9.14. Boiling-point diagram for the system benzene/methanol at 1 atmosphere. The composition of the azeotropic mixture is 0.609 mole fraction methanol, and it boils at 57.6°C. The point • indicates boiling point and composition of minimum-boiling azeotrope.

The converse is also observed. If the total vapor pressures of mixtures go through a minimum at intermediate composition (negative deviations from Raoult's law), the boiling point of the mixture whose composition corresponds to the vapor-pressure minimum will be higher than that of either pure component (or any other mixture), and the vapor in equilibrium with this mixture will have the same composition as the liquid. Chloroform/acetone is such a system; its vapor pressure–composition diagram is given in Figure 9.15, and its distillation diagram in Figure 9.16.

Since the composition of the vapor in equilibrium with these minimum- or maximum-boiling mixtures is identical with the composition of the liquid, separation of the components of such a mixture by distillation is impossible. Such mixtures are called *azeotropic*.

Azeotrope

It is invariably true that if the vapor pressure versus composition diagram shows a maximum at an intermediate composition, a minimum-boiling azeotrope will exist at that composition, at the same temperature and pressure. That the

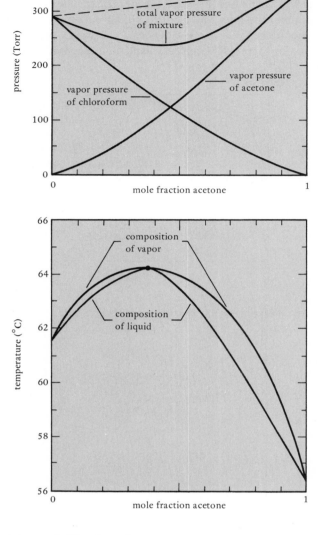

Figure 9.15. Vapor pressure–composition diagram for the system chloroform/acetone at 35.17°C. The vapor-pressure minimum occurs at 0.383 mole fraction acetone.

Figure 9.16. Boiling-point diagram for the system chloroform/acetone at 1 atmosphere. The composition of the azeotropic mixture is 0.360 mole fraction acetone, and it boils at 64.4°C.

mixture is minimum boiling is easily seen, since high vapor pressure corresponds to low boiling point. That the mixture must be azeotropic can best be deduced by seeing from the boiling-point diagram (Figure 9.14) that the situation cannot be otherwise. If the upper curve (representing vapor composition) does not touch the lower curve (representing liquid composition) at the minimum point, the implication would be that there is no vapor in equilibrium with this liquid. Therefore, the curves must touch at the minimum point, which is to say that the same point represents the composition of both liquid and vapor. A similar argument can be applied to the case of a maximum-boiling azeotrope as well.

The practical consequences of distillation of mixtures that form either minimum- or maximum-boiling azeotropes are best deduced from the boiling-point diagrams. If a pair of liquids can form a minimum-boiling azeotrope, a perfectly efficient fractionating column will produce a distillate of the composition of the azeotrope, no matter what the initial composition of the mixture.

The residue in the boiler after complete removal of the other components as the azeotropic mixture will be pure *A* or pure *B*, depending upon whether the mixture initially contained relatively more *A* than the azeotropic composition, or

Table 9.2. Azeotropic mixtures of maximum boiling point 57

SECTION 9

| Component A | b.p. (°C) | Component B | b.p. (°C) | —Azeotropic Mixture— | |
				b.p. (°C)	Weight % B
Water	100	Hydrofluoric acid	19.5	111.4	35.6
Water	100	Hydrochloric acid	−80	108.6	20.22
Water	100	Hydrobromic acid	−73	126	47.5
Water	100	Hydriodic acid	−34	127	57
Water	100	Nitric acid	86	120.5	68
Water	100	Sulfuric acid	dec.	338	98.3
Water	100	Perchloric acid	110	203	71.6
Water	100	Formic acid	100.8	107.2	22.6
Water	100	Ethylenediamine	116	118	20–25
Acetic acid	118.2	Pyridine	115.5	140	53
Acetone	56.1	Chloroform	61.2	64.4	78.5

relatively more B than the azeotropic composition. If, on the other hand, a pair of liquids can form a maximum-boiling azeotrope, a perfectly efficient fractionating column can produce a distillate of either pure A or pure B, depending upon whether the mixture initially contained relatively more A or relatively more B than the azeotropic composition. The residue, after complete removal of that amount of either A or B, which was initially present in excess over the material of azeotropic composition, will be the pure azeotrope.

Possibly the easiest way to interpret the boiling-point diagrams of mixtures involving azeotropes is to consider them to be made up of two simple distillation diagrams side by side, one of which involves pure A and the azeotrope of A and B, and the other the azeotrope of A and B, and pure B. Then you can apply the same stepping process to either side of the diagram as was used in the case of the ideal mixture.

Table 9.3. Azeotropic mixtures of minimum boiling point

| Component A | b.p. (°C) | Component B | b.p. (°C) | —Azeotropic Mixture— | |
				b.p. (°C)	Weight % B
Water	100	Acetonitrile	81.5	76.0	85.8
Water	100	Ethanol	78.3	78.1	96
Water	100	n-Propanol	97.3	87	71.7
Water	100	i-Propanol	82.3	80.3	87.4
Water	100	Propionic acid	141.4	99.1	17.8
Water	100	t-Butanol	82.5	79.9	88.2
Water	100	Pyridine	115	94	43
Water	100	Dioxane	101.3	87.8	82
Carbon tetrachloride	76.8	Methanol	64.7	55.7	20.6
Carbon tetrachloride	76.8	Ethanol	78.3	65.1	15.8
Carbon tetrachloride	76.8	Acetone	56.2	56.1	88.5
Chloroform	61.2	Methanol	64.7	53.4	12.6
Chloroform	61.2	Ethanol	78.3	59.4	7
Chloroform	61.2	Hexane	69.0	60.0	28
Methanol	64.7	Acetone	56.2	55.5	88
Methanol	64.7	Benzene	80.1	57.5	60.9
Methanol	64.7	Cyclohexane	80	54	72
Ethanol	78.3	Ethyl acetate	77.1	71.8	69
Ethanol	78.3	Benzene	80.1	68.2	67.4
Ethanol	78.3	Cyclohexane	80	64.9	69.5
Acetone	56.2	Hexane	69	49.7	46.5
Benzene	80.1	Cyclohexane	80.6	77.7	48.2

The existence of maximum-boiling azeotropes (negative deviations from Raoult's law) can be interpreted in terms of energetically more favorable interactions (lower potential energy) between the unlike molecules than between the like molecules. It is tempting to extend this idea to that of the formation of a "complex" or weakly bonded compound. However, since the compositions of most azeotropes do not correspond to nice molecular ratios and, even worse, show slight variations with temperature and pressure, this interpretation must be ruled out. Table 9.2 lists some maximum-boiling azeotropes and some of their properties.

The existence of minimum-boiling azeotropes (mixtures of maximum vapor pressure) can be interpreted in terms of energetically less favorable interactions (higher potential energy) between the unlike molecules than between the like molecules. This has been described as "incipient insolubility" in which the like molecules tend to squeeze out the unlike. The extreme would be when the two liquids are immiscible (see Steam Distillation, Section 11) and each exerts its own vapor pressure independently. Table 9.3 lists some minimum-boiling azeotropes and some of their properties.

9.6 TECHNIQUE OF DISTILLATION

Heating

A constant rate of heating is essential for optimum performance of any distillation apparatus. Of the methods of heating described in Section 32.1, an electrically heated oil bath or an electric heating mantle gives the most constant and most easily regulated rate of heat input. If the separation is very easy and high efficiency is not needed, a steam bath may be used at lower temperature with flammable materials (for example, to remove the volatile solvent from a high-boiling product) or a flame may be used with higher-boiling substances (for example, to complete the distillation after the solvent has been removed by heating with the steam bath).

Rate of distillation

The rate of distillation is controlled by the rate of heat input. The higher the temperature of the bath above the boiling point of the mixture, the greater the rate of distillation. The lower the rate of distillation, the greater the efficiency of any distillation apparatus. In the case of an easy separation, a collection rate for the distillate of one to three drops per second (three to ten ml per minute) will often be a reasonable compromise between speed and efficiency. In careful fractionations, the reflux ratio may be quite high, and the collection rate much lower than this. Since the required rate of heating for a given rate of distillation depends upon how well the column is insulated against heat loss, it is best determined by trial.

Bumping

Even with a constant source of heat, the rate of boiling will not always be constant. Sometimes alternate periods of superheating followed by vigorous boiling will occur. In addition to being annoying, this phenomenon, called *bumping*, will prevent equilibrium from being established within the fractionating column. Even boiling can usually be promoted by adding several *boiling stones*. These may be bits of unglazed porous clay plate, pieces of carborundum, or bits of a specially prepared anthracite coal. The tiny air bubbles trapped in the pores of such materials prevent superheating by providing the nuclei for bubble formation. The use of several boiling stones is recommended whenever a liquid is to

Boiling stones

be boiled, in recrystallizations as well as in distillations. Should you forget to add the boiling stones, let the mixture cool a bit before you drop them in; if you do not allow a brief cooling time, you run the risk of having the mixture boil over.

In order for a distillation column to operate at maximum efficiency, *it must be insulated against heat loss*. The most usual form of insulation is a high-vacuum jacket (like a Thermos bottle) that is an integral part of the column. Sometimes provision is made for electrical heating, with or without the vacuum jacket. No insulation at all is provided with the simple columns illustrated in Figure 9.10. The simplest way of providing some insulation for these columns is to wrap them, not too smoothly, with a couple of layers of aluminum foil. Asbestos tape, glass wool, and many other materials have been used, either singly or in combination. The higher boiling a substance is, the more easily it will be condensed by the column. If the column is tall and poorly insulated, it may be impossible to put as much heat in at the boiler as can be lost in the column. In this case, it will not be possible to drive the material over.

With flame heating, it is easy to superheat the vapor, upset the equilibria within the column, and lower column efficiency. Superheating can be minimized by heating through an asbestos board, and by wrapping the upper part of the boiling flask and the column in a layer of aluminum foil.

The flask used as the boiler in a distillation should not be more than half full to begin with. If a large amount of solvent must be removed before distilling a small amount of a high-boiling compound, it is best to distill most of the solvent and then transfer the residue to a smaller flask for the last part of the distillation. Losses in transfer can be prevented by rinsing with a small amount of the solvent that was just removed. If you attempt to complete the distillation from a large flask, not only will the losses be greater due to the larger vapor volume and surface area, but the walls of the flask will act as a condenser and may make it impossible to drive the material over.

There are several reasons that material is lost upon distillation. The vapor

Insulation

Avoiding superheating

Size of boiling flask

Avoiding loss of material

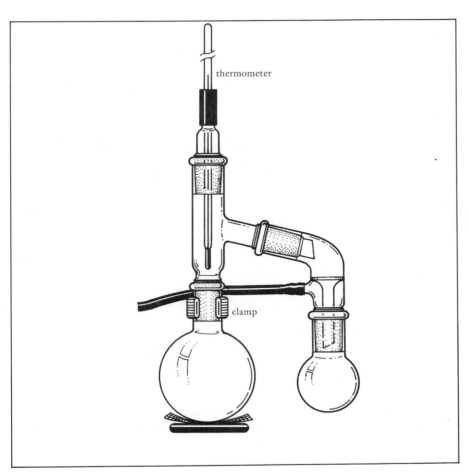

thermometer

clamp

Figure 9.17. An apparatus for distillation. The receiver and vacuum adapter should each be held on with a spring clamp or a rubber band. The hose should lead over the edge of the desk in order to keep flammable fumes from the burner, or to the vacuum pump in the case of a vacuum distillation. A small piece of stainless steel sponge may be placed just below the thermometer as a packing for better fractionation. This set-up, which does not make use of a condenser, should not be used with low-boiling compounds.

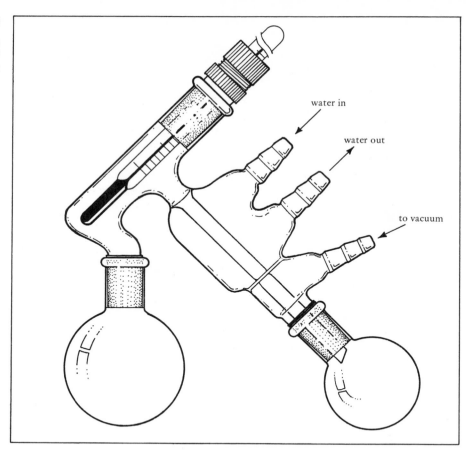

water in

water out

to vacuum

Figure 9.18. A small-scale apparatus for distillation.

volume of the boiling flask will result in a loss of 0.3 to 0.6 g per 100 ml; the walls of the boiling flask, adapters, and condenser will be wet with liquid at the end of the distillation; and the column will hold up a certain amount. The distillation apparatus illustrated in Figure 9.3 will result in losses of approximately 1 to 3 ml, depending on the size of the flask, and if a fractionating column is used, the total may be between 3 and 5 ml. It should be obvious from this that if the volume of material to be distilled is less than 15 or 20 ml, a smaller apparatus should be used. If the separation problem is very easy, the apparatus of Figure 9.17 may be used. If a substance is to be distilled without a condenser, it must boil at a relatively high temperature, or there must be very little of it, or the distillation must go slowly; the receiver may be cooled occasionally in a cold water bath. A nicer but more expensive apparatus is shown in Figure 9.18. It is very easy to superheat the vapors in a small apparatus if the procedures described above are not used.

Position of thermometer   Since the temperature of the vapor is very often taken to be the boiling point of the corresponding distillate, care must be taken (1) to avoid superheating the vapor and (2) to make certain that the entire mercury-containing bulb of the thermometer is below the bottom of the side arm leading to the condenser. A very common error is to position the thermometer so high that the bulb is not entirely heated to the temperature of the vapor; too low a temperature will then be recorded.

If the desired separation requires more than a few theoretical plates, a fractionating column more elaborate than the types shown in Figure 9.10 is usually desirable. References 1 and 2 describe the details of operation of a number of highly efficient fractionating columns of different designs.

1. Explain what effect each of the following mistakes would have on the success of a distillation:
   a. You forgot to add boiling stones.
   b. You attempted to distill a low-boiling, flammable liquid (such as ether) using a flame but without using a condenser.
   c. Your thermometer was positioned too high in the distilling adapter. That is, the top of the mercury bulb was above the bottom of the opening of the side-arm of the adapter that leads to the condenser.
   d. You distilled too fast.
   e. When isolating carvone (boiling point around 230°C; Section 46) from either oil of spearmint or oil of caraway you collect as product the material boiling from 110°C to 231°C.

Problems

1. **a.** Estimate from Figure 9.12 the boiling point of a water/methanol mixture 0.6 mole fraction in water.
   b. What would be the composition of the vapor in equilibrium with this mixture at the boiling point?
2. **a.** Estimate from Figure 9.14 the boiling point of a mixture of benzene and methanol that is 0.1 mole fraction in methanol. What would be the composition of the vapor in equilibrium with this mixture at the boiling point?
   b. Answer the same questions as in **a.**, for the case of a mixture that is 0.1 mole fraction in benzene.
3. Referring to Figure 9.12, and assuming the use of a magic fractionating column with an infinite number of theoretical plates and zero holdup:
   a. What would be the boiling point and the composition of the initial distillate, starting with an equimolar mixture of methanol and water?
   b. What mole percent of the total mixture would distill with this composition?
   c. What would be the boiling point and composition of the remainder?
   d. Present the results of your calculations for **b.** and **c.** as a plot of boiling point of of the distillate (ordinate) versus mole percent distilled (abscissa).
4. Do the same as in Problem 3, but substitute benzene and methanol (Figure 9.14).
5. Do the same as in Problem 3, but substitute acetone and chloroform (Figure 9.16).
6. Interpret the existence of a minimum-boiling azeotrope in the system acetone/chloroform in terms of the attractive forces between the molecules. Does your hypothesis explain the composition of the azeotrope?

Exercises

1. Distillation of a pure substance. Distill 50 ml of water (or methanol or carbon tetrachloride) in an apparatus similar to that illustrated in Figure 9.3. Keep a record of the temperature of the vapor versus the volume of the distillate. When the distillation is complete, make a plot of the temperature (ordinate) versus the volume (abscissa).
2. Distillation of an azeotropic mixture. Distill (as described in Exercise 1) 50 ml of a mixture of *n*-propyl alcohol and water that is 0.57 mole fraction water. How can you tell whether you are distilling a pure substance or an azeotropic mixture?
3. Distillation of a mixture of methanol and water. Distill 100 ml of an equimolar mixture of methanol and water, using a simple fractionating column as illustrated in Figure 9.10. Keep a record of the temperature of the vapor at the top of the column versus the volume of the distillate collected. When the distillation is complete, plot your data as in Problem 3 and compare the actual result with the ideal.
4. Isolation of (R)-(−)- or (S)-(+)-carvone from oil of spearmint or oil of caraway by distillation (Section 46).

1. K. B. Wiberg, *Laboratory Technique in Organic Chemistry*, McGraw-Hill, New York, 1960, p. 24.

2. *Technique of Organic Chemistry*, Vol. IV, A. Weissberger, editor, Interscience, New York, 1951.

Sources from which more information about simple distillation and fractional distillation may be obtained include References 1 and 2.

§10 REDUCED-PRESSURE DISTILLATION

A liquid will boil when its vapor pressure equals the pressure on its surface. This means that you can make a substance boil at a temperature lower than its normal boiling point by distilling it in an apparatus in which the surface pressure can be reduced. Because many substances undergo noticeable decomposition at elevated temperatures, distillation under reduced pressure is desirable for these substances in order to minimize decomposition. In the case of substances whose thermal stability is not known, distillation is usually carried out under reduced pressure if it appears that the normal boiling point will be greater than 125–175°C.

10.1 ESTIMATION OF THE BOILING POINT AT REDUCED PRESSURE

A useful generalization is that the boiling point of a compound will decrease by about 20–30°C each time the external pressure is reduced by a factor of two. Thus, the boiling point of a compound in an apparatus in which the pressure is maintained at 45–50 Torr (1/16 atmosphere) would be expected to be 80–100°C lower than its normal (atmospheric-pressure) boiling point.

The nomograph in Figure 10.1 is useful for estimating both expected reduced-pressure boiling points from normal boiling points and normal boiling points from observed reduced-pressure boiling points. The nomograph applies to liquids which are not associated in the liquid phase. The variation of boiling

Figure 10.1. A nomograph for estimating the boiling point as a function of pressure. The nomograph relates the normal boiling point of a substance (scale B) to boiling points at reduced pressures (scales A and C). A line connecting points on two scales will intersect the third scale at some point. Thus, from values for A and C, you can estimate B. Knowing B (or having estimated B), you can estimate the boiling point A′ at a reduced pressure C′. A transparent plastic ruler works as well as anything for connecting points and noting intersections.

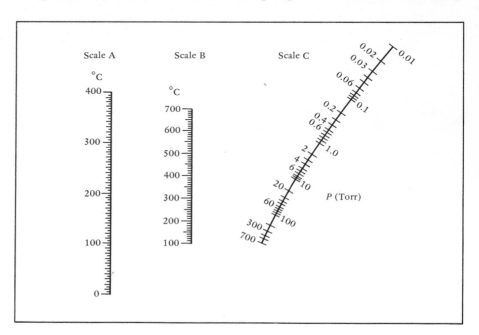

point with pressure for associated liquids, such as alcohols, is 10–20 percent less

than for nonassociated liquids.

Reference 1 describes a more accurate method of calculating changes in boiling point with variations in pressure.

10.2 APPARATUS

It is possible to carry out distillations at reduced pressure in a regular distillation apparatus such as those illustrated in Figures 9.3 and 9.17 by connecting a source of vacuum to the side arm of the distillation adapter. Other apparatus that are

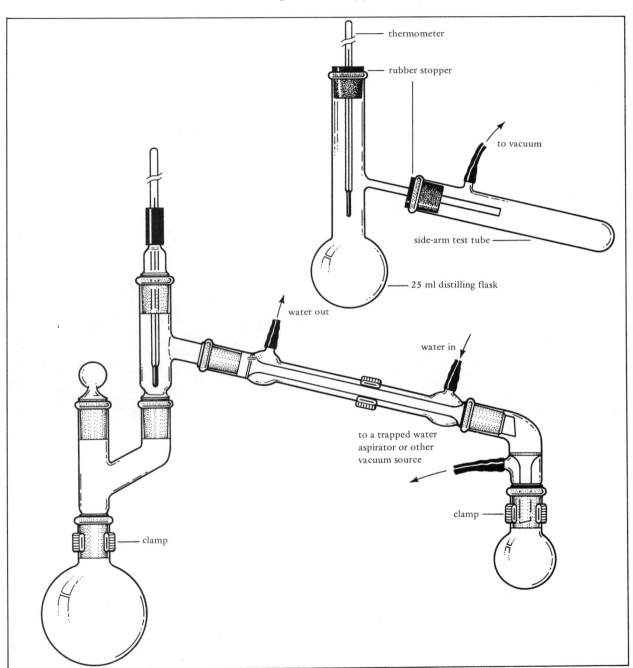

Figure 10.2. Two examples of apparatus for vacuum distillation.

suitable for vacuum distillation if the separation requirements are modest are illustrated in Figure 10.2. For smaller scale work, an arrangement such as that shown in Figure 9.18 could be used.

For separations that require more than simply removing a volatile solvent and then vacuum distilling the product, or vacuum distilling the product from a nonvolatile residue, a more specialized apparatus must be used. This apparatus must allow the collection of fractions of different boiling ranges without variation of the internal pressure or interruption of the distillation process. The simplest *Fraction collectors* type of so-called *fraction collector* uses a distillation adapter that may be rotated in order to collect the distillate in different receivers. Some examples are illustrated in Figure 10.3. The apparatus illustrated in Figure 9.18 may also be used for vacuum distillation of small samples when a rotating fraction collector of this type is attached to it as shown in Figure 10.4. The disadvantages of this type of fraction collector are the limited number of fractions allowed and the fact that the more volatile fractions can distill into the less volatile ones in the course of the distillation.

Another approach to the problem of how to collect fractions without disturb-*Fraction cutter* ing the equilibria in the apparatus is to use a distillation adapter (*fraction cutter*), which will allow removal and replacement of a receiver without breaking the vacuum in the distillation apparatus. An example of an adapter of this type is

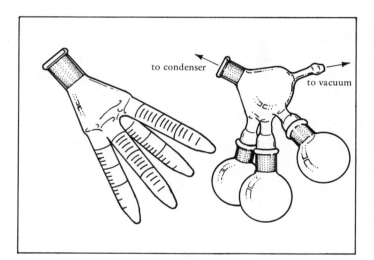

Figure 10.3. Two simple fraction collectors ("cows").

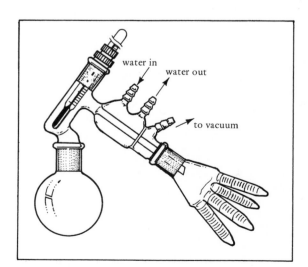

Figure 10.4. Small-scale vacuum distillation apparatus with rotating fraction collector.

shown in Figure 10.5. Distillation apparatus have been designed that include a short vacuum-jacketed Vigreux column, condenser, and fraction cutter in one compact piece. An example is illustrated in Figure 10.6.

10.3 SOURCE OF VACUUM

Water aspirator

The most common and convenient source of vacuum is the water aspirator. The ultimate pressure attainable with a water aspirator equals the vapor pressure of water at temperature of flow. Between 5 and 30°C, the vapor pressure of water, in Torr, is numerically equal to the temperature, in degrees C, within 2.5 Torr. Thus, in the winter, you might hope to be able to attain a pressure below 15 Torr, while in the summer, 25 Torr might be the ultimate pressure. An arrangement such as that illustrated in Figure 10.7 is suitable for reduced-pressure distillation when the desired pressure is equal to or above the pressure attainable with the water aspirator. The suction flask not only serves as a trap to prevent water from being sucked back into the apparatus in the event that the water pressure decreases momentarily, but also buffers the system against changes in pressure resulting from variations in the flow rate of the water or from changing distillation receivers. The Bunsen burner valve can serve to provide a controlled rate of leakage of air into the system in case you wish to operate at a pressure above the ultimate pressure that you can attain with the aspirator.

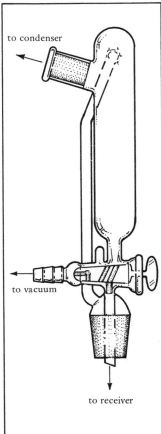

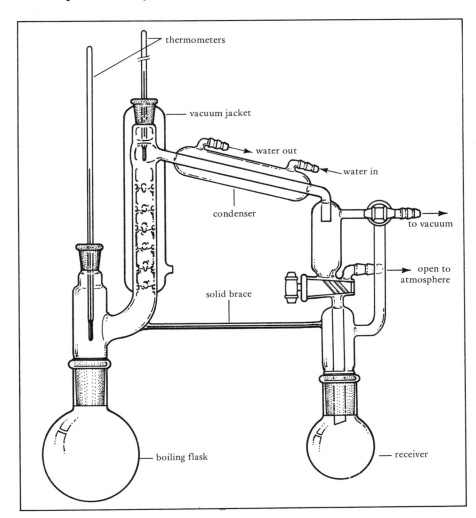

Figure 10.5. Fraction cutter for vacuum distillation. Rotation of the stopcock to different positions allows changing of receivers without breaking the vacuum to the system.

Figure 10.6. Apparatus for fractional distillation under vacuum.

Oil pump
Cold trap

For pressures below 10–25 Torr, a mechanical vacuum pump is generally used. A pump in average condition can give an ultimate pressure of 3–5 Torr, and 1 Torr if in good condition. A mechanical vacuum pump should always be connected to the system through a *cold trap*, as shown in Figure 10.8, in order to keep volatile (or corrosive) materials from reaching the pump, since the ultimate pressure attainable by the pump will be no lower than the vapor pressure of the liquid (oil) in the pump. Often, the performance of a mechanical pump can be restored merely by changing the oil. If it is necessary to carry out a reduced-pressure distillation with a mechanical vacuum pump at a pressure greater than the ultimate attainable by the pump, a system involving a *manostat*, such as that described in *Laboratory Technique in Organic Chemistry* (2), should be used, and *not* an air leak. Drawing air through the hot oil of the pump would lead to degradation and an increase in vapor pressure.

10.4 PRESSURE MEASUREMENT

Manometer; McLeod gauge

For measuring pressures in excess of 5–10 Torr, you will find a simple manometer such as that illustrated in Figure 10.9 to be quite adequate. For determination of pressures down to 1 Torr, a tilting McLeod gauge is useful (2).

10.5 TECHNIQUE OF DISTILLATION UNDER REDUCED PRESSURE

The vapor volume resulting from the vaporization of a given amount of substance is many times larger at a reduced pressure than at 1 atmosphere (100 times larger at 7.6 Torr; 760 times larger at 1 Torr). This means that maintaining an even rate of boiling and prevention of bumping are even more important in vacuum distillations than in distillations at atmospheric pressure: a burst of vapor produced in a bump during vacuum distillation can easily splash material over into the receiver, as well as upset the vapor–liquid equilibria.

Boiling stones are often used to prevent bumping in a brief distillation, but they seem to lose their effectiveness fairly quickly. If for some reason the pressure in the apparatus rises momentarily, the boiling stones often cease to function afterwards. Microporous carbon boiling chips (available from Fisher Scientific Company) are often recommended as being most suitable.

One of the most convenient methods of heating the boiling flask during distillation is to use a magnetically stirred and electrically heated oil bath (see Figure 33.1). If this method is used, bumping can usually be eliminated by placing a small magnetic stirring bar in the boiling flask and adjusting the level of the heating bath until it is somewhat higher than the level of the liquid in the flask. If the temperature of the bath is not too much above the boiling point of the liquid, and the liquid and bath are well stirred (the stirring bar in the bath turns the smaller one in the flask), boiling will usually be very smooth.

Experimental measurement of the boiling point of a substance in a reduced-pressure distillation is much less certain than in distillations at atmospheric pressure, and extrapolations of the reduced-pressure boiling points to atmospheric pressure boiling points will contain these uncertainties. Since, for a given rate of heat input, the vapor velocity in a reduced-pressure distillation may be several hundred times greater (due to the greater volume) than in an atmospheric pressure distillation, it is much easier for superheated vapors to reach the thermometer. In addition, the rapid flow of large amounts of vapor through the apparatus may

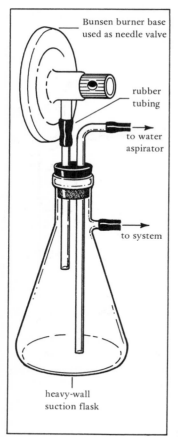

Bunsen burner base
used as needle valve

rubber
tubing

to water
aspirator

to system

heavy-wall
suction flask

Figure 10.7. Pressure regulation using the water aspirator.

result in a considerable pressure drop. That is, the pressure may be considerably higher in the boiler than at the manometer. Both of these effects may be minimized by using a heating bath, and using it at a temperature that provides a minimum rate of distillation. An additional uncertainty is introduced when an air bleed is used to promote even boiling, because the air will contribute an unknown amount to the total pressure of the system.

Other techniques of distillation, which also apply to distillation under reduced pressure, are described in Section 9.6.

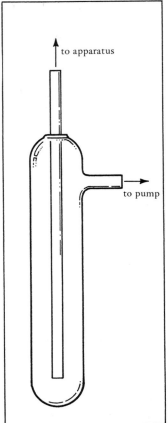

Figure 10.8. A cold trap. The lower part of the trap must be immersed in a Dry Ice/acetone bath or liquid nitrogen bath in order to condense vapors to keep them from getting to the pump.

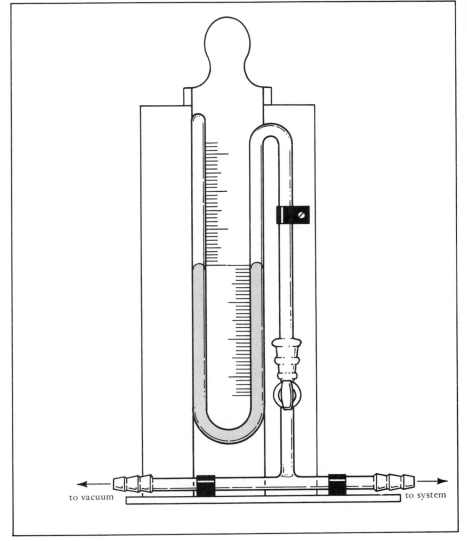

Figure 10.9. A simple manometer. The pressure equals the difference in height of the two columns of mercury. This height difference may be read by means of the movable scale.

Problems

1. a. Benzyl alcohol is reported to boil at 93°C at 10 Torr. At what temperature would it be expected to boil at atmospheric pressure?

   b. 1,2-Dibutoxybenzene is reported to boil at 135–138°C at 12 Torr. What would you expect its normal boiling point to be?

2. a. α-Phenylethylamine boils at 187°C at atmospheric pressure. At what temperature would you expect it to distill under the vacuum that can be obtained with a water aspirator (assume 20 Torr)?

b. The normal boiling point of nitrobenzene is 211°C. At what temperature would it distill under a pressure of 15 Torr?

3. a. Diethyl ether has a normal boiling point of 34.6°C. Calculate the vapor pressure of ether at 0°C.

b. Diethyl phthalate is reported to have a normal boiling point of 296°C. How low would the pressure have to be in order to distill this substance at a temperature below 100°C?

Exercises

1. In the isolation of eugenol from oil of cloves (Section 44), the procedure may be extended to include vacuum distillation of the eugenol.
2. (R)-(−)- or (S)-(+)-carvone may be isolated from oil of spearmint or oil of caraway by distillation under reduced pressure (Section 46).
3. In the preparation of dibenzylketone (Section 74.3), the final distillation of the product can be carried out much more easily under reduced pressure.

References

1. H. B. Hass, *J. Chem. Educ.* **13**, 490 (1936).
2. K. B. Wiberg, *Laboratory Technique in Organic Chemistry*, McGraw-Hill, New York, 1960.

---

§11 DISTILLATION OF MIXTURES OF TWO IMMISCIBLE LIQUIDS; STEAM DISTILLATION

In a mixture of two completely immiscible liquids, each exerts its own vapor pressure independently of the other. As the temperature of such a mixture in an apparatus open to the atmosphere is raised, the vapor pressure of each substance increases until the total vapor pressure equals the pressure of the atmosphere. Since the total vapor pressure is the sum of the individual vapor pressures, the total vapor pressure must become equal to atmospheric pressure at a temperature below the boiling point of either pure substance. The mixture thus distills at a temperature below the boiling point of either pure component.

Since organic compounds are generally miscible with one another, this phenomenon is usually observed only when one of the liquids is water; in these cases, the distillation process is called *steam distillation*. Steam distillation is a useful technique for effecting certain separations and a method for distilling certain high-boiling compounds at a temperature no greater than 100°C.

11.1 THEORY OF STEAM DISTILLATION

In Section 9.5, the occurrence of minimum-boiling azeotropes was interpreted in terms of incipient insolubility. If insolubility persists to high enough temperatures, a situation involving the distillation of two insoluble liquids will be attained. The gradual change from the one situation to the other can be imagined to occur according to the process outlined in Figures 11.1 and 11.2.

Composition of the Distillate

The composition of the vapor at the boiling point, and therefore of the distillate, can be calculated by realizing that the ratio of the number of molecules of each

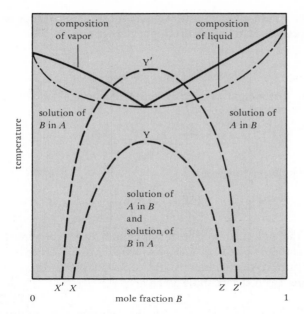

composition of vapor   composition of liquid

solution of B in A

solution of A in B

solution of A in B and solution of B in A

X′ X   Z Z′

0   mole fraction B   1

temperature

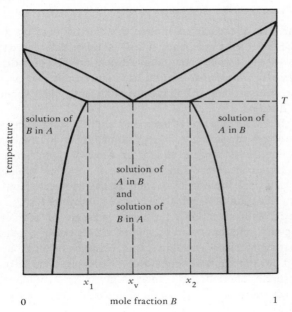

solution of B in A

solution of A in B

solution of A in B and solution of B in A

x_1 x_v x_2

0   mole fraction B   1

temperature

T

Figure 11.1. A hypothetical boiling point diagram for a system involving a minimum-boiling azeotrope. If A and B are not completely miscible at all temperatures below the boiling point, certain mixtures of A and B will exist as two phases below certain temperatures. The area of the diagram under the broken curve XYZ represents the temperatures and compositions at which two phases exist in this hypothetical example. If you imagine that immiscibility could persist to temperatures above boiling point (XYZ moves up to X′Y′Z′), the boiling point diagram could be imagined to change to that shown in Figure 11.2.

Figure 11.2. A hypothetical boiling-point diagram of a system involving two "immiscible" liquids. At all compositions between x_1 and x_2, two phases will exist at the boiling point. The boiling point of all such mixtures will be $T°$, and the composition of the vapor will be x_v.

type in the vapor must equal the ratio of the vapor pressures of the pure substances at that temperature:

$$\frac{P_B}{P_A} = \frac{\text{molecules of } B}{\text{molecules of } A} = \frac{\text{moles of } B}{\text{moles of } A} = \frac{n_B}{n_A} \qquad (11.1\text{-}1)$$

For example, a mixture of the immiscible liquids bromobenzene and water steam distills at 95.3°C at 760 Torr. The vapor pressure of water at this temperature is 641 Torr, and therefore the vapor pressure of bromobenzene must be $760 - 641 = 119$ Torr. The vapor pressure ratio of 119/641 must equal the mole ratio of the vapor and thus of the distillate, as in Equation 11.1-2:

$$\frac{119}{641} = \frac{n_{\text{brom.}}}{n_{\text{water}}} = 0.186 \text{ mole bromobenzene/mole water} \qquad (11.1\text{-}2)$$

Since water is usually one of the substances involved in the distillation of mixtures of two immiscible liquids, we may substitute water for one of the substances in Equation 11.1-1. Substituting water for substance A gives us:

$$\frac{P_B}{P_{H_2O}} = \frac{n_B}{n_{H_2O}} \tag{11.1-3}$$

Remembering that weight divided by molecular weight equals moles, we may write:

$$\frac{P_B}{P_{H_2O}} = \frac{\text{wgt}_B/M_B}{\text{wgt}_{H_2O}/M_{H_2O}} = \frac{\text{wgt}_B}{\text{wgt}_{H_2O}} \cdot \frac{18}{M_B} \tag{11.1-4}$$

Solving for $\text{wgt}_B/\text{wgt}_{H_2O}$, we get:

$$\frac{\text{wgt}_B}{\text{wgt}_{H_2O}} = \frac{M_B}{18} \cdot \frac{P_B}{P_{H_2O}} \tag{11.1-5}$$

where M_B is the molecular weight of compound B, which is insoluble in water, and P_B/P_{H_2O} is the ratio of the vapor pressure of B to the vapor pressure of water at the temperature at which the sum of the two vapor pressures equals atmospheric pressure. Using again the example given above:

$$\frac{\text{wgt}_{\text{brom.}}}{\text{wgt}_{\text{water}}} = \frac{157}{18} \cdot \frac{119}{641} = 1.62 \text{ g bromobenzene/gram water}$$

The higher the ratio of organic compound to water the better, since less distillate will have to be collected (and hence it will take less time) in order to collect a given amount of the organic compound. Equation 11.1-5 indicates that for a compound of a given molecular weight, the higher the vapor pressure, the more efficient the steam distillation. Solids with sufficiently high vapor pressures can also be steam distilled.

Because all real liquids are at least very slightly soluble in one another, the analysis given above can be only an approximation to any real situation. In the example involving bromobenzene and water, the two "immiscible" liquid layers would be a solution of a very small amount of water in bromobenzene and a solution of a very small amount of bromobenzene in water. However, according to Raoult's law (Section 9.3), which accurately accounts for the vapor pressure of the *solvent* in dilute solutions, the vapor pressure of bromobenzene in the bromobenzene-rich solution would be approximately equal to that of pure bromobenzene at that temperature, and the vapor pressure of water in the water-rich solution would be approximately equal to the vapor pressure of pure water at the same temperature.* Therefore, the analysis given for "completely immiscible" liquids will apply quite well if the two liquids are not very soluble in one another.

11.2 TECHNIQUE OF STEAM DISTILLATION

Internal generation of steam

A substance can be distilled with steam by adding water to a mixture in which the substance (for example, a crude product from a reaction) is present, and boiling the resulting suspension. If the desired compound has a significantly higher vapor pressure at about 100°C than any of the other components of the mixture, it can be selectively removed by this process. An apparatus suitable for this kind of operation is illustrated in Figure 11.3.

Sometimes the presence of a solid in a mixture will cause unavoidable

---

* It is interesting to realize that since both solutions are in equilibrium with vapor of the same composition, and thus must be in equilibrium with each other, the vapor pressure of water in the bromobenzene-rich solution (less than 1% water) must be equal to the vapor pressure of water in the water-rich solution (more than 99% water). A similar statement can be made about the vapor pressure of bromobenzene in the two solutions.

superheating and violent bumping if the flask is heated directly. In these cases, the steam must be passed in from an external source. When steam is supplied externally, a satisfactory rate of steam distillation can be attained without any danger that the contents of the flask will be overheated by the vigorous heating necessary to supply the large molar heat of vaporization of water. If a large amount of water will be required to distill the substance, this can be an important consideration, since a maximum rate of distillation will probably be desired. An arrangement for the production and use of external steam is shown in Figure 11.4 (see page 72). When an external source of steam is used, water may condense in the distillation flask. This can be prevented by gently heating the flask with a Bunsen burner, an oil bath, or a heating mantle.

Test for Completion of Distillation

There are two useful ways to determine when a steam distillation is complete. The first is to determine that the distillate currently being produced contains no water-insoluble material. This can be done by collecting a sample of the distillate

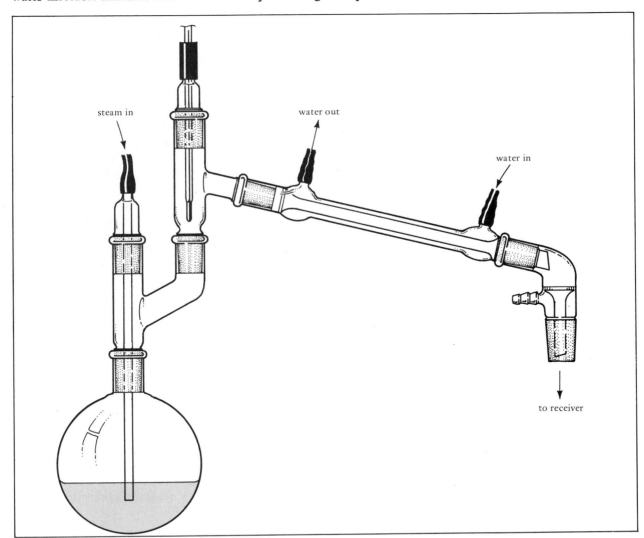

Figure 11.3. An apparatus for steam distillation. Replace the adapter and steam inlet tube with a separatory funnel if steam is generated internally.

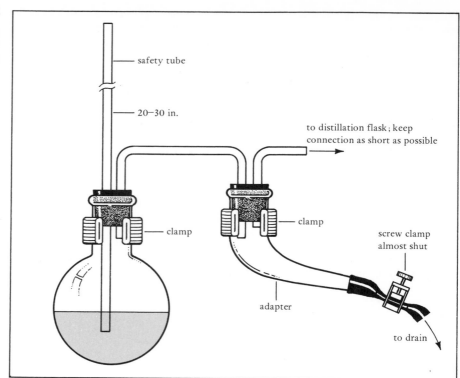

Figure 11.4. An apparatus for generating steam.

Look for oily droplets

separately, or by observing either that the distillate flowing down the condenser no longer contains little oily droplets or that solid is no longer collecting in the cold condenser. (If solid tends to collect and block the condenser, the cooling water should be turned off until the solid has melted and run into the receiver.) No harm is done in allowing the steam distillation to continue somewhat past the point at which insoluble material can be observed in the newly formed distillate.

Another way of estimating when a steam distillation can be terminated is by calculating how much water should be collected in order to distill a given amount of substance. The amount of water collected can be estimated by collecting the distillate in a graduated cylinder or an approximately calibrated Erlenmeyer flask. According to Equation 11.1-5, if you know the molecular weight of the substance and the ratio of the vapor pressure of the substance to the vapor pressure of water at the temperature of distillation, the weight of water required to distill a given amount of the substance can be calculated. In the example worked out above, it was calculated that 1.62 grams of bromobenzene could be distilled by 1 gram of water. Since most steam distillations occur at approximately 100°C, it is satisfactory to use the ratio of vapor pressures at 100°C, rather than at the (usually) unknown temperature of distillation:

$$\frac{\text{wgt}_B}{\text{wgt}_{H_2O}} = \frac{M_B}{18} \cdot \frac{P_B^{100°}}{760} \tag{11.2-1}$$

The vapor pressure of B at 100°C may be estimated according to the nomograph of Figure 10.1 as long as the boiling point at atmospheric pressure is known. Employing the case of bromobenzene again as an example, and using 140 Torr as the value of the vapor pressure that can be estimated for bromobenzene at 100°C, we get:

$$\frac{\text{wgt}_{\text{brom.}}}{\text{wgt}_{\text{water}}} = \frac{175}{18} \cdot \frac{140}{760}$$

$$= 1.6 \text{ g bromobenzene per gram water}$$

This is the same value as that calculated using the ratio of vapor pressures at the actual temperature of distillation. The less the mutual solubilities of the substances and the closer the temperature of distillation to 100°C, the better this approximation.

Problems

1. In Experiment 43, oil of cloves is isolated from cloves by steam distillation. Assuming oil of cloves to be composed mostly of eugenol, which has a normal boiling point of 252°C.

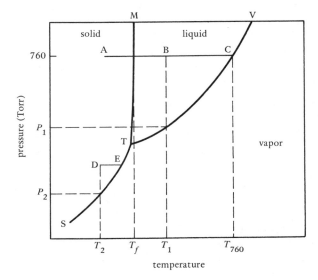

eugenol

a. Estimate the vapor pressure of eugenol at 100°C.
b. Calculate how many grams of water would be required to steam distill a gram of eugenol (oil of cloves).

Exercises

1. Isolation of oil of cloves from whole cloves (Section 43).
2. Other experiments that involve a purification by steam distillation include the oxidation of cyclohexanol to cyclohexanone (Section 53, procedure **a**), the preparation of n-butyl bromide (Section 55.1), the preparation of aniline (Section 59), the preparation of chlorobenzene and the chlorotoluenes (Section 64), and the synthesis of 1-bromo-3-chloro-5-iodobenzene (Section 76.4).

§12 SUBLIMATION

12.1 THEORY OF SUBLIMATION

The processes of sublimation and distillation are closely related. They may be compared with reference to the pressure–temperature diagram in Figure 12.1. In the process of *distillation of a liquid*, the liquid is heated until it reaches the temperature at which its vapor pressure equals the total pressure of the system; the

Figure 12.1. A general pressure–temperature diagram showing equilibria between solid, liquid, and vapor phases.

state of the system may be represented during this process by a point moving from B to C in Figure 12.1.

The vapors that expand to a cooler part of the apparatus are then recondensed, the point now moving from C back to B. The distillation of a solid involves first melting and then boiling, condensation, and resolidification (a point moving from A to C and back again); it is just like distillation of a liquid except for melting and solidification. However, if a solid is heated under a total pressure less than that corresponding to the triple point (T in Figure 12.1), it will reach a temperature at which its vapor pressure equals the total pressure and will "distill" without melting, a point moving from D to E and back. Such behavior is called *sublimation*. The loss of water from food that has been stored in a freezer for a long time takes place by sublimation ("freezer burn"). Similarly, wet clothes hung out to dry on a winter day first freeze and then dry by sublimation.

It is very rare for a substance to have a vapor pressure of 760 Torr or greater at the temperature of the triple point, which, as Figure 12.1 indicates, is approximately the same temperature as the normal melting point T_f, but many substances have a large enough vapor pressure near their melting point to be sublimed at the reduced pressures easily obtained in the laboratory. Table 12.1 lists a number of substances and their vapor pressures at the melting point.

Table 12.1. Some substances that are easily sublimed under laboratory conditions

Compound	Melting Point $(T_f, °C)$	Vapor Pressure at T_f Vapor Pressure at T[a]
Hexachloroethane	185	780 Torr
Perfluorocyclohexane	59	950 Torr
Camphor	179	370 Torr
Anthracene	218	41 Torr
Naphthalene	80	7 Torr
Benzene	5	36 Torr
p-Dibromobenzene	87	9 Torr
p-Dichlorobenzene	53	8.5 Torr
Phthalic anhydride	131	9 Torr
Benzoic acid	122	6 Torr
Carbon dioxide	−57	5.2 atm
Iodine	114	90 Torr

Many aromatic amines and phenols are easily sublimed.

[a] Since the volume change upon melting is very small, the melting point of a substance is practically independent of pressure; therefore, the temperature at the triple point will be practically the same as at the normal melting point. Thus, since the temperature of the solid at the triple point and at the melting point is practically the same, the vapor pressure of the substance at the melting point should be the same as at the triple point.

Sublimation is a useful procedure for purification if the impurities are essentially nonvolatile and the desired substance has a vapor pressure of at least a few Torr at its melting point. If the impurities have about the same vapor pressure as the desired compound, very little separation will be achieved and it would be much better to attempt to purify the substance by recrystallization. Sublimation is most useful in the case of small samples, as mechanical losses can be kept very low. Several apparatus for sublimation are illustrated in Figures 12.2 and 12.3. Substances like camphor whose vapor pressure at the melting point is substantial, although less than 760 Torr, can be sublimed successfully at atmospheric pressure.

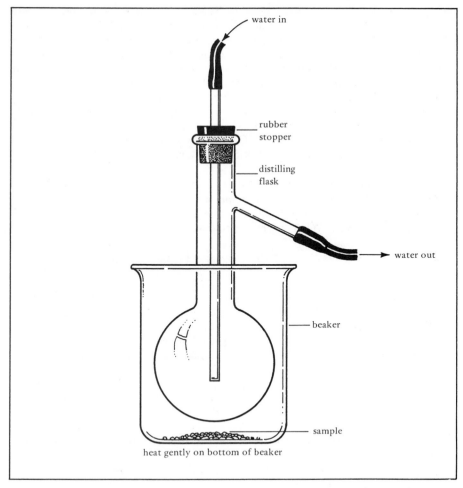

water in

rubber stopper

distilling flask

water out

beaker

sample

heat gently on bottom of beaker

Figure 12.2. An apparatus for sublimation at atmospheric pressure.

The rate of sublimation will be relatively low and will depend upon the rate of diffusion of the vapor to the cold condensing surface.

12.2 TECHNIQUE OF SUBLIMATION

In a sublimation at atmospheric pressure, spread out the material to be sublimed in the container to be heated and arrange for the condensation of the vapors (Figure 12.2). Heat the container cautiously without melting the solid until crystals can be seen to appear on the cold surface of the condenser. At this point, only a small amount of heat is needed to keep the sublimation going. If you are heating with a burner, the flame will probably have to be applied intermittently. The rate of sublimation is much easier to control if an oil bath is used for heating (Section 32.1). When only a nonvolatile residue remains, carefully remove the condenser and scrape off the product.

Heat gently

In sublimation at reduced pressure, the procedure is the same except that a closed system in which the pressure can be reduced must be used (Figure 12.3). The source of vacuum can be either the water aspirator or the oil pump (Section 10.3).

Problems

1. Whether or not a substance can be sublimed depends upon its vapor pressure at the triple point. To be easily and conveniently sublimed, a substance should have a

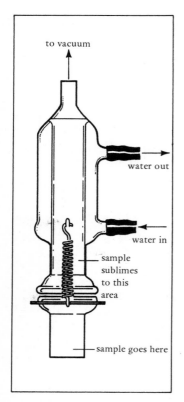

Figure 12.3. An
arrangement for sublimation
under reduced pressure.

vapor pressure of at least 5–25 Torr. In Table 12.1, it was pointed out that the vapor pressure at the melting point should be a good approximation of the vapor pressure at the triple point. Since at the melting point the vapor pressure of the solid is equal to the vapor pressure of the liquid, and since it is possible to estimate the vapor pressure of a liquid at any temperature (even at its freezing point) when its vapor pressure at any other temperature is known (for example, by using the nomograph of Figure 10.1), knowledge of the boiling point of the substance at any pressure and the melting point of the substance should make it possible to estimate the vapor pressure at the triple point.

  a. Knowing that benzene boils at 80°C, calculate the vapor pressure to be expected for benzene at its freezing point of 5°C. Compare your result with the value in Table 12.1.

  b. Do the same for benzoic acid, given that its normal boiling point is 249°C.

2. The vapor pressure, in Torr, of a substance at its melting point (P_f) can be related to its heat of vaporization ΔH_{vap}, its boiling point T_b, its melting point T_f, and atmospheric pressure (760 Torr) by the following equation:

$$\log P_f = \log 760 - \frac{\Delta H_{vap}}{2.3RT_b}\left(\frac{T_b - T_f}{T_f}\right)$$

You can see that compounds with relatively high melting points and short liquid ranges should be relatively easily sublimed.

  a. Which of the isomeric octanes would be expected to sublime most easily? (See Table 17.2.)

  b. In Table 12.1, it was stated that in general aromatic amines and phenols are relatively easily sublimed. Why does this make sense in terms of the structures of these molecules?

Exercises

1. Sublime a sample of hexachloroethane, using an arrangement such as that shown in Figure 12.2.

2. Sublime a sample of camphor. This may be done at atmospheric pressure or in an apparatus such as that illustrated in Figure 12.3.

3. Benzoic acid has a vapor pressure of 6 Torr at the melting point and may be purified by vacuum sublimation as well as by crystallization. A mechanical vacuum pump should be used so that the process will not take too long.

4. In the procedure of Section 42, caffeine is purified by recrystallization from ethanol. An alternative method should be purification by sublimation, since caffeine is reported to sublime slowly at atmospheric pressure at temperatures above 120°C, to sublime quickly at 1 Torr at 160–165°C, and to sublime at 89°C and 15 Torr.

---

§13 EXTRACTION BY SOLVENTS

The fact that different substances may differ in solubility in various liquids can serve as a basis for separating them. Purification by recrystallization involves one application of this phenomenon, and solvent extractions and chromatographic methods (Section 14) are others.

13.1 THEORY OF EXTRACTION

If a mixture of substances, say A and B, is dissolved in an organic solvent such as diethyl ether and the solution is mixed thoroughly with the immiscible solvent water, an essentially complete separation of A and B may be effected if one of the substances, say A, is relatively more soluble in water than in ether compared to

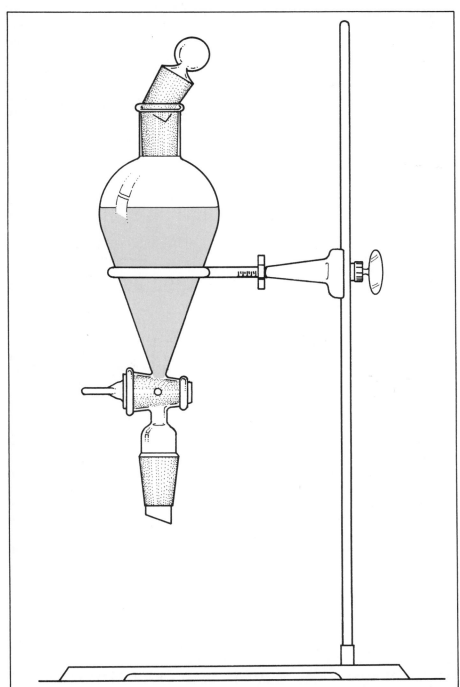

Figure 13.1. A separatory funnel.

the other. If at equilibrium most of the *A* molecules are in the water phase and most of the *B* molecules are in the ether phase, a physical separation of the layers by means of a separatory funnel (Figure 13.1) will result in a separation of the two kinds of molecules. The material in each layer may be recovered typically by distillation or evaporation of the solvent.

Table 13.1 indicates in a general way the relative solubility of different types of compounds in water and organic solvents.

Table 13.1. Estimated relative solubility of different types of compounds in organic solvents and water

Solubility in all solvents decreases with increasing molecular weight

Type of Compound	Estimated Ratio of Solubility in Organic Solvent to Solubility in Water
Covalent substances containing only carbon, hydrogen, and halogen	Very much greater than 1
Covalent substances containing oxygen and/or nitrogen in addition to carbon, hydrogen, and halogen	
a. 5 carbon atoms per functional group	10:1
b. 2 carbon atoms per functional group	1:1
c. 1 carbon atom per functional group	1:10
Salt of an organic acid	Very much less than 1
Salt of an organic base; amine salt	Very much less than 1
Inorganic salts	Very much less than 1

13.2 EXTRACTION OF ACIDS AND BASES

The extraction of an acidic or basic substance, either the product of the reaction or an undesired side-product, from an organic solvent into water can be effected by mixing (shaking) the solution in the separatory funnel with an aqueous solution of base or acid, respectively.

Extraction of acids by bases

In the case of the extraction of an acidic material with aqueous base, the acid in the organic solution dissolves in the water and is immediately converted to its salt. The salt is very much more soluble in water than in the organic solvent. Thus, the concentration of free acid in the aqueous phase will remain very low as long as the aqueous phase remains sufficiently basic to convert the acid to its salt. After equilibrium has been established, the *ratio* of free acid dissolved in the organic solvent to free acid dissolved in water may be much greater than 1 (due to greater solubility in the organic solvent than in water). Still, the fact that the vast majority of acid molecules are in the form of their corresponding salt dissolved in water means that only a very small *amount* of the acid originally present in the organic solvent is still there. From this analysis, you can see that the degree of completeness with which an acid can be extracted from an organic solvent depends upon the basicity of the extracting aqueous solution: the solution must be basic enough to convert the acid "completely" to its salt. A useful approximation is that the pH of the extracting solution should be at least four pH units more basic than the pK_a of the acid to be extracted. Thus, aqueous sodium carbonate solution (pH $\approx$ 11) should "completely" extract acids stronger than $pK_a \approx 7$ (see Table 13.2).

Extraction of bases by acids

Similarly, you could expect to "completely" extract an amine from an organic solvent by means of an aqueous acid solution that is at least four pH units more acidic than the pK_a of the conjugate acid of the amine (see the data of Table 13.2).

It should also be apparent that if a basic aqueous solution of an organic acid is acidified with a mineral acid (pH $\approx$ 1; $[OH^-] \approx 10^{-13}$), the acid should be extractable from water into an organic solvent if the free acid is less soluble in water than in the organic solvent. Of course, if the acid is not very soluble in water, it will separate or precipitate from the acidified solution before the organic solvent is added.

Table 13.2. pH required for extraction of different acids and bases

Approximate pH *of Aqueous Solution Needed to Extract an Acid from Organic Solvent*

Compound	pKa	pH *as Basic or More Basic Than:*
Mineral acids	> 1	~ 4
Ar—C(=O)—OH	~ 4	~ 8
R—C(=O)—OH	~ 5	~ 9
Phenols	~ 10	~ 14

Approximate pH *of Aqueous Solution Needed to Extract a Base from an Organic Solvent*

Compound	pKa	pH *as Acidic or More Acidic Than:*
Anilines	~ 5	~ 1
Pyridines	~ 6	~ 2
Aliphatic amines	~ 11	~ 7

Approximate pH *of Aqueous Solutions, 5–10% by Weight*

Compound	*Approximate* pH
HCl; H_2SO_4	0
Acetic acid	3
$NaHCO_3$	8
Na_2CO_3; K_2CO_3	11
$NaOH$; KOH	14

In a similar way it is possible to extract an organic base from an acidic aqueous solution by making the solution basic and then extracting the mixture with an organic solvent.

13.3 TECHNIQUE OF EXTRACTION

The objective of a simple extraction is to partition one or more substances between two immiscible solvents. This is usually accomplished with the use of a separatory funnel (Figure 13.1). If the separatory funnel has a glass stopcock, prepare the funnel for use by making sure that the stopcock is lightly greased and will turn without difficulty; Teflon stopcocks need not be greased. With the separatory funnel supported in a ring (Figure 13.1), check to make sure that the stopcock is closed and pour in the solution to be extracted. Then add the extracting solvent (the funnel should not be filled to more than about three-fourths of its height), replace the stopper after wetting it with water (to keep the organic solvent from creeping out around the stopper), and swirl or shake the contents to mix them. With vigorous shaking, a total mixing period of ten to thirty seconds is usually considered adequate to establish equilibrium. After allowing the mixture to stand in the separatory funnel until the two immiscible layers have separated cleanly, remove the stopper at the top and draw off part of the lower

Close the stopcock

Mix thoroughly

Remove stopcock for storage

Relief of pressure

Warning: boiling of solvent

Warning: CO_2 evolution

Choice of solvent

Extraction in batches

layer through the stopcock at the bottom. Wait a little while for the remainder of the lower layer to drain down (gentle swirling of the separatory funnel can speed this up), and draw this off also. Then pour the upper layer out the top.

If a glass stopcock plug is used, it should be removed from the separatory funnel, cleaned, and stored out of the funnel. If this is not done, you may find it impossible to turn the stopcock next time you use the funnel. (Teflon stopcock plugs are much better suited for separatory funnels than are glass plugs since they will not freeze in place, do not need to be greased, and do not need to be removed for storage.)

When a volatile solvent is involved in an extraction, the establishment of the equilibrium vapor pressure of the solvent will cause the pressure to rise inside the stoppered separatory funnel. The pressure is best released by turning the funnel upside down (with the stopper held in place with the palm of the hand) and cautiously opening the stopcock. When a very volatile solvent such as ether is being used, the first mixing should consist only of a slow inversion of the separatory funnel followed by release of the pressure. After alternate cautious sloshing of the contents of the separatory funnel and then release of pressure, the sound of the escaping vapors will indicate that the pressure is not being built up so fast and the periods of mixing can be longer and more vigorous. If you neglect to release the pressure inside the funnel, the stopper may be forced out. If the contents of the funnel are forced out as well, the escape of a volatile and flammable solvent can result in a dangerous fire.

It should be obvious that it is dangerous to attempt to extract a solution if its temperature is near or above the boiling point of the extracting solvent. This means that if an extraction is to be carried out with pentane, ether, or dichloromethane, it may be necessary to cool the solution to below room temperature. When these solvents are used, it is also a good idea to hold the separatory funnel by the ends so that the contents will not be warmed by your hands.

If a strong acid is to be extracted with carbonate or bicarbonate solution, the carbon dioxide produced can cause a large buildup in pressure unless mixing is done very cautiously with frequent release of pressure. In cases where much carbon dioxide production is anticipated, it is best to do the mixing in a flask or beaker and then to transfer the mixture to the separatory funnel for separation.

If an organic product is to be purified by dissolution in an organic solvent followed by extraction of the solution with two or more portions of aqueous solution, the whole process will be much faster and easier, and will involve less loss, if the organic solution is less dense than water. In this case, the water layer can be drawn off through the stopcock, and the organic solution is retained in the funnel ready for the next extraction. If the organic phase is heavier than water, it will have to be drawn off through the stopcock, the aqueous layer poured out, and the organic layer returned to the separatory funnel for the next extraction. Each such transfer will take time and may involve a loss of material. Conversely, if it is necessary to extract an aqueous solution with several portions of solvent in order to achieve the maximum recovery of a substance, it will be more convenient to extract with a solvent heavier than water so that the solvent can simply be drawn off each time through the stopcock without removal of the water layer first. The densities of a number of solvents are listed in Table 8.1.

If it is necessary to extract a larger volume of material than will fit into the available separatory funnel, the extraction may, of course, be done in batches. Small amounts of water insoluble material may be efficiently removed from large amounts of water by adding a little ether (or other solvent less dense than water) to the flask containing the water/product mixture, swirling the mixture well, and then adding it in portions to the separatory funnel, drawing off most of the water layer after each addition. The converse of this procedure can be used to wash a large amount of an organic solution with a little water, if the organic solution is heavier than water.

Often it is possible to tell which layer in the separatory funnel is organic and which is aqueous from knowledge of the relative volumes used or relative densities of the two solvents. Sometimes, however, the transfer of material from one layer to the other or the presence of several substances of different densities can make the identification of the layers uncertain. It is not possible, of course, to tell by smell which layer is which, since the vapor pressure of each component in each phase will be the same. But sometimes you can determine which is which by adding a little of the organic solvent or water and seeing which layer it goes into. Or you can withdraw a little of the lower layer and see if it is miscible with water. If doubt still remains, each layer should be carried on in the procedure as if it were the desired one until it becomes obvious that one of the two cannot be the right one. It is very common to discard the product layer through error or ignorance; it is always advisable to *save everything* until the product has been safely isolated.

Which layer is which?

Don't throw away the wrong layer

In most extractions, at least a trace of insoluble material collects at the interface between the two immiscible layers. It is often impossible to separate the layers without taking along some of this insoluble material. This is not a crucial matter, since whatever is picked up can always be removed by filtration at the end of the extraction or at some later stage in the purification. For example, in the very common case in which the organic product of a reaction is isolated by dissolving it in an organic solvent, extracting ("washing") the solution with one or more aqueous solutions in order to remove certain undesired materials, drying it (Section 15.2), and removing the drying agent by gravity filtration, any insoluble impurity that may have been carried along in the organic layer will be removed along with the drying agent in the gravity filtration.

If the distribution coefficient for a substance to be extracted from water is much less than one—that is, if the ratio at equilibrium of the concentration in the organic solvent to the concentration in water is much less than one—a simple extraction process will not give a satisfactory recovery. The distribution coefficient can sometimes be increased by adding sodium chloride or sodium sulfate to the aqueous solution, since the solubility of most organic compounds is less in salt solutions than in water. (The interpretation for this phenomenon, known as *salting out*, is given in Section 22.1.) Alternatively, the distribution coefficient may be increased by using an organic solvent that is a better solvent for the type of compound being extracted. In the case of substances with oxygen-containing functional groups, ethyl acetate and *n*-butyl alcohol are probably better solvents than non-oxygen-containing solvents. Chloroform is an especially good solvent for amines. (The factors that determine solubility are also discussed in Section 22.1.)

Salting out

$CHCl_3$

Sometimes the mixture in the separatory funnel does not separate into two phases, but forms a homogeneous solution. This happens when a good mutual solvent for the two solvents you are using in the extraction is present in a relatively large amount. A low-molecular-weight alcohol, tetrahydrofuran, or dioxane is usually involved when this happens. The situation can be cured by adding more of the two immiscible solvents so as to reduce the relative amount of the mutual solvent that is bringing about the total miscibility. It is sometimes possible to effect the formation of a second layer by adding a saturated salt solution. It is best to avoid this problem entirely by removing the bulk of the mutual solvent, possibly by distillation, before doing the extraction.

Only one layer!

Occasionally, when the organic substance being purified is a solid, it will begin to crystallize out of the organic solvent during an extraction process. This may be because the amount of organic solvent has been reduced below that required to dissolve all the compound, owing to its small but finite solubility in the aqueous solutions being used for washing the solution. Or this may happen because another substance, perhaps an alcohol, that helped to dissolve the product in the organic layer has itself been extracted out. In either case, addition of more of the same or a better organic solvent should bring the material back into solution.

Emulsions

A common problem in extraction is failure of the immiscible solutions to separate completely and cleanly into two layers; a certain volume of the mixture at the interface is very likely to consist of droplets of one solution suspended in the other (an *emulsion*). In some cases, the separation becomes clean and complete if the separatory funnel and its contents are allowed to stand undisturbed for a few minutes. In others, the emulsion may persist for hours or days. If the volume of the emulsion is relatively small, it can sometimes be temporarily ignored and the extraction procedure continued in the hope that the emulsion will disappear in later extractions. But if most of the mixture is emulsified, it must either be allowed to stand (if the time is available) or be broken in some other way.

Breaking emulsions   If the emulsion is caused by too small a difference in density between the two layers, the addition of solvent to one or both layers may produce a larger density difference. Pentane will most efficiently decrease the density of an organic solution, while carbon tetrachloride will most efficiently increase the density. The addition of water may either increase or decrease the density of the aqueous phase, depending upon its composition. If the aqueous phase contains appreciable amounts of organic solvents (such as alcohols), it may have a density of less than one gram per milliliter; if, on the other hand, it is a solution of inorganic materials, its density will be greater than one gram per milliliter. The addition of salt or saturated sodium chloride solution can also increase the density of the aqueous phase.

Emulsions are most commonly encountered in extractions involving basic solutions. Presumably this is because traces of higher-molecular-weight organic acids are converted to their salts, and the resulting soap causes emulsification. The tendency of an extraction with a "neutral" aqueous solution to emulsify can sometimes be overcome by adding a few drops of acetic acid, thus suppressing soap formation.

Stubborn emulsions can sometimes be broken by centrifugation or suction filtration of the emulsified material. Suction filtration of an emulsion is a very messy operation and should be done only in desperation.

Avoiding emulsions   In the case of emulsion formation, prevention is far better than cure. If the mixture is sloshed or shaken only gently at first and then allowed to stand, the tendency toward emulsion formation may be estimated. When it appears that emulsion formation may be a problem, it is wise to mix the layers more gently for a longer time. In extreme cases, it may be desirable to carry out the mixing in a round-bottom flask by slow stirring with a magnetic stirrer. Since in this case the area of the interface will be much less than if the mixture were broken up into a suspension of tiny bubbles by vigorous shaking or stirring, it may take 30–60 minutes to approach equilibrium.

Problems

1. Estimate whether or not each of the following acids could be "completely extracted" with one portion of an excess of (a) aqueous sodium bicarbonate solution, (b) aqueous sodium carbonate solution, and (c) aqueous sodium hydroxide solution:

Trichloroacetic acid, $Cl_3C-\overset{\displaystyle O}{\overset{\|}{C}}-O-H$ $K_a = 2 \times 10^{-1}$

Butyric acid, $CH_3CH_2CH_2\overset{\displaystyle O}{\overset{\|}{C}}-O-H$ $K_a = 1.5 \times 10^{-5}$

Phenol, **I** $K_a = 1.3 \times 10^{-10}$

Picric acid, **II** $K_a = 1.6 \times 10^{-1}$

I   **II**

2. Using extraction procedures only, how would you separate a mixture of:
  a. Naphthalene, benzoic acid, and α-naphthylamine?
  b. Naphthalene, benzoic acid, and α-naphthol?

naphthalene   α-naphthylamine   α-naphthol

3. The solubility of adipic acid in water is 1.5 g per 100 ml at 15°C, and 0.6 g per 100 ml in ether at the same temperature.
  a. What fraction of a sample of adipic acid could not be extracted from water into ether with one extraction with a volume of ether equal to that of the aqueous solution?
  b. What fraction of a sample of adipic acid could not be extracted from ether into water with one extraction with a volume of water equal to that of the ethereal solution?
  c. As in **a.**, but consider using three portions of ether, each equal to one-third of the volume of the aqueous solution.
  d. As in **b.**, but consider using three portions of water, each equal to one-third of the volume of the ethereal solution.
  e. Calculate the limiting fraction, which could not be extracted with an equal volume of ether.
  f. Calculate the limiting fraction, which could not be extracted with an equal volume of water.
  g. As in **a.**, **c.**, and **e.**, but using a total volume of ether equal to five times the volume of the aqueous solution.

Exercises

1. Extraction of caffeine from tea or NoDoz (Section 42).
2. Extraction of eugenol from oil of cloves (Section 44).
3. Separation of a mixture of an acid A—H, a base B:, and a neutral substance N.

Procedure. Dissolve about 5 grams of the mixture in 50 ml of diethyl ether and transfer the solution to a 125-ml separatory funnel. Add to the funnel 30 ml of 1 *M* HCl solution. Shake the mixture well in order to extract the basic substance as its hydrochloric acid salt into the water layer:

$$B: + H_3\overset{+}{O} + Cl^- \longrightarrow \overset{+}{B}-H + Cl^- + H_2O$$

Draw off the lower aqueous layer and save it.

Add to the ether solution in the separatory funnel about 30 ml of 1 M NaOH solution. Shake the mixture well in order to extract the acidic substance as its sodium salt into the water layer:

$$A\text{---}H + HO^- + Na^+ \longrightarrow A:^- + Na^+ + H_2O$$

Draw off the lower aqueous layer and save it. The ether layer should now contain only the neutral substance N.

Isolation of the neutral substance N. Wash the ether layer by adding to the separatory funnel about 25 ml of water, shaking the mixture, allowing the layers to separate, and then drawing off and discarding the lower, aqueous, layer. Transfer the ethereal solution of the neutral compound to a small Erlenmeyer flask, dry it over 1–2 grams of anhydrous sodium sulfate for a few minutes (Section 15.2), remove the drying agent by gravity filtration (Section 7.1), and remove the ether by evaporation or distillation on the steam bath (Section 36). The infrared spectrum of the residue may be determined (Section 23); or, if the residue is a solid, it may be recrystallized (Section 8) and its melting point determined (Section 17).

Isolation of the basic substance B:. Transfer the aqueous solution of the hydrochloric acid salt of the basic substance to a clean 125 ml separatory funnel. Make the solution strongly basic by adding about 2 ml of 50% aqueous sodium hydroxide. Make sure the solution is well mixed. The basic substance will now be present as the free base:

$$\overset{+}{B}\text{---}H + Cl^- + Na^+ + HO^- \longrightarrow B: + Na^+ + Cl^- + H_2O$$

Add 25 ml of ether to the separatory funnel. Shake the mixture well so as to extract the free base into the ether layer. Draw off the lower aqueous layer (which may now be discarded) and transfer the ethereal solution of the basic substance to a small Erlenmeyer flask. Dry the solution over 1–2 grams of anhydrous potassium carbonate for a few minutes, remove the drying agent by gravity filtration, and remove the ether by evaporation or distillation on the steam bath. The infrared spectrum of the residue may be determined; or, if the residue is a solid, it may be recrystallized and its melting point determined.

Isolation of the acidic substance A—H. Transfer the aqueous solution of the sodium salt of the acidic substance to a clean 125 ml separatory funnel. Make the solution strongly acidic by adding about 3 ml of conc. HCl. Make sure the solution is well mixed. The acidic substance will now be present as the free acid:

$$A:^- + Na^+ + H_3O^+ + Cl^- \longrightarrow A\text{---}H + Na^+ + Cl^- + H_2O$$

Add 25 ml of ether to the separatory funnel. Shake the mixture well so as to extract the free acid into the ether layer. Draw off the lower, aqueous, layer (which may now be discarded) and transfer the ethereal solution of the acidic substance to a small Erlenmeyer flask. Dry the solution over 1–2 grams of anhydrous sodium sulfate for a few minutes, remove the drying agent by gravity filtration, and remove the ether by evaporation or distillation on the steam bath. The infrared spectrum of the residue may be determined; or, if the residue is a solid, it may be recrystallized and its melting point determined.

The accompanying flow chart (Figure 13.3) summarizes these procedures.

4. Separation of the components of a commercial mixture of aspirin, phenacetin, and caffeine.

Some brands of headache or cold tablets, the so-called "APC tablets," contain a mixture of acetylsalicylic acid (aspirin), phenacetin, and caffeine. It is possible to take advantage of the acid-base properties of these compounds so as to separate them by an extraction procedure. Caffeine, whose conjugate acid has a pKa of −0.16, can just be extracted as the conjugate acid from chloroform into 4 M HCl. After the acid extract has been neutralized, caffeine can be reextracted from the water with additional chloroform and isolated by evaporation of the chloroform.

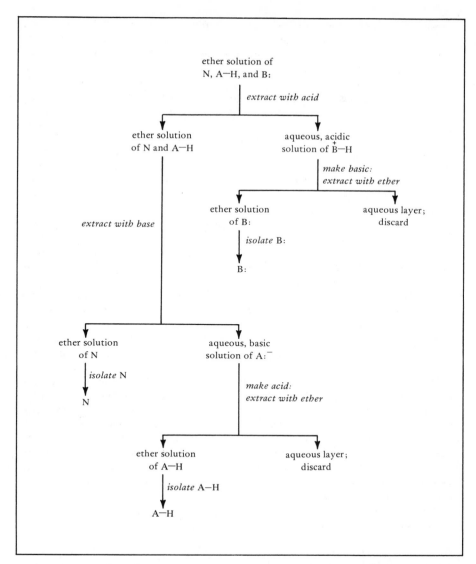

Figure 13.2. Flow chart for the separation of a mixture of an acid A—H, a base B:, and a neutral substance N.

Aspirin, having a pKa of 3.49, can be extracted from chloroform by 0.5 *M* aqueous sodium bicarbonate solution. After the aqueous extract is neutralized by addition of HCl, the precipitated aspirin can be isolated by reextraction with more chloroform and recovered by evaporation of the chloroform. Phenacetin, a substance that is neither acidic nor basic, remains in the original chloroform solution and is recovered by evaporation of the chloroform after the other two substances have been removed.

Procedure.

Isolation of caffeine: Crush 3 APC tablets (Note 1) and add them to a separatory funnel that contains 25 ml of chloroform. Then add to the funnel 20 ml of 4 *M* HCl (80 milliequivalents of acid). Shake the funnel in order to thoroughly mix the contents (Note 2). Allow the funnel to stand so that the layers will separate. Draw off and save the lower chloroform layer that contains the unextracted aspirin and phenacetin (Note 3).

Recovery of caffeine: Neutralize the aqueous acidic extract of caffeine in the separatory funnel by adding 7.0 grams of solid sodium bicarbonate (83 meq. of base). Since a lot of carbon dioxide will be produced, you should add the sodium bicarbonate in portions. Swirl the contents of the funnel after each portion has appeared to react. After all the base has been added and the reaction appears to be complete add 10 ml of chloroform, stopper the funnel, invert it, and release the

caffeine

acetylsalicylic acid

(aspirin)

p-ethoxyacetanilide

(phenacetin)

pressure by opening the stopcock. Cautiously mix the contents of the funnel and frequently release the pressure by opening the stopcock of the inverted separatory funnel. Finally, when the pressure does not build up any more, shake the funnel to thoroughly mix the contents of the flask (Note 2), allow the layers to separate (Note 4) and draw off the lower, chloroform, layer into a small Erlenmeyer flask labeled "caffeine." Reextract the neutralized acid solution with a second 5 ml portion of chloroform and add this further chloroform extract to the flask labeled "caffeine." Dry the combined extracts with a small portion of anhydrous magnesium sulfate (Section 15.2) and filter the mixture by gravity (Section 7.1) into a small, weighed, Erlenmeyer flask containing a boiling stone (Note 5). Remove the chloroform by distillation on the steam bath (Note 6) and determine the weight of the residual solid by reweighing the flask. The recovered caffeine (Note 7) can be recrystallized from 10 ml of carbon tetrachloride, if desired. The melting point of caffeine is reported to be 238°C, with sublimation starting at 170°C, and its infrared and NMR spectra are shown in Figures 42.1 and 42.2.

Isolation of aspirin: Place the original chloroform solution that was saved from the first extraction in a clean separatory funnel. Then add 25 ml of 0.5 M sodium bicarbonate solution (12.5 meq. of base) and thoroughly mix the contents of the funnel (Note 2) so as to extract the aspirin into the basic water layer. Since carbon dioxide will be formed the funnel must be vented occasionally in order to prevent too great an increase in pressure. After mixing allow the funnel to stand undisturbed so that the layers can separate. Draw off the lower, chloroform, layer into a small Erlenmeyer flask labeled "phenacetin" and add to this a small amount of anhydrous magnesium sulfate.

Recovery of aspirin: Add to the basic aqueous solution of aspirin in the separatory funnel 5 ml of 4 M HCl (20 meq. of acid). Mix the contents of the funnel by swirling it gently. Carbon dioxide will be evolved and the aspirin will separate as a solid. Recover the aspirin by extracting first with a 15 ml portion of chloroform and then with a 5 ml portion. Combine these two extracts in a small Erlenmeyer flask labeled "aspirin." Dry the aspirin extracts with a small portion of anhydrous magnesium sulfate and then filter the mixture by gravity into a small, weighed, Erlenmeyer flask containing a boiling stone (Note 5). Remove the chloroform by distillation on the steam bath (Note 6) and determine the weight of the residue. The recovered aspirin (Note 8) can be recrystallized by adding 5 ml of water, heating the mixture on the steam bath, and adding slightly more than the minimum amount of 95% ethanol required to dissolve the solid; about 1.5 ml should be added. Aspirin is reported to melt at 135°C with rapid heating. Its IR and NMR spectra are shown in Figures 56.8 and 56.9.

Recovery of phenacetin: Remove the magnesium sulfate from the chloroform solution of phenacetin by gravity filtration, collecting the filtrate in a small, weighed, Erlenmeyer flask containing a boiling stone (Note 5). Remove the chloroform by distillation on the steam bath (Note 6) and determine the weight of the residual solid. The recovered phenacetin (Note 9) can be recrystallized from a very small amount of 95% ethanol. The melting point of phenacetin is reported to be 134–135°C, and its IR and NMR spectra are shown in Figures 60.2 and 60.3.

The flow chart (Figure 13.3) summarizes these procedures.

Notes

1. Three APC tablets typically weigh 1.5 grams and contain 10.5 grains aspirin (680 mg), 7.5 grains phenacetin (486 mg), and 1.5 grains caffeine (97 mg).

2. One full minute of continuous shaking is sufficient.

3. If you have a second separatory funnel, this extract can be added directly to it and two parts of the experiment can be carried out simultaneously.

4. If an emulsion is formed, it can be broken by adding several drops of glacial acetic acid to the separatory funnel.

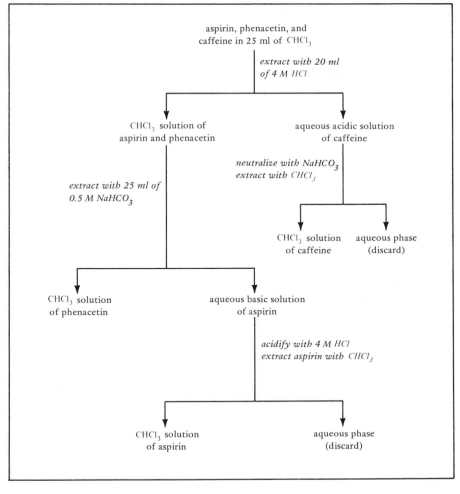

Figure 13.3. Flow chart for the separation of a mixture of aspirin, phenacetin, and caffeine.

5. Weigh the flask with the boiling stone in it.

6. Put the recovered chloroform into the container reserved for it. The last of the chloroform may have to be removed by heating the flask without the adapters and condenser attached.

7. Only about half the caffeine is recovered, and NMR analysis shows that it is contaminated with phenacetin.

8. About three-fourths of the aspirin is recovered, and it appears to be pure by NMR analysis.

9. Almost all the phenacetin is recovered, but it contains a small amount of a less soluble material, perhaps part of the binder. Analysis by NMR shows it to contain between 5 and 10% caffeine, and something that contributes a single peak that overlaps the highest field component of the methyl triplet (2).

References

1. P. Haddad and M. Rasmussen, *J. Chem. Educ.* **53**, 731 (1976).

2. D. P. Hollis, *Analytical Chemistry* **35**, 1682 (1963).

§14 CHROMATOGRAPHY

Chromatographic methods of separation are similar in basic concept to extraction, in that they take advantage of the fact that different substances are partitioned differently between two phases. But in practical techniques, the two methods vary greatly. Since the techniques of chromatography are so different from those of extraction, chromatographic methods are discussed separately in this section.

14.1 THEORY OF COLUMN CHROMATOGRAPHY

In *column chromatography*, a solid phase (adsorbent) is held in a vertical tube, the mixture to be separated (*A* plus *B*) is placed on top of the column of adsorbent, and a solvent (eluant) is allowed to flow down through the column. At all times, a certain fraction of each component of the mixture will be adsorbed by the solid and the remainder will be in solution. Any given molecule will spend part of the time sitting still on the adsorbent and the remainder flowing down the column with the solvent. A substance that is relatively strongly adsorbed (say, *A*) will have a greater fraction of its molecules adsorbed at any one time, and thus any given molecule of *A* will spend more time sitting still and less time moving. By contrast, a weakly adsorbed substance (*B*) will have a smaller fraction of its molecules adsorbed at any one time, and hence any given molecule of *B* will spend less time sitting and more time moving. Thus, the more weakly a substance is adsorbed, the faster it will get to the bottom of the column and flow out with the eluant. Since the eluant is collected in small portions (*fractions*), the early fractions will contain *B* and the later ones will contain *A*. The name *chromatography* was given to this process because the substances to which this method of separation was first applied were colored plant pigments.

There are a number of factors that determine the efficiency of a chromatographic separation of this type. The adsorbent should show a maximum of selectivity toward the substances being separated so that the differences in rate of elution will be large. For the separation of any given mixture, some adsorbents may be too strongly adsorbing (holding all components near the top of the adsorbent) or too weakly adsorbing (allowing all components to move through the adsorbent almost as fast as the eluting solvent). Table 14.1 lists a number of adsorbents in order of adsorptive power.

Selectivity of adsorbent

The eluting solvent should also show a maximum of selectivity in its ability to dissolve or desorb the substances being separated. The fact that one substance is relatively soluble in a solvent can result in its being eluted faster than another substance. However, a more important property of the solvent is its ability to be itself adsorbed on the adsorbent. If the solvent is more strongly adsorbed than the substances being separated, it can take their place on the adsorbent and all the substances will flow along rapidly together. If the solvent is less strongly adsorbed than any of the components of the mixture, its contribution to different rates of elution will be only through its difference in solvent power toward them. If, however, it is more strongly adsorbed than some components of the mixture and less strongly than others, it will greatly speed the elution of those substances that it can replace on the column, without speeding the elution of the others.

Selectivity of eluting solvent

Table 14.2 lists a number of common solvents in approximate order of increasing adsorbability, and hence in order of increasing eluting power. The order can be only approximate since it is not independent of the nature of the adsorbent. Mixtures of solvents can be used, and, since increasing eluting power results mostly from preferential adsorbtion of the solvent, addition of only a little (0.5–2%, by volume) of a more strongly adsorbed solvent will result in a large increase

Table 14.1. Chromatographic adsorbents

Most Strongly Adsorbent[a]	
Alumina	Al_2O_3
Charcoal	C
Florisil	MgO/SiO_2 (anhydrous)
Silica gel	SiO_2
Lime	CaO
Magnesia	MgO
Magnesium carbonate	$MgCO_3$
Calcium phosphate	$Ca_3(PO_4)_2$
Calcium carbonate	$CaCO_3$
Potassium carbonate	K_2CO_3
Sodium carbonate	Na_2CO_3
Talc	MgO/SiO_2 (hydrous)
Sucrose	Carbohydrate (polyhydroxylic)
Starch	Carbohydrate (polyhydroxylic)
Least Strongly Adsorbent	

[a] The order in the table is approximate since it depends upon the substance being adsorbed and the solvent used for elution.

in the eluting power. Because water is among the relatively strongly adsorbed solvents, the presence of a little water in a solvent can greatly increase its eluting power. For this reason, solvents to be used in column chromatography should be quite dry. (See Section 15.3 for methods of drying solvents.)

Solvents should be dry

The particular combination of adsorbent and eluting solvent that will result in the acceptable separation of a particular mixture can be determined only by trial. Alumina is the most commonly used adsorbent because it is readily available at relatively low cost and has a wide range of adsorptive power, depending upon how much water it has adsorbed. The adsorptive power of alumina can be decreased by adding up to 15 percent by weight of water. If alumina has become too heavily hydrated in storage, it may be "activated" by being heated at 200°C for about three hours. A trial to determine the conditions for a chromatographic separation might be made by preparing a small column using partially

Table 14.2. Eluting solvents for chromatography

Least Eluting Power (alumina as adsorbent)[a]
Petroleum ether (hexane; pentane)
Cyclohexane
Carbon tetrachloride
Benzene
Dichloromethane
Chloroform
Ether (anhydrous)
Ethyl acetate (anhydrous)
Acetone (anhydrous)
Ethanol
Methanol
Water
Pyridine
Organic acids
Greatest Eluting Power (alumina as adsorbent)

[a] The order of eluting power of the solvents listed in the table will generally be observed with other highly polar adsorbents.

hydrated alumina, adding a sample of the mixture, and starting elution with a solvent of weak eluting power (hexane, for example). Solvents of successively greater eluting power (benzene, dry ether, dry acetone, methanol; see Table 14.2) can then be tried until one is found that will move the material down the column. Further, more sensitive, trials can be made using different solvents, or mixtures of solvents, of similar eluting power. If the trial adsorbent is too strongly adsorbing, a more completely hydrated alumina can be tested or a weaker adsorbent can be tried; if it is too weakly adsorbing, a less hydrated grade of alumina can be tested.

If the substances in the mixture differ greatly in adsorbability, it will be much easier to separate them. Often in this case a succession of solvents of increasing eluting power is used. One substance may be eluted easily while the other stays at the top of the column, and then the other can be eluted with a solvent of greater eluting power. Table 14.3 indicates an approximate order of adsorbability by functional group.

Table 14.3. Adsorbability of organic substances by functional group

Comparison of this table with Table 14.2 indicates that there is a relationship between the eluting power of a solvent and the tendency of the solvent to be adsorbed.

Least Strongly Adsorbed [a]
Saturated hydrocarbons; alkyl halides
Unsaturated hydrocarbons; alkenyl halides
Aromatic hydrocarbons; aryl halides
Polyhalogenated hydrocarbons
Ethers
Esters
Aldehydes and ketones
Alcohols
Acids and bases (amines)
Most Strongly Adsorbed

[a] The order depends upon the adsorbent.

14.2 TECHNIQUE OF COLUMN CHROMATOGRAPHY

The column

There are several types of tubes that may be used to support the adsorbent. One of the most satisfactory and least expensive is illustrated in Figure 14.1a. A chromatographic column like this, essentially a giant medicine dropper, may be made by heating, drawing down, cutting off, and fire-polishing a long piece of glass tubing 10–25 mm in diameter. The column should be at least 50 cm long so that the column of eluting solvent above the adsorbent can be high enough to contribute a sufficient hydrostatic head to achieve an acceptably large flow rate. If the tip is not too narrow, the flow of solvent through the column may be stopped by slipping a rubber policeman over it. Otherwise, a piece of rubber tubing and a clamp can be used. Burets and other columns with stopcocks (Figure 14.1b and c) are sometimes used, but they have the disadvantage that the stopcock grease will be eluted along with the other substances in the system and will contaminate the material being purified. Small trial chromatograms may be run in medicine droppers.

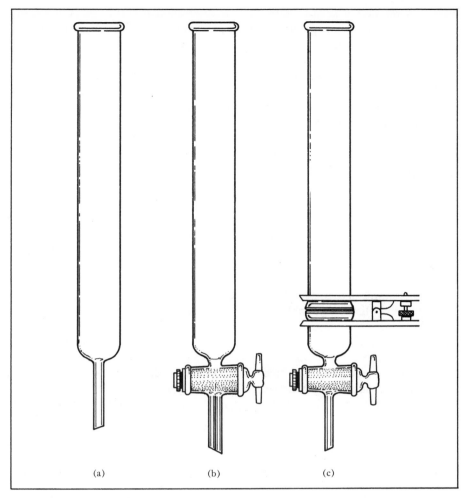

(a)   (b)   (c)

Figure 14.1.
Chromatographic columns.

Packing the Column

You may pack the column with adsorbent by partially filling it with a liquid of
low adsorbability (petroleum ether, for example) and slowly adding the dry
powdered adsorbent down the top so that it settles evenly and uniformly.
Alternatively, the adsorbent may be mixed with the liquid to form a thin slurry
and then poured and rinsed into the column. The objective is to form a uniform
bed of adsorbent without holes, channels, or air bubbles. The separation will be
best if the bottom of the bed of adsorbent is flat. Therefore, if you are using a
column that is tapered or has some other odd shape at the bottom, you should
first partially fill the column with solvent, then push a plug of glass wool down to
fill up the odd volume, and finally add a little sand, rinsing it down to form a flat
base on which the adsorbent can come to rest. After all traces of adsorbent have
been rinsed down (solvent will usually have to be drained off during this process)
and the top of the bed of adsorbent has been flattened by jiggling the tube, a disk of
filter paper just a little smaller than the diameter of the tube should be dropped
flat on the adsorbent, and then enough sand to form a layer 2 or 3 mm deep
should be sprinkled on the paper. The idea here is to protect the top of the column
of adsorbent so that it will not be disturbed when the sample or the eluting solvent is poured onto it. Figure 14.2 shows a column ready for use.
     The ratio of the weight of adsorbent to the weight of sample may vary greatly,
but usually will not be much outside the range of 25 or 50 to 1. The ratio of the

Adsorbent:sample ratio

height of the bed of adsorbent to its diameter should normally be between 3:1 and 10:1. If the ratio is greater than this, the rate of flow of eluant may be too low. Thus, when larger samples are to be chromatographed, large-diameter columns must be used and not just deeper beds of adsorbent. The finer the adsorbent, the greater its surface area and hence its adsorptive capacity, but the lower the flow rate. Sometimes it is necessary to use a mixture of adsorbent plus diatomaceous earth (Celite) as a nonadsorbing diluent in order to attain a minimum acceptable flow rate.

Adding the Sample

When the column has been prepared, the solvent should be allowed to drain until it is just level with the top of the sand that covers the adsorbent. If the column is allowed to drain dry, the bed of adsorbent usually will crack, and its ability to separate will be greatly diminished because the solution will be able to flow through the cracks. It is a waste of time to try to use a column with cracked or channeled adsorbent.

The sample is then added. A liquid can be added directly; a solid should be dissolved in as little as possible of a solvent of relatively low eluting power. After the addition, the column is allowed to drain again until the level of liquid has fallen just to the top of the adsorbent. If necessary, rinse the sample down from the walls of the column with additional *small* portions of solvent, draining the column each time. You will find that a medicine dropper is useful for these additions. The idea is to get the sample adsorbed in a minimum layer of adsorbent. The narrower the sample band, the better the separation, since narrow bands will overlap less.

Elution

After the sample has been added, the column is eluted using a solvent or series of solvents as recommended or as determined by trial. In a new situation, you should start with a solvent of low eluting power (petroleum ether, hexane) and work up through solvents of increasing eluting power (benzene, ether, etc.) either as mixtures or as pure solvents. When the level of liquid in the column is low, the eluting solvent should be added very cautiously to keep from disturbing the top of the bed of adsorbent. During elution, the solvent should not be allowed to flow through any sort of rubber or plastic tubing, since material can be eluted from the tubing as well.

Collecting the Fractions

If the substances to be separated are colored, their progress down the column can be followed visually and the eluant that contains each component can be collected separately. If the substances are colorless, the eluant must be collected in successive fractions and the presence of components of the original mixture determined by some analytical procedure. The most common method is to collect fractions of approximately equal volume in tared (previously weighed) flasks, evaporate or distill the solvent (Section 36), and reweigh to determine the weight of the residue. If the residue is a solid, it will usually crystallize and can then be easily seen. The residue can be identified by determining one or more physical properties such as melting point or the infrared absorption spectrum. In the case of substances that

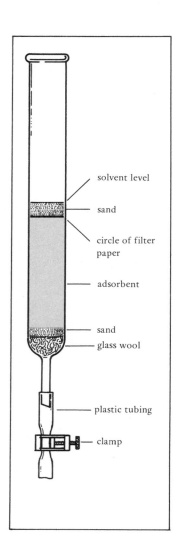

solvent level

sand

circle of filter paper

adsorbent

sand
glass wool

plastic tubing

clamp

Figure 14.2. A simple chromatographic column ready for use.

fluoresce (many aromatic compounds do), their progress down the column can be followed by illuminating the column with ultraviolet light ("black light"). Often, the presence or absence of colorless substances in the eluant can be established by determining its ultraviolet absorption; negative fractions need not be evaporated.

Sometimes elution is carried out only until the components of the sample appear to be separated on the column. The adsorbent is then pushed out of the tube (you must use a column such as (c) in Figure 14.1), and the parts of the adsorbent containing the various components are separated and stirred with a solvent of high eluting power. Removal of the adsorbent by filtration and evaporation of the solvent yields the purified sample of the material.

The rate of flow through the column can be increased by filling the column with eluting solvent to a higher level. If the flow is still too slow, it can be hastened by applying a little air pressure to the top of the column with an atomizer bulb connected through a relatively large ballast volume (Figure 14.3). Connecting the column to a water aspirator by means of a suction flask is not recommended because it is very easy to evaporate the solvent from the bottom of the column and thus form cracks and spoil the column.

Increasing flow rate

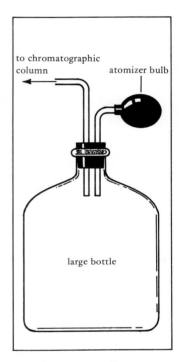

Column chromatography is especially suited for purifying small amounts of material. Normally, it effects separations far more completely than distillation or recrystallization can. Its disadvantages are that it is a highly empirical method, often requiring several trials to establish acceptable conditions, and that it is not suited to the purification of large quantities since most columns of reasonable size have only a small capacity.

Figure 14.3. A way to apply pressure to the top of a chromatographic column.

Exercises

1. Purification of technical-grade anthracene.

   Procedure.
   Column: 1.5 to 2 cm diameter.
   Adsorbent: alumina. Use enough to make a bed of adsorbent 8–10 cm deep. Pack the column using hexane or petroleum ether.
   Sample solution: 0.3 g of technical anthracene in 150–200 ml of hexane.
   Eluting solvent: hexane. Examination of the column in ultraviolet light will show a narrow, deep-blue fluorescent zone near the top, due to carbazole. Immediately below this should appear a nonfluorescent yellow band due to naphthacene. Anthracene forms a broad blue-violet fluorescent zone in the lower part of the column. Elution with hexane should be continued until anthracene begins to come off the column; the eluant collected before the anthracene begins to come off should be discarded. At this point, the elution should be continued with hexane: benzene, 1:1 (by volume), until the yellow band reaches the bottom of the column (1).

2. Isolation of (R)-(−)- or (S)-(+)-carvone from oil of spearmint or oil of caraway (Section 46).

14.3 THEORY OF THIN-LAYER CHROMATOGRAPHY

The theoretical basis for the separation of substances by *thin-layer chromatography* is exactly the same as that for column chromatography. The difference between the methods is that in thin-layer chromatography the adsorbent is used in the form of a thin layer (about 0.25 mm thick) on a supporting material, usually a sheet of glass or plastic. The sample is applied to the layer of adsorbent, near one edge, as a small spot of a solution. After the solvent has evaporated, the adsorbent-coated sheet is propped more or less vertically in a closed container, with the edge to which the spot was applied down. The solvent, which is in the bottom of the container, creeps up the layer of adsorbent, passes over the spot, and, as it continues up, effects a separation of the materials in the spot ("develops" the chromatogram) in the same way as the eluting solvent does in column chromatography. When the solvent front has nearly reached the top of the adsorbent, the sheet is removed from the container.

Since the amount of adsorbent involved is relatively small, and the ratio of adsorbent to sample must be high, the amount of sample must be very small, usually much less than a milligram. For this reason, thin-layer chromatography (TLC) is usually used as an analytical technique rather than a preparative method, although with thicker layers (about 2 mm) and large plates with a number of spots or a stripe of sample, it can be used as a preparative method. The separated substances are recovered by scraping the adsorbent off the plate (or cutting out the spots if the supporting material can be cut) and extracting the substance from the adsorbent.

Because the distance traveled by a substance relative to the distance traveled by the solvent front depends upon the molecular structure of the substance, TLC can be used to identify substances as well as to separate them. The relationship *Identification through* between the distance traveled by the solvent front and the substance is usually *R_f value* expressed by the so-called R_f value:

$$R_f \text{ value} = \frac{\text{distance traveled by substance}}{\text{distance traveled by solvent front}}$$

The R_f values are greatly dependent upon the exact nature of the adsorbent and solvent. Therefore, the comparison of experimental R_f values with reported R_f values involves much uncertainty. In order to determine whether an unknown substance is the same as a known substance with the same R_f value, it is necessary to run the two substances side by side in the same chromatogram, preferably at the same concentration.

14.4 TECHNIQUE OF THIN-LAYER CHROMATOGRAPHY

Preparation of Microscope Slide Plate

The two most widely used adsorbents in TLC are silica gel and alumina. Adsorbent-coated TLC plates may be either obtained commercially or prepared in the laboratory. Ready-made plates are more convenient but are relatively expensive. The preparation of large plates requires the use of fairly expensive apparatus, but small plates using microscope slides as the support are easily prepared in the laboratory and are very widely used for small-scale work. The procedure consists of preparing a suspension of the adsorbent in a volatile solvent, dipping in a pair of clean microscope slides held tightly together until all but a handle of about 1 cm of their length is immersed in the slurry, and then drawing the slides out in one smooth, continuous motion. When the solvent has evaporated (in about a minute), the *Activation* two slides may be separated. After the adsorbent has been activated by heating (Figure 14.4), the plates are ready for use.

Application of Sample

The sample to be separated is generally applied as a small spot (1 to 2 mm diameter) of solution about 1 cm from the end of the plate opposite the handle. The addition may be made with a microsyringe or with a micropipet prepared

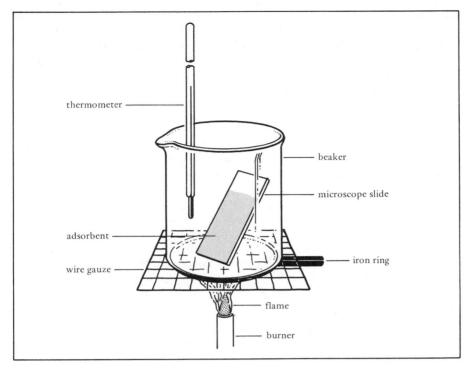

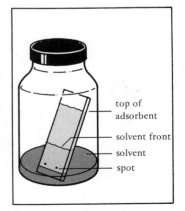

Figure 14.4. Activation of a plate for TLC.

by heating and drawing out a melting-point capillary. (The disposable micro-pipets made by the Drummond Scientific Company are very useful.) As small a sample as possible should be used, since this will minimize tailing and overlap of spots; the lower limit is the ability to visualize the spots in the developed chro-matogram. If the sample solution is very dilute, make several small applications in the same place, allowing the solvent to evaporate between additions. Do not disturb the adsorbent when you make the spots, since this will result in an uneven flow of the solvent. The starting position may be indicated by making a small mark near the edge of the plate.

Development

The chamber used for development of the chromatogram may be as simple as a beaker covered with a watch glass, or a cork-stoppered bottle. The developing solvent (an acceptable solvent or mixture of solvents must be determined by trial) is poured into the container to a depth of a few millimeters. The spotted plate is then placed in the container, spotted end down; the solvent level must be below the spots (see Figure 14.5). The solvent will then slowly rise in the adsorbent by capillary action.

In order to get reproducible results, the atmosphere in the development chamber must be saturated with the solvent. This may be accomplished by sloshing the solvent around in the container before any plates have been added. The atmosphere in the chamber is then kept saturated by keeping your container closed all the time except for the brief moment during which a plate is added or removed.

Visualization

When the solvent front has moved to within about 1 cm of the top end of the adsorbent (after 15 to 45 minutes), the plate should be removed from the develop-ing chamber, the position of the solvent front marked, and the solvent allowed to

Figure 14.5. Development of a plate in TLC.

evaporate. If the substances that were in the sample are colored, they may be observed directly. If not, they can sometimes be visualized by shining ultraviolet light on the plate or by allowing the plate to stand for a few minutes in a closed container in which the atmosphere is saturated with iodine vapor. Sometimes the spots may be visualized by spraying the plate with a reagent that will react with one or more of the components of the sample.

Exercises

Silica gel-coated TLC plates may be obtained commercially from a number of sources, or may be prepared in the laboratory. Two procedures for coating microscope slides are described here.

a. Prepare a suspension of 35 g of silica gel G (Merck) in 100 ml of a 2:1 (by volume) mixture of chloroform:methanol. Coat the plates as described in Section 14.4.

b. Prepare a suspension of 5 g of silica gel (Biosil A-30B) in 12 ml of water. Pour 1 ml of the slurry on a clean microscope slide and spread it out with a stirring rod. Tap the slide gently to settle the slurry into an even film. After allowing it to set for a couple of minutes, dry it in an oven at 110°C for 15 minutes or longer.

The following separations have been described using silica gel TLC plates.

1. Separation of leaf pigments.

Sample solution. Crush green leaves in a mortar with a few ml of ethanol or acetone and twice this volume of petroleum ether or hexane. Filter the mixture, saving the filtrate. Extract the leaves again, filter, combine the two filtrates and transfer them to a separatory funnel. Wash the extract with three small portions of water and then dry it (Section 15.2) over anhydrous sodium sulfate.

Developing solvent. Benzene:acetone, 7:3 (by volume). Ligroin:acetone, 7:3 by volume) is said to work better as a developing solvent (private communication from reviewers).

Visualization. The carotenes move most rapidly, then chlorophyll a, chlorophyll b, and the xanthophylls (2).

2. Analysis of mixtures of aspirin, phenacetin, and caffeine.

Sample solution. Use methanol; known solutions should be prepared to contain 5–10 mg per ml. Unknown mixtures may be prepared from materials available at drugstores.

Developing solvent. Anhydrous methanol:glacial acetic acid:diethyl ether: benzene, 1:18:60:120 (by volume).

Visualization. Spray with 0.1 N potassium permanganate in 0.05 N sulfuric acid (3).

3. Analysis of 2,4-dinitrophenylhydrazones.

Sample solution. 2,4-Dinitrophenylhydrazones of known aldehydes or ketones and mixtures of unknown aldehydes or ketones must be prepared as described in Section 29.5. Low-molecular-weight aliphatic compounds are recommended. Dissolve the 2,4-dinitrophenylhydrazones in ethyl acetate.

Developing solvent. Benzene:ethyl acetate, 19:1 (by volume), or benzene: petroleum ether (b.p. 67–75°C), 3:1 (by volume). Activate the plate by heating at 110°C for 10 minutes in an oven or as shown in Figure 14.4 (4).

14.5 THEORY OF PAPER CHROMATOGRAPHY

The process of *paper chromatography* is very similar to that of thin-layer chromatography (TLC) except that a strip of paper replaces adsorbent and support. As with TLC, a very small amount of a dilute solution of the substance is applied as a spot near one end of a strip of filter paper, and the strip is supported in a closed container with the end containing the spot hanging down into a solvent. As the solvent rises up the paper by capillary action, it effects the separation ("develops" the chromatogram) as does the eluant in column chromatography.

Under ordinary conditions, about 18 percent by weight of filter paper consists of adsorbed water. This means that when the molecules of a substance are sitting still on the paper, it is possible that they should be considered to be dissolved in the adsorbed water rather than adsorbed by the paper.

As in the case of TLC, the amounts of sample must be so small that paper chromatography is used almost entirely as an analytical method in which you can demonstrate the homogeneity of a sample or qualitatively determine the composition of a mixture through determination of R_f values.

14.6 TECHNIQUE OF PAPER CHROMATOGRAPHY

Arrangement of Paper Strip

The following simple procedure can be used for the paper chromatographic analysis of a single sample spot. As a container for the development of the paper chromatogram, use a 25 mm × 125 or 150-mm test tube. Cut a half-inch-wide strip of filter paper, and thumbtack one end of it to the bottom of the cork used to stopper the test tube. The strip should be just long enough that the lower end will dip into the half-inch layer of solvent in the bottom of the test tube. During the development of the chromatogram, the paper must not touch the walls of the test tube. A couple of staples fastened into the lower end of the strip will help it to hang straight. Figure 14.6 illustrates the suggested arrangement.

Several samples may be run simultaneously by using a sheet of filter paper bent around to form a short tube and held that way (without the edges touching) by two staples. Development is carried out with the tube standing on end in a beaker covered with a watch glass. The grain of the paper—that is, the direction of the longer axis of the ellipse formed when a drop of water is allowed to spread on the paper—must be vertical. This arrangement is illustrated in Figure 14.7.

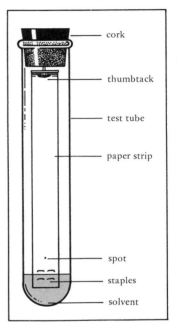

Figure 14.6. An arrangement for paper chromatography.

Application of Sample

The solution of the sample of the material to be separated should be applied in as small a spot as possible. Use a micropipet, which can be made by heating and drawing out a melting-point capillary or can be obtained commercially. If the solution is so dilute that a relatively large amount must be used, several small applications should be made, allowing time for the previous one to dry before the next one is added. As with TLC, the less sample used, the smaller the spots and the better the resolution. The only limitation is the ability to see the spots of separated material. The sample spot should be made far enough from the end of the paper that it will not dip into the solvent. Mark the starting line lightly with a pencil so that later on you will be able to more easily determine the R_f values.

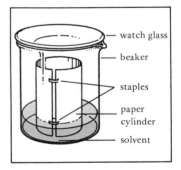

Figure 14.7. An arrangement for paper chromatography allowing several samples to be run at once.

Development

The chromatogram should be developed until the solvent front has risen about 10 cm. The paper should then be removed from the developing chamber, the line of the solvent front marked with a pencil, and the solvent rinsed off or allowed to evaporate.

If the substances are colored, the spots can be observed directly. Ultraviolet light may sometimes be used to visualize the spots, or the paper may be sprayed with a reagent that will react with some or all of the substances of the sample in such a way as to transform them to a colored derivative.

R_f values

As in the case of TLC, R_f values are fairly sensitive to the exact experimental conditions. Comparison of experimental R_f values of unknown substances with those of substances whose structures are known are best made by running a sample of the known substance alongside the unknown sample on the same strip of paper.

Exercises

1. Analysis of spinach leaves.

   Sample solution.   Grind a few grams of fresh or frozen spinach leaves with about three volumes of acetone. Filter the mixture with suction and discard the filtrate. Grind the residue with a minimum volume of acetone and use the resulting solution.

   Developing solvent.   Petroleum ether.

   Visualization.   Carotenes move most rapidly, followed by xanthophylls and chlorophylls; a gray area due to decomposed chlorophylls may appear after the carotenes (6).

2. Analysis of carrots. Analysis of carrots as in Exercise 1 shows carotene to be predominant (6).

3. Analysis of food coloring.

   Sample solution.   Dilute the food coloring by a factor of five with isopropyl alcohol:water, 1:2 (by volume).

   Developing solvent.   Isopropyl alcohol:water, 1:2 (by volume) (7).

4. Separation of ink pigments.

   Sample solution.   Dilute 1 ml of "Script" "washable" writing fluid in 5 ml of 95% ethanol. Use 0.005 ml of solution.

   Developing solvent.   95% ethanol (5).

5. Separation of *cis*- and *trans*-azobenzene.

   Sample solution.   Use cyclohexane as solvent.

   Developing solvent.   Cyclohexane:benzene, 3:1 (by volume) (4).

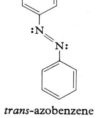

trans-azobenzene

14.7 THEORY OF VAPOR-PHASE CHROMATOGRAPHY

In *vapor-phase chromatography* (VPC), the stationary phase is a high-boiling liquid that is present as a coating upon an inert granulated support, and the mobile phase is a gas, usually helium. As in the other chromatographic methods, separations are possible when there are differences in the way in which different substances are partitioned between the stationary and mobile phases.

If the liquid phase does not preferentially dissolve molecules with certain functional groups, the order of elution is most volatile to least volatile (order of increasing boiling point). One would expect this order always to be observed

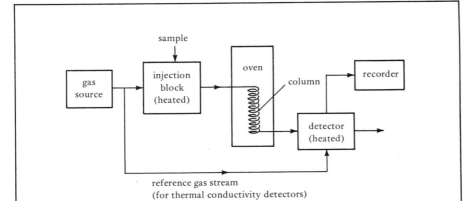

Figure 14.8. Schematic diagram of a vapor-phase chromatograph.

with molecules of an homologous series and with structural isomers involving the same functional group. Some liquids do appear to preferentially dissolve molecules with certain functional groups. They therefore display a relatively high selectivity toward these molecules and are especially well suited for the analysis of mixtures containing them. Table 14.4 lists some liquids that are commonly used as the stationary phase in VPC.

The apparatus required for VPC is considerably more complex and expensive than that needed for the other chromatographic methods. The essential parts, shown schematically in Figure 14.8, consist of a source of the carrier gas under pressure, a port through which the sample may be injected, the column (usually

Table 14.4. Stationary phases most commonly used in vapor-phase chromatography

Stationary Phase	Maximum Useful Operating Temperature	Selectivity; Suitability
Silicone oils $R—Si—(—O—Si—)_n—O—Si—R$ with R groups	250°C	According to volatility; Generally useful
Apiezon grease $CH_3—(—CH_2—)_n—CH_3$	250°C	According to volatility; Not suitable for hydroxylic compounds (tailing)
Polyethylene glycol $HO—(—CH_2—CH_2—O—)_n—CH_2—CH_2—OH$	150°C	Relatively selective toward polar compounds (alcohols, amines, aldehydes, ketones); Best suited for polar compounds
Diisodecyl phthalate	175°C	According to volatility; Generally useful

Hot wire detector

Flame ionization detector

a 5 to 10-foot length of ¼-inch-diameter metal tubing packed with the liquid-coated support), and the detector, and its recorder. The detector senses the presence of material in the carrier gas, and the recorder in effect records the output of the detector as a function of time.

A commonly used type of detector is an electrically heated wire (filament) positioned in the gas stream at the exit to the column. The wire is cooled to an equilibrium temperature by the flowing carrier gas, and at that temperature the wire has a certain resistance. When some substance other than the gas itself is present in the stream, the wire is not cooled as efficiently (helium has a very much larger thermal conductivity than any other gas but hydrogen); its temperature rises and so does its resistance. The wire is part of a Wheatstone bridge circuit and, as the resistance of the wire varies, the resulting imbalance voltage developed in the bridge is compensated for by the recorder, which is an automatic recording null potentiometer. The position of the pen of the recorder corresponds to the position of the slide on the slide-wire of the potentiometer. The recorder output, which is a record of the pen position on the slide-wire as a function of time, is thus related to the variation of the thermal conductivity of the gas at the exit to the column. When nothing but helium is coming off the column, the pen of the recorder will draw a straight line (the base line). As a substance comes off the column, the pen will deflect and then return to the base line, thus drawing a "peak" (see Figure 14.9).

A more sensitive type of detector is the flame ionization detector. In a system using a flame ionization detector a portion of the eluant from the column is fed to a hydrogen/air flame. When a substance other than the carrier gas is present in the eluant the electrical conductivity of the hot gas of the flame will increase due to the formation of ions as the substance in the carrier gas burns. This change in conductivity is sensed as a change in voltage between a wire in the flame and the burner, and this change in voltage is followed by the recorder operating as a potentiometer, just as with the hot wire detector. Nitrogen gas may be used as the carrier gas when a flame ionization detector is used, rather than the much more expensive helium gas required when using a thermal conductivity detector.

Quantitative analysis by VPC

It seems reasonable that the larger the amount of a given substance that comes off the column, the higher its concentration in the carrier gas. Thus the area under the peak should be a measure of the amount of that substance present in the sample. If the sample is a mixture, the ratio of the areas of any two peaks should be a measure of the relative amounts of the two substances.

It is easy to determine which peak corresponds to which substance if authentic samples are available. A portion of the mixture being analyzed can be spiked with one of the substances known to be present in the mixture. Analysis of this sample will show that the area of one peak has increased relative to the others, thus indicating to which component of the mixture it corresponds. The more difficult problem is to determine exactly in what way the area ratio is a measure of the relative amounts of the two substances. That is, does the area ratio correspond to the weight ratio or to the mole ratio or to neither? The only way to know for sure is to analyze mixtures of known composition. In cases where it has been determined, the area ratio generally corresponds approximately to the weight ratio, the agreement being better in the cases of similar compounds. If pure samples of the substances being analyzed are not available, you can only hope that the detector response is approximately the same for each. The more similar the substances are in molecular structure, the more likely it is that this will be the

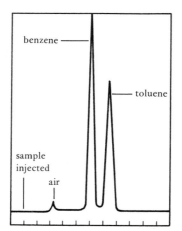

Figure 14.9. Vapor-phase chromatogram of a mixture of benzene and toluene.

case. If the compounds are structural isomers with the same functional groups, **101**
the area ratio will probably be very close to the weight ratio (which is the same SECTION 14
as the mole ratio).

Qualitative analysis by VPC

Vapor-phase chromatography is often used to roughly estimate the purity of a substance by comparing the area of the main peak, which corresponds to the compound of interest, to the area of the other peaks. It is also often used to determine whether or not a certain substance is present in a mixture. This is done by running first a sample of the mixture and then a sample of the mixture plus a little of the compound of interest. If a peak present in the first chromatogram is larger in the second, the substance added probably has the same structure as a component of the original mixture. If the added substance results in a new peak in the second chromatogram, you can set a relatively low upper limit on its possible concentration in the original mixture. Similarly, as implied above, it is possible to identify the substance responsible for a given peak in the chromatogram of a mixture by spiking portions of the mixture with small amounts of various known substances and comparing the resulting chromatograms of the spiked mixtures with the chromatogram of the original mixture. Although this method of identification is fairly reliable, especially if the result is the same when several different columns are used, it requires that you be able to guess what the substance responsible for the peak is and that you have a pure sample of that material available. In some cases, it may be desirable or necessary to attempt to collect the substance responsible for a certain peak as it leaves the gas chromatograph, determine some physical property such as its infrared or mass spectrum, and deduce from that the identity of the substance.

Preparative VPC

So far, the applications of VPC that have been described have been analytical. As in other chromatographic methods, it is possible to use VPC for preparative purposes by leading the carrier gas stream from the detector through a cooled tube in which the sample components of the gas stream can condense. If the tube is changed as each peak comes through, the components of the mixture may be collected separately. In order to purify larger amounts of material this way, it is not possible just to inject larger samples, since peak width increases with increasing sample size and soon the point is reached where the impurity peaks overlap the peak of the desired substance. It is also not possible just to use a larger-diameter column, because the separating power of the column falls off very rapidly with increasing column diameter. It is therefore usually necessary to process successive samples of 0.01 to 0.25-ml volume, depending upon the difficulty of the separation, using quarter-inch or sometimes half-inch columns. Instruments are available commercially in which the repetitive processes of sample injection and fraction collection are carried out automatically.

Vapor-phase chromatography is one of the most widely used methods for separation or analysis of small amounts of reasonably volatile substances. While a very good fractionating column may have an efficiency corresponding to about 100 theoretical plates, an ordinary gas chromatographic column may have an efficiency corresponding to more than 1000 theoretical plates, and columns are available with efficiencies of greater than 10,000 theoretical plates. Holdup is essentially zero, but throughput is very small.

14.8 TECHNIQUE OF VAPOR-PHASE CHROMATOGRAPHY

In most cases, the gas chromatograph will be all set up ready to go. The things that will have been done for you are:

1. The appropriate *column* has been installed, and the inlet port, column, and detector have been brought to their operating temperatures by their corresponding *heaters*.

2. The *carrier gas flow rate* has been set by adjustment of the *pressure regulator* and a *needle valve*.

3. The *filament current* has been set to the proper operating value.

4. The *sensitivity* has been set so that the pen of the recorder will not go off scale on the largest peak. If the pen does go off scale, the sensitivity or the sample size should be decreased; if the largest peak involves less than 50 percent of full-scale deflection, the sensitivity or the sample size should be increased.

5. The *balance* and *recorder zero* have been set to put the pen at the position desired for the base line.

6. The *recorder speed* has been set at the desired speed. One inch per minute is normal; if the sample comes through fast, a greater recorder speed is probably desirable.

Injection of Sample

What remains is for you to start the recorder and inject the sample. Liquid samples are injected as is; solids must be dissolved in a volatile solvent such as ether. Normally, the special (expensive) microsyringe used for injection will be wet with the previous sample. If so, you can rinse it by drawing in some of your sample and then squirting it out into a waste solvent container. Two or three repetitions should be adequate. The next step is to draw your sample into the syringe without any air bubbles. Sometimes the air can be removed by dipping the needle into the sample and pumping the plunger a few times. Sometimes you may need to draw in some liquid, hold the syringe with the needle up, and tap the barrel with your finger until the bubbles rise to the needle end. They may then be forced out by pushing the plunger in, and then the rest of the syringe can be filled. The desired amount of sample—0.001 to 0.005 ml (1 to 5 microliters)—is taken by adjusting the end of the plunger so that the desired volume fills the barrel of the syringe. Now you are ready to inject the sample. Turn the recorder from Standby to On. Insert the needle of the syringe all the way into the rubber septum of the inlet port, slide in the plunger, and pull the needle straight back out.

The progress of the chromatogram is followed by watching the recorder. A chromatogram may take less than a minute or more than half an hour; a typical time is a few minutes. When the chromatogram is complete, turn the recorder from On to Standby. If the pen goes off scale or the peaks are too small, change either the sensitivity setting or the sample size and try it again.

It is not unusual to decide that a chromatogram is complete when there is still material in the column. If the gas chromatograph is not used again immediately, this residue may go undetected. If, however, it is used again soon, very shallow, broad peaks can sometimes be observed under the relatively sharp peaks of the new sample. The broad peaks can be attributed to the previous sample, since the longer a substance remains in the column, the more time it has to diffuse and the shorter and wider will be its peak, for a given area.

Determination of Peak Areas

If the recorder has either a mechanical or an electronic integrator built into it, it is possible to measure the area of a peak automatically. Otherwise, peak areas may be measured with a planimeter or by counting squares. If the peak is symmetrical, the product of the height of the peak times the width of the peak at half-height equals 93 percent of the area of the peak. If area ratios are to be used, as is usually the case, the product of peak height times width at half-height is just as useful as the area. Experimental error in measuring the width of the peak may be minimized by using a higher chart speed. If the peaks are not completely resolved but overlap somewhat, it is much more difficult to estimate their area. Resolution is increased by using a longer or narrower column, smaller sample size, and lower operating temperature. Two substances that cannot be resolved on one column may be resolved on another.

Exercises

1. Chlorination of 2,4-dimethylpentane: determination of the relative reactivity of primary, secondary, and tertiary hydrogens in 2,4-dimethylpentane toward chlorine atoms. The experimental observations in the free-radical chlorination of alkanes are generally interpreted in terms of the following mechanism:

1) $Cl—Cl \longrightarrow 2\,Cl\cdot$ } initiation

2) $Cl\cdot + H—R \longrightarrow Cl—H + R\cdot$
3) $R\cdot + Cl—Cl \longrightarrow R—Cl + Cl\cdot$ } chain propagation

4) $2\,R\cdot \longrightarrow R—R$
5) $2\,Cl\cdot \longrightarrow Cl—Cl$
6) $R\cdot + Cl\cdot \longrightarrow R—Cl$ } chain termination

If the alkane has more than one kind of hydrogen atom, as in propane, products of more than one structure may be obtained. With propane, both 1-chloropropane and 2-chloropropane may be formed, depending upon which type of hydrogen is removed in Step 2:

2 (primary)

$$Cl\cdot + CH_3—CH_2—CH_3 \longrightarrow Cl—H + CH_3—CH_2—CH_2\cdot$$

3 (primary)

$$CH_3—CH_2—CH_2\cdot + Cl_2 \longrightarrow CH_3—CH_2—CH_2—Cl + Cl\cdot$$
$$\text{1-chloropropane (45\%)}$$

2 (secondary)

$$Cl\cdot + CH_3—CH_2—CH_3 \longrightarrow Cl—H + CH_3—\overset{\cdot}{C}H—CH_3$$

3 (secondary)

$$CH_3—\overset{\cdot}{C}H—CH_3 + Cl_2 \longrightarrow CH_3—\overset{\overset{\textstyle Cl}{|}}{C}H—CH_3 + Cl\cdot$$
$$\text{2-chloropropane (55\%)}$$

The relative amounts of the products are explained in terms of the relative rates of formation of the radicals that can be produced by removal of the different kinds of hydrogen atoms in Step 2. In the case of the light-catalyzed chlorination of propane at 25°C, the product is 45% 1-chloropropane and 55% 2-chloropropane.

The ratio of the rate of Steps 2 (secondary) and 2 (primary) is inferred to be 55:45 or 1.22:1. Since there are three times as many primary hydrogen atoms as secondary hydrogen atoms, the ratio of the rate of Step 2 (secondary) to the rate of Step 2 (primary) *per hydrogen atom* is:

$$\frac{1.22/1}{1/3} = 3.66:1$$

In the case of the free-radical chlorination of 2,4-dimethylpentane, three products are possible, each corresponding to the abstraction of a hydrogen atom from a primary, secondary, or tertiary position:

primary:

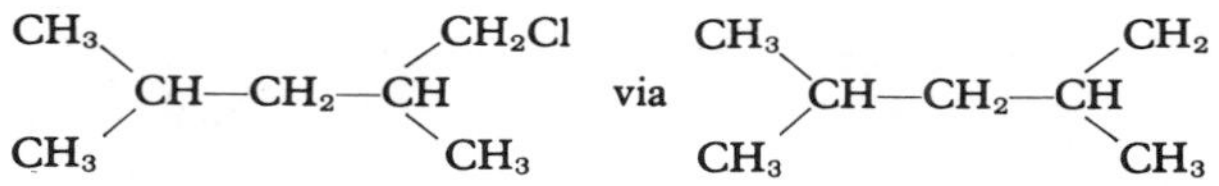

secondary:

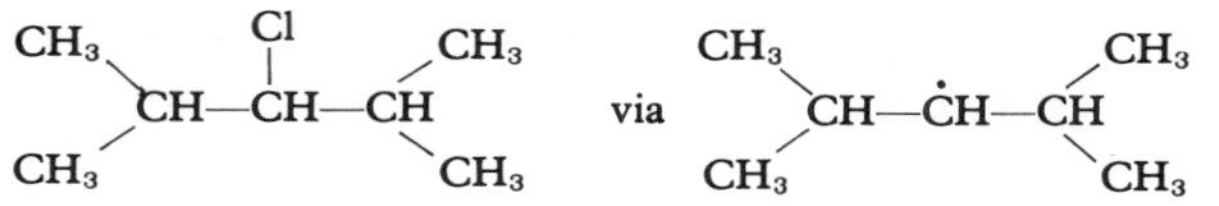

tertiary:

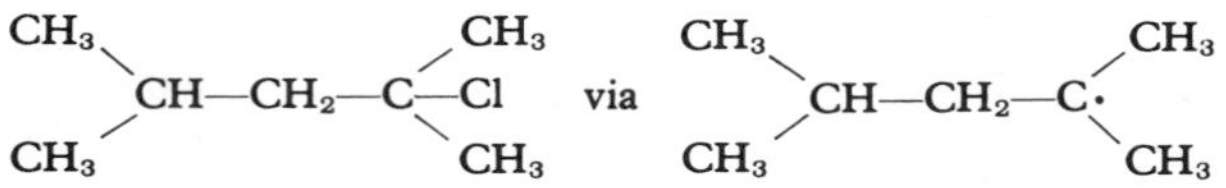

Procedure. Pass a slow stream of chlorine gas through about 5 ml of 2,4-dimethylpentane in a 5-ml Erlenmeyer flask while irradiating the mixture with a weak ultraviolet light for about 5 minutes, as shown in Figure 14.10, on page 105. The mixture may be analyzed by means of vapor-phase chromatography under the following conditions: 5-ft × ¼-in Carbowax 6000 column (20% on 30/60 firebrick) in an Aerograph A-90-P2 gas chromatograph using a Sargent Model-SR 1-millivolt recorder; injector temperature, 70°C; column temperature, 74°C; detector temperature, 157°C; sample size, 5 microliters; helium gas pressure, 15 lb; attenuation, factor of 1; chart speed, 1 inch per minute. The analysis takes about 14 minutes.

In addition to the large peak corresponding to unreacted 2,4-dimethylpentane, three peaks will be observed that correspond to the three isomeric monochlorination products. Assuming that the detector is equally sensitive to each isomer, the ratio of peak areas may be taken as the mole ratio of the isomers in the mixture. The order of appearance of the peaks is: tertiary, secondary, primary (8).

2. The products of the acid-catalyzed dehydration of 2-methyl-cyclohexanol (Section 49.2) may be determined by vapor-phase chromatography.

14.9 BATCHWISE ADSORPTION; DECOLORIZATION

Sometimes it is possible to adsorb an undesired colored compound by treating a solution with an adsorbent such as activated charcoal or alumina and then filtering the mixture by gravity to remove the adsorbent and the adsorbed material (Section 8.3). The ideal conditions are that the impurity is much more strongly adsorbed than the desired substance, and that an amount of adsorbent just sufficient to adsorb the impurity is used.

Occasionally it may be possible to decolorize a solution by filtering it through a short column of alumina, thus preferentially adsorbing the colored substance. A poorly adsorbed solvent (hydrocarbons, ether; not alcohols) must be used so that the solvent will not take the place of the colored material.

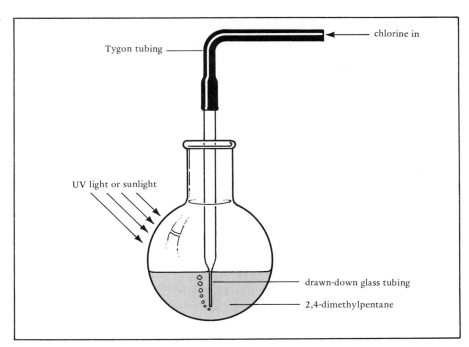

Figure 14.10. Arrangement for chlorination of 2,4-dimethylpentane.

Labels in figure: Tygon tubing — chlorine in — UV light or sunlight — drawn-down glass tubing — 2,4-dimethylpentane

References

1. A. I. Vogel, *A Textbook of Practical Organic Chemistry*, 3rd edition, Wiley, New York, 1956.

2. C. Rollins, *J. Chem. Educ.* **40**, 32 (1963).

3. K. A. Conners and S. F. Eriksen, *American Journal of Pharmaceutical Education* **28**, 161 (1964).

4. J. Baldwin, *Experimental Organic Chemistry*, 2nd edition, McGraw-Hill, New York, 1970.

5. L. F. Druding, *J. Chem. Educ.* **40**, 536 (1963).

6. A. R. Patton, *J. Chem. Educ.* **27**, 574 (1950).

7. E. S. and D. Kritchevsky, *J. Chem. Educ.* **30**, 370 (1953).

8. A. Ault and W. Lambrecht, unpublished results; G. A. Russell and P. G. Haffley, *J. Org. Chem.* **31**, 1869 (1966).

Sources from which more information about chromatographic methods may be obtained include:

9. E. Lederer and M. Lederer, *Chromatography*, Elsevier, Amsterdam and New York, 1967.

10. E. Heftmann, *Chromatography*, 3rd edition, Reinhold, New York, 1975.

11. R. P. W. Scott, Contemporary Liquid Chromatography, in *Techniques of Chemistry*, Vol. XI, Wiley, New York, 1976.

§15 REMOVAL OF WATER, DRYING

One very common separation problem is that of removing water. Since the products of many reactions are separated by procedures that use water or aqueous solutions, water usually must be removed during the purification process. Occasionally, the presence of even small amounts of water in a reaction mixture is undesirable. If so, it is necessary to remove even the water that is present in the reagents and solvents because of exposure to water vapor in the air. The following sections describe some of the methods for removal of water from solids, liquids, and gases.

15.1 DRYING OF SOLIDS

A damp solid, such as that which may be obtained when isolating a solid by suction filtration, may often be dried simply by spreading it out on a sheet of filter paper. If the solid is only slightly damp and the crystals are not so small that the material is a damp powder or paste, it may dry within an hour to the point that the only water associated with the solid is that which must be adsorbed in order for it to be in equilibrium with the water vapor present in the air.

If the material is quite damp or pasty, the bulk of the water can often be removed by pressing the material between sheets of adsorbent paper (filter paper) or by spreading it out and pressing it down on a piece of unglazed porcelain plate or a block of plaster of Paris.

Increasing Rate of Drying

Drying oven; heat lamp

The overall *rate* of drying may be increased by increasing the rate of evaporation (by raising the temperature) and by minimizing the rate of recondensation (by decreasing the partial pressure of water in the atmosphere around the sample). To increase the rate of evaporation, the sample may be heated in an oven to a temperature somewhat below its melting point (but usually not above 110°C for organic compounds, in order to minimize the rate of reaction of the substance with oxygen in the air). Another very convenient method of heating is to shine an infrared heat lamp on the sample. In all cases, the thinner the layer in which the sample is spread, the faster it will dry. The rate of recondensation of water vapor on the sample can be decreased by providing for circulation of air over it, or by carrying out the drying process in a container in which the pressure can be reduced. Figure 15.1 shows two simple arrangements for drying relatively small

Vacuum desiccator

samples. Figure 15.2 illustrates a vacuum desiccator. This apparatus may be

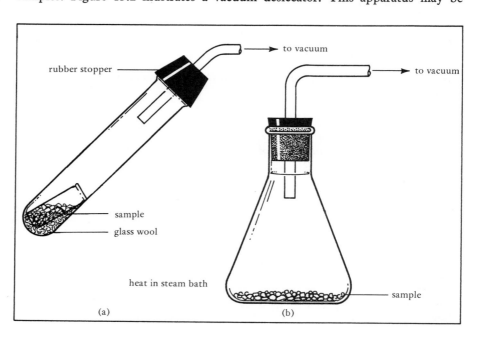

Figure 15.1. Arrangements for drying small samples.

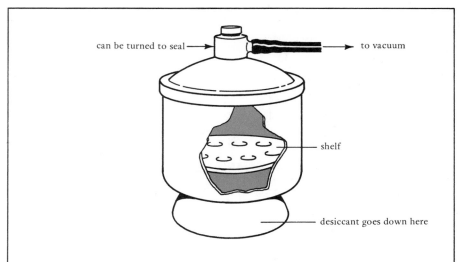

can be turned to seal → ← to vacuum

shelf

desiccant goes down here

Figure 15.2. A vacuum desiccator.

evacuated and also may contain a desiccant, a material that combines readily with water and has a low equilibrium vapor pressure of water in the hydrated form. The drying process inside an evacuated vacuum desiccator involves evaporation of water from the sample, diffusion to the desiccant, and essentially irreversible combination with the desiccant. The presence of the vacuum greatly increases the rate of diffusion, which is usually the slow step. Vacuum ovens are useful for drying larger quantities under vacuum at elevated temperature.

Increasing Extent of Drying

The *extent* to which a substance can be dried depends upon the partial pressure of water in the atmosphere surrounding it. In the open air, it is determined by the partial pressure of water in the air. For example, at 25°C and 60 percent relative humidity, the partial pressure of water will equal 0.60 times the equilibrium vapor pressure of water at this temperature, or about 14 Torr. For many substances, the amount of adsorbed or combined water that will result in an equilibrium vapor pressure of this magnitude for the substance will be quite small, and drying in air will serve to remove practically all of the water. If a substance will adsorb or combine with an objectionable amount of water at the partial pressure of water in the atmosphere, it must be dried in a closed chamber that contains a desiccant. The desiccant can be chosen to provide a sufficiently low equilibrium vapor pressure of water (see Table 15.1). Figure 15.3 illustrates an

Use of desiccant

Table 15.1. Desiccants commonly used in desiccators

Substance	Hydrated Form	Equilibrium Vapor Pressure of Hydrated Form
Aluminum oxide (alumina)	1% by weight of water (estimate)	0.001 Torr
Conc. sulfuric acid	95% sulfuric acid	0.001 Torr
	80% sulfuric acid	0.6 Torr
Potassium hydroxide	$KOH \cdot H_2O$	1.5 Torr
Sodium hydroxide	$NaOH \cdot H_2O$	0.7 Torr
Calcium chloride	$CaCl_2 \cdot H_2O$	0.04 Torr
Drierite	$CaSO_4 \cdot \frac{1}{2}H_2O$	0.004 Torr

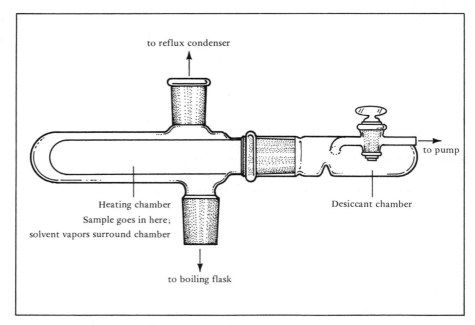

Figure 15.3. Apparatus for drying a small sample under vacuum in the presence of a desiccant: Abderhalden drying pistol. The temperature at which the sample is dried is determined by the boiling point of the solvent used.

apparatus that is used to dry small samples quickly and thoroughly by providing for heating, evacuation, and the use of a desiccant.

A less generally useful method of drying a solid that has been collected by suction filtration is to wash it with a solvent in which water is at least somewhat soluble but the substance is not. Anhydrous methanol, ethanol, or ether may be used for this purpose, but the problem is that many organic substances are appreciably soluble in these solvents.

Removal of Organic Solvents

Although removal of water has been the objective of the methods previously described, they may also be used for the *removal of other solvents* from solids. Removal of other solvents by air drying is usually complete, since the partial pressure in air of solvents other than water should be zero. Of course, the drying process can always be speeded up by increasing the temperature or using a vacuum desiccator, or both. Occasionally, a substance will hold solvent quite tenaciously and will require relatively long periods of drying at an elevated temperature. High-boiling solvents, acetic acid especially, are best removed directly after collection by suction filtration: the sample is washed on the funnel with small portions of a relatively volatile solvent that will dissolve the high-boiling solvent but not the compound.

15.2 DRYING OF SOLUTIONS

When an organic substance is isolated by extraction with an organic solvent, the aqueous phase from which the substance was extracted or the aqueous solutions with which the extracts are washed will leave the extract saturated with water. It is usually most convenient to remove the water by treating the extract with a solid that will combine with the water:

$$\underset{\text{anhydrous salt}}{\text{Solid} + n\text{H}_2\text{O}} \; \underset{\longleftarrow}{\overset{\longrightarrow}{}} \; \underset{\text{hydrated salt}}{\text{Solid} \cdot n\text{H}_2\text{O}}$$

The hydrated salt is then removed by gravity filtration. The completeness of water removal by this method depends both upon how long the drying agent is left in contact with the extract and upon how low the equilibrium vapor pressure of water is for the particular hydrated salt.

The ideal material for this purpose is a solid that is insoluble in the organic solvent, is inert toward both the solvent and the substance in solution, can be used in small amounts, and will combine quickly and completely with the dissolved water to give a solid (the hydrate of the original material) that may be removed by gravity filtration. Table 15.2 lists the drying agents that are most commonly used in this way. The drying agents are characterized by their capacity (the amount of water that can be removed by a given weight of drying agent), speed (rate at which water combines to form the hydrate), intensity (degree of dryness ultimately achieved), suitability for use with different classes of compounds, and cost. Since the criteria are not independent of one another (for instance, the intensity depends upon the degree of hydration allowed, and thus upon the capacity) and since they also depend upon the solvent being dried (ether solutions dry faster than benzene or ethyl acetate solutions), the ratings can be only approximate and qualitative.

Drying agents

Table 15.2. Drying agents commonly used for drying solutions in organic solvents

Substance	Capacity	Speed	Intensity	Cost	Convenience	Suitability
Calcium chloride	H	M	H	L	H	a
Calcium sulfate						
(Drierite)	L	H$^+$	H$^+$	M	H	b
Magnesium sulfate	H	H	M,H	L	M	c
Molecular sieves, 4Å	H	H	H		H	c
Potassium carbonate	M	M	M		M	d
Sodium sulfate	H$^+$	L	L	L	M	e

H: high   M: medium   L: low

a. Combines with alcohols, phenols, amines, amino acids, amides, ketones, and some aldehydes and esters. It should not be used to dry solutions containing compounds of these types unless it is desired to remove them also. Some calcium hydroxide may be present that will combine with acids. The hexahydrate is unstable above 30°C.
b. Generally useful. The hemihydrate is stable to at least 100°C.
c. Generally useful.
d. Combines with acids and phenols. It should not be used to dry solutions containing acids unless it is desired to remove them also.
e. Generally useful. The decahydrate is unstable above 32°C.

The drying of solutions involving solvents in which water is relatively soluble (ether, ethyl acetate) can be achieved more economically by carrying out the drying process in two stages, using a relatively inexpensive, high-capacity, low-intensity drying agent in the first stage, and a high-intensity drying agent in the second. The first stage can be to wash the solution with a saturated sodium chloride solution, in which the water will be more soluble than it is in the organic solvent, or to dry with anhydrous sodium sulfate, which will combine with 127 percent of its weight of water to form a decahydrate, $Na_2SO_4 \cdot 10H_2O$. The second stage can then be to dry with anhydrous calcium sulfate, which has a very high intensity but a low capacity since it combines with only 6 percent of its weight of water to form a hemihydrate, $CaSO_4 \cdot \frac{1}{2}H_2O$.

The procedure for drying a solution usually consists of adding a portion of the solid drying agent and allowing the mixture to stand for 5–15 minutes with occasional swirling. If a second liquid phase appears (a saturated aqueous solution of the drying agent), or if the drying agent clumps together, more should be added.

Procedure

It is easy to tell when an excess of anhydrous magnesium sulfate is present, since the suspension that forms when the mixture is swirled is quite cloudy and settles relatively slowly.

15.3 DRYING OF SOLVENTS AND LIQUID REAGENTS

Pure organic liquids (solvents and reagents) can be dried by treatment with various materials much in the way that solutions are dried. The drying agent may act either by forming a relatively stable hydrate or by entering into an essentially irreversible chemical reaction with the water. Table 15.3 indicates which drying agents are useful for drying certain compounds or classes of compounds. If drying is the only thing to be done, you may simply add the drying agent to the liquid in the bottle in which it is stored, providing for venting if hydrogen gas will be evolved. Drying may be practically complete in less than half an hour, but in some cases it may take several days.

Table 15.3. Suitability of various drying agents for drying pure solvents

You must take care not to try to dry a liquid or solution with something that will react with it. Sodium, potassium, and hydrides must be kept away from acids, alcohols, phenols, and other active hydrogen compounds; sodium and potassium must be kept away from alkyl or aryl halides.

| | Drying Agent | | | | | |
Solvent	P_2O_5	KOH, NaOH	BaO, CaO	K_2CO_3	$CaCl_2$	$MgSO_4$
Alcohols	−		+			
Aldehydes, ketones	−	−	−	+		
Alkanes	+				+	
Alkenes	+				+	
Alkyl halides	+				+	+
Amines	−	+	+	+	−	
Aromatic hydrocarbons	+				+	
Aryl halides	+				+	+
Ethers	−		+		+	+[a]
Nitriles	+			+		

(+) *Recommended for use*   (−) *Advised against using*

[a] Magnesium sulfate is satisfactory for drying ethyl ether for use in the Grignard reaction [*J. Chem Educ.* **39**, 578 (1962)].

Drying by Azeotropic Distillation

In the special case of solvents in which water is sparingly soluble and form minimum-boiling azeotropes with water, the solvent can be dried by distillation. The initial distillate will contain the water as the minimum-boiling azeotrope; the remainder of the distillate will be the dry solvent. Solvents that can be dried in this way include benzene, toluene, xylene, hexane, heptane, petroleum ether, carbon tetrachloride, and 1,2-dichloroethane. As long as the condensate appears cloudy, water is being removed; after about 10 percent of the volume of the solvent has been distilled, all the dissolved water and the water adsorbed on the walls of the flask and condenser should be gone.

When any of these solvents must be dry for use as a solvent (or reagent) in a reaction, it is often most convenient to add a 10–15 percent excess to the reaction flask and then to remove the excess by distillation. If this is done, both the dissolved water and the water adsorbed on the inside of the apparatus will be removed as a minimum-boiling azeotrope. Sometimes it is possible to dry one or more of the reagents as well by adding it before the distillation.

The presence of water in a liquid may sometimes be detected by adding a small amount of anhydrous cobaltous chloride or anhydrous cobaltous bromide. If water is present, the blue color of the anhydrous salt gives way to the pink color of the hydrate. One form of anhydrous calcium sulfate commercially available (Drierite) comes as granules whose surface is impregnated with cobaltous chloride. A few granules of this material is useful for testing for the presence of water.

Test for presence of water

15.4 DRYING OF GASES

Gases are most conveniently dried by passing them through a tube packed with a granular adsorbent. Table 15.4 indicates which absorbents have been recommended for a number of different gases. The size of the column of adsorbent and the flow rate should be such that a contact time of at least five seconds is attained. Figure 15.4 illustrates a gas-drying tube.

Table 15.4 also indicates certain gases that may be dried by bubbling them through concentrated sulfuric acid.

Table 15.4. Suitability of various drying agents for drying gases

			Drying Agent		
Gas	P_2O_5	$CaCl_2$	$CaSO_4$	Al_2O_3	conc. H_2SO_4
Air		+	+	+	+
Nitrogen		+	+	+	+
Hydrogen	+	+		+	+
Carbon dioxide	+	+		+	+
(+) Recommended for use					

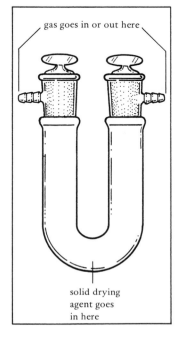

gas goes in or out here

solid drying agent goes in here

Figure 15.4. Gas-drying tube.

Reference

1. D. R. Burfield, K.-H. Lee, and R. H. Smithers, Desiccant Efficiency in Solvent Drying. A Reappraisal by Application of a Novel Method for Solvent Water Assay, *J. Org. Chem.* **42**, 3060 (1977). As the authors say in their abstract, the results range from the expected to the highly surprising.

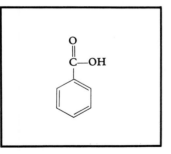

Determination of Physical Properties

It is currently believed that the physical and chemical properties of a substance are determined entirely by the molecular structure of the substance. Different physical states of a substance are believed to differ not in molecular structure but only in the relationships between the molecules. The different intermolecular relationships are determined by the structure of the molecules.

It follows from this assumption that samples that have identical physical and chemical properties are identical at the molecular level. *If substances are discovered that differ in at least one chemical or physical property, they must have different molecular structures. Thus, you can decide whether two or more substances are composed of identical molecules by comparing their physical and chemical properties. If a substance is judged to be pure because there is no method known that will separate it into components, its physical and chemical properties can serve as the standards by which its kind of molecule may be identified at another time. In a similar way, the properties of a mixture can serve to establish that mixture's identity. If a sample is thought to be a mixture, the progress of its purification may be followed by comparison of its physical or chemical properties before and after the application of separation procedures.*

In the preparative experiments in this book, the identity and purity of the products of the reactions may be established by comparison of certain of their physical and chemical properties with the properties of substances whose molecular structure and degree of purity are assumed to be known. The physical properties most often used in this way are the melting or boiling point and the infrared or nuclear magnetic resonance spectrum.

Since the physical and chemical properties of a substance are thought to be determined entirely and uniquely by the molecular structure of the substance, it follows that we can infer the molecular structure of the substance from the physical and chemical properties of the substance. *After all, the properties of a substance are the only experimental knowledge we will ever have of that substance, and hence any theoretical description must be inferred from these observable properties. It is impossible to explain briefly exactly how empirical evidence has in the past been translated into*

112

structural theory and how, in general, theory is arrived at, but in the following discussion of physical properties and their rationalization in terms of molecular structure, many examples will be given of how conclusions about the molecular structures of substances may be inferred from their physical properties.

The physical properties most often used in the past to characterize substances were the boiling point, melting point, density, index of refraction, and optical rotation (Sections 16 through 20). For substances of unknown structure, determinations of molecular weight and solubility characteristics were often helpful (Sections 21 and 22). More recently, spectrometric methods have become very widely used, both for the characterization of substances and for structure determination. Of these, the most common are infrared, ultraviolet-visible, nuclear magnetic resonance, and mass spectrometry (Sections 23 through 26).

---

§16 BOILING POINT

The boiling point is one of the most often reported physical properties of a liquid, since it can usually be determined while purifying a liquid by distillation. It is often used along with the density and index of refraction to establish the identity and to estimate the purity of a liquid.

16.1 EXPERIMENTAL DETERMINATION OF BOILING POINT

If sufficient material is available, the boiling point of a liquid may be determined by distillation (Section 9.6). The temperature of the vapor should be observed during the course of the distillation, and the temperature range over which most of the material distills should be taken as the boiling point. Since the boiling point is a function of the pressure, the barometric pressure or, in a distillation under reduced pressure, the pressure of the system should be recorded.

If less than a milliliter of liquid is available—as, for instance, when the sample has been isolated by vapor-phase chromatography—a *small-scale method* may be used. In this procedure, a 10- to 15-cm length of glass tubing 3–5 mm in inside diameter is sealed shut at one end by heating in a flame. Two or three drops of the liquid are added to this sample tube, and a length of melting-point capillary, sealed about 5 mm from one end, is dropped in with the sealed end down. (This small tube is most conveniently made by melting shut a melting-point capillary near the middle and cutting off all but 5 mm on one side of the seal.) The sample tube is fastened to a thermometer by a 2- to 3-mm slice of rubber tubing used as a small rubber band (see Figure 16.1). The thermometer is then supported in a melting-point bath (Figure 17.4) and heated until a very rapid, steady stream of bubbles issues from the sealed capillary. The bath is then allowed to cool slowly, and the temperature at which a bubble just fails to come out of the capillary and the liquid starts to enter it is taken as the boiling point of the liquid.

If the barometric pressure is not 760 Torr, the observed boiling point may by corrected to the temperature that would be expected at 760 Torr. This correction amounts to about 0.5°C for each 10-Torr deviation of atmospheric pressure from 760 Torr; the observed boiling point will be low if the atmospheric pressure is low.

In addition to errors introduced by experimental variables such as the rate

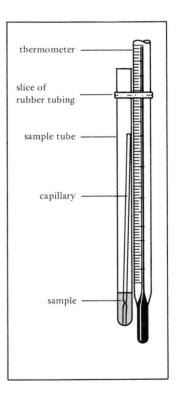

Figure 16.1. Apparatus for small-scale determination of boiling point.

of heating, superheating, and the presence of impurities, the observed boiling point may be in error for two other reasons. The first is that the thermometer may not be correctly calibrated (the scale of degrees may not be in exactly the right place on the stem of the thermometer). The extent of any error of calibration may be determined by using the thermometer to experimentally determine the melting point of pure samples of solids of known melting point (Section 17.1). The second possible error involves the intended use for which the thermometer was designed. Many thermometers are calibrated with the understanding that they will be immersed only to a certain line engraved upon the stem. Some thermometers, however, are calibrated with the understanding that the entire thermometer will be at the temperature that is being determined. If only part of the thermometer is at this temperature, the extent of this error, which is the result of unequal coefficients of thermal expansion of mercury and glass, may be calculated according to the following equation:

$$\text{Correction to be added to } t_1 = KN(t_1 - t_2)$$

where K = the apparent coefficient of thermal expansion of mercury in glass: 0.000159 for mercury in soft glass at 200°C; 0.000167 for borosilicate (Pyrex) glass at 200°C

   N = the length, in degrees, of the part of the mercury column not heated to the temperature of the bulb: the length of the exposed column of mercury

   t_1 = the observed temperature: the temperature of the mercury bulb

   t_2 = the temperature at the middle of the exposed mercury column; this may be determined by a second thermometer

A typical emergent stem correction for an observed temperature reading of 200°C might be +5°C.

One of the best ways to estimate (and thus take into account) thermometer error and many experimental errors is to determine the boiling point of a pure sample of a substance of known boiling point under the same experimental conditions used to determine the unknown boiling point. Table 16.1 lists some liquids and their boiling points, which have been recommended for use as standards for boiling-point determinations.

Table 16.1. Reference liquids for boiling-point determinations

Compound	Normal b.p. (°C)	ΔT/10 Torr (°C)[a]
Acetone	56.1	0.39
Water	100.0	0.37
Bromobenzene	156.2	0.53
Nitrobenzene	210.9	0.48
Quinoline	237.5	0.59
Benzophenone	305.9	0.6

[a] Variation of boiling point with pressure.

16.2 BOILING POINT AND MOLECULAR STRUCTURE

The variation of boiling point with the structure of covalently bonded molecules may be resolved into three contributing factors: the molecular weight, the nature of the functional group, and the degree of branching of the molecule. In order to see the regularity in the effect of any one factor, the other two must be held constant.

Boiling Point and Molecular Weight

As an illustration of the effect of increasing molecular weight, the boiling points of the first members of the homologous series of straight-chain alkanes and primary alkyl chlorides are listed in Tables 16.2 and 16.3. From these data it appears that *boiling points increase with increasing molecular weight* (the nature of the functional group and degree of branching being held constant). It is also apparent that the increase in boiling point per additional methylene group is not constant, but decreases with increasing molecular weight. For this reason, a graph of boiling point versus molecular weight (or number of carbon atoms) is a curved line.

Table 16.2. Normal alkanes: boiling points and enthalpies and entropies of vaporization

Compound	—Boiling Point— °C	°K	Molecular Weight	—Increase— b.p.	m.w.	ΔH_{vap} (cal/mole)	ΔS_{vap} (cal/mole-degree)
C_1	−161.49	111.66	16.042			1,955	17.51
C_2	−88.63	184.52	30.068	72.86	14	3,517	19.06
C_3	−42.07	231.08	44.094	46.56	14	4,487	19.42
C_4	−0.50	272.65	58.120	41.57	14	5,352	19.63
C_5	36.07	309.22	72.146	36.57	14	6,160	19.92
C_6	68.74	341.89	86.172	32.67	14	6,896	20.17
C_7	98.47	371.62	100.198	29.72	14	7,575	20.38
C_8	125.67	398.82	114.224	27.20	14	8,214	20.60
C_9	150.80	423.95	128.250	25.13	14	8,777	20.70
C_{10}	174.12	447.27	142.276	23.32	14	9,390	20.99

Table 16.3. Primary, straight-chain alkyl chlorides: boiling points and enthalpies and entropies of vaporization

Compound	—Boiling Point— °C	°K	Molecular Weight	—Increase— b.p.	m.w.	ΔH_{vap} (cal/mole)	ΔS_{vap} (cal/mole-degree)
C_1	−24.22	248.93	50.5			5,126	20.59
C_2	12.27	285.42	64.5	36.49	14	5,832	20.43
C_3	46.60	319.75	78.5	34.33	14	6,512	20.37
C_4	78.44	351.59	92.6	31.84	14	7,174	20.40
C_5	107.76	380.91	106.6	29.32	14	7,824	20.54
C_6	134.50	407.65	120.6	26.74	14	8,458	20.75
C_7	159.1	432.3	134.6	24.65	14	9,091	20.98
C_8	182.0	455.2	148.7	22.9	14	9,673	21.25
C_9	203.4	476.6	162.7	21.4	14	10,250	21.51
C_{10}	223.4	496.6	176.7	20.0	14	10,800	21.75

Boiling Point and the Nature of the Functional Group

As an illustration of the effect of the nature of the functional group, the boiling points of a number of straight-chain compounds with approximately the same molecular weight but with different functional groups are presented in Table

Table 16.4. Boiling points and enthalpies and entropies of vaporization of straight-chain compounds with approximately the same molecular weight but with different functional groups

Compound	Molecular Weight	—Boiling Point — °C	°K	ΔH_{vap} (cal/mole)	ΔS_{vap} (cal/mole-degree)
1-Pentene	70	30.0	303.2	6,021	19.86
1-Fluorobutane	76	32.5	305.7	6,264	20.49
Diethyl ether	74	34.3	307.5	6,355	20.67
Pentane	72	36.1	309.3	6,157	19.91
1-Chloropropane	78	45.7	318.9	6,594	20.68
Diethylamine	73	55.5	328.7	6,888	20.96
Methyl acetate	74	57.8	331.0	7,454	22.52
n-Butylamine	73	77.8	351.0	7,678	21.88
Butyraldehyde	72	74.6	347.8	7,880	22.66
2-Butanone	72	79.6	352.8	7,837	22.21
Propionitrile	69	117.4	390.6	7,937	20.32
1-Butanol	74	117.6	390.8	10,505	26.88
Propionic acid	74	139.3	412.5	7,318	17.74
Dimethylformamide	73	149.6	422.8	9,164	21.67
N-Methylacetamide	73	206	479		
Propionamide	73	213	486	14,860	30.56

16.4. It is apparent that *the more polar the functional group, the higher the boiling point.* (Actually, the degree of polarity of the functional group is inferred from the b.p.'s of the substances.)

Boiling Point and Degree of Branching

As an illustration of the effect of the degree of branching, the boiling points and molecular structures of the eight isomers of molecular formula $C_5H_{11}Cl$ are presented in Table 16.5. The general result is that *the more highly branched the molecule (the functional group and the molecular weight being the same), the lower the boiling point.* A comparison of the last three entries in the table illustrates the

Table 16.5. Boiling points of the isomeric alkyl chlorides of molecular formula $C_5H_{11}Cl$

Compound	b.p. (°C)	Compound	b.p. (°C)
C—C—C—C—C—Cl	108	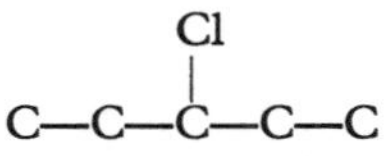 C—C—Ċ—C—C (Cl on third C)	97
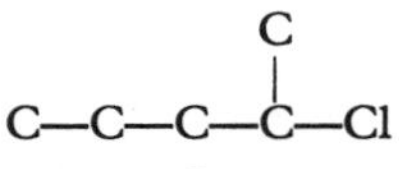 C—C—C—Ċ—Cl (C branch)	97	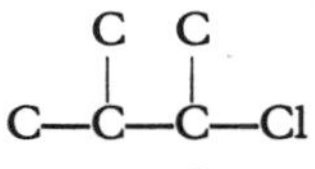 C—C—Ċ—Cl (two C branches)	91
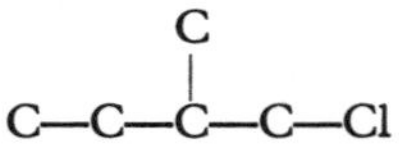 C—C—Ċ—C—Cl (C branch)	98	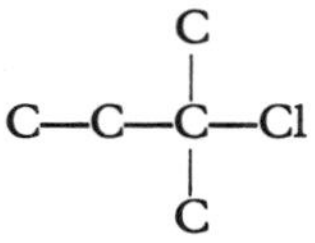 C—C—Ċ—Cl (C branch below)	86
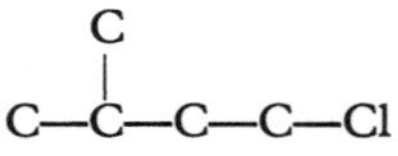 C—Ċ—C—C—Cl (C branch)	100	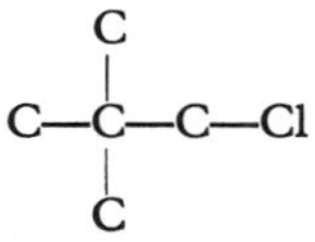 C—Ċ—C—Cl (C branches above and below)	84

general observation that if the branches are on the same carbon atom rather than on different ones, the boiling point will be a little lower.

For processes that take place at a constant temperature, such as vaporization or freezing of a liquid, the entropy change (ΔS) is equal to the enthalpy change (ΔH) divided by the temperature at which the process occurs:

$$\Delta S = \frac{\Delta H}{T}$$

At the normal boiling point, then,

$$T_{vap} = \frac{\Delta H_{vap}}{\Delta S_{vap}}$$

Thus, the high boiling points are the result of large heats of vaporization and small entropies of vaporization. It remains to see how the factors of molecular weight, nature of the functional group, and degree of branching contribute to these quantities.

The heat of vaporization is interpreted as a measure of the amount of energy required to separate the molecules against an attractive intermolecular force in the change, at constant temperature, from the liquid state to the vapor state. The stronger the intermolecular forces, the higher the heat of vaporization. Four different kinds of intermolecular forces may be distinguished: (1) van der Waals' forces proportional to the square of the polarizability, (2) dipole–dipole forces, proportional to the fourth power of the dipole moments, (3) induced dipolar forces, proportional to the polarizability and the square of the dipole moment, and (4) hydrogen-bonding forces, usually due to the presence of O—H or N—H groups in the molecule (1).

ΔH_{vap}

The increase of boiling point with increasing molecular weight (the nature of the functional group and the degree of branching remaining constant) may be interpreted in terms of increased van der Waals' forces. The higher the molecular weight (the more CH_2 units), the larger the total intermolecular attraction due to van der Waals' forces. This results in an increase in heat of vaporization with increasing molecular weight. Data for the first members of the series of normal alkanes and alkyl chlorides are presented in Tables 16.2 and 16.3.

The decrease in boiling point with increasing branching (the nature of the functional group and the molecular weight remaining constant; see Table 16.5) may also be interpreted in terms of the influence of structure on the magnitude of the van der Waals' forces. A more highly branched and thus more compact molecule will have less surface area and therefore a smaller total intermolecular attraction due to van der Waals' forces.

When functional groups are present, which cause the molecule to have a dipole moment, dipolar and induced dipolar forces can contribute to the intermolecular attraction in addition to the van der Waals' forces. The higher boiling points of ethers, aldehydes, ketones, esters, nitriles, and nitro compounds (and to a certain extent alkyl halides) may be interpreted in terms of the presence of additional forces of these types. The presence of O—H or N—H (or, to a smaller extent, S—H) groups in a molecule adds a further particularly effective, localized type of dipole–dipole attraction (the "hydrogen bond"), which is due almost uniquely to these types of functional groups. The relatively high boiling points (for a given molecular weight and degree of branching) of alcohols, thiols, primary and secondary amines, phenols, carboxylic acids, and unsubstituted and monosubstituted amides can be interpreted mainly in terms of this additional relatively large force. Tables 16.4 and 16.6 present data that illustrate the relative effectiveness of different functional groups in contributing to a high boiling point.

Table 16.6. Boiling points of substituted butanes

Compound[a]	Boiling point (°C)	Molecular Weight	Intermolecular Forces	ΔH_{vap} (cal/mole)	ΔS_{vap} (cal/mole-degree)
R—H	0	58	van der Waals' only	5,342	19.63
R—F	33	76	van der Waals' and dipolar	6,264	20.49
R—OCH₃	70	88	,,		
R—Cl	78	93	,,	7,174	20.40
R—Br	102	137	,,	7,613	20.32
R—CHO	103	86	,,	8,550	22.7
R—COOCH₃	127	116	,,		
R—COCl	128	121	,,		
R—I	131	184	,,	7,983	19.77
R—C≡N	141	83	,,	8,669	20.82
R—NO₂	152	103	,,	10,000	23.5
R—NH₂	78	73	van der Waals', dipolar, and hydrogen bonds	7,678	21.88
R—SH	99	90	,,	7,700	20.72
R—OH	118	74	,,	10,505	26.88
R—COOH	187	102	,,	11,891	25.89

[a] R = n-Butyl

The boiling point increase resulting from the introduction of a single halogen atom is due partly to the creation of a dipole moment (about the same for —F, —Cl, and —Br, and slightly less for —I) and partly to the introduction of a polarizable atom (the polarizability is least for fluorine and increases through chlorine to bromine to iodine; polarizability generally increases with atomic number or number of electrons). The effect of an iodine atom is mostly through its polarizability, and the effect of fluorine is almost entirely due to any dipole moment resulting from its presence. It is reasonable to account for the relatively low boiling points of fluorocarbons in terms of the low polarizability of fluorine.

ΔS_{vap} The entropy of vaporization is interpreted as a measure of the increase in disorder that results when a collection of molecules is changed from a relatively confined state to a relatively free state (or from a small volume to a large volume, or from a state with relatively widely spaced energy levels to a state with relatively closely spaced energy levels). In general, the entropy of vaporization for members of an homologous series increases only slowly with increasing molecular weight (Tables 16.2 and 16.3) and is independent of the degree of branching. Thus, the effect of molecular weight and degree of branching upon boiling point is almost entirely through the heat of vaporization and not through the entropy of vaporization. In fact, for most substances of moderate boiling point, the entropy of vaporization is equal to 20–22 calories/mole-degree (Trouton's rule).

However, there are some interesting exceptions. Alcohols (and water) generally have an unusually large entropy of vaporization. This is interpreted by saying that the increase in disorder is unusually great when an alcohol or water is vaporized, and the reason for this is that the liquid state is relatively ordered or structured compared to other liquids. This extra degree of structure, presumably due to the formation of chains or networks of hydrogen bonds, is lost upon vaporization. Thus, from the point of view of the entropy change upon vaporization, we would expect alcohols and water to have unusually low boiling points. This is not the case, however; alcohols (and water) have relatively high boiling points compared to other compounds of the same molecular weight and degree of branching. The reason is that additional energy is required to break up the hydrogen bonds, which results in an unusually high heat of vaporization for alcohols and water. In fact, the heat of vaporization more than compensates for the increased entropy of vaporization. Acetic acid has an unusually low heat of vaporization, less than that for pentane. On the basis of heats of vaporization, acetic acid might be expected to boil below pentane. However, it also has an unusually low entropy of vaporization—about 14.5 cal/mole-degree. If acetic acid had a "normal" entropy of vaporization, it would boil at $-3°C$, rather than at $118°C$. The low entropy of vaporization is interpreted by saying that acetic acid must retain some order or structure in the vapor phase. Vapor density measurements support this by indicating an average molecular weight considerably higher than 60, and both phenomena are interpreted in terms of partial dimerization through hydrogen bond formation in the vapor phase. Since not all the molecules are separated upon vaporization, the low heat of vaporization may thus be explained as well.

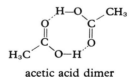

acetic acid dimer

Boiling points might be expected to vary much more widely than they do, except that differences in heat of vaporization and entropy of vaporization tend to balance each other. An increase in the intermolecular force would be expected to increase the order of the liquid, and thus lead to a larger entropy of vaporization, as well as to increase the heat of vaporization. Table 16.7 lists some compounds which have about the same boiling points (and therefore the same ratio of heat of vaporization to entropy of vaporization), but which show a considerable variation in these quantities.

Table 16.7. Enthalpies and entropies of vaporization of selected compounds boiling near 120°C

| Compound | Molecular Weight | —Boiling Point— | | ΔH_{vap} | ΔS_{vap} |
		°C	°K	(cal/mole)	(cal/mole-degree)
Acetic acid	60	118.5	391.7	5,662	14.46
Butyronitrile	71	117.4	390.6	7,937	20.32
n-Butanol	74	117.6	390.8	10,505	26.88
2-Hexanone	100	127.4	400.6	8,243	20.38
Ethyl butyrate	114	118.9	392.1	8,673	22.12
n-Octane	114	125.8	398.8	8,214	20.60
n-Butyl iodide	184	130.5	403.7	7,983	19.77

Problems

1. A liquid was observed to distill between 206 and 207.5°C when the atmospheric pressure was 743 Torr. The thermometer used was a total-immersion thermometer, and the exposed length was equivalent to 225°C. Room temperature was about 27°C. Calculate the expected normal boiling point of the substance. What factors contribute to uncertainty in this value?

2. A liquid was observed to distill between 121.5 and 122.5°C under the conditions described in Problem 1. Under the same conditions, toluene distilled between 109 and 109.5°C. Calculate the expected normal boiling point of the substance. What factors contribute to the uncertainty of this value?

3. What correction should be applied when a partial-immersion thermometer is used as a total-immersion thermometer?

4. The boiling points in degrees C, of a series of analogous chlorine and fluorine compounds are presented in the following table. Rationalize the different trends shown by the different halogens.

	CH$_3$X	CH$_2$X$_2$	CHX$_3$	CX$_4$
Fluorine	−78	−51	−82	−129
Chlorine	−24	40	62	76

Which trend would you expect the analogous bromine compounds to follow?

5. The following table presents values for the enthalpy and entropy of vaporization for several aliphatic carboxylic acids. How do you interpret these data?

Compound	ΔH_{vap} (cal/mole)	ΔS_{vap} (cal/mole-degree)	b.p.
Formic acid	5,318	14.26	101
Acetic acid	5,558	14.19	118
Propionic acid	9,998	24.14	141
n-Butyric acid	10,780	24.70	163
i-Butyric acid	10,630	24.92	153
Valeric acid	11,890	25.90	183

6. Rationalize the differences in boiling points of members of each group of compounds.

a.   CH$_3$CH$_2$—O—CH$_2$CH$_3$
34°C   66°C

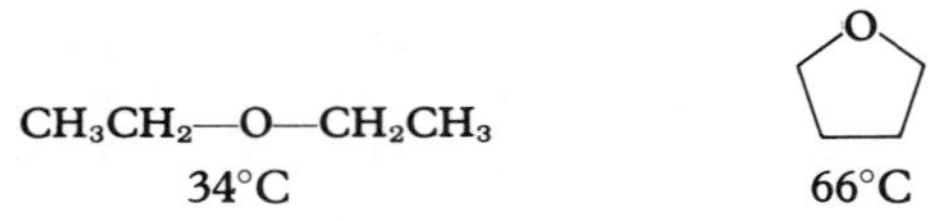

b.   CH$_3$—C—O—C—CH$_3$
140°C   261°C

c.   CH$_3$—C—O—CH$_2$CH$_3$
77°C   206°C

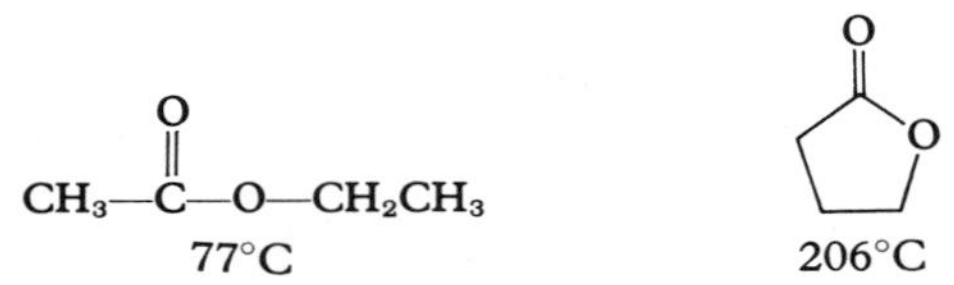

d.
80°C   87°C   102°C   105°C

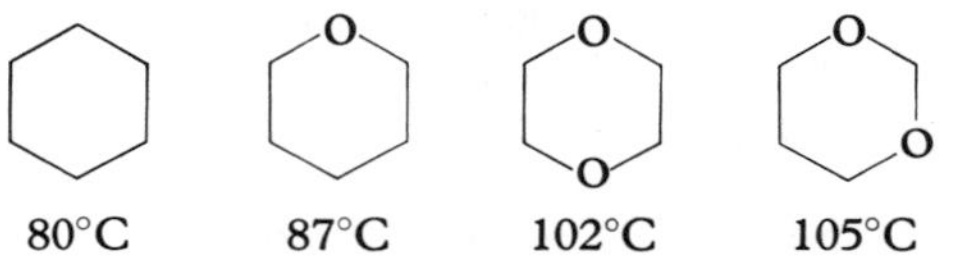

e.
166°C   220°C

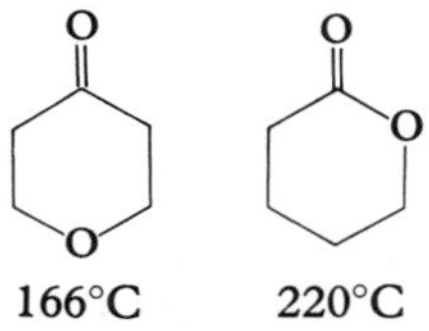

f.
134°C   160°C   184°C   182°C

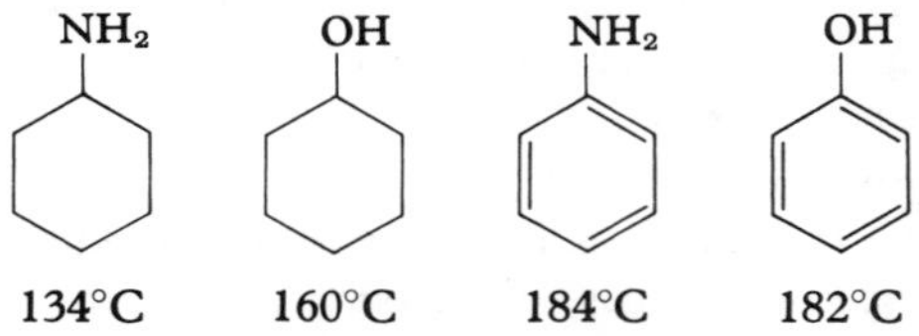

g.
146°C   135°C

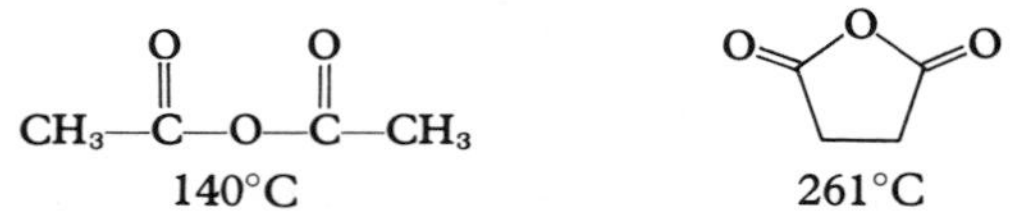

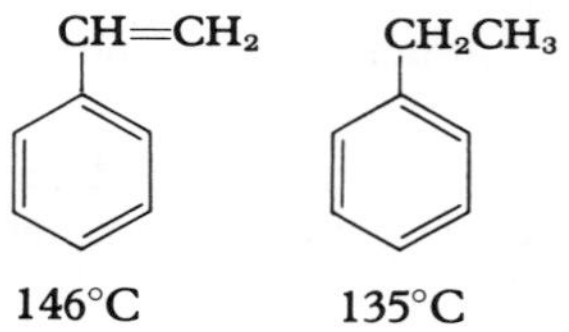

h.

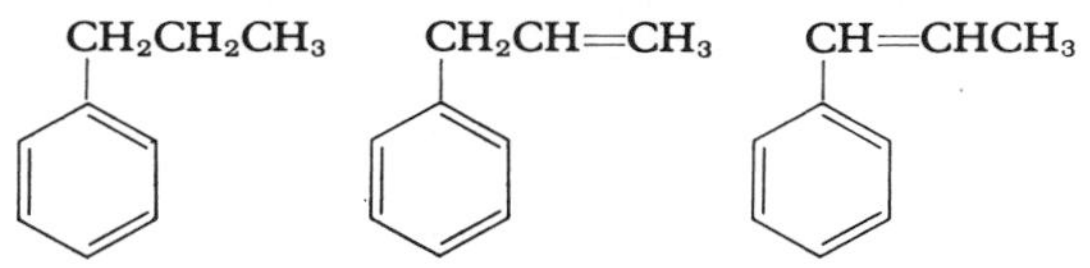

CH₂CH₂CH₃   CH₂CH=CH₃   CH=CHCH₃

158°C   157°C   177°C

i.

$$CH_3-\overset{\overset{\textstyle O}{\|}}{C}-CH_2CH_3 \qquad CH_3-\overset{\overset{\textstyle O}{\|}}{C}-CH=CH_2$$

80°C   80°C

$$CH_3-\overset{\overset{\textstyle O}{\|}}{C}-CH_2-\overset{\overset{\textstyle CH_3}{}}{\underset{\textstyle CH_3}{C}}-H \qquad CH_3-\overset{\overset{\textstyle O}{\|}}{C}-CH=\overset{\overset{\textstyle CH_3}{}}{\underset{\textstyle CH_3}{C}}$$

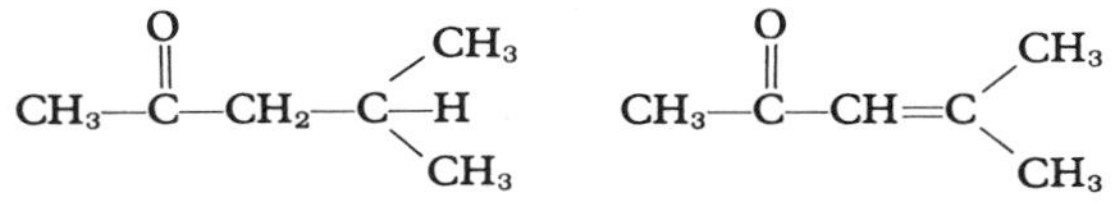

119°C   130°C

7. Assign structures to the following substances on the basis of their boiling points.
  a. Isomers of molecular formula C_3H_9N. *Boiling points:* 49°, 35°, 35°, and 3°C.
  b. Isomers of molecular formula $C_4H_{10}O$. *Boiling points:* 118°, 108°, 100°, 83°, 39°, 34°, and 31°C.
  c. Methyl ethers of molecular formula $C_5H_{12}O$. *Boiling points:* 70°, 61°, 61°, and 55°C.

Reference

1. A discussion of the origin of these forces is presented by G. M. Barrow in *Physical Chemistry*, 3rd edition, McGraw-Hill, New York, 1973, pp. 512–519.

---

§17 MELTING POINT

The normal melting point of a solid is defined as the temperature at which the solid and liquid are in equilibrium at a total pressure of 1 atmosphere (the vapor pressure of the solid is usually very much less than 1 atmosphere at the melting point). In contrast to the volume change that accompanies the vaporization of a liquid, the change in volume that takes place upon the melting of a solid is very small. Hence, the melting point of a solid, unlike the boiling point of a liquid, is practically independent of any ordinary pressure change. Since the melting point of a solid can be easily and accurately determined with small amounts of material, it is the physical property that has most often been used for the identification and characterization of solids.

17.1 EXPERIMENTAL DETERMINATION OF THE MELTING POINT

Melting Points from Cooling Curves

If large amounts of the solid are available (a gram or so), the most accurate method for the determination of the melting point as defined above is to heat the sample until it is melted (probably by means of an oil bath, Section 32.1) and then allow it to cool slowly for crystallization, keeping track of the temperature of the sample as a function of time by means of an immersed thermometer or thermistor. At first, the temperature will be observed to fall as the liquid loses heat to the surroundings. When crystallization begins, however, the heat evolved during

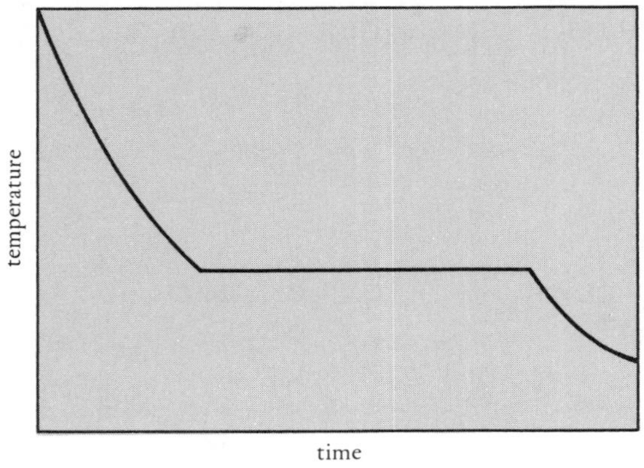

DETERMINATION OF
PHYSICAL PROPERTIES

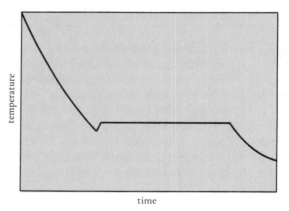

Figure 17.1. Expected
cooling curve for a pure
substance.

this process (ΔH_f, the heat of fusion) will maintain the temperature at a constant
value until crystallization is complete. At this point, the temperature will again
fall as the solid loses heat to the surroundings. If the material is pure, the temperature of the sample remains constant during the entire process of solidification;
this temperature is the melting point. Figure 17.1 illustrates this expected cooling
curve for a pure substance. It is not unusual for the temperature to fall a little
below the melting point before crystallization begins. When this happens, the

Supercooling   sample is said to have *supercooled*. The heat evolved (as crystallization takes place)
will then warm the sample to the melting point (see Figure 17.2).

This procedure may also be used for determining the freezing point of substances that are liquids at room temperature. The liquids are cooled in a cold
or refrigerated bath. During the cooling time, the temperature of the bath should
be only a little below the expected freezing point. (Several trials can be made
with the same sample, and the temperature of the bath can be adjusted according
to the results of the previous trial.)

Figure 17.2. Expected
cooling curve, with
supercooling, for a pure
substance.

This procedure is the one that should be used for calibrating a thermometer
or checking the calibration of a thermometer by means of solids of known melting
point. In this case, any disagreement between the reading of the thermometer
and the true melting point, which is assumed to be known, is attributed to an

Thermometer calibration   error in the calibration of the thermometer. Table 17.1 lists a number of substances of known melting point that have been recommended for use in this
method of calibration.

Capillary Melting Points

The method most often used for the determination of the melting point of a solid
requires only a very small sample. A few crystals of the sample are placed in a
thin-walled capillary tube 10–15 cm long and about 1 mm in inside diameter,

Table 17.1. Reference substances for calibration of thermometers by melting-point determination

Compound	Melting Point (°C)
Ice	0.0
m-Dinitrobenzene	89.7
Benzoic acid	121.7
Salicylic acid	160.4
3,5-Dinitrobenzoic acid	205
Sym.-di-*p*-tolyl urea	268

which has been sealed shut at one end. The capillary containing the sample and the thermometer are then suspended in an oil or air bath that can be heated slowly and evenly. The temperature (range) over which the sample is observed to melt is taken as the melting point. Obviously, if you are to be able to determine at what temperature the crystals are melting, the thermometer and sample must be at the same temperature at the melting point. This means that the rate of heating of the bath should be very low as the melting point is approached (about 1 degree per minute). Otherwise, the temperature of the mercury in the thermometer bulb and the temperature of the crystals in the capillary will each be an unknown amount below the temperature of the bath and probably not equal to each other. This is because of the slowness with which heat energy is transferred by conduction.

If the approximate temperature at which the sample will melt is not known, a preliminary melting-point determination should be made in which the temperature of the bath is raised quickly. Then the more accurate determination should be carried out, with a low rate of heating near the melting point. A preliminary melting point can be determined within ten minutes, which is the time it would take to raise the temperature of the bath 10 degrees at 1 degree per minute. It should be obvious that if you cannot estimate the melting point within about 20 degrees, then two melting-point determinations (one fast, one slow) will take less time than one determination that is slow over a wide range.

Usually, the melting-point capillary can be filled by pressing the open end into a small heap of the crystals of the substance, turning the capillary open end up, and vibrating it by drawing a file across the side to rattle the crystals down into the bottom. If filing does not work, drop the tube, open end up, down a length of glass tubing about 1 cm in diameter (or a long condenser) onto a hard surface such as a porcelain sink, stone desk top, or the iron base of a ringstand. The solid should be tightly packed to a depth of 2–3 mm. If the sample sublimes rapidly at the melting point, it will be necessary to seal the capillary before attempting to determine the melting point; in an unsealed tube, the sample would sublime to the cooler part of the capillary above the bath. The capillary may be melted shut about 2 cm above the sample by briefly holding that part of the capillary in a small burner flame. The entire sealed portion should be immersed in the bath during heating. When an oil bath is used, the capillary may be fastened to the thermometer by means of a small slice of rubber tubing used as a rubber band (see Figure 17.3); if the capillary is straight, it may stick to the thermometer by the capillary action of the bath oil without the help of a rubber band. If a compound begins to decompose near the melting point, the capillary with the sample should be placed in the bath after the temperature has been raised to within 5 or 10 degrees of the expected melting point, so as to minimize the length of time that the sample is heated.

Trial run

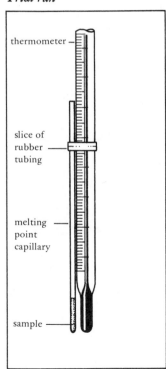

thermometer

slice of rubber tubing

melting point capillary

sample

Figure 17.3. Arrangement of sample and thermometer for melting-point determination.

Melting-point baths

There are many types of oil baths that can be used for this type of melting-point determination, as well as for micro boiling-point determination. The simplest involve heating with a burner flame and depend upon convection for mixing; the more elaborate are heated electrically by an immersion heater and are stirred. For greater accuracy and convenience, the latter is necessary. It is very easy to heat at a low and steady rate with an electric immersion heater, but almost impossible with a flame. Several heating baths are illustrated in Figure 17.4. It is dangerous to exceed 200° with a bath oil such as mineral oil. Other liquids such as silicone oils, which have higher flash points, can be used at these higher temperatures.

Thermometer correction

As in the case of the determination of the boiling point, the emergent stem correction of the thermometer must be taken into account (see Section 16.1), as well as any correction for error in calibration of the thermometer.

Capillary melting points are properly compared with one another, but they are occasionally considerably different from melting points determined from experimental cooling curves.

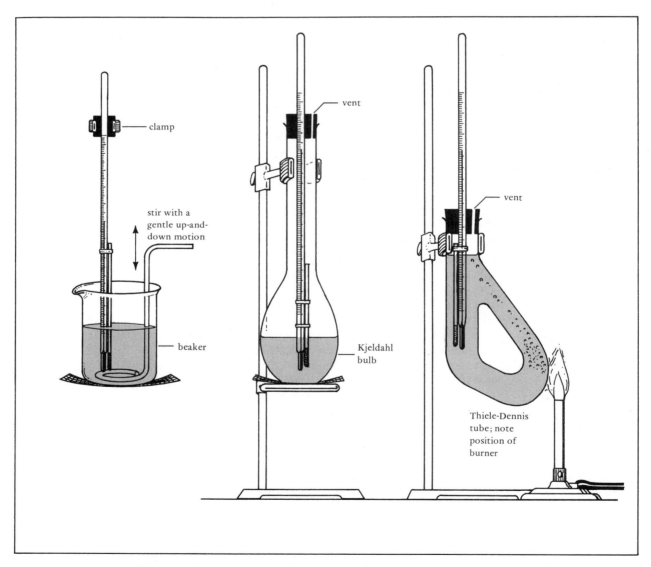

Figure 17.4. Heating baths for melting- or boiling-point determination.

A very fast and easy method of melting-point determination involves heating a few crystals of the sample between a pair of microscope cover glasses on an electrically heated metal block while observing them with the aid of a magnifying glass. This method has the advantages of requiring as little as a single crystal, permitting good temperature control, and being very convenient. However, complete thermal equilibrium between the sample, block, and thermometer is not possible, since the thermometer is inside the block and the sample is on the surface, exposed to the cooler atmosphere. For this reason, observed block melting points are often higher than capillary melting points; the higher the melting point, the greater the difference. However, a melting point quickly determined on a block can serve as an approximate melting point for the determination of a capillary melting point.

Although a pure solid would be expected to have a single temperature at which solid and liquid are in equilibrium, most samples are observed to melt over a finite temperature range. With capillary or block melting points, this may be due in part to the fact that the bath or block temperature slowly rises during the time it takes the sample to melt. A melting range can also be due to the presence of impurities in the sample. Thus, the melting point will usually be reported as two temperatures between which the sample was observed to melt, a *melting range*.

Melting range

17.2 THE MELTING POINT AS A CRITERION OF PURITY

A dilute solution of a liquid begins to freeze at a temperature somewhat lower than the freezing point of the pure liquid. The lowering of the freezing point (assuming that the material that separates out is the pure solvent) is given by

$$\Delta T = \frac{RT_f^2}{\Delta H_f} x_B$$

where ΔT is the difference between the freezing points of the dilute solution and the pure solvent, T_f is the freezing point of the pure solvent, ΔH_f is the heat of fusion of the pure solvent, and x_B is the mole fraction of the solute (impurity) (1). Thus, the presence of an impurity makes itself known by lowering the initial freezing point of the sample. As the pure solvent crystallizes from solution, the concentration of the impurity must increase (x_B increases), and the freezing point of the solution must fall. The result is that the cooling curve for such a solution will appear as in Figure 17.5: freezing not only starts at a lower temperature, but as Figure 17.5 shows, it also becomes complete only over a range of temperatures.

Thus, a "sharp" melting point (actually, a melting *range* of less than 1°C) is often taken as evidence that the sample is fairly pure, and a wide melting range is evidence that it is not pure. If the identity of the substance has been established by other experiments, the degree of purity can be estimated not only from the melting range, but from the difference between the actual temperature at which melting is complete (or freezing begins) and the known melting point of the substance, or ΔT in the equation. Since ΔT is inversely proportional to ΔH_f, as well as proportional to x_B, a given level of impurity, x_B, may result in a large depression of the freezing point (when ΔH_f is small) or a small depression of the freezing point (when ΔH_f is large).

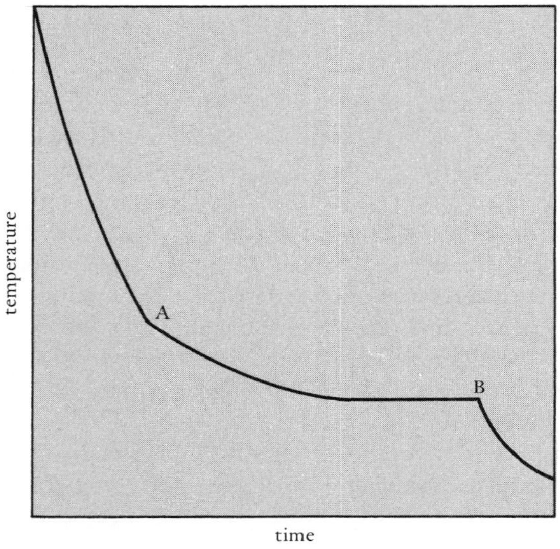

Figure 17.5. Expected cooling curve for an impure substance. Both solid and liquid are present between A and B.

17.3 THE MELTING POINT AS A MEANS OF IDENTIFICATION AND CHARACTERIZATION

If two substances have different melting points (outside of experimental uncertainty), they must have different molecular structures or different configurations, as with *cis* and *trans* isomers or other diastereomers. If the melting points of two substances are the same (within experimental uncertainty), the molecular structures may be, but are not necessarily, the same; even with the same structure, the configurations may be enantiomeric. These statements hold, of course, only for pure substances. Recall that "sharp" melting points as empirically determined are often taken as satisfactory evidence of purity.

For capillary melting points, experimental uncertainty is derived not only from any impurities in the sample and the procedure by which the melting point is determined, but also from the subjective interpretation of what is seen. The precise point at which melting starts is often difficult to distinguish from shrinking or sintering. Even with good technique, the uncertainty of the temperature at which melting begins may be several degrees; agreement between different determinations by different people on the same sample by the same method may not be within a degree.

17.4 MIXTURE MELTING POINTS

Since mixtures of different substances generally melt over a range of temperatures and are usually completely melted at a temperature below the melting point of at least one of the components, the nonidentity of two substances can often be established by determining the melting point of mixtures of the two. If each individual sample melts "sharply" (and at the same temperature, of course), and if an intimate mixture of the two made by rubbing approximately equal amounts together melts over a wide range, the two substances are not identical in structure and configuration. Usually you wish to establish the identity rather than the nonidentity of two samples, so it is unfortunate that the converse is not always true: the absence of a depression of the melting point or of a wide melting range of the mixture is not certain evidence that the two substances are identical in molecular structure and configuration. Examples of the various types of behavior that may be observed for mixtures of different substances, and their interpretation in terms of phase diagrams, are given in References 2 and 3.

Double melting points and the behavior of polymorphic forms are also **127**
described in *Technique of Organic Chemistry* (3).

SECTION 17

17.5 MELTING POINT AND MOLECULAR STRUCTURE

Systematic variations of melting point with changes in structure are not as
obvious or predictable as is the case with boiling points. Although melting points
do generally increase with increasing molecular weight, the first members of
homologous series often have melting points that are considerably different from
what would be expected on the basis of the behavior of the higher homologs. In
some homologous series of straight-chain aliphatic compounds, melting points
alternate: the melting point of successive members of the series is higher or
lower than that of the previous member, depending on whether the number of
carbon atoms is even or odd. Sometimes, as in the case of the normal alkanes,
the melting points of successive members of the series always increase, but by a
larger or smaller amount depending upon whether the number of carbons is even
or odd. As with boiling points, compounds with polar functional groups generally
have higher melting points than compounds with nonpolar functional groups,
but in contrast to the case with boiling points, highly branched or cyclic molecules
(relatively symmetrical molecules) tend to have higher melting points than their
straight-chain isomers. The combined effects of branching or the presence of
rings, then, are to reduce the range of temperature over which the liquid can
exist at a vapor pressure of less than 760 Torr. In extreme cases, a liquid range
may not be possible at a vapor pressure of less than 760 Torr; at atmospheric
pressure, the substance may sublime without melting. Hexachloroethane and
perfluorocyclohexane behave in this way.

Alternation of m.p. in homologous series

Polarity; Symmetry

As with the boiling point of a liquid, the absolute melting point of a substance equals the heat of fusion ΔH_f, in calories per mole, divided by the entropy
of fusion, ΔS_f, in calories per mole-degree:

$$T_f = \frac{\Delta H_f}{\Delta S_f} \qquad (17.5\text{-}1)$$

That this should be so can be seen most quickly by realizing that at the melting
point the liquid and solid phases are in equilibrium, which means that at this
temperature no change in free energy will accompany the conversion of liquid
to solid (or solid to liquid), i.e., $\Delta G_f = 0$. Since

$$\Delta G_f = \Delta H_f - T_f \Delta S_f = 0$$

then

$$\Delta H_f = T_f \Delta S_f$$

and Equation 17.5-1 follows from this. Another way of looking at this is to say
that, given definite values for ΔH_f and ΔS_f, there is only one temperature at which
ΔH_f can equal $T \Delta S_f$ (that ΔG_f can equal 0; that the two phases can be in equilibrium). This temperature is called the melting point, T_f.

The increase in boiling point with molecular weight, and the influence of the
nature of the functional group and degree of branching on the boiling point, are
interpreted in terms of the effect of structure on the heat of vaporization; to a
first approximation, the entropy of vaporization has a constant value of 20–22
calories per mole-degree. In the case of melting points, however, both the heat of
fusion and the entropy of fusion can vary widely, even for apparently closely

Table 17.2. Melting points, heats of fusion, and entropies of fusion of selected compounds

Compound	Melting Point °C	°K	ΔH_f (cal/mole)	ΔS_f (cal/mole-degree)
n-Butyl alcohol	−87.3	185.9	2,130	11.48
t-Butyl alcohol	25.2	298.4	1,620	5.44
Diethyl ether	−116.2	157.0	1,649	10.50
o-Dichlorobenzene	−16.7	256.5	3,080	12.02
m-Dichlorobenzene	−26.2	247.0	3,010	12.19
p-Dichlorobenzene	53	326	4,340	13.31
o-Dibromobenzene	1.8	275.0	2,030	7.38
m-Dibromobenzene	−6.9	266.3	3,150	11.84
p-Dibromobenzene	87	360	4,780	13.27
o-Xylene	−25.2	248.0	3,250	13.11
m-Xylene	−47.9	225.3	2,765	12.27
p-Xylene	13.3	286.4	4,090	14.28
Cyclohexane	6.6	279.7	673	2.28
Methylcyclohexane	−126.6	146.6	1,613	11.01
Cyclohexanol	23.5	296.7	406	1.37
Phenol	41	314	2,771	8.82
n-Octane	−56.8	216.4	4,957	22.91
2-Methylheptane	−109.0	164.1	2,451	14.94
2,3,3-Trimethylpentane	−100.7	172.5	370	2.12
2,2,3,3-Tetramethylbutane	100.7	373.9	1,802	4.82

The other isomeric octanes (values for four isomers are not reported) have melting points between −126 and −91°C, heats of fusion between 3,070 and 1,700 cal/mole, and entropies of fusion between 17.8 and 10.69 cal/mole-degree.

related compounds. Table 17.2 presents the melting points, heats of fusion, and entropies of fusion of several sets of related compounds.

The higher melting point of n-butyl alcohol compared to diethyl ether can easily be interpreted in terms of the larger heat of fusion of the more polar alcohol. Similarly, the higher melting points of the p-disubstituted benzenes compared to the *ortho* and *meta* isomers must be interpreted in terms of a larger heat of fusion ΔH_f for the *para* isomer, since the differences in the entropy of fusion among the isomers are either small or in the wrong direction to explain the facts.

The unusually high melting point of cyclohexane compared to methylcyclohexane is primarily the result of an unusually low entropy of fusion for cyclohexane. Similarly, the relatively high melting point of *tert*-butyl alcohol compared to its isomers is the result of a relatively low entropy of fusion. The ΔS_f entropy of fusion is usually interpreted primarily in terms of increased freedom of rotation of the molecule as a whole, and of parts of the molecule with respect to other parts, in the liquid phase compared to the solid phase. The increase in vibrational and translational entropy upon melting is relatively small; the large increase in translational entropy occurs upon vaporization, and there is considerable vibrational motion possible in the crystal. A small entropy of fusion, then, implies either that the molecule has relatively little rotational freedom in either the solid or liquid state (cyclic, polycyclic molecules), or that it can gain certain degrees of rotational freedom, with their associated increase in entropy, in the solid state without melting (highly branched, highly symmetrical molecules). Cyclohexane is known to have a transition at 185.9°K with an associated entropy change of 8.63 cal/mole-degree, and similar transitions have been observed for other highly symmetrical molecules.

The data for the isomeric octanes show how complex and variable the relationships can be between structure and melting point, heats of fusion, and entropies of fusion.

Problems

1. **a.** Calculate the freezing point depression, ΔT, for a sample of *tert*-butyl alcohol containing 0.01 mole fraction of an impurity.
   b. Do the same for a similar solution of cyclohexanol.

2. **a.** A sample of a substance of known molecular structure starts to freeze at 57.00°C and is half frozen at 56.80°C. What would be the melting point of the pure substance?
   b. The sample in part **a.** is remelted and sufficient solute of known molecular weight is added to give a solution containing 1 mole percent of this solute (you must assume that the solvent is pure). The sample now starts to freeze at 56.90°C. What mole fraction of impurity was present in the original sample?

References

1. G. M. Barrow, *Physical Chemistry*, 3rd edition, McGraw-Hill, New York, 1973, p. 605.
2. A. I. Vogel, *A Textbook of Practical Organic Chemistry*, 3rd edition, Wiley, New York, 1957, p. 21.
3. E. L. Skau, J. C. Arthur, and H. Wakeham, *Technique of Organic Chemistry*, Vol. I, Part I, 3rd edition, A. Weissberger, editor, Interscience, New York, 1959, p. 287.

§18 DENSITY; SPECIFIC GRAVITY

Before the advent of the spectroscopic methods, density was one of the most important physical properties by which liquids were characterized. Since the density is often reported and tabulated, especially in the older literature, and is fairly easy to determine, it is still often used to characterize and identify liquids.

18.1 EXPERIMENTAL DETERMINATION OF THE DENSITY

The *density* is defined as the *mass per unit volume*; the units usually used by chemists are grams/ml. The determination of density, then, involves the determination of the weight and volume of a sample of the substance. Since density is a function of temperature, the temperature at which the density is determined should also be recorded. For most organic liquids, the density decreases by about 0.001 g/ml per degree increase in temperature.

Temperature dependence

The weight of the sample is determined by weighing the container when it is empty and then when it is filled (completely or to the mark) with the liquid. The difference gives the weight of the sample at the temperature T, or w_{sample}^T. The volume of the sample is determined by cleaning and refilling the container with water and reweighing. Subtracting the weight of the empty container gives the weight of an equal volume of water at the temperature T, or $w_{H_2O}^T$. From the known density of water at the temperature of the samples, $d_{H_2O}^T$, the volume of the two samples can be calculated:

$$\text{Density of water at } T°C = \frac{w_{H_2O}^T}{\text{volume of sample}}$$

or

$$\text{Volume of sample} = \frac{w_{H_2O}^T}{d_{H_2O}^T}$$

The density d of the sample at $T°C$ may now be calculated:

$$d_{sample}^T = \frac{\text{wgt of sample}}{\text{vol of sample}} = \frac{w_{sample}^T}{w_{H_2O}^T/d_{H_2O}^T} = \frac{w_{sample}^T}{w_{H_2O}^T} \times d_{H_2O}^T \qquad (18.1\text{-}1)$$

Table 18.1 gives the density of water at several temperatures.

Table 18.1. Density of water between 20 and 30°C

Temperature	Density
20	0.9982
21	0.9980
22	0.9978
23	0.9975
24	0.9973
25	0.9970
26	0.9968
27	0.9965
28	0.9962
29	0 9959
30	0 9956

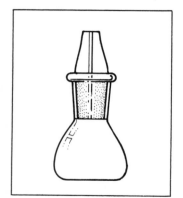

Figure 18.1. A specific gravity bottle. As the stopper is put in, the air escapes through the capillary. "Full" means filled to the top of the capillary.

The container used for density determinations may be as simple as a volumetric flask. A pycnometer, which is essentially a U-shaped tube, is often used with large samples, although small pycnometers may be constructed from a piece of glass tubing. The pycnometer is most easily filled by drawing in the liquid with gentle suction until it is filled past the mark, and adjusting to the mark by touching a piece of filter paper to the capillary tip; the excess will be drawn out into the filter paper by capillary action. A 1- or 2-ml specific gravity bottle may be used for small samples. The stopper for such a bottle has a capillary that allows it to be filled (to the top of the capillary) without trapping any air bubbles inside the bottle. (See Figure 18.1.) It is possible to use a 0.05-ml syringe as a container if the sample is very small. In this case, the weighings must be done to 0.1 milligram.

Sometimes it is desirable to determine the density at a particular temperature, rather than at the temperature of the laboratory. If so, the container and sample must be brought to the desired temperature by storage in a bath maintained at this temperature. The final small adjustment to the desired volume must be done after thermal equilibrium is attained.

Specific gravity

Specific gravity, which is related to density and sometimes confused with density, is defined as the ratio of the weight of a certain volume of liquid to the weight of an equal volume of water:

$$\text{Specific gravity} = \frac{w_{sample}^T}{w_{H_2O}^T} \text{ (dimensionless)}$$

Comparison of this expression with Equation 18.1-1 will show that the density equals the specific gravity times the density of water. Since the density of water is practically equal to 1, the density and specific gravity have almost the same value.

The density of a 0.5-ml sample may conveniently be determined by means of the *Fisher-Davidson gravitometer*. In effect, the density of the unknown liquid is determined by comparing the height to which a column of unknown liquid may be raised by a slight vacuum with the height to which a liquid of known density (ethylbenzene) may be raised by the same slight vacuum. Since the heights are inversely proportional to the densities, and the density of ethylbenzene is known, the unknown density may be calculated. The instrument is actually calibrated in units of grams/ml.

18.2 DENSITY AND MOLECULAR STRUCTURE

The density of a substance is determined mainly by the atomic weight of its constituent atoms: high atomic weight results in high density.

Most organic liquids have densities between 0.8 and 1.1 at room temperature. The structural possibilities for liquids that are more or less dense than this are relatively limited. Aliphatic acyclic hydrocarbons, saturated and unsaturated, and aliphatic acyclic ethers and amines generally have densities less than 0.8. Aromatic hydrocarbons usually have densities between 0.86 and 0.9. Iodides and bromides have densities greater than 1.1, while alkyl chlorides have densities less than 1.

Compounds of density greater than 1.2 usually have at least one iodine or bromine atom, or two or more chlorine atoms. Compounds with two or more functional groups, especially compounds with relatively large intermolecular forces (relatively high boiling for their molecular weight), generally have densities greater than 1.

As the ratio of —CH$_2$— units per functional group increases, the densities tend toward a value between 0.8 and 0.9.

Problems

1. The weight of a certain volume of a liquid at 25°C was determined to be 2.339 grams, and the weight of an equal volume of water at the same temperature was found to be 2.013 grams.
   a. Calculate the density (d^{25}) of the liquid.
   b. Calculate the specific gravity (d_{25}^{25} and d_4^{25}) of the liquid. *Note:* This notation means, in the first case, that the temperature of both the sample and the equal volume of water is assumed to be 25°C, and, in the second case, that the temperature of the sample is 25°C but that of the water is 4°C.

2. How would densities determined on the moon differ from those determined on earth?

§19 INDEX OF REFRACTION

Along with the boiling point and density, the *index of refraction* is one of the physical properties most often used to identify and characterize liquids.

19.1 EXPERIMENTAL DETERMINATION OF THE INDEX OF REFRACTION

The index of refraction of a substance is the ratio of the speed of light in a vacuum to the speed of light in the substance. It is often measured by making use of the fact that the critical angle of refraction of light passing from one medium to

*Wavelength dependence and
temperature dependence*

Abbe refractometer

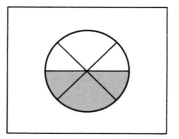

Figure 19.1. View through
the eyepiece of the Abbe
refractometer. When the
refractometer is set to be
read, the intersection of the
cross-hairs falls on the border
between the light and dark
fields; the border should be
as sharp and free from color
as possible.

another is a function of the refractive indexes of the two media. The Abbe refractometer in effect measures the critical angle for refraction of light passing from a liquid of unknown refractive index to a glass prism of known refractive index. It is calibrated in units of index of refraction (a dimensionless ratio). Since the index of refraction is a function of both the wavelength of light and the temperature, these must both be specified along with the measured refraction. The D line of sodium is the wavelength for which the index of refraction is often reported, and at the temperature T, the index of refraction determined at this wavelength would be reported as n_D^T. The index of refraction of most organic liquids decreases between 3.5 and 5.5×10^{-4} per degree increase in temperature.

With the Abbe refractometer, two drops of sample are required in the space between the prisms, and the instrument is adjusted until the field seen through the eyepiece appears as shown in Figure 19.1. The intersection of the cross-hairs should be on the border between the light and dark sections of the field, and the compensator should be set to sharpen and achromatize the border between the light and dark sections until the difference is as sharp and as near black and white as possible. When the cross-hairs and border are lined up as shown in Figure 19.1, the index of refraction is read from the scale. The value is reliable to ± 0.0002, provided that the instrument is properly calibrated. The calibration may be checked by determining the index of refraction of water, which is 1.3330 at 20°C and which decreases 0.0001 unit per degree increase in temperature between 20° and 30°C. With the compensator, the index of refraction determined with ordinary (white) light is very close to that which would be obtained using the light from a sodium lamp.

If no sharp boundary can be observed, insufficient sample was used, or the sample evaporated before the adjustment was completed. In making determinations on very volatile liquids, you can introduce the sample with a fine dropper through a little channel that leads to the space between the prisms, without having to open them up.

After use, the sample should be removed by means of a soft tissue and a little acetone or ethanol.

The thermometer indicates the temperature of the sample space. It may be maintained at a particular value by circulating water from a constant-temperature bath through the prism housings, using the hose connections provided.

19.2 INDEX OF REFRACTION AND MOLECULAR STRUCTURE

The index of refraction of a substance is determined mainly by the polarizability of its constituent atoms and functional groups: the presence of atoms of higher atomic number and of conjugated unsaturation results in a higher index of refraction. Alkyl iodides and aromatic compounds usually have an index of refraction greater than 1.5000; most other compounds give a value lower than this.

As with the boiling point and density, the main use of experimental values for the index of refraction is for comparison with values reported in the literature for compounds whose structure is known. It is possible, however, to calculate the expected index of refraction for a substance from its molecular structure through the following relationship:

Molecular refraction

$$\text{Molecular refraction} = M n_D^T = \text{molecular weight} \times n_D^T$$

Solving for n_D^T shows that the index of refraction equals the molecular refraction divided by the molecular weight. Contributions of individual structural features (atoms or bonds) to the molecular refraction may be found in Reference (1).

In a similar way, a related property, the molecular refractivity, may be estimated from independent contributions of structural features (1). The estimated value can then be compared with the experimental value, which depends upon the index of refraction and the density in this way:

$$\text{Molecular refractivity} = R_{\text{D}} = \frac{n^2 - 1}{n^2 + 2} \cdot \frac{mw}{d}$$

where n is the index of refraction at the wavelength of the D line of sodium, d is the density of the liquid at the same temperature at which the index of refraction was determined (preferably near 20°C), and mw is the molecular weight of the substance.

The fact that the molecular refraction and molecular refractivity can be estimated from the structural formula of a compound makes it possible to use the index of refraction (or index of refraction and density) as criteria of identity even though literature values may not be available.

Reference

1. A. I. Vogel et al., *J. Chem. Soc.* **1952** (514).

---

§20 OPTICAL ACTIVITY

20.1 EXPERIMENTAL DETERMINATION OF OPTICAL ROTATION

An additional physical property that is of use in the identification and characterization of optically active solids or liquids is the *optical rotation*. The optical rotation is found by preparing a solution of the substance and then determining, by means of a *polarimeter*, the direction and amount by which the plane of polarization of a beam of plane-polarized light is rotated upon passage through the solution. The polarimeter consists essentially of a polarizing prism (polarizer), which transmits from the monochromatic light source a beam of plane-polarized light, a trough to hold the sample tube, and a second polarizing prism (analyzer) that may be rotated so as to exactly compensate for the rotation that the solution induces in the polarized light beam. The correct position of the analyzer is determined by looking through the central eyepiece at the light transmitted by the instrument. Figure 20.1 indicates the appearance of the field for possible positions of the analyzer with respect to the plane of polarization of the beam. The angular displacement of the analyzer is read in degrees from the graduated circle with the aid of an auxiliary eyepiece and vernier, and the difference between the angular position of the analyzer with the sample in and out of the trough is the observed rotation, α.

The sample is prepared by making up a solution of known concentration by dissolving a weighed amount of the substance in a volumetric flask. A concentration of 0.5–2 grams per 100 ml is usually recommended. Solvents most commonly used are water, methyl and ethyl alcohols, and chloroform. A dilute solution is most desirable, but if the rotatory power of the substance is low, a higher concentration may be necessary. If the solution is not free from dust, it should be filtered. The volume of solution required varies with the length and diameter of the sample tube. Twenty-five milliliters should be enough in any case, and tubes are available that require 1 ml or less. With a liquid, the rotation may be determined using the neat liquid.

The polarimeter

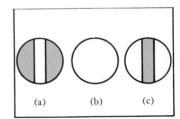

(a)   (b)   (c)

Figure 20.1. View through the eyepiece of the polarimeter. The analyzer should be set so that the intensity of all parts of the field is the same (b). When the analyzer is displaced to one side or the other, the field will appear as in (a) or (c).

The sample tube is usually a 1- or 2-decimeter-long glass tube with a fitting for a brass screw cap at each end. It is filled by placing one end glass against one end of the tube (glass-to-glass) and then screwing on the cap, using a rubber washer between the end glass and the cap. With the tube standing vertically on this cap, the solution is added through the other end (possibly by means of a dropper or pipet) until the rounded meniscus stands above the end of the tube. The other end glass is then slid across the end of the tube, without leaving an air bubble in the tube; if air is trapped in the tube, remove the glass, add more solution if necessary, and try sliding it on again. The other brass cap is then screwed on, using a rubber washer as with the first cap. Screw the ends on firmly but not too tight, as strain in the glass of the end caps may cause them to rotate the beam slightly. Some polarimeter tubes are constructed so that they can be filled through a port in the middle with both ends capped. After placing the tube in the trough of the instrument, close the cover.

The image may be brought into focus, if necessary, by moving the eyepiece in or out. Distortion of the image may be due to an air bubble in the light path, or to the presence of density gradients in the sample caused by either concentration or temperature gradients.

The magnitude of the observed rotation, α, depends upon several variables that must therefore be recorded along with the observed rotation.

- *Concentration:* It is reasonable that the greater the concentration of molecules in the path of the beam, the greater should be the rotation. However, in many cases, the increase in rotation is not exactly proportional to the concentration. This is explained by postulating that solute–solute interactions, which are relatively more prevalent at higher concentrations, affect the rotation differently from solute–solvent interactions, which are (approximately) the only ones present at low concentrations.

- *Length of sample tube:* The rotation of the beam will be proportional to the number of molecules in the path; for a given concentration, the number of molecules in the path will be proportional to the length of the path. The path length is conventionally expressed in decimeters.

- *Solvent:* The observed rotation of the same substance at the same concentration in different solvents may be very different, even of opposite sign!

- *Temperature:* A change in concentration accompanies the expansion and contraction that a solution undergoes with change in temperature. This experimental variable can be eliminated by making up the solution at the temperature of the room containing the polarimeter or the temperature at which the optical rotation will be determined. (Polarimeter tubes are available that can be thermostatted by passing water from a constant-temperature bath through a jacket surrounding the tube.) It can also be calculated from the coefficient of thermal expansion of the solvent. More important proposed explanations of the rather large changes in the observed rotation that occur with changes in temperature are (1) the change with temperature of the relative populations of conformational isomers, which have different rotatory powers, and (2) the change with temperature of the solvation of the optically active molecules.

- *Wavelength of light:* The magnitude (and also the sign, in many cases) of the rotation depends on the wavelength of light used. Since the D line of sodium was used early, most rotations are reported at this wavelength. Generally, the magnitude of rotation is larger at shorter wavelengths, and for this reason and the fact that the human eye (especially the dark-adapted eye) is more sensitive to green light, the green line of the mercury arc (5461 Å) has been recommended for use.

The optical rotation is usually reported as the specific rotation, $[\alpha]$:

for solutions, $$[\alpha]_D^T = \frac{100\,\alpha_D^T}{lc}$$

for pure liquids, $$[\alpha]_D^T = \frac{\alpha_D^T}{ld}$$

where $[\alpha]_D^T$ = specific rotation, at the D line of sodium at the temperature T

α_D^T = observed rotation—positive if clockwise; negative if counterclockwise—at the D line of sodium at the temperature T

l = length of sample tube, in decimeters

c = concentration of solution, in grams per 100 ml

d = density of the pure liquid at the temperature T, in grams/ml

The specific rotation $[\alpha]$ is a measure of the rotary power per unit path length per unit weight concentration. In taking your experimental data, you must report the solvent and the concentration as well, since the variation of rotation with concentration is not necessarily linear and is not usually independent of the nature of the solvent. And when comparing an experimental value of the specific rotation with a literature value, you must compare solvent and concentration as well as temperature and wavelength.

20.2 OPTICAL ACTIVITY AND MOLECULAR STRUCTURE

A compound that is optically active must be composed of molecules for which no conformation that can be attained by the molecules is the same as the mirror image of that conformation; *all* conformations must be chiral; *all* conformations must belong to point groups $\mathbf{C}_n$ or $\mathbf{D}_n$. Thus, if in any conformation the molecules of a substance possess a symmetry element whose corresponding symmetry operation involves reflection through a plane (a plane of symmetry, a center of symmetry, or a higher even improper axis of rotation), the substance cannot be optically active.

The presence of a chiral center in the molecules of a substance will often result in optical activity. The most common type of chiral center is a carbon atom to which are bonded four different atoms or groups of atoms (a chiral carbon). In some rare cases the expected effect will be too small to observe, and in others the presence of a second chiral center similar to the first may give at least one conformation of the molecule a plane or center of symmetry (*meso* form). Of course, if the sample is a racemic mixture (an equimolar mixture of enantiomers), no optical activity will be observed.

The presence of a chiral center in the molecules of a substance is not required for molecular chirality, as illustrated by the existence of certain optically active allenes and biphenyls. Excellent discussions of the details of the relationships between molecular structure, molecular symmetry, and optical activity are presented in References 1 through 3.

If a sample is observed to be optically active, then, it is not a racemic mixture and it must be composed of chiral molecules. The absence of optical activity indicates that the sample is a racemic mixture or that its molecules possess a plane of symmetry, a center of symmetry, or a higher even improper axis of rotation in at least one conformation.*

---

* An exceptional case is discussed in Reference 2, p. 26.

The sign and magnitude of the rotation are a function of both the configuration of the molecule and the wavelength of the light, and a graph of the specific rotation of a compound versus wavelength is called its *optical rotatory dispersion curve* (ORD curve). Since enantiomers have equal but opposite optical rotations at all wavelengths, their ORD curves are exactly similar in shape but opposite in sign. A more useful observation is that similar or homologous compounds of the same relative configuration have similar ORD curves of the same sign. This makes it possible in certain cases to establish the relative configurations of optically active molecules without interconversion. If, in addition, the absolute configuration of one has been established, the absolute configuration of all is thereby established. A more complete discussion of the interpretation of ORD curves may be found in Reference 1.

(S)-(-)-α-phenylethylamine

Problems

1. The observed rotation at 25°C of a sample of α-phenylethylamine (Section 47) in a 10-cm tube, using a sodium lamp, is −35.6°. The density, d^{25}, of the amine is reported to be 0.953 g/ml. Calculate the specific rotation, $[\alpha]_D^{25}$.

2. The observed rotation α_D^{20} of a solution of 232 mg of cholesterol (Section 39) in 10 ml of chloroform in a 1-dm tube was −0.73°. Calculate the specific rotation $[\alpha]_D^{20}$.

3. How can a clockwise rotation of 10° be distinguished from a counterclockwise rotation of 350°?

4. For each of the following molecular structures, determine (1) the total number of stereoisomers possible, and (2) which of the stereoisomers should be optically active.

   a. 1-bromohexane   **b.** 2-bromohexane
   c. 1,2-dibromohexane   **d.** 2,3-dibromohexane
   e. 1,2-dibromocyclohexane   **f.** 1,3-dibromocyclohexane
   g. 1,4-dibromocyclohexane   **h.** chloroallene (CHCl=C=CH₂)
   i. 1,1-dichloroallene   **j.** 1,3-dichloroallene
   k. 3-hexene   **l.** 2-chloro-3-hexene
   m. 2,5-dichloro-3-hexene   **n.** 2,5-dichloro-3-hexyne

References

1. K. Mislow, *Introduction to Stereochemistry*, W. A. Benjamin, New York, 1965.
2. H. H. Jaffe and M. Orchin, *Symmetry in Chemistry*, Wiley, New York, 1965, Chapters 1 and 2.
3. M. Orchin and H. H. Jaffe, *J. Chem. Educ.* **47**, 246 (1970).

§21 MOLECULAR WEIGHT

One of the bits of information most useful in establishing the molecular structure of a new substance is the molecular weight. If the molecular weight is known with confidence to 1 atomic mass unit, all but a small fraction of the total structural possibilities may be eliminated on this basis alone.

If the problem is simply to establish the identity of a substance with a known compound, the determination of the molecular weight of the unknown is much less useful, since it provides only a single data point for comparison. The infrared spectrum, for example, would provide many more points for comparison and would usually be easier to determine than the molecular weight. Of course, the molecular weight of any substance of known structure is also automatically known and thus may be compared to that of an unknown substance, whereas this is not true of the infrared spectrum and other physical and chemical properties.

Mass spectrometry is currently the most commonly used method of molecular weight determination. It has the great advantages of being applicable to most substances that are at least slightly volatile (a vapor pressure of 10^{-2} Torr at a temperature of $100°C$), of requiring a very small amount of sample, and of routinely giving the molecular weight to 1 atomic mass unit. The disadvantages of the mass spectrometric method are that the required instrumentation is relatively expensive and difficult to maintain in top operating condition. Section 26.2 describes how the mass spectrum of a substance may be determined and how it can provide information as to the molecular structure, as well as the molecular weight, of a substance.

21.2 MOLECULAR WEIGHT DETERMINATION BY OTHER METHODS

Other methods of molecular weight determination include measurement of freezing-point depression, boiling-point elevation, or osmotic pressure. In these methods, the molality of a solution containing the unknown substance is determined. Knowing the weight of the solute added, you can then calculate its molecular weight.

The equivalent weight of acidic or basic substances can be determined by titration. From the equivalents of titrant consumed, the equivalents of sample are known. Knowing the weight of sample used, you can calculate the weight per equivalent.

A particularly convenient and relatively accurate method is that of isothermal distillation. In this procedure, a solution of a known weight of a compound of known molecular weight is placed in one leg of an H-shaped tube, and a solution of an equal weight of the compound of unknown molecular weight is placed in the other. If the loaded tube is held in a bath at a constant temperature, solvent will slowly distill from the leg with the more dilute solution to the leg with the more concentrated solution until the concentrations are equal. In this state, the ratio of the heights of the solution in the two legs is the inverse of the ratio of the molecular weights. Of course, the weights of the two samples need not be equal, and if they are not, the calculation is slightly more complicated.

Methods for the determination of molecular weights by freezing-point depression, using camphor and tertiary butyl alcohol as solvents, are given in References 1 and 2.

References

1. E. J. Cowles and M. T. Pike, *J. Chem. Educ.* **40**, 422 (1963).
2. M. J. Bigelow, *J. Chem. Educ.* **45**, 108 (1968).

---

§22 SOLUBILITY

The solubility of a substance is significant for two reasons. First, before it is possible to start working with a compound in the laboratory, it is necessary to know or to be able to predict whether it will be soluble in water, aqueous acid, aqueous base, or organic solvents such as ether. Second, if the molecular structure of a substance is unknown, knowledge of its solubility characteristics can be the basis for certain deductions about its structure.

It must be emphasized that *solubility is a matter of degree*; that is, some things are more soluble than others, and the decision of where you wish to draw the line that says that substances less soluble than this will be called "insoluble" (in a particular solvent) and substances more soluble than this will be called "soluble" (in a particular solvent) is completely arbitrary. Very often, the line is drawn at 30 mg of substance per ml of solvent (3 g per 100 ml of solvent). Obviously there will be borderline cases, and one must look out for the use of different definitions of solubility and insolubility in the chemical literature.

22.1 SOLUBILITY OF LIQUIDS IN LIQUIDS

Mixing is normal

Although the mutual insolubility of "oil" and water is probably the most familiar example of the solubility behavior of two liquids, it is much more satisfactory to approach the phenomena of solubility or insolubility of liquids from the point of view that mutual solubility is the expected or normal behavior for two liquids, and that insolubility is exceptional or unusual behavior.

The two liquids benzene and toluene are mutually soluble in all proportions (miscible) for the same reason that red and white marbles will mix if you shake them together: the free energy of the mixed state is less than the free energy of the unmixed state because of the increase in entropy associated with mixing. The mixed state has the same potential energy as the unmixed state, but it is more probable than the unmixed state. If the intermolecular forces between unlike molecules and like molecules are the same, the heat of mixing, ΔH_{mix}, will be zero. Since the entropy of mixing, ΔS_{mix}, is always positive, the free-energy change upon mixing will, in this case, be negative.

$$\Delta G_{mix} = \Delta H_{mix} - T\Delta S_{mix} \qquad (22.1\text{-}1)$$

In the special case when ΔH_{mix} equals zero this equation becomes

$$\Delta G_{mix} = -T\Delta S_{mix} \qquad (22.1\text{-}2)$$

The phenomenon of mixing is one of a very few in which the free-energy change can be considered to be primarily a function of the change in entropy rather than primarily a function of the change in enthalpy (energy). Thus, the expected or normal miscibility of liquids is observed when the heat of mixing, ΔH_{mix}, is negative (mixing is exothermic), zero, or only slightly positive (mixing is only slightly endothermic).

Insolubility

If ΔH_{mix} is large and positive (highly endothermic), it can more than compensate for $T\Delta S_{mix}$ and will cause ΔG_{mix} to be positive (Equation 22.1-1). If this is the case, the unmixed state will have a lower free energy than the mixed state, and mixing will not occur. The enthalpy of mixing, ΔH_{mix}, can be positive if the forces between like molecules are larger than the forces between unlike molecules, because in this case it will take more energy (work) to separate the like molecules (of both solvent and solute) to the larger average intermolecular distance of the mixed state than will be regained upon moving the unlike molecules together in the mixed state:

$$\Delta H_{mix} = E_{separation} + E_{association} \qquad (22.1\text{-}3)$$

where $E_{separation}$ is the energy required to separate like molecules (always positive), and $E_{association}$ is the energy released upon moving unlike molecules together (always negative). The larger the intermolecular attractions, the greater the magnitude of each term of this equation.

Thus, the nonmiscibility of water with most organic liquids is the result of the large intermolecular attraction between water molecules (hydrogen bonding), which must be lost and remain uncompensated for in the mixed state when the organic liquid cannot form hydrogen bonds with water. It is in this way that the immiscibility of "oil" (or hexane, benzene, carbon tetrachloride, etc.) and water can be rationalized. On the other hand, methanol, ethanol, acetone, and acetic acid are miscible with water because the intermolecular attractions between the molecules of water and solvent are comparable in magnitude to the intermolecular attractions between water and water, and solvent and solvent. This results in a relatively small heat of mixing and therefore, by Equation 22.1-1, a negative free-energy change upon mixing. The often-quoted generalization that "like dissolves like" can be understood by realizing that substances whose molecules have similar intermolecular forces (type and magnitude) tend to mix.

acetone

"Like dissolves like"

Partial Solubility

Complete immiscibility, however, is an extreme behavior actually approached by relatively few pairs of liquids; mercury and water appear to approach this extreme very closely. It is more usual for pairs of liquids that are not completely miscible to have a measurable solubility in one another. For example, 100 ml of water will dissolve 7.5 g of diethyl ether, and 100 ml of ether will dissolve 1.3 g of water at 25°C; 100 ml of water will dissolve 8.5 g of *n*-butyl alcohol, and 100 g of *n*-butyl alcohol will dissolve 25 g of water at 20°C.

Partial solubility can be understood by realizing that the increase of entropy due to mixing is largest for the first material that dissolves. That is, ΔS_{mix} for going from mole fraction solvent $= 1.0$ to mole fraction solvent $= 0.9$ (or $x_{\text{solute}} = 0$ to $x_{\text{solute}} = 0.1$) is greater than ΔS_{mix} for going from $x_{\text{solvent}} = 0.9$ to $x_{\text{solvent}} = 0.8$ (or $x_{\text{solute}} = 0.1$ to $x_{\text{solute}} = 0.2$). Since ΔH_{mix} might be expected to be approximately the same in both cases, ΔG_{mix}, which equals $\Delta H_{\text{mix}} - T\Delta S_{\text{mix}}$, can be negative for the formation of a dilute solution (when the average ΔS_{mix} per increase in concentration of solution is larger) and positive for the formation of a concentrated solution (when the average ΔS_{mix} per increase in concentration of solute is smaller). The larger and more positive the enthalpy of mixing ΔH_{mix}, the sooner ΔH_{mix} will equal $T\Delta S_{\text{mix}}$; the larger and more positive the enthalpy of mixing, the sooner the solution should be expected to become saturated with respect to the solute, the point at which ΔG_{mix} is equal to zero. Table 22.1 presents a qualitative summary of the solubility in water of organic compounds containing different functional groups.

Since both the heat of vaporization of a liquid (and thus the boiling point of a liquid) and the heat of mixing are functions of intermolecular forces, it is understandable that there is a correspondence between the solubility of certain organic compounds in water and their boiling points. For example, for the four isomeric butyl alcohols, the order of decreasing boiling point is the order of increasing solubility in water. In Section 16.2, lower boiling point was interpreted in terms of smaller intermolecular forces. Similarly, ΔH_{mix} would be expected to be more negative (or less positive) when less energy is required to separate the molecules from their average distance in the unmixed state to their average distance in the mixed state. Table 22.2 gives the boiling points and solubilities of the isomeric butyl alcohols and several other compounds.

Solubility and b.p.

It is generally true that the lower homologs corresponding to a given functional group are more soluble in water than are the higher homologs (see Table 22.1). This can be understood by realizing that the larger the hydrocarbon part

Table 22.1. Solubility in water of straight-chain organic compounds with different functional groups

Branching of the alkyl group increases the solubility (see Table 22.2).

Type of compound	\u2014 Number of Carbon Atoms \u2014						
	1	2	3	4	5	6	7
Alcohol	· ·						
Aldehyde	· – – – – – – – – – – –						
Alkane	———————————————————————————						
Alkene	———————————————————————————						
Alkyl halide	———————————————————————————						
Amide	· – – – – – – –						
Amine	· – – – – – – –						
Carboxylic acid	· – – – – – – – – – – – – –						
Methyl ester	· – – – – – – –						
Methyl ether	· · · · · · · · · · · · · · · · – – – – – – – – – – –						
Methyl ketone	· – – – – – – –						
Nitrile	· · · · · · · · · · · · · · · · – – – – – – – – – – –						
Thiol	· · · · · · ———————————————————						

Solubility is greater than 5 grams per 100 ml of water: · · · · · ·
Solubility is between 1 and 5 grams per 100 ml of water: – – – – – –
Solubility is less than 1 gram per 100 ml of water: ————

of the molecule relative to the functional group that can hydrogen-bond to water, the smaller the net intermolecular attraction between water molecules and the molecules of the organic liquid. This will result, according to Equation 22.1-3, in a more positive heat of mixing. Another way of looking at this is to say that the larger the alkyl group, the more the molecule will behave like the water-insoluble hydrocarbons.

The general phenomenon of increasing solubility with increasing temperature can be rationalized in terms of Equation 22.1-1. Greater solubility means that the same amount of material may be dissolved in a smaller volume of solvent to form a more concentrated solution. Although the formation of a relatively concentrated solution involves a smaller ΔS_{mix} than does the formation of a dilute solution, the term $T\Delta S_{mix}$ can maintain the same value despite a decrease in ΔS_{mix} if T is correspondingly greater.

Solubility increases with increasing T

Table 22.2. Solubility in water and boiling point of isomeric compounds

Compound	Boiling Point (°C)	Solubility in Water
n-Butyl alcohol	118	8.3[a]
Isobutyl alcohol	108	9.6[a]
Sec-butyl alcohol	100	13.0[a]
Tert-butyl alcohol	83	Miscible in all proportions
Diethyl ether	35	7.5[a]
Methyl *n*-butyl ether	70	1.00[b]
Methyl isobutyl ether	58	1.24[b]
Methyl *sec*-butyl ether	59	1.79[b]
Methyl *tert*-butyl ether	54	5.89[b]

[a] Grams per 100 ml water at 20°C.
[b] Weight percent of ether in saturated aqueous solution at 20°C.

The fact that most organic compounds are less soluble in aqueous salt solutions than in water can be understood by realizing that ΔH_{mix} for a salt solution should be more positive than for water; mixing with salt solution requires the input of additional energy to separate the charged ions. Thus, relatively soluble organic substances can often be *salted out* of aqueous solution by saturating the solution with sodium chloride or sodium sulfate.

Since ΔS_{mix} must always be positive, Equation 22.1-1 predicts that anything ought to be miscible with anything else if the temperature is high enough. This will certainly be true when all the substances are above their critical temperatures, because they will then behave as gases. As vapors, mercury and water are completely miscible.

22.2 SOLUBILITY OF SOLIDS IN LIQUIDS

The process of dissolution of a solid in a solvent may be considered to be the sum of two consecutive processes: the melting of the solid (to give the supercooled liquid) followed by the mixing of the liquid with the solvent. Thus, the free-energy change for the dissolving of a solid may be considered to be the sum of the free-energy changes for these two processes:

$$\Delta G_{dis} = \Delta G_f + \Delta G_{mix} \qquad (22.2\text{-}1)$$

Since

$$\Delta G_f = \Delta H_f - T\Delta S_f \qquad \text{and} \qquad \Delta G_{mix} = \Delta H_{mix} - T\Delta S_{mix}$$

it follows that:

$$\Delta G_{dis} = \qquad \Delta G_f \qquad + \Delta H_{mix} - T\Delta S_{mix} \qquad (22.2\text{-}2)$$

$$\Delta G_{dis} = \Delta H_f - T\Delta S_f + \Delta H_{mix} - T\Delta S_{mix} \qquad (22.2\text{-}3)$$

Solids are generally more soluble in a given solvent at higher temperatures than at lower temperatures (this is what makes recrystallization possible). At a higher temperature, less solvent will be needed to dissolve a given amount of material to give a saturated solution ($\Delta G_{dis} = 0$) because, although ΔS_{mix} will be smaller since a more concentrated solution is being formed, a larger value for T will compensate partially through the $T\Delta S_{mix}$ term, just as in the mixing of liquids that was discussed in the previous section, and additionally through the $T\Delta S_f$ term.

At the melting point of a solid, T_f,

$$\Delta G_f = \Delta H_f - T_f\Delta S_f = 0 \qquad (22.2\text{-}4)$$

or:

$$\Delta H_f = T_f\Delta S_f \qquad \text{or} \qquad \Delta S_f = \frac{\Delta H_f}{T_f} \qquad (22.2\text{-}5)$$

By means of these last two relationships, ΔG_f at any temperature T may be expressed as a function of ΔH_f alone or ΔS_f alone:

$$\Delta G_f = \Delta H_f - T\left(\frac{\Delta H_f}{T_f}\right) = \Delta H_f\left(1 - \frac{T}{T_f}\right) \qquad (22.2\text{-}6)$$

$$\Delta G_f = T_f\Delta S_f - T\Delta S_f = \Delta S_f(T_f - T) \qquad (22.2\text{-}7)$$

(Remember from Section 17.5 that ΔG_f will equal zero only when $T = T_f$; that is, the solid and liquid will be in equilibrium only at a unique temperature, the melting point. Below the melting point, $T_f > T$, and therefore ΔG_f will be positive.) These relationships may each be substituted into Equation 22.2-2 to give:

$$\Delta G_{dis} = \Delta H_f\left(1 - \frac{T}{T_f}\right) + \Delta H_{mix} - T\Delta S_{mix} \qquad (22.2\text{-}8)$$

$$\Delta G_{dis} = \Delta S_f(T_f - T) + \Delta H_{mix} - T\Delta S_{mix} \qquad (22.2\text{-}9)$$

These last two relationships make it possible to understand why two (or more) substances that might be expected to have the same tendency to mix in a solvent can have different solubilities. For example, the *cis/trans* isomers maleic and fumaric acid might be expected to mix equally well with water, but the *cis* isomer, maleic acid, is 100 times as soluble as the *trans* isomer at room temperature (see Table 22.3). Similarly, the three position isomers of dinitrobenzene might be expected to mix equally well with benzene, but their solubilities range over a factor of 10 at 50°C (see Table 22.3). These phenomena can be understood by observing that *in such cases solubility decreases with increasing melting point of the compound.* If the difference in melting point is due entirely to a difference in the enthalpies of fusion ΔH_f, the entropies of fusion ΔS_f are thus assumed to be the same for all the isomers of the set and, according to Equation 22.2-9, the positive term $\Delta S_f(T_f - T)$ will be greater, the larger the T_f. Thus, the ΔS_{mix} term must remain larger for the higher-melting isomers, and saturation must occur at higher dilutions. Similarly, if the difference in melting point is due entirely to a difference in the entropies of fusion, the enthalpies of fusion are thus assumed to be the same for all the isomers of the set and, according to Equation 22.2-8, the positive term $\Delta H_f(1 - T/T_f)$ will be greater, the larger the T_f. Thus, again the ΔS_{mix} term must remain larger for the higher-melting isomers; that is, only relatively dilute solutions are possible for the higher-melting isomers. It seems reasonable that the result should be qualitatively the same if the differences in melting point are due to differences in both the enthalpy and entropy of fusion. More examples of this phenomenon are given in Table 22.3.

maleic acid
m.p. 143°C

Solubility and m.p.

fumaric acid
m.p. 286°C

22.3 CLASSIFICATION OF COMPOUNDS BY SOLUBILITY: RELATIONSHIPS BETWEEN SOLUBILITY AND MOLECULAR STRUCTURE

In the following sections, structural features of molecules that lead to solubility under various conditions in different solvents will be described. This information will often enable you to predict the solubility behavior of a compound of known structure and to exclude certain possibilities for the structure of an unknown substance on the basis of its solubility or insolubility under different conditions. Again, a compound will be classified as "soluble" if it is soluble to the extent of 30 or more milligrams per milliliter of solvent (3 or more grams per 100 ml of solvent).

Solubility in Water

Covalent substances are soluble in water at room temperature only if the molecules have a functional group that can form hydrogen bonds with water—a functional group, in other words, that contains a nitrogen, oxygen, or sulfur atom. The higher

Table 22.3. Solubilities and melting points of isomeric compounds

Compound	m.p. (°C)	Water	Alcohol	Benzene	Hexane
Maleic acid (cis)	143	60[a]	51[a]		
Fumaric acid (trans)	286	0.6[a]	5[a]		
d-Tartaric acid	170	139[a]	27[b]		
l-Tartaric acid	170	139[a]	27[b]		
dl-Tartaric acid	206	20[a]	2[b]		
o-Dinitrobenzene	116			17.5[c]	
m-Dinitrobenzene	90			37.6[c]	
p-Dinitrobenzene	170			3.1[c]	
Phenanthrene	100			18.6[b]	4.2[b]
Anthracene	217			0.6[b]	0.2[b]
o-Nitrobenzoic acid	148			0.35[d]	
m-Nitrobenzoic acid	142			0.80[d]	
p-Nitrobenzoic acid	240			0.08[d]	
o-Dihydroxybenzene	105	14[e]		1.15[e]	
m-Dihydroxybenzene	109	24[e]		0.40[e]	
p-Dihydroxybenzene	173	1.4[e]		0.04[e]	

[a] Grams per 100 ml solvent at 20°C.
[b] Grams per 100 ml solvent at 25°C.
[c] Grams per 100 ml solvent at 50°C.
[d] Grams per 100 ml solvent at 30°C.
[e] Mole percent of solute in saturated solution at 30°C.

homologs are decreasingly soluble in water, and with monofunctional compounds they become "insoluble" at about five carbon atoms (see Table 22.1). If two or more functional groups are present, especially if they include the amino or hydroxy group, compounds with more than five or six carbon atoms may be water soluble.

Ionic substances (salts) are often very soluble in water. Although these compounds are solids and may have very large heats of fusion, solvation of the ions is often sufficiently exothermic that they are soluble in water.

If an organic liquid is soluble in water, it is most likely to be a relatively low-molecular-weight mono- or difunctional compound. Higher-molecular-weight and polyfunctional compounds are much more likely to be solids.

If an organic solid is soluble in water, it is most likely to be either a salt (the alkali metal salt of an acid or the hydrogen halide or sulfate salt of an amine), a polyfunctional compound that happens to be a solid (polyhydric phenols, sugars), or an amino acid (which exists as an inner salt).

Acidification of the aqueous solution of the salt of an organic acid will convert it to the free acid, which will separate from solution if it is not itself soluble in water or aqueous acid; similarly, making the solution of the salt of an amine basic will liberate the free base, which will separate from solution if it is not soluble in water or aqueous base. Salts of weak acids will hydrolyze in aqueous solution to form hydroxide ion, and salts of amines will hydrolyze to produce hydronium ion; the decrease or increase in acidity of the water in which the salt is dissolved may be detected by means of pH paper or a pH meter.

α-D-glucose
(a sugar)

Solubility in Diethyl Ether

Almost all organic liquids are soluble in ether. A great many organic solids are soluble in ether. Those that are not are compounds with a relatively large heat of fusion and thus include salts and relatively high-molecular-weight, high-melting covalent compounds.

Solubility in 5% Aqueous Sodium Hydroxide

Since the sodium salts of acids are generally very soluble in water, water-insoluble organic compounds that can be converted to their conjugate bases by 5% aqueous sodium hydroxide will dissolve in this solvent. Such compounds include carboxylic acids, sulfonic acids, sulfinic acids; phenols; sulfonamides of primary amines, imides; some β-diketones and β-keto esters; mercaptans and thiophenols; some primary and secondary nitro compounds; and oximes. Acidification of the basic solution of such compounds should precipitate the sample.

It should be obvious that substances that are soluble in water would be expected to be soluble in aqueous base, with the exception of the salts of water-insoluble amines.

Some acid halides and some easily hydrolyzed esters will react to give soluble products upon standing with 5% aqueous sodium hydroxide.

Solubility in 5% Aqueous Sodium Bicarbonate

OH
NO$_2$

NO$_2$

2,4-dinitrophenol

Of the substances listed as soluble in 5% aqueous sodium hydroxide, only carboxylic acids, sulfonic acids, sulfinic acids, and certain phenols with multiple electronegative substituents, such as 2,4,6-tribromophenol or 2,4-dinitrophenol, are sufficiently acidic to be converted to their conjugate bases by 5% aqueous sodium bicarbonate solution and thus to dissolve in this solvent. Dissolution of acids in sodium bicarbonate solution will result in the evolution of carbon dioxide gas.

Solubility in 5% Aqueous Hydrochloric Acid

H
N

carbazole

Five percent aqueous hydrochloric acid will convert many amines to their water-soluble hydrochloride salts and thus cause them to dissolve. Most aliphatic amines (primary, secondary, and tertiary) will be converted to their conjugate acids under these conditions, as well as most aromatic amines in which no more than one aromatic ring is attached directly to the nitrogen. Diphenylamine, carbazole, 2,4,6-tribromoaniline, and the nitroanilines are not sufficiently basic to be converted to their conjugate acids under these conditions.

N,N-disubstituted amides and N-benzylacetamide can be converted to their conjugate acids under these conditions, but not N-unsubstituted or most N-monosubstituted amides.

The hydrochloride salts of some basic substances are not soluble in 5% hydrochloric acid, although the salt may be soluble in water. Thus, if a solid remains after treatment with 5% hydrochloric acid, it should be separated from the supernatant liquid by suction filtration or centrifugation and its solubility in water determined. If doubt still remains as to whether the solid is the salt or the original compound, its melting point or infrared spectrum can be compared to that of the original material. Basification of the supernatant liquid should give a precipitate if the hydrochloride salt was partially dissolved.

It should be obvious that substances that are soluble in water would be expected to be soluble in aqueous acid, with the exception of the salts of water-insoluble acids.

Compounds containing an oxygen or nitrogen atom will dissolve in or react with cold concentrated sulfuric acid. Alkenes will react to form either a soluble alkyl hydrogen sulfate or an insoluble polymer, or will oxidize, usually with the evolution of heat. Polyalkylbenzenes will undergo sulfonation to give a product that is soluble in the mixture.

All saturated compounds containing only carbon, hydrogen, or halogen will be insoluble in concentrated sulfuric acid, as will those aromatic hydrocarbons or their halogenated derivatives that cannot be easily sulfonated.

Solubility in 85% Phosphoric Acid

Certain oxygen-containing compounds such as alcohols, aldehydes, methyl and cyclic ketones, and esters are soluble in 85% phosphoric acid, provided that they contain fewer than nine carbon atoms. Ethers are somewhat less soluble: di-*n*-propyl ether is soluble, whereas di-*n*-butyl ether and anisole are not.

anisole

22.4 TECHNIQUES FOR DETERMINATION OF SOLUBILITY

A sample size of 30 mg of a solid or one or two drops of a liquid to 1 ml of solvent may be satisfactory in most cases. The solid should be powdered. All solubility determinations should be made at room temperature, without heating.

In determining solubility in water or ether, add the solvent in portions to the sample in a small test tube (8-mm diameter by 50-mm length). Mix the contents of the tube well after each addition of solvent. Adding the solvent in portions will allow you to distinguish between substances that are freely soluble, barely soluble, and relatively insoluble. If the substance is insoluble in water, the resulting suspension can be used to determine solubility in hydrochloric acid or sodium hydroxide: add the calculated amount of concentrated hydrochloric acid or 50% sodium hydroxide solution.

In determining solubility in 5% sodium hydroxide, 5% sodium bicarbonate (only for substances found to be soluble in 5% sodium hydroxide), or 5% hydrochloric acid (only for compounds that contain nitrogen), add the sample in portions to the solvent, mixing well after each addition. If the first portion does not dissolve completely, the fact that the material is of limited solubility is established, and the remainder need not be added. The solution should be neutralized after the removal of any insoluble residue.

When determining solubility in concentrated sulfuric acid, add the sample to the acid. Color changes, evolution of heat or a gas, or other evidence of reaction can be as significant as simple dissolving.

Solubility in 85% phosphoric acid should be determined only if the compound is soluble in concentrated sulfuric acid.

Problems

1. Perfluoroheptane (C_7F_{16}) dissolves in heptane only to the extent of 0.27 mole per mole of heptane. Account for the immiscibility of these two substances.

2. Account for the miscibility with water of polyethyleneglycol, a high-boiling substance whose formula is

$$HO—(CH_2CH_2O—)_nCH_2CH_2OH$$

3. Account for the fact that, although diethyl ether boils lower than the isomeric butyl alcohols, it is less soluble in water (see Table 22.2).

4. Account for the lower solubility and higher melting point of the equimolar mixture of d- and l-tartaric acids compared to either pure enantiomer (see Table 22.3).

5. Evaluate the relative effectiveness of nitrogen, oxygen, and sulfur atoms in hydrogen bonding to water (see Table 22.1).

6. What is the significance of the way Equations 22.2-8 and 22.9-9 reduce when $T = T_f$?

7. Rationalize the trends in the solubilities of the following isomeric compounds:

Compound	b.p.	m.p.	Solubility in Water (grams per 100 g)
Valeric acid	187°		3.4
Isovaleric acid	177°		4.2
Trimethylacetic acid		36°	2.2
n-Propyl acetate	101°		1.9
Isopropyl acetate	88°		3.2

8. Rationalize the trends in the solubilities of the dibasic acids of the following type:

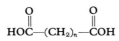

Acid	n	m.p.	Solubility in Water (grams per 100 g; 20°C)
Oxalic	0	189°	9.5
Malonic	1	135°	73.5
Succinic	2	185°	6.9
Glutaric	3	98°	63.9
Adipic	4	152°	1.5
Pimelic	5	105°	4.9
Suberic	6	142°	0.15
Azelaic	7	106°	0.24
Sebacic	8	134°	0.10

9. The solubility of caffeine in several solvents is given; rationalize the relatively large solubility in chloroform.

Solvent	Solubility (grams per 100 g solvent; 18°C)
Chloroform	11.2
Carbon tetrachloride	0.09
Ether	0.12
Ethyl acetate	0.73
Benzene	0.91

10. Explain the fact that some organic compounds, for instance many proteins, are more soluble in salt solutions than in pure water; that is, they may be "salted in" to aqueous solution.

References

Sources from which more information about the relationships between solubility and molecular structure may be obtained include:

1. R. L. Shriner, R. C. Fuson, and D. Y. Curtin, *The Systematic Identification of Organic Compounds*, 5th edition, Wiley, New York, 1965, p. 67.

2. N. D. Cheronis and J. B. Entrikin, *Identification of Organic Compounds*, Interscience, New York, 1963, p. 77.

3. A. I. Vogel, *A Textbook of Practical Organic Chemistry*, 3rd edition, Wiley, New York, 1957, p. 1045.

Absorption spectroscopic methods are relatively recent developments. Ultraviolet-visible spectroscopy (Section 24) has been known the longest. It was followed by infrared spectroscopy and, most recently, by nuclear magnetic resonance spectroscopy (Section 25). Since equipment is widely available for the determination of IR, UV, and NMR spectra, these methods are now commonly and routinely used for the analysis and characterization of pure substances and mixtures.

Absorption spectroscopy involves the determination of the degree to which electromagnetic radiation (light; light energy) is absorbed by a substance over a range of wavelengths. The record of energy absorption versus wavelength is the absorption spectrum. The usual wavelength range for IR spectroscopy is from 2 microns to 16 microns, although instruments are available that scan up to 200 microns. Table 23.1 indicates the relationship between the IR region and other regions of the electromagnetic spectrum, including the UV-visible region and the NMR region.

Table 23.1. The electromagnetic spectrum: energy, frequency, and wavelength

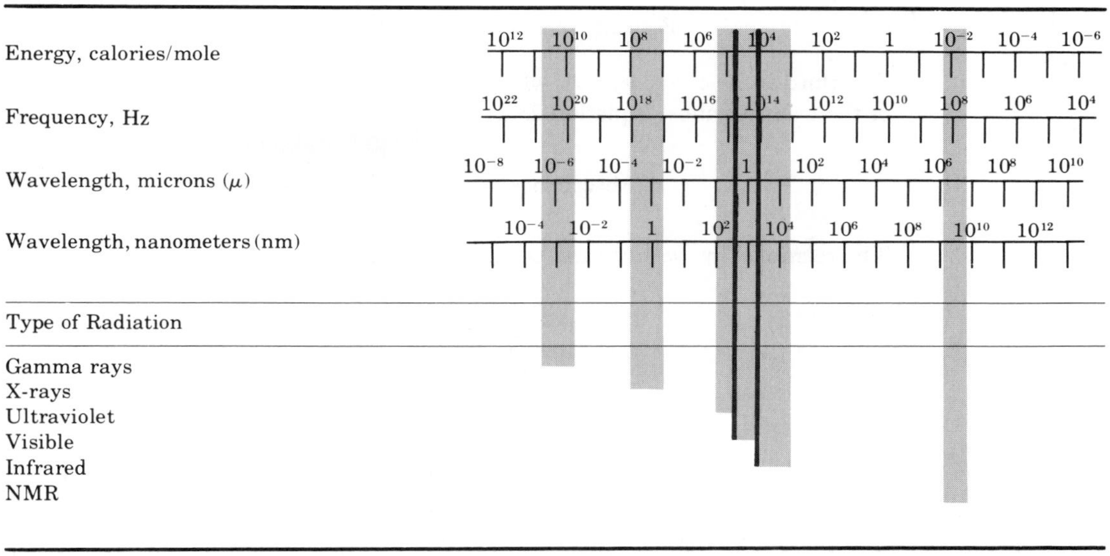

Energy, calories/mole	10^{12}	10^{10} 10^8	10^6	10^4	10^2	1	10^{-2}	10^{-4}	10^{-6}
Frequency, Hz	10^{22}	10^{20} 10^{18}	10^{16}	10^{14}	10^{12}	10^{10}	10^8	10^6	10^4
Wavelength, microns (μ)	10^{-8}	10^{-6} 10^{-4}	10^{-2}	1	10^2	10^4	10^6	10^8	10^{10}
Wavelength, nanometers (nm)		10^{-4} 10^{-2}	1	10^2	10^4	10^6	10^8	10^{10}	10^{12}

Type of Radiation

Gamma rays
X-rays
Ultraviolet
Visible
Infrared
NMR

23.1 WAVELENGTH, FREQUENCY, AND ENERGY OF ELECTROMAGNETIC RADIATION

Electromagnetic radiation (light) can be described in terms of wavelength, frequency, or energy. The product of wavelength (λ; length) and frequency (ν; time^{-1}) equals the speed of light (c; length per time):

$$\lambda\nu = c \qquad (23.1\text{-}1)$$

According to Equation 23.1-1, wavelength and frequency are inversely proportional to one another. That is, short wavelengths correspond to high frequencies, and vice versa.

Light energy is described in terms of *quanta*, or the least increment of energy. The size of the least increment is directly proportional to the frequency of the light, according to Equation 23.1-2:

Light as energy

$$E = h\nu \qquad (23.1\text{-}2)$$

where E is the energy of the quantum, ν is the frequency, and h is the proportionality constant, Planck's constant. Thus, large quanta are associated with high frequency and, by inverse proportionality, with short wavelength.

The relationships between wavelength, frequency, and energy are also illustrated in Table 23.1.

23.2 UNITS OF LIGHT ABSORPTION

The degree to which light is absorbed by a sample at a particular wavelength may be expressed in several ways. The first is as *transmittance*, T, which is defined in Equation 23.2-1 as the ratio of the intensity of light that passes through the sample to the intensity incident upon the sample (that is, the ratio that expresses the fraction of light that gets through the sample):

$$\text{Transmittance} = T = \frac{I_{\text{transmitted}}}{I_{\text{incident}}} = \frac{I_t}{I_i} \qquad (23.2\text{-}1)$$

The second is the *percent transmittance*, $\% T$, which is defined as the transmittance $\times$ 100, Equation 23.2-2:

$$\text{Percent transmittance} = \% T = T \times 100 \qquad (23.2\text{-}2)$$

The third is the *absorbance*, A, which is defined as the logarithm of the reciprocal of the transmittance, Equation 23.2-3:

$$\text{Absorbance} = A = \log \frac{1}{T} = \log \frac{I_i}{I_t} \qquad (23.2\text{-}3)$$

The relationships between these units are illustrated in Table 23.2.

Table 23.2. Relationships between transmittance, percent transmittance, and absorbance units

$\% T$	T	$1/T$	Absorbance
100	1	1	0
75	3/4	1.33	0.125
50	1/2	2	0.301
25	1/4	4	0.602
10	1/10	10	1.000
5	1/20	20	1.301
2	1/50	50	1.699
1	1/100	100	2.000
0	0	infinite	infinite

It is too bad to have to use "complicated" units like absorbance, but *only* absorbance is directly proportional to sample concentration and sample thickness (path length). The relationship between absorbance and concentration is given by Equation 23.2-4:

$$A = \epsilon c l \qquad (23.2\text{-}4)$$

Extinction coefficient

where A is the absorbance; c is the concentration, in moles/liter; l is the path length, in cm; and ϵ, the extinction coefficient, is the proportionality constant. The practical consequences of this relationship between concentration and path length will be brought out in the discussions of sample preparation for infrared and ultra-violet spectroscopy.

Most IR and UV spectrometers are calibrated in both percent transmittance and absorbance units.

23.3 INFRARED LIGHT ABSORPTION AND MOLECULAR STRUCTURE

In all forms of absorption spectroscopy, an interpretation of a spectrum involves a discussion of the ways in which light energy is absorbed by the molecules of the sample. In the infrared range, energy absorption is interpreted in terms of an increase in the amplitude of various molecular vibrations. The frequency of the light absorbed equals the frequency of the molecular vibration, and thus the higher-frequency vibrations will absorb the larger quanta or the shorter-wavelength light.

Mechanism of IR energy absorption

For the purpose of interpreting infrared spectra, bonds between atoms in molecules can be thought of as springs, and the atoms as masses at the ends of the springs. According to this model, then, parts of the molecule can stretch and bend with respect to the rest of the molecule. The frequency of stretching and bending can be related to the atomic masses and the force constants of the springs by Equation 23.3-1.

$$\nu \propto \sqrt{\frac{k}{\mu}} \qquad (23.3\text{-}1)$$

Here, ν is the frequency of stretching or bending, k is the force constant (a measure of the stiffness of the spring), and μ is the reduced mass. The reduced mass is defined by Equation 23.3-2.

$$\mu = \frac{M_1 M_2}{M_1 + M_2} \qquad (23.3\text{-}2)$$

where M_1 and M_2 are the total masses at the two ends of the spring. If one part of the molecule is much lighter than the remainder ($M_1 \ll M_2$) then μ approximately equals the mass of the light part, M_1. In this case, Equation 23.3-1 reduces to Equation 23.3-3

$$\nu \propto \sqrt{\frac{k}{M_1}} \qquad (23.3\text{-}3)$$

and the frequency of bending or stretching of a small part of the molecule with respect to the remainder is proportional to the square root of the stiffness of the spring and inversely proportional to the square root of the mass of the small part. The fact that the absorption of energy by the increase in amplitude of the stretching vibration of C—H, C—F, C—Cl, C—Br, and C—I bonds occurs at lower and lower frequencies (longer wavelength; lower energy) can thus be interpreted by saying that the force constant k appears to remain about the same as M_1 increases. Similarly, the fact that the stretching frequency of a multiple bond is higher than that of the corresponding single bond can be interpreted by saying that the force constant is larger for the multiple bond than for the single bond. It is harder to stretch a double or triple bond than a single bond.

23.4 INTERPRETATION OF INFRARED SPECTRA

Establishment of Identity of Samples

The simplest use of infrared spectra is to determine whether two samples are identical or not. If the samples are the same, their IR spectra (obtained under identical conditions) must be the same. If the samples are different, their spectra (obtained under identical conditions) will theoretically be different. If the two samples are both pure substances very similar in structure, the differences in the spectra may be so small that it is not easy to see them; it may even be beyond the power of the instrument to detect them. The absorption peaks of the spectrum of an impure sample should be less intense than those of a pure sample, assuming equal amounts of sample, and will contain additional peaks. The infrared spectra of enantiomers should be identical, and those of diastereomers should be different. Figures 46.1 and 46.2 show the IR spectra of a pair of enantiomers, the *R* and *S* isomers of carvone; the spectra are identical within experimental uncertainty. Figures 48.1 and 48.3 illustrate the IR spectra of a pair of diastereomers, the *trans* and *cis* isomers of 1,2-dibenzoylethylene. The spectra are obviously different.

Comparison of IR spectra is a convenient and relatively sensitive way to establish the identity of a substance with a sample of known structure and purity. Comparison of the IR spectra of different fractions obtained in a fractional distillation or of material before and after recrystallization is a good way to determine and follow the progress of a purification. Most IR spectra are obtained for use in these ways.

Determination of Molecular Structure

A more sophisticated level of interpretation of an IR spectrum is to use it to establish the structure of an unknown material. Comparison of IR spectra of substances of known structure has led to the empirical establishment of a great many correlations between wavelength (or frequency) of IR absorption and features of molecular structure. The presence or absence of absorption at certain wavelengths can thus be interpreted in terms of the presence or absence of certain structural features. Table 23.3 presents a number of these correlations.

Certain structural features can be established fairly easily. For example, if a substance contains only C, H, and O, the oxygen can be present only as C=O, O—H, or C—O—C (or a combination of these, such as the ester or carboxylic acid group). The presence or absence of absorption in the carbonyl region (~ 5.8–6.0 microns; ~ 1730–1670 cm^{-1}) or O—H region (~ 2.7–3.0 microns; ~ 3700–3300 cm^{-1}) can serve to eliminate or establish some of these possibilities. The

nature of a functional group involving nitrogen can be inferred in a compound containing only C, H, and N in a similar way. If the compound contains both O and N and/or other atoms other than C and H, the problem is more difficult but can be approached in the same way.

It is also fairly easy to establish whether a compound is primarily aromatic or primarily aliphatic. The IR spectrum of an aromatic compound (without a long aliphatic side chain) generally has only weak absorption in the C—H stretching region (~ 3.3 microns; ~ 3000 cm^{-1}), sharp bands in the 6–7-micron region (1660–1430 cm^{-1}), and strong absorption at wavelengths greater than 12 microns (less than 840 cm^{-1}), and gives the general impression of having sharp and sym-

$$CH_3CH_2CH_2CH_2CH_2CH_2\!-\!\overset{\displaystyle O}{\overset{\|}{C}}\!-\!O\!-\!CH_3$$

methyl heptanoate
(aliphatic)

methyl benzoate
(aromatic)

Table 23.3. Infrared absorption–structure correlations

151

SECTION 23

		Range (microns)	Intensity	Range (cm⁻¹)
C—H stretching vibrations				
Alkane		3.38–3.51	m–s	2962–2853
Alkene		3.23–3.32	m	3095–3010
Alkyne		3.03	s	3300
Aromatic		3.30	v	3030
Aldehyde		3.45–3.55	w	2900–2820
	and	3.60–3.70	w	2775–2700
C—H bending vibrations				
Alkane		6.74–7.33	v	1485–1365
Alkene				
monosubstituted (vinyl)		7.04–7.09	s	1420–1410
		7.69–7.75	w–s	1300–1290
		10.05–10.15	s	995–985
	and	10.93–11.05	s	915–905
disubstituted, *cis*		14.5	s	690
disubstituted, *trans*		7.64–7.72	m	1310–1295
	and	10.31–10.42	s	970–960
disubstituted, *gem*		7.04–7.09	s	1420–1410
	and	11.17–11.30	s	895–885
trisubstituted		11.90–12.66	s	840–790
Aromatic				
5 adjacent hydrogen atoms		13.3	v,s	750
	and	14.3	v,s	700
4 adjacent hydrogen atoms		13.3	v,s	750
3 adjacent hydrogen atoms		12.8	v,m	780
2 adjacent hydrogen atoms		12.0	v,m	830
1 isolated hydrogen atom		11.3	v,w	880
N—H stretching vibrations				
Amine, not hydrogen bonded		2.86–3.03	m	3500–3300
Amide		2.86–3.2	m	3500–3140
O—H stretching vibrations				
Alcohols and phenols				
not hydrogen bonded		2.74–2.79	v,sh	3650–3590
hydrogen bonded		2.80–3.13	v,b	3750–3200
Carboxylic acids				
hydrogen bonded		3.70–4.00	w	2700–2500
C—O stretching vibrations				
Esters				
formates		8.33–8.48	s	1200–1180
acetates		8.00–8.13	s	1250–1230
propionates and higher homologs		8.33–8.70	s	1200–1150
benzoates and phthalates		7.63–8.00	s	1310–1250
	and	8.69–9.09	s	1150–1100
Carbon–halogen stretching vibrations				
C—F		7.1–10.00	s	1400–1000
C—Cl		12.5–16.6	s	800–600
C—Br		16.6–20.0	s	600–500
C—I		~20	s	~500

(continued)

		Range (microns)	*Intensity*	*Range (cm⁻¹)*
C═C stretching vibrations				
Isolated alkene		5.99–6.08	v	1669–1645
Conjugated alkene				
C═C conjugated		6.25	m–s	1600
C═O conjugated		6.07–6.17	m–s	1647–1621
phenyl conjugated		6.15	m–s	1625
Aromatic		6.25	v	1600
		6.33	m	1580
		6.67	v	1500
	and	6.90	m	1450
C═O stretching vibrations				
Aldehydes				
saturated, aliphatic		5.75–5.81	s	1740–1720
α,β-unsaturated, aliphatic		5.87–5.95	s	1705–1680
Ketones				
saturated, acyclic		5.80–5.87	s	1725–1705
saturated, 6-membered ring				
and larger		5.80–5.87	s	1725–1705
saturated, 5-membered ring		5.71–5.75	s	1750–1740
α,β-unsaturated, acyclic		5.94–6.01	s	1685–1665
aryl, alkyl		5.88–5.95	s	1700–1680
diaryl		5.99–6.02	s	1670–1660
Carboxylic acids				
saturated, aliphatic		5.80–5.88	s	1725–1700
aromatic		5.88–5.95	s	1700–1680
Carboxylic acid anhydrides				
saturated, acyclic		5.41–5.56	s	1850–1800
	and	5.59–5.75	s	1790–1740
Acyl halides				
chlorides		5.57	s	1795
bromides		5.53	s	1810
Esters and lactones (cyclic esters)				
saturated, acyclic		5.71–5.76	s	1750–1735
saturated, 6-membered ring				
and larger		5.71–5.76	s	1750–1735
α,β-unsaturated and aryl		5.78–5.82	s	1730–1717
vinyl esters		5.56–5.65	s	1800–1770
Amides and lactams (cyclic amides)		5.88–6.14	s	1700–1630
Triple-bond stretching vibrations				
C≡N		4.42–4.51	m	2260–2215
C≡C		4.42–4.76	v,m	2260–2100

Intensity: w weak absorption, m medium absorption, s strong absorption, v variable intensity of absorption, sh sharp absorption, b broad absorption.

metrical absorption bands. The IR spectrum of an aliphatic compound generally has relatively strong absorption in the C—H region ($\sim$3.4 microns; $\sim$2950 cm⁻¹) and little or no absorption at wavelengths greater than 12 microns (less than 840 cm⁻¹), and gives the impression of having broad and not particularly symmetrical bands. Compare, for example, the spectra of methyl heptanoate and methyl benzoate, Figures 23.1 and 23.2. Cyclic and conformationally rigid compounds generally give spectra that contain a relatively small number of sharp, symmetrical bands; the spectrum of cyclohexanone, Figure 53.1 is an example.

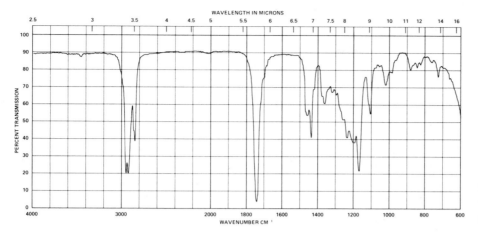

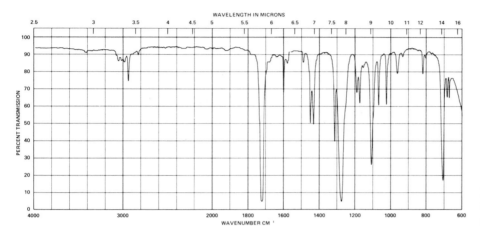

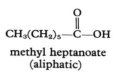

$$CH_3(CH_2)_5-\overset{\overset{\displaystyle O}{\|}}{C}-OH$$

methyl heptanoate
(aliphatic)

Figure 23.1. IR spectrum of the aliphatic compound methyl heptanoate; thin film.

methyl benzoate
(aromatic)

Figure 23.2. IR spectrum of the aromatic compound methyl benzoate; thin film.

23.5 SAMPLE PREPARATION

For infrared absorption spectroscopy, the sample whose absorption is to be determined must be placed in the beam of infrared radiation. The container (cell) or support for the sample must be transparent to infrared radiation and must therefore be made of one of a small number of materials, not including glass. The most commonly used material is sodium chloride, which is transparent between 2 and 16 microns (5000 and 600 cm^{-1}). Certain other metal halide salts are transparent over other wavelength ranges.

Liquid Samples

The spectrum of a pure liquid is most easily determined as a *film* between a pair of salt plates. One salt plate is positioned in the holder, a drop of the liquid is placed in the center, and the second plate is put on top. The top of the holder is then pressed or screwed on gently and the holder placed in the sample (near) beam of the instrument. Nothing need be placed in the reference (far) beam, since the only absorbing material in the sample beam is the liquid whose absorption spectrum is being determined. If the strongest peak in the spectrum absorbs more than about 98 percent of the light, tighten the top of the holder to make the film thinner and rerun the spectrum. The spectra shown in Figures 23.1 and 23.2, for example, were obtained from thin films of liquid.

Thin film method

The IR spectrum of a pure liquid may also be determined in solution as with solid samples.

Solid Samples: The Solution Method

Solution method

The IR spectra of solids are most often determined *in solution*. Table 23.4 suggests appropriate concentrations for an average sample; the concentration should be adjusted, if necessary, so that the strongest peak will absorb between 90 and 98 percent of the light. As Equation 23.2-4 indicates, the absorbance of a sample is proportional to both its concentration and its thickness. Table 23.4 is based on the assumption that the sample thickness (the path length of the infrared cell) will be 0.2 mm. If a cell of 0.4-mm path length will be used instead, the amount of sample required will be only half that specified by the table. It is assumed, of course, that 0.5 ml of solution will be prepared in either case.

Table 23.4. Milligrams of sample required for 0.50 ml of solution

A cell path length of 0.2 mm is assumed. For a different cell path length, adjust the amount of sample used so as to keep the number of molecules in the beam the same as for the examples given.

Molecular weight	100	200	300	400	500
Mg of sample	9	18	27	36	45

Double beam operation

The ideal solvent would be transparent over the entire infrared range of wavelength, but no liquids attain this ideal. For this reason, a so-called double-beam operation must be used for obtaining the IR spectrum of solids in solution. The solution cell is placed in the sample beam. An identical cell containing pure solvent must be placed in the reference beam; its absorption is subtracted electronically by the instrument from the absorption of the sample beam, with the result that the spectrum recorded is that of the net absorption due to sample

Solvents

molecules in the sample beam. Solvents most often used for infrared spectroscopy are carbon tetrachloride, chloroform, and carbon disulfide (Figures 23.3, 23.4, 23.5). Water is *never* used (except with special cells with water-resistant windows).

Cl
|
Cl—C—Cl
|
Cl
carbon tetrachloride

Figure 23.3. IR spectrum of carbon tetrachloride; 0.05-mm path length.

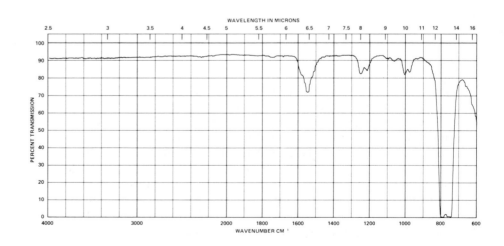

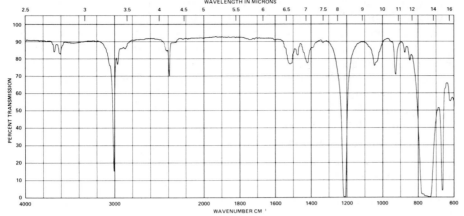

chloroform

Figure 23.4. IR spectrum of chloroform; 0.05-mm path length.

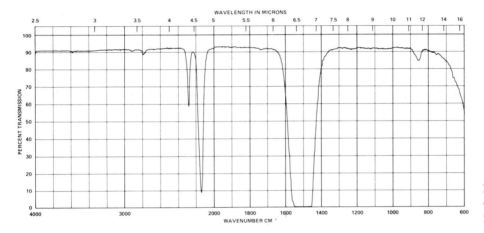

S═C═S
carbon disulfide

Figure 23.5. IR spectrum of carbon disulfide; 0.05-mm path length.

In those wavelength regions where the solvent absorbs strongly (more than about 95 percent of the light), the absorption recorded by the spectrophotometer is meaningless. This is because as the solvent absorption increases, the net absorption due to sample becomes an increasingly smaller fraction of the total absorption, and the difference in intensity of the two beams finally falls below the limit detectable by the instrument. You can verify this by placing your hand in the sample or reference beam when the instrument is scanning a range of strong solvent absorption (for carbon tetrachloride, between 12.3 and 13.7 microns; 810 and 730 cm^{-1}; Figure 23.3) and seeing that the spectrum is unaffected. Between carbon tetrachloride or chloroform and carbon disulfide, sample absorption may be determined at all wavelengths between 2 and 16 microns (5000 and 600 cm^{-1}; see Figures 23.3, 23.4, and 23.5). If the recorder pen runs either up- or downscale in a range of intense solvent absorption, the balance control is not adjusted correctly and the peaks of the rest of the spectrum will be distorted.

Balance-control setting

Solid Samples: The KBr Method

The infrared spectrum of a solid may also be determined in the solid state. One way is to grind 1–2 mg of sample with about 100–400 mg of anhydrous potassium bromide in a clean mortar and to press the resulting mixture into a translucent wafer, using a die. The wafer is then mounted in the sample beam and an attenuator is placed in the reference beam. The function of the attenuator is to compensate for the loss of sample beam intensity due to scattering. The screen of the attenuator is set so that the recorder pen indicates 100 percent transmission at

Attenuator

the wavelength at which the sample is most transparent. The appropriate wavelength may be found by a trial scan of the spectrum or by a shortcut appropriate to the individual instrument. Use of the attenuator gives a more normal-looking spectrum, since it in effect moves the spectrum to the part of the chart where a given amount of absorbance corresponds to the largest movement of the pen; compare, for example, the distance of pen travel from 0.0 to 0.1 and from 1.0 to 1.1 absorbance units.

—OH Absorption

Absorption in the O—H region in a spectrum obtained by the KBr pellet method must be interpreted with great caution, since it is hard to make sure that no water is in the KBr or gets into the sample during preparation.

Solid Samples: The Mull Method

A second method of determining the IR spectrum of a solid in the solid state is to grind (mull) 2–5 mg with a drop of mineral oil (paraffin oil) or hexachlorobutadiene. The spectrum of the mull is then determined as a liquid film. Since the recorded spectrum will be that of the compound plus the mulling agent, the spectrum of the mulling agent must be mentally subtracted from the recorded spectrum. The spectra of mineral oil and hexachlorobutadiene are presented in Figures 23.6 and 23.7.

In Section 23.4, it was stated that the spectra of identical materials should be the same *if* the spectra are obtained under the same conditions. The reason for this qualification is that for some compounds the appearance of the spectrum depends upon the nature of the sample. For example, the appearance of the O—H

$CH_3(CH_2)_nCH_3$
mineral oil

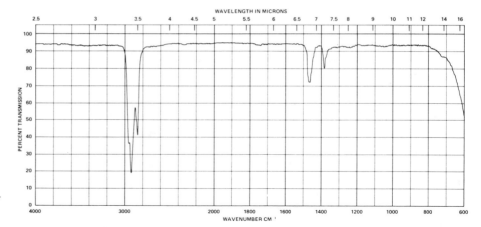

Figure 23.6. IR spectrum of mineral oil; thin film.

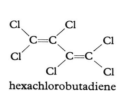

hexachlorobutadiene

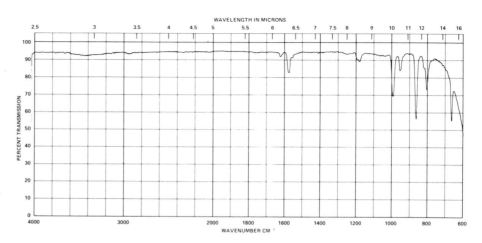

Figure 23.7. IR spectrum of hexachlorobutadiene; thin film.

part of the spectrum of an alcohol depends upon the concentration of the solution.

The reason is that in dilute solution a smaller fraction of the molecules are hydrogen bonded than in a concentrated solution, and these two molecular species—hydrogen bonded and not hydrogen bonded—absorb at different wavelengths in the O—H region. Also, the appearance of the spectrum of a sample obtained by the KBr technique often depends upon the exact details of its preparation, presumably because of differences in particle size and distribution. The spectra of solids obtained in the solid state sometimes differ from spectra obtained in solution, and spectra of the same compound in different solvents will almost always look different because of different regions of strong solvent absorption.

Reference 11 presents an excellent introduction to practical aspects of infrared spectroscopy, especially with respect to sample preparation.

Problems

1. Figures 23.8 and 23.9 show the infrared spectra of anisole and benzyl alcohol. Which is which? Justify your choice.

$$CH_3 \qquad OH$$
$$| \qquad |$$
$$O \qquad CH_2$$

anisole   benzyl alcohol

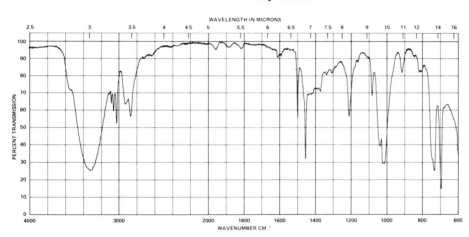

Figure 23.8.

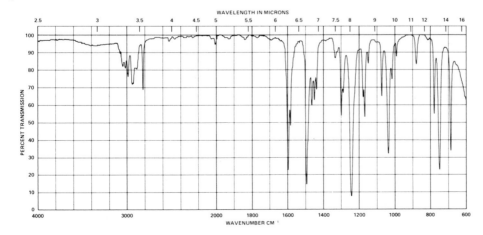

Figure 23.9.

2. The three figures for Problem 2 (Figures 23.10, 23.11, 23.12) show the infrared spectra of ethyl acetate, methyl propionate, and butyric acid. Which is which? Justify your choice.

$$CH_3-\overset{\overset{\displaystyle O}{\|}}{C}-O-CH_2-CH_3 \qquad CH_3-CH_2-\overset{\overset{\displaystyle O}{\|}}{C}-O-CH_3 \qquad CH_3-CH_2-CH_2-\overset{\overset{\displaystyle O}{\|}}{C}-O-H$$

ethyl acetate   methyl propionate   butyric acid

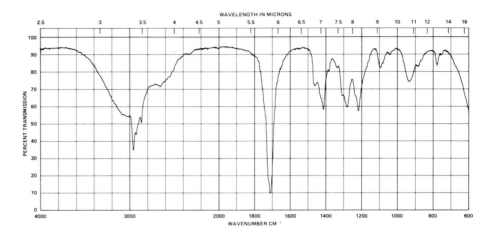

Figure 23.10.

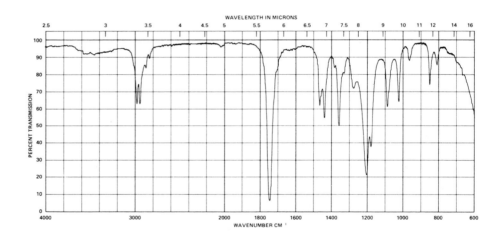

Figure 23.11.

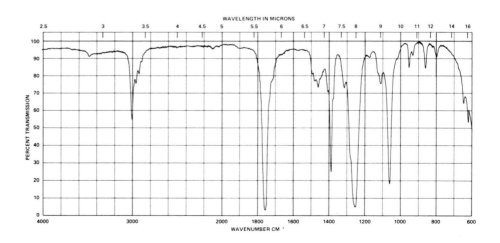

Figure 23.12.

3. The three figures for Problem 3 (Figures 23.13, 23.14, 23.15) show the infrared spectra of the three isomeric dichlorobenzenes. Which is which? Defend your choice.

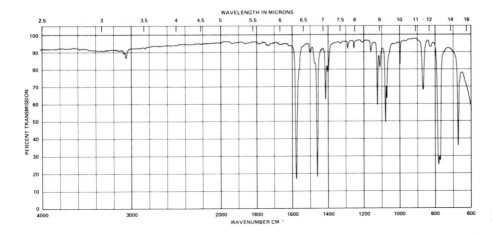

Figure 23.13.

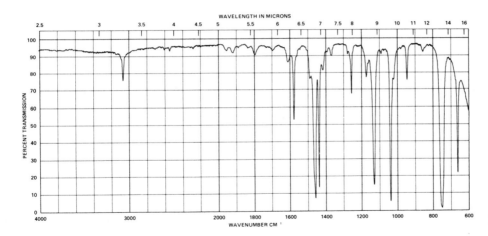

Figure 23.14.

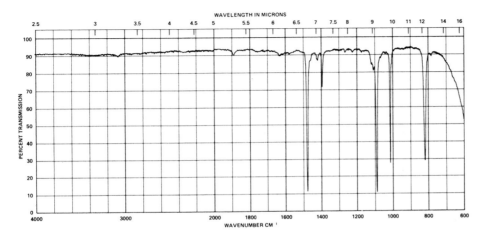

Figure 23.15.

4. The five figures for Problem 4 (Figures 23.16, 23.17, 23.18, 23.19, 23.20) show the infrared spectra of the following five isomers of molecular formula $C_9H_{10}O$. Assign the structures to the spectra. Defend your choices.

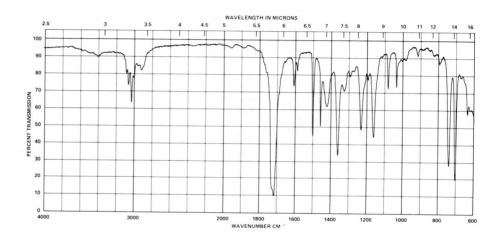

I

II

III

IV

V

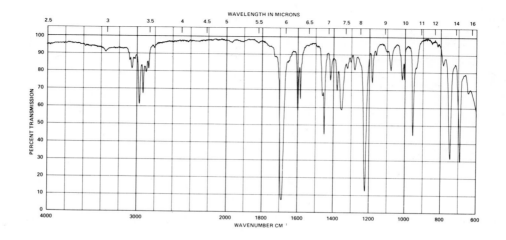

Figure 23.16.

Figure 23.17.

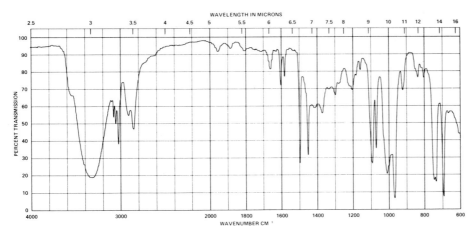

Figure 23.18.

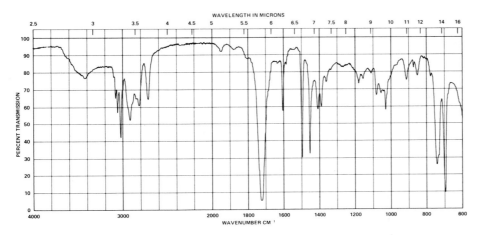

Figure 23.19.

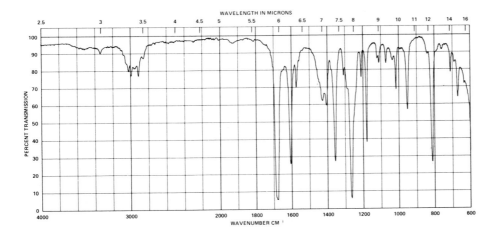

Figure 23.20.

5. Saturated aliphatic esters (Structure **I**) absorb at about 5.71–5.76 microns. α,β-Unsaturated esters (**II**) absorb at longer wavelength, while vinyl esters (**III**) absorb at shorter wavelength. Explain.

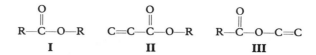

6. Interpret the variation in wavelength of the carbonyl absorption in compounds of the type R—C—X as —X is —Cl, —O—R, —H, —R, and —NH₂.

$$R-\overset{\overset{\displaystyle O}{\|}}{C}-X$$

References

A brief introduction is:

1. P. R. Jones, *Chemistry* **38**, 5 (1965).

Introductory discussions that include spectra–structure correlations are:

2. R. M. Silverstein, G. C. Bassler, and T. C. Morrill, *Spectrometric Identification of Organic Compounds*, 3rd edition, Wiley, New York, 1974, pp. 73–157.

3. J. R. Dyer, *Applications of Absorption Spectroscopy of Organic Compounds*, Prentice-Hall, Englewood Cliffs, N.J., 1965, pp. 22–57.

4. D. J. Pasto and C. R. Johnson, *Organic Structure Determination*, Prentice-Hall, Englewood Cliffs, N.J., 1969, pp. 109–158.

5. A. D. Cross and R. A. Jones, *Introduction to Practical Infrared Spectroscopy*, 3rd edition, Plenum, New York, 1969.

6. R. T. Conley, *Infrared Spectroscopy*, 2nd edition, Allyn and Bacon, Boston, 1972.

An extensive presentation of spectra–structure correlations is made in:

7. L. J. Bellamy, *The Infrared Spectra of Complex Molecules*, 3rd edition, Halsted, New York, 1975.

Indexes of published infrared spectra include:

8. H. M. Hershenson, *Infrared Absorption Spectra: Index for 1945–1957*, Academic Press, New York, 1959.

9. H. M. Hershenson, *Infrared Absorption Spectra: Index for 1958–1962*, Academic Press, New York, 1964.

A catalog of about 8000 infrared spectra is:

10. C. J. Pouchert, *The Aldrich Library of Infrared Spectra*, Aldrich Chemical Co., Inc., Milwaukee, Wisconsin, 1970 (2nd edition, 1975).

Finally, an extensive discussion of methods of sample preparation is presented in:

11. J. S. Swinehart, *Organic Chemistry*, Appleton-Century-Crofts, 1969, Appendix III.

§24 ULTRAVIOLET-VISIBLE ABSORPTION SPECTROSCOPY

Ultraviolet-visible spectroscopy involves the determination of the degree to which light is absorbed by a substance over certain ranges of wavelength of light; the output of a UV-visible spectrometer is a record of energy absorption versus wavelength, the UV-visible spectrum. Usually the wavelength ranges are from 200 nanometers to 360 nm (near ultraviolet range) and from 360 nm to 800 nm (visible range) (see Table 23.1). Absorption at shorter wavelengths is more difficult to measure because air absorbs in this region, and absorption at longer wavelengths involves a different mechanism of energy absorption (Section 23.3). Since absorption of light energy in these wavelength ranges is interpreted in terms of the promotion of electrons in lower energy levels to higher energy levels, UV-visible spectroscopy is often called *electronic absorption* spectroscopy.

Mechanism of UV energy absorption

24.1 ULTRAVIOLET-VISIBLE LIGHT ABSORPTION AND MOLECULAR STRUCTURE

Ultraviolet-visible spectroscopy, the oldest of the spectroscopic methods, found its greatest use in the determination of the structure of molecules containing conjugated systems of double bonds. The presence of conjugated unsaturation was found always to result in one or more intense absorption maxima at wavelengths greater than 200 nm. Very useful correlations were deduced for the positions and intensities of these peaks as a function of structure for conjugated dienes (Table 24.1), α,β-unsaturated ketones (Table 24.2), and substituted aromatic compounds.

Table 24.1. Conjugated dienes: correlations between structure and wavelength of absorption maximum. From A. Ault, *Problems in Organic Structure Determination*, McGraw-Hill, New York, 1967, p. 16. Reprinted with permission.

In many cases, the agreement between calculated values and observed values is within 5 nanometers.

	Nanometers
Base values	
Acyclic, conjugated dienes, and conjugated dienes contained in two nonfused 6-membered-ring systems	217
Conjugated dienes contained in two fused 6-membered-ring systems (heteroannular dienes)	214
Conjugated dienes contained in a single ring (homoannular dienes)	253
Increments	
C=C extending conjugation	+30
Each alkyl substituent, including ring residues	+5
Each ring, 6-membered or less, to which the diene double bonds are exocyclic	+5
Each —Cl or —Br substituent	+17
Alkoxy or acyloxy substituent	0
Solvent corrections	
No solvent corrections are necessary.	

Table 24.3 summarizes the wavelength maximum and intensity of absorption to be expected for certain other types of compounds. From this, you can see that saturated compounds with O, N, S, or halogen atoms can absorb light in the ultraviolet range because of the presence of unshared pairs of electrons, *n* electrons, on these atoms. Nonconjugated aldehydes and ketones absorb weakly due to the possibility of excitation of a nonbonded pair of electrons into an antibonding π orbital, an $n \rightarrow \pi^*$ transition. However, nonconjugated alkenes cannot absorb light by this mechanism. With the exception of the tail end of an intense absorption band whose maximum falls at a wavelength shorter than 200 nm ("end absorption"), they are transparent in the ultraviolet region.

Table 24.2. α,β-unsaturated aldehydes and ketones: correlations between structure and wavelength of absorption maximum. From A. Ault, *Problems in Organic Structure Determination*, McGraw-Hill, New York, 1967, p. 15. Reprinted with permission.

In many cases, the agreement between calculated values and observed values is within 5 nanometers.

	Nanometers
Base values	
α,β-Unsaturated ketones	215
α,β-Unsaturated aldehydes	210
Cyclopentenones	205
Increments	
C=C extending conjugation	+30
α-Alkyl substituent, including ring residue	+10
β-Alkyl substituents, including ring residues	+12
γ, δ, or further alkyl substituents, including ring residues	+18
Each ring, 6-membered or less, to which a C=C is exocyclic	+5
C=C contained in a 5-membered ring (except cyclopentenones)	+5
Enolic α or β—OH	+35
(Alkoxy and acyloxy substituents are treated as alkyl groups.)	
Solvent corrections	
Water	−8
Methanol and ethanol	0
Chloroform	+1
Dioxane	+5
Ether	+7
Hexane	+11

Table 24.3. Correlations between ultraviolet light absorption and molecular structure

Type of Compound	*Transition*	*Wavelength*[a]	ϵ[b]
Alkane	$\sigma \rightarrow \sigma^*$	<200	
Nonconjugated alkene	$\pi \rightarrow \pi^*$	<200	
Nonconjugated alkyne	$\pi \rightarrow \pi^*$	<200	
Alcohols	$n \rightarrow \sigma^*$	185	10^2
Ethers	$n \rightarrow \sigma^*$	185	10^3
Amines	$n \rightarrow \sigma^*$	200	10^3
Mercaptans	$n \rightarrow \sigma^*$	200; 230	$10^3; 10^2$
Sulfides	$n \rightarrow \sigma^*$	210; 240	$10^3; 10^2$
Alkyl chlorides	$n \rightarrow \sigma^*$	175	10^2
Alkyl bromides	$n \rightarrow \sigma^*$	210	10^2
Alkyl iodides	$n \rightarrow \sigma^*$	260	10^2
Nonconjugated aldehydes	$n \rightarrow \pi^*$	290	10
Nonconjugated ketones	$n \rightarrow \pi^*$	280–290	10
Carboxylic acid derivatives	$n \rightarrow \pi^*$	200–210	$10–10^2$
Conjugated alkenes	$\pi \rightarrow \pi^*$	Table 24.1	
Conjugated aldehydes and ketones	$\pi \rightarrow \pi^*$	Table 24.2	
Aromatic compounds	$\pi \rightarrow \pi^*$	200–visible	

[a] Approximate wavelength maximum, in nanometers.
[b] Approximate extinction coefficient; see Section 23.2.

24.2 INTERPRETATION OF UV-VISIBLE SPECTRA

The two features of an ultraviolet-visible spectrum that can give information concerning molecular structure are (1) the wavelengths at which the maximum light is absorbed, and (2) the intensity of absorption at that wavelength. The way in which the degree of absorption is expressed and measured experimentally is explained in Section 23.2. Certain correlations of these two parameters with molecular structure are summarized in the three preceding tables.

Since the development of infrared and nuclear magnetic resonance spectroscopy, ultraviolet spectroscopy has become less important for the determination of structure of organic molecules. For this purpose, the newer methods are simply more convenient and more reliable. For example, the establishment of the presence of a carbonyl group is much easier and more certain through infrared spectroscopy. The principal application of UV spectroscopy to determination of molecular structure lies at present in the study of more complex molecules containing conjugated systems.

Ultraviolet spectroscopy is, however, still widely used as a quantitative method of analysis: to determine concentrations. This is because many compounds absorb strongly even at low concentration, and because UV spectrometers can determine the degree of absorption of a sample with great accuracy.

24.3 COLOR AND MOLECULAR STRUCTURE

Color is the result of a selective reflection or transmission of light in the visible range of the electromagnetic spectrum. A white or colorless substance reflects or transmits all incident light equally well. For any substance to appear colored, then, it must selectively absorb light of certain wavelengths in the visible part of the spectrum. For most organic compounds, the mechanism of light absorption in this part of the spectrum involves a π electronic system that usually contains four or more conjugated double bonds. Thus, colored compounds are most often found to be aromatic rather than aliphatic compounds.

A substance that absorbs light near the blue end of the visible spectrum appears yellow. As the wavelength of visible absorption moves toward the red end of the spectrum, the observed color becomes orange, red, and finally purple or blue. Colored organic compounds that are not yellow or orange are very unusual (see Section 78).

The most common types of colored compounds are nitro-substituted phenol or aniline derivatives. Less common types are quinones and compounds containing the azo (—N=N—) system. Many dyes and colorings combine these two structural features (see Section 65).

24.4 SAMPLE PREPARATION

In most cases, a suitable sample for UV-visible spectroscopy will be a solution of the substance in a solvent transparent to UV and visible radiation. Thus, choices of sample concentration and solvent must be made, in addition to a choice of cell type.

Concentration

As stated in Section 23.2, absorbance is proportional to the concentration of the sample (c, in moles per liter) and the sample thickness, or path length (l, in cm): $A = \epsilon cl$.

A knowledge of the value of the extinction coefficient ϵ at the absorption maximum is of use in the determination of the molecular structure of the sample. Therefore, you will want to determine the spectrum at a concentration that will give an absorbance of about 1 or a little less, because the value of ϵ can be estimated most accurately under these conditions. Since the standard sample cell has a path length of 1.00 cm, the concentration c, in moles per liter, will equal $1/\epsilon$ when $A = 1$; and for A not to exceed 1, c should be equal to or less than $1/\epsilon$. Table 24.4 summarizes the concentration, in moles per liter, required for an absorbance of 1 for absorption bands of various extinction coefficients. Table 24.5 presents some data that may be useful in preparing samples of various con-

Table 24.4. Concentration c required for an absorbance of 1 for absorption bands of various extinction coefficients ϵ

	ϵ	c
Very weak band	10	0.1 M
Weak band	100	0.01 M
Medium band	1,000	0.001 M
Strong band	10,000	0.0001 M

Table 24.5. Milligrams of sample required for 5 ml of solution as a function of molecular weight and desired concentration[a]

	Molecular Weight				
Concentration	100	200	300	400	500
0.1 M	50	100	150	200	250
0.01 M	5	10	15	20	25
0.001 M	0.5	1	1.5	2	2.5
0.0001 M	0.05	0.1	0.15	0.2	0.25

[a] The most dilute solutions must be made by dilution of a more concentrated solution.

centrations. Of course, if the molecular weight of a substance is not known with certainty, the calculated value of the extinction coefficient will contain a corresponding uncertainty.

There are two important practical consequences of the fact that values for ϵ for different functional groups may differ by several powers of 10. First, observing that a sample of 0.0001 M concentration appears to be completely transparent ($A \approx 0$) does not necessarily mean that absorption maxima with $\epsilon \approx 100$ or less are absent. This is because at a concentration this low they would show an absorbance of only 0.01 or less, which could easily be overlooked. Similarly, a strong absorption band can completely obscure a weak one. The second consequence is that for the accurate determination of extinction coefficients that differ by much more than a power of 10, spectra obtained at different concentrations will be needed. The UV spectra presented in Figures 67.3 and 67.4, obtained at concentrations that differ by a factor of 100, illustrate this very nicely.

Solvent

The ideal solvent for UV-visible spectroscopy is completely transparent over the entire spectral range. In contrast to the case with IR spectroscopy, many solvents do approach the ideal fairly closely. These include cyclohexane, ethanol, methanol, and water. In order to compensate for light absorption by the solvent and the cell (and any absorption by impurities in the solvent), UV-visible spectrometers are generally run in the double-beam mode wherein the absorption of pure solvent in a second cell is automatically subtracted from the absorption of the sample solution. The spectrum recorded is the difference spectrum, and it represents the net absorption due to the solute.

Since small concentrations of many compounds result in very intense absorption, a small amount of an impurity introduced during careless sample preparation may dominate the spectrum and completely obscure the spectrum of the compound you are interested in.

Beware of contamination

Cell Type

The standard UV-visible cell is a rectangular cell having a 1-cm path length and approximately a 3-ml volume. Circular cells and cells of different path lengths are also available.

For visible spectra, Pyrex glass cells are appropriate. For spectra in the UV range, the much more expensive quartz cells will be needed, since Pyrex absorbs strongly at wavelengths shorter than 320 nm.

Problems

1. Four derivatives of cholesterol—A, B, C, and D—have the following spectral properties:

    A IR: strong absorption near 5.95 microns
       UV: $\lambda_{max} = 230$ nm; $\epsilon = 10,700$

    B IR: strong absorption near 5.95 microns
       UV: $\lambda_{max} = 241$ nm; $\epsilon = 16,600$

    C IR: no absorption near 5.95 microns
       UV: $\lambda_{max} = 234$ nm; $\epsilon = 20,000$

    D IR: no absorption near 5.95 microns
       UV: $\lambda_{max} = 315$ nm; $\epsilon = 19,800$

    To each of the compounds, assign one of the following molecular structures **I**, **II**, **III**, or **IV**—where R is:

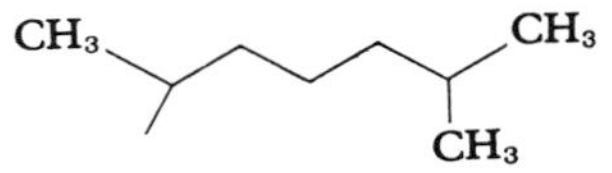

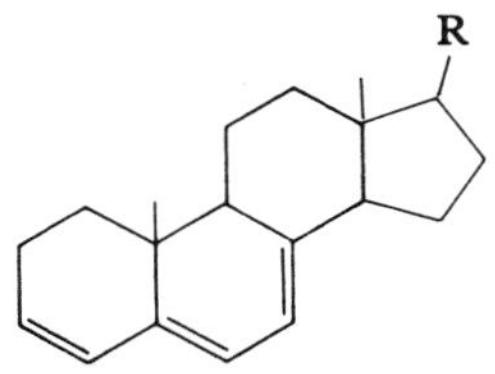

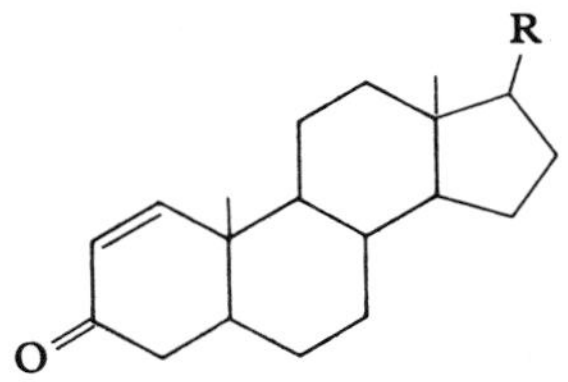

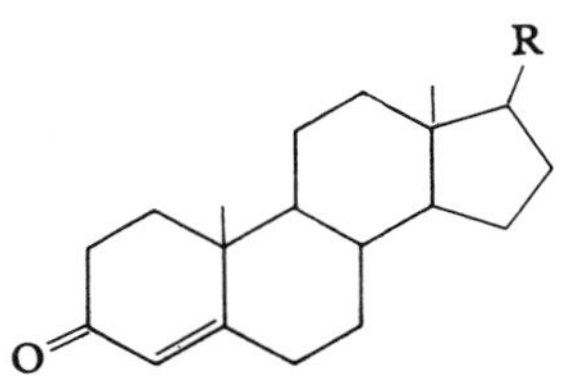

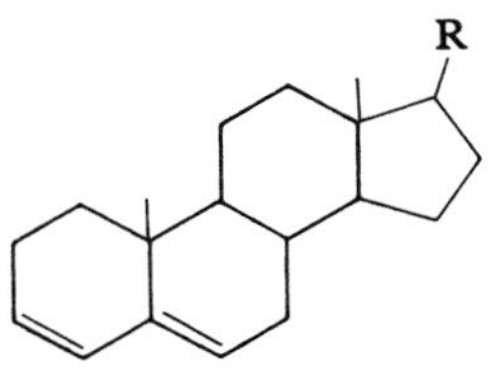

2. Absolute ethanol (anhydrous ethanol) can be prepared by adding benzene to 95%
ethanol and distilling. The first distillate is a ternary azeotrope of benzene, water,
and ethanol; the next is a binary azeotrope of benzene and ethanol. Benzene has a
maximum extinction coefficient of 200 at 256 nm. What is the minimum concentration of benzene (mg/liter) that could be detected by UV spectroscopy, assuming
that an absorbance as little as 0.02 could be determined?

3. LSD has a maximum extinction coefficient of less than 1000. What are the chances
of detecting 10 micrograms of LSD by means of UV spectrometry?

4. Calculate the wavelength to be expected for the $\pi \to \pi^*$ transition for 2-methyl-2-
pentenal. Compare your calculated value with the position of the peak in spectrum
shown in Figure 67.3.

$$CH_3CH_2 \quad \overset{\displaystyle O}{\underset{\displaystyle C-H}{\|}}$$

$$\underset{H}{\overset{}{\diagdown}} C = C \underset{CH_3}{\overset{}{\diagup}}$$

2-methyl-2-pentenal

References

1. R. M. Silverstein, G. C. Bassler, and T. C. Morrill, *Spectrometric Identification of
Organic Compounds*, 3rd edition, Wiley, New York, 1974, pp. 231–257.

2. J. R. Dyer, *Applications of Absorption Spectroscopy of Organic Compounds*, Prentice-
Hall, Englewood Cliffs, N.J., 1965, pp. 4–21.

§25 NUCLEAR MAGNETIC RESONANCE SPECTROSCOPY

Nuclear magnetic resonance spectroscopy (NMR) involves measuring the absorption of "light" energy in the radiofrequency portion of the electromagnetic
spectrum. The mechanism of energy absorption is the reorientation of magnetic
nuclei with respect to a large external magnetic field, H_{ext}, in which the sample is
placed. Unlike a compass needle in the earth's magnetic field, which can have a
great many different orientations, a nuclear magnet can have only a few. The pro-
Mechanism of energy ton, for example, can have only two orientations with respect to the field: one
absorption "with" the field (of lower energy) and one "against" the field (of higher energy).
NMR spectroscopy is based on the fact that the energy of the quantum of absorbed *radiofrequency* (rf) *radiation* goes into changing the orientation of the
magnetic nucleus from being "with" the magnetic field to being "against" the
field.

25.1 SHIELDING; CHEMICAL SHIFT

Since the difference in energy of the proton in the two states "with" and "against"
depends upon the size of the magnetic field that it experiences (H_{nuc}), and since
the energy for transitions between the two states ($h\nu$) is supplied by the rf field,
the following relationship holds:

$$h\nu \propto H_{nuc}$$

where $h\nu$ is, according to Planck's relationship, the energy of the quanta of
frequency ν. Protons experiencing different magnetic fields H_{nuc} will therefore
absorb rf energy at different frequencies. For an NMR spectrometer operating
Resonance condition at a frequency of 60 Megahertz (60 MHz), H_{nuc} must equal 14,092 gauss for
energy to be absorbed (for *resonance* to occur).

The usefulness of NMR spectrometry to the organic chemist stems from
the fact that *protons in different structural environments experience a different H_{nuc}
for the same H_{ext}*. For a particular proton, the difference between H_{nuc} and H_{ext} is
called the *shielding*:

$$H_{nuc} = H_{ext} - H_{shielding}$$

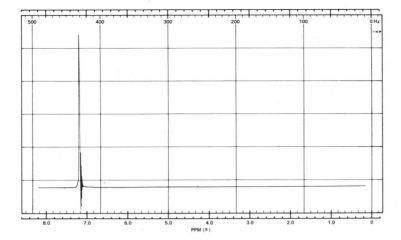

benzene

Figure 25.1. NMR
spectrum of benzene; neat.

p-xylene

*Chemical shift reference
compound*

Scales of chemical shift

Thus, if the sample in the magnetic field H_{ext} is irradiated with a constant rf field of exactly 60 MHz and H_{ext} is slowly increased, energy will be absorbed by the sample each time H_{nuc} (which equals $H_{ext} - H_{shielding}$) becomes equal to 14,092 gauss. The graph of energy absorption versus H_{ext} is the NMR spectrum. The NMR spectrum of benzene shows only a single instance of energy absorption (a single resonance; see Figure 25.1), whereas *p*-xylene shows two resonances, one for the methyl protons and one for the ring protons (Figure 25.2). It is important to remember that shielding is proportional to H_{ext} and that differences in shielding, or chemical shift differences, will also be proportional to H_{ext}.

Chemical shifts are usually reported as the amount by which the H_{ext} required for the resonance of the sample substance varies from the H_{ext} required for the resonance of a *reference substance, tetramethylsilane (TMS)*. A unit of convenient size is one part per million (ppm) of H_{ext}, and the TMS resonance is taken sometimes as 0 (δ scale) and sometimes as 10 (τ scale). For comparison with coupling constants (see Section 25.2), chemical shift differences must be expressed in frequency units; 1 ppm of 60 MHz is 60 Hz. The relationships between the various units and scales are shown in Table 25.1.

25.2 SPIN-SPIN SPLITTING

Although the presence of resonances at different chemical shifts gives information about the numbers of different functional groups in which protons are present,

Table 25.1. Units of chemical shift

Increasing H_{ext}:   →
Increasing Frequency: ←
Increasing Shielding: →

11	10	9	8	7	6	5	4	3	2	1	0	−1	δ
−1	0	1	2	3	4	5	6	7	8	9	10	11	τ
330	300	270	240	210	180	150	120	90	60	30	0	−30	Hz[a]
660	600	540	480	420	360	300	240	180	120	60	0	−60	Hz[b]
1100	1000	900	800	700	600	500	400	300	200	100	0	−100	Hz[c]
											↑		
											TMS resonance		

[a] 30-MHz instrument.
[b] 60-MHz instrument.
[c] 100-MHz instrument.

$$CH_3—Si—CH_3$$

tetramethylsilane (TMS)

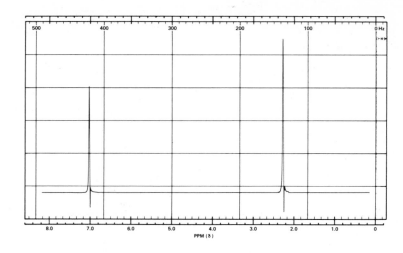

CH₃ / p-xylene structure

Figure 25.2. NMR spectrum
of *p*-xylene; neat.

the features of the NMR spectra from which the most detailed information concerning the molecular structure of the sample can be obtained are those that are the result of the effect upon one another of protons in different groups at different chemical shifts (spin-spin coupling). While the NMR spectrum of methyl iodide (Figure 25.3) consists of a single peak, the NMR spectrum of ethyl iodide (Figure 25.4) shows two *sets* of peaks. The resonances of the two different kinds of protons in ethyl iodide, the methyl protons and the methylene protons, are not each a single peak but rather a set of peaks, or *multiplet*. The methylene protons

Multiplets

methyl iodide

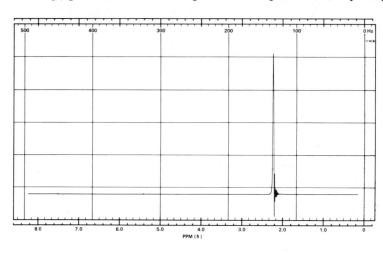

Figure 25.3. NMR spectrum
of methyl iodide; neat.

ethyl iodide

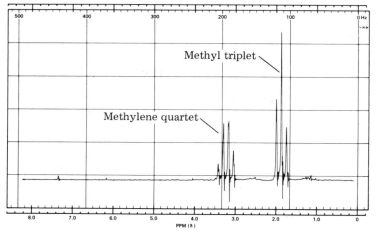

Methyl triplet

Methylene quartet

Figure 25.4. NMR spectrum
of ethyl iodide; neat.

$$
\begin{array}{c}
\text{H} \\
| \\
\text{H}-\text{C}-\text{I} \\
| \\
\text{H}
\end{array}
\qquad
\begin{array}{c}
\text{H}\quad\text{H} \\
|\quad| \\
\text{H}-\text{C}-\text{C}-\text{I} \\
|\quad| \\
\text{H}\quad\text{H}
\end{array}
$$

methyl iodide   ethyl iodide

have caused the methyl resonance to appear as a triplet of approximate relative intensity $1:2:1$, and the methyl protons have caused the resonance of the methylene protons to appear as a quartet of approximate relative intensity $1:3:3:1$.

In general, it is necessary to use the methods of quantum mechanics to calculate the expected multiplicity of a resonance. It is fortunate that in many cases a very much simpler analysis is adequate.

Interpretation of Splitting of the CH$_3$ Resonance of CH$_3$CH$_2$I

Consider first the appearance of the methyl resonance of ethyl iodide as a $1:2:1$ triplet due to the presence of the two adjacent methylene protons. Because of the small energy difference for a proton in either of the two states, *a particular proton in a magnetic field of 14,092 gauss at room temperature has an almost equal probability of being in either the lower or the upper spin state*—of being oriented, that is, either "with" or "against" the external magnetic field. Therefore, for the purpose of this analysis, you can estimate that of the ethyl iodide molecules one-fourth will have both the methylene protons oriented "with" H_{ext} ($\uparrow\,\uparrow$), one-fourth will have both the methylene protons "against" H_{ext} ($\downarrow\,\downarrow$), and one-fourth will be in each of the two states possible where one proton is "with" and one "against" H_{ext} ($\frac{1}{4}$ as $\uparrow\,\downarrow$ and $\frac{1}{4}$ as $\downarrow\,\uparrow$).

The effect of having both of the two adjacent methylene protons "with" the external field is to bring the methyl protons of those molecules into resonance at a slightly lower value of H_{ext}, since the two protons "with" the field will augment H_{ext} for the neighboring methyl group. This effect of the adjacent protons may be expressed by Equation 25.2-1, where H_{coupling} in this case is positive.

$$H_{\text{nuc}} = H_{\text{ext}} - H_{\text{shielding}} + H_{\text{coupling}} \qquad (25.2\text{-}1)$$

Thus, the methyl resonance for one-fourth of the molecules will occur at slightly lower field (smaller H_{ext}) than would be expected in the absence of spin-spin coupling.

The methyl resonance for half of the molecules will occur at the same chemical shift as would be expected in the absence of spin-spin coupling. These are the molecules in which the two methylene protons are oriented in opposite directions, cancelling one another, so that in this case in Equation 25.2-1, H_{coupling} has a value of zero.

In the remaining fourth of the molecules, the adjacent methylene protons will have their spins oriented so that both oppose H_{ext}, diminishing the effect of H_{ext} on the methyl groups of these molecules. These molecules will thus have their methyl resonances at a slightly higher H_{ext} than would be expected in the absence of spin-spin coupling; in this case in Equation 25.2-1, H_{coupling} is negative. In ethyl iodide, as in most ethyl groups, the magnitude of this effect, the spin-spin coupling constant J, is about 7 Hz or 7/60 of 1 ppm in a 60-MHz machine.

It is important to notice in this analysis that the multiplicity of the methyl resonance is interpreted in terms of the effect of the two *neighboring* protons. In this case, the protons of the methyl group have no apparent effect upon the multiplicity of their own resonance.

Interpretation of the Splitting of the CH$_2$ Resonance of CH$_3$CH$_2$I

The multiplicity of the resonance of the methylene group of ethyl iodide may be explained in a similar way. One-eighth of the molecules would be expected to have the protons of the adjacent methyl group all "with" H_{ext} ($\uparrow \uparrow \uparrow$), three-eighths would be expected to have two "with" and one "against" ($\frac{1}{8}\uparrow \uparrow \downarrow$, $\frac{1}{8}\uparrow \downarrow \uparrow$, and $\frac{1}{8}\downarrow \uparrow \uparrow$), three-eighths would be expected to have one "with" and two "against" ($\frac{1}{8} \uparrow \downarrow \downarrow$, $\frac{1}{8} \downarrow \uparrow \downarrow$, and $\frac{1}{8} \downarrow \downarrow \uparrow$), and one-eighth would be expected to have all three "against" ($\downarrow \downarrow \downarrow$). Thus, the molecules are divided into four

Splitting is due to neighbors

groups of relative population $1:3:3:1$ in which the methylene protons are expected to go into resonance at slightly different values of H_{ext} because of the effect of the adjacent methyl protons. Again, the multiplicity of the resonance of a group of protons is interpreted in terms of the effect of the adjacent protons; the multiplicity of a resonance does not depend upon the number of protons in the group corresponding to that resonance.

In general, when a simple analysis such as this applies, the number of peaks in a multiplet is one more than the number of adjacent protons; the "N + 1

N + 1 Rule

Rule": a single adjacent proton will split the resonance of a group of neighboring protons into a $1:1$ doublet; two adjacent protons will split the resonance of a group of neighboring protons into a $1:2:1$ triplet; etc. The relative intensities of the peaks of a multiplet follow the coefficients of the binomial expansion: $1:1$, $1:2:1$, $1:3:3:1$, $1:4:6:4:1$, $1:5:10:10:5:1$, etc.

Conditions for N + 1 Rule spectra

Magnetic equivalence

This type of analysis is appropriate *if two conditions are met*. First, the average coupling constant J between each proton in one group and each proton in the other must be the same, and second, the coupling constant J must be small relative to the chemical shift difference between the two groups, $\Delta\delta$. Magnetic equivalence must prevail.

Figures 25.5 through 25.10 present the NMR spectra of some compounds that can be readily interpreted according to this simple method of analysis. Figure 25.5 shows the spectrum of 1,1,2-trichloroethane, in which the methylene protons appear as an approximately $1:1$ doublet and the single proton as an

```
        Cl  H                      Cl  H  Cl
        |   |                      |   |   |
    H—C—C—Cl               H—C—C—C—H
        |   |                      |   |   |
        Cl  H                      Cl  Cl  Cl
  1,1,2-trichloroethane    1,1,2,3,3-pentachloropropane
```

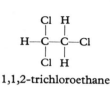

```
        Cl  H
        |   |
    H—C—C—Cl
        |   |
        Cl  H
  1,1,2-trichloroethane
```

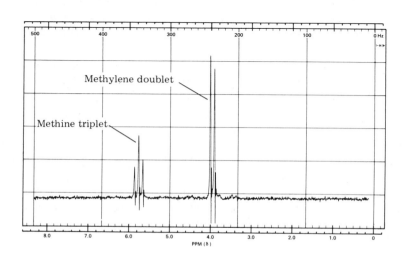

Figure 25.5. NMR spectrum of 1,1,2-trichloroethane; CCl$_4$ solution.

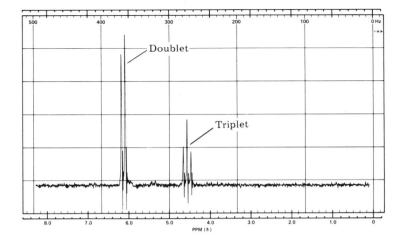

173

SECTION 25

1,1,2,3,3-pentachloropropane

Figure 25.6. NMR spectrum of 1,1,2,3,3-pentachloropropane; CCl_4 solution.

approximately 1:2:1 triplet. In Figure 25.6, the two equivalent end protons of 1,1,2,3,3-pentachloropropane appear as an approximately 1:1 doublet, and the middle proton appears as an approximately 1:2:1 triplet.

Figure 25.7 shows the NMR spectrum of 1,1-dichloroethane, in which the

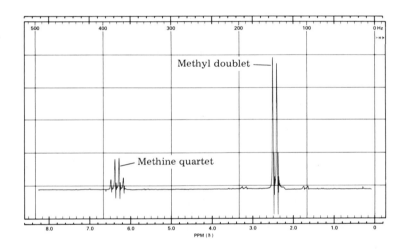

1,1-dichloroethane

Figure 25.7. NMR spectrum of 1,1-dichloroethane; neat.

methyl group appears as an approximately 1:1 doublet, and the single proton as an approximately 1:3:3:1 quartet. Figure 25.8 shows the NMR spectrum of ethyl chloride, which illustrates against the characteristic resonance of the ethyl group: an approximately 1:3:3:1 quartet for the methylene protons and an approximately 1:2:1 triplet for the methyl protons. Notice that the actual relative intensities of the peaks within a multiplet differ slightly from the predicted 1:1, 1:2:1,

1,1-dichloroethane

ethyl chloride

etc., relative intensities. The distortion is such that the stronger peaks are on the side nearer the resonance of the nuclei to which they are coupled. In this sense the peaks of a multiplet can be said to "point" toward the resonance of the nuclei with which they are coupled. Thus in Figure 25.8, the methylene quartet is said

"Pointing"

DETERMINATION OF
PHYSICAL PROPERTIES

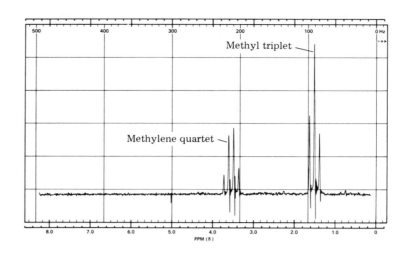

H H
| |
Cl—C—C—H
| |
H H

ethyl chloride

Figure 25.8. NMR spectrum of ethyl chloride; CCl_4 solution.

Methyl triplet

Methylene quartet

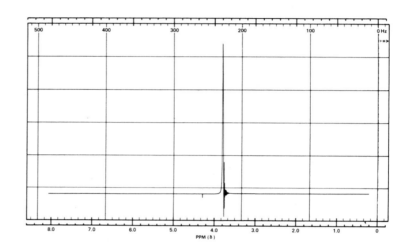

H H
| |
Cl—C—C—Cl
| |
H H

1,2-dichloroethane

Figure 25.9. NMR spectrum of 1,2-dichloroethane; neat.

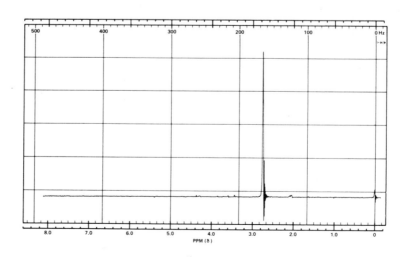

Cl H
| |
Cl—C—C—H
| |
Cl H

1,1,1-trichloroethane

Figure 25.10. NMR spectrum of 1,1,1-trichloroethane; CCl_4 solution.

to point toward the methyl resonance, and the methyl triplet is said to point toward the methylene quartet. By now you should expect that the NMR spectra
of 1,2-dichloroethane and 1,1,1-trichloroethane should each consist of a single
peak. They do; see Figures 25.9 and 25.10.

$$
\begin{array}{ccc}
\text{H} & \text{H} & \\
| & | & \\
\text{Cl}-\text{C}-\text{C}-\text{Cl} \\
| & | & \\
\text{H} & \text{H} & \\
\end{array}
\qquad
\begin{array}{ccc}
\text{Cl} & \text{H} & \\
| & | & \\
\text{Cl}-\text{C}-\text{C}-\text{H} \\
| & | & \\
\text{Cl} & \text{H} & \\
\end{array}
$$

1,2-dichloroethane   1,1,1-trichloroethane

You should also notice from this series of spectra that protons bonded to a
carbon bearing a chlorine atom are deshielded; two chlorine atoms appear to
deshield more than one. The dependence of shielding upon molecular structure
will be discussed in more detail in Section 25.4.

25.3 THE INTEGRAL

The third feature of the NMR spectrum from which information about the
molecular structure of the sample may be obtained is the area under the absorption peaks, or the *integral*. While chemical shift data provide information about
the number and type of functional groups in which sets of protons occur, and the
spin-spin splitting patterns indicate the structural and geometrical relationships
between protons, the total area under the peaks of a multiplet corresponding to
any set of protons provides information on the number of protons in that set: the
integral is proportional to the number of protons. The degree to which energy is
absorbed by a set of protons is independent of its structural environment.

The area under the peaks of an NMR spectrum is determined electronically by
use of the NMR spectrometer in a separate operation after obtaining the spectrum. Figure 25.11 shows the NMR spectrum of ethyl iodide and its integral.
The vertical displacement of the second (uppermost) trace is proportional to the
area under the corresponding peaks. The total area of the methylene quartet is
seen to be two-thirds that of the methyl triplet. You can also estimate in the
spectra of Figures 25.4 through 25.8 that the relative area of each multiplet corresponds to the relative number of protons involved.

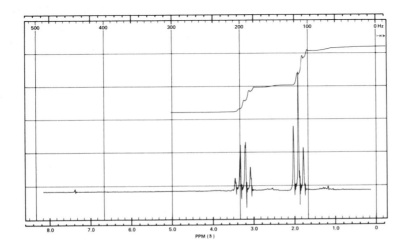

$$
\begin{array}{ccc}
\text{H} & \text{H} & \\
| & | & \\
\text{I}-\text{C}-\text{C}-\text{H} \\
| & | & \\
\text{H} & \text{H} & \\
\end{array}
$$

ethyl iodide

Figure 25.11. NMR
spectrum and integral of
ethyl iodide; neat.

25.4 NUCLEAR MAGNETIC RESONANCE AND MOLECULAR STRUCTURE

Structure and chemical shift

The amount of energy absorbed by a proton is independent of its structural environment, and therefore the integral has the same value for every proton. In contrast, the shielding or chemical shift (shielding relative to a standard) does depend upon the structural environment of a proton. Table 25.2 presents some of the correlations between chemical shift and molecular structure. From the table, you can see two trends. (1) Methyl protons are more shielded than analogous methylene protons, which, in turn, are more shielded than methine protons. (2) Electron-withdrawing groups tend to deshield adjacent protons.

Structure and coupling constants

Coupling constants are related in a very sensitive way to the structural and geometrical relationships between nuclei. The functional relationship is illustrated in Table 25.3. The table shows a few correlations between values of coupling constants and the relationships between nuclei. A study of the table begins to suggest that, in general, coupling does not extend over more than three bonds, but that sp^2 hybridized atoms are better "conductors" of coupling than are sp^3.

25.5 INTERPRETATION OF NMR SPECTRA

If the NMR spectrum is that of a pure substance of unknown molecular structure, the goal of interpretation is to decide what possible molecular structures could account for the spectrum. In contrast to the case with infrared spectroscopy, it is sometimes possible to deduce the structural formula of a substance from its molecular formula and NMR spectrum alone, without the use of reference spectra. For example, either isomer of molecular formula $C_2H_4Cl_2$ can be distinguished on the basis of its NMR spectrum alone (Figures 25.7 and 25.9).

The steps of such an interpretation are as follows.

1. Determine the relative number of protons responsible for each peak or set of peaks (multiplet) from the integral. If the molecular formula is known, then the absolute number of protons can be calculated.

Table 25.2. Correlations between chemical shift and molecular structure

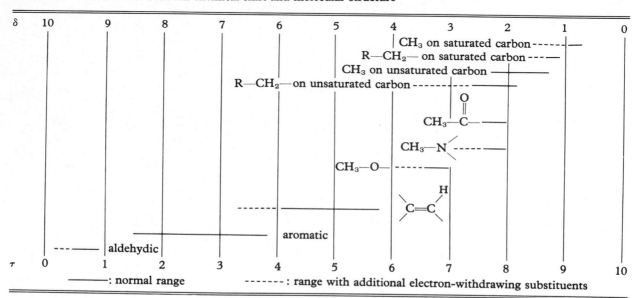

2. Determine the chemical shift of each multiplet. The spectrum must have been determined with TMS added so that the difference between the shielding of the sample peaks or multiplets and the TMS resonance can be read from the scale on the chart. When the spectrum is run, the instrument is normally adjusted so that the TMS resonance falls on the zero of the δ scale (10 on the τ scale).

3. Determine the coupling constant between nuclei in different sets at different chemical shifts. In the examples illustrated by the figures up to this point, the spacing between peaks of a multiplet equals the coupling constant.

4. The final and most difficult step is to think of a molecular structure that accounts for the information determined in steps 1, 2, and 3. Correlation tables such as those illustrated in Tables 25.2 and 25.3 help out here.

Unfortunately, few spectra will be as easy to interpret as the examples that have been presented. The first reason is that in the examples the chemical shift difference was always large compared to the coupling constant. Figure 25.12

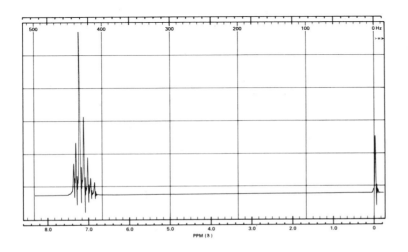

1,2,3-trichlorobenzene

Figure 25.12. NMR spectrum of 1,2,3-trichlorobenzene; CCl_4 solution.

Table 25.3. Correlations between coupling constants and molecular structure

J(Hz)		J(Hz)	
H—C—C—H	~7	$C=C$ with H and $C—H$	~7
H—C—C—C—H	~0	$C=C$ with H and $C—H$	~2
$C=C$ with H and H	0–2	$C=C$ with H and $C—H$	~1.5
$C=C$ with H and H (cis)	~10	H—C—C(=O)—H	2–3
$C=C$ with H and H (trans)	~17	benzene ring H	o ~9 m ~3 p ~0

shows the NMR spectrum of 1,2,3-trichlorobenzene, which, according to the simple analysis, should show a 1:1 doublet for the B protons, and a 1:2:1 triplet for the A proton. The NMR spectrum actually consists of seven lines.

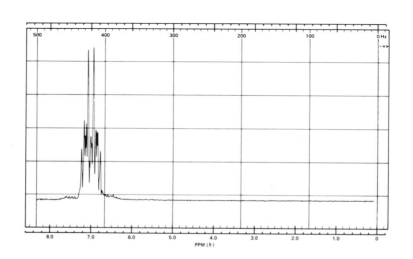

1,2,3-trichlorobenzene   1-chloro-4-bromobenzene

The second reason for the difficulty in interpreting spectra is that in the examples described so far, the coupling constants between each proton in one set and every proton in the other set were exactly equal (the interset coupling constants were equal; the two sets of chemical shift equivalent nuclei were magnetically equivalent). Figure 25.13 shows the NMR spectrum of 1-chloro-4-bromobenzene. Here, the interset coupling constants are not equal: $J_{AB(ortho)}$ is not equal to $J_{AB(para)}$.

1-chloro-4-bromobenzene

Figure 25.13. NMR spectrum of 1-chloro-4-bromobenzene; CCl₄ solution.

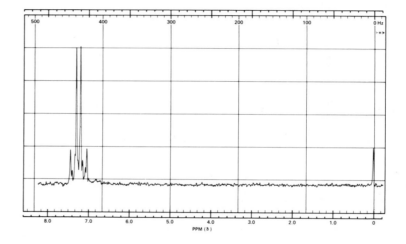

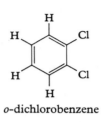

o-dichlorobenzene

Figure 25.14. NMR spectrum of o-dichlorobenzene; neat.

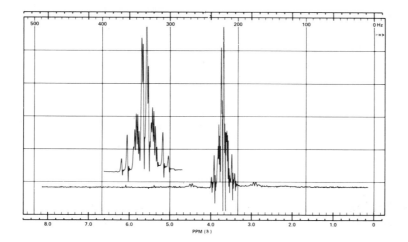

$$Br-\underset{\underset{H}{|}}{\overset{\overset{H}{|}}{C}}-\underset{\underset{H}{|}}{\overset{\overset{H}{|}}{C}}-Cl$$

1-bromo-2-chloroethane

Figure 25.15. NMR spectrum of 1-bromo-2-chloroethane; neat.

Similarly, as shown in Figures 25.14 and 25.15, the spectra of *o*-dichlorobenzene and 1-bromo-2-chloroethane are complex because of the inequality of the interset coupling constants. The analysis of the NMR spectra of compounds such as these in which magnetic nonequivalence is involved is discussed in References 1 and 2.

Magnetic nonequivalence

o-dichlorobenzene   1-bromo-2-chloroethane

Reference 3 considers the concepts presented in this section in more detail, and discusses the interpretation of the proton NMR spectra of a number of additional compounds of known structure.

25.6 SAMPLE PREPARATION

Nuclear magnetic resonance spectra are determined on pure liquids or solutions. Carbon tetrachloride is a very good NMR solvent in that it contains no protons itself. However, some compounds are not very soluble in it. Chloroform ($CHCl_3$) is a better solvent for many substances than is carbon tetrachloride, but it contains a proton whose resonance may obscure some of those of interest in the sample. Deuterochloroform ($CDCl_3$) has the solvent properties of chloroform, and the deuterium resonance is far from the resonances of the protons. For routine work where solubility is not a problem, carbon tetrachloride should be used. If the compound is not very soluble in carbon tetrachloride, deuterochloroform should be tried. Other deuterated solvents such as hexadeuteroacetone are available but expensive. Water-soluble samples can be run in deuterium oxide (D_2O).

Solvent

Deuterochloroform

Sample concentration should be about 10–30% by weight or volume. A total volume of 0.5 ml is all that is needed.

Concentration

The spectrum is determined with the sample in an *NMR tube*, a thin-walled glass tube 5 mm in diameter and 7 in long, sealed at the bottom. A tight-fitting plastic cap closes the top of the tube.

NMR tube

The reference standard, TMS, must be added if it is not already present (1%) in the NMR solvent as purchased. Since TMS boils at 27°C, it is best added by means of a microsyringe that has been stored in the freezer. Only about 5–10 microliters is needed. Recap the NMR tube immediately after adding the TMS; invert the tube several times to mix. Before attempting to insert the sample tube into the probe of the instrument, wipe the outside clean and insert the tube into the white Teflon turbine ("spinner"), which allows the tube to be spun in the probe. Position the tube in the spinner with the depth gauge, and then place the tube and spinner in the probe.

Problems

1. Which isomer will give a single peak in its NMR spectrum?
   a. 1,1-Dichloroethane or 1,2-dichloroethane?
   b. Cyclobutane or methylcyclopropane?
   c. CH_3—O—CH_3 or CH_3CH_2—O—H?
   d. *ortho*-, *meta*- or *para*-dichlorobenzene?
   e. Isomers of molecular formula C_5H_{12}?
   f. Isomers of molecular formula C_4H_9Br?

2. Predict the appearance of the NMR spectra of the following compounds:

   a. CH_3—$\overset{\overset{\displaystyle O}{\|}}{C}$—$CH_3$ (acetone)

   b. CH_3—O—$\overset{\overset{\displaystyle O}{\|}}{C}$—O—$CH_3$ (dimethylcarbonate)

   c. CH_3—$\overset{\overset{\displaystyle O}{\|}}{C}$—O—$CH_3$ (methyl acetate)

   d. CH_3CH_2—O—CH_2CH_3 (diethyl ether)

   e. CH_3—$\overset{\overset{\displaystyle O}{\|}}{C}$—$CH_2CH_3$ (2-butanone)

   f. CH_3—$\overset{\overset{\displaystyle O}{\|}}{C}$—O—$CH_2CH_3$ (ethyl acetate)   and

   CH_3—O—$\overset{\overset{\displaystyle O}{\|}}{C}$—$CH_2CH_3$ (methyl propionate)

3. Figure 25.16 shows the NMR spectrum of a compound of molecular formula $C_6H_{12}O_2$. What molecular structure is most consistent with the spectrum?

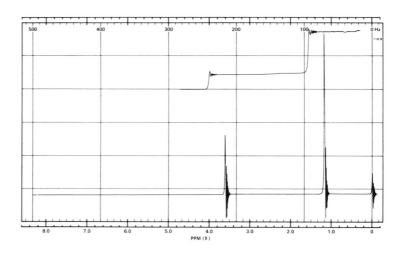

Figure 25.16. NMR spectrum of a compound of molecular formula $C_6H_{12}O_2$; CCl_4 solution.

4. Figure 25.17 shows the NMR spectrum of a compound of molecular formula $C_7H_{14}O$. What molecular structure is most consistent with the spectrum?

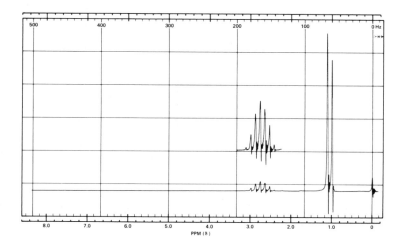

Figure 25.17. NMR spectrum of a compound of molecular formula $C_7H_{14}O$; CCl_4 solution.

5. Figure 25.18 shows the NMR spectrum of a compound of molecular formula C_5H_8. What molecular structure is most consistent with the spectrum?

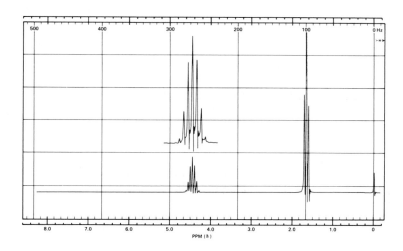

Figure 25.18. NMR spectrum of a compound of molecular formula C_5H_8; CCl_4 solution.

6. Figure 25.19 shows the NMR spectrum of a compound of molecular formula $C_6H_{14}O_2$. What molecular structure is most consistent with the spectrum?

7. Figure 25.20 is the NMR spectrum of a compound of molecular formula $C_{10}H_{12}O$. The portion of spectrum at the left is a continuation of the rest of the spectrum. The inset is an amplification of the resonance at $\delta = 3$. The relative areas under the multiplets are as follows:

δ	Relative Area
1.15	6
3.0	1
7.55	4
9.95	1

What molecular structure is most consistent with the spectrum?

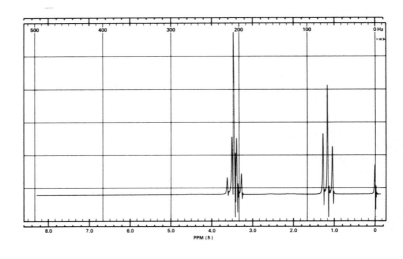

Figure 25.19. NMR spectrum of a compound of molecular formula $C_6H_{14}O_2$; CCl_4 solution.

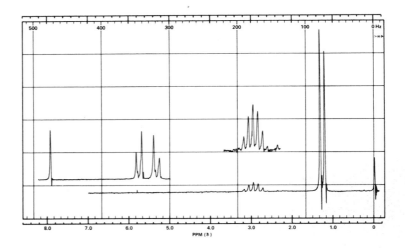

Figure 25.20. NMR spectrum of a compound of molecular formula $C_{10}H_{12}O$; CCl_4 solution.

8. Figure 25.21 shows the NMR spectrum of a compound of molecular formula $C_{11}H_{14}O$. The relative areas under the peaks are as follows:

δ	Relative Area
2.15	6
2.20	3
2.30	3
6.70	2

What molecular structure is most consistent with the spectrum?

9. Figure 25.22 shows the NMR spectrum of a compound of molecular formula $C_{11}H_{20}O_4$. What molecular structure is most consistent with the spectrum?

10. Figure 61.4 shows the NMR spectrum of 2,4-dinitrobromobenzene. The doublet on the left is the resonance of one hydrogen; the pair of doublets in the middle is the resonance of a second hydrogen; and the doublet on the right is the resonance of the third hydrogen. Assuming that the spacings between lines of a multiplet are equal to coupling constants involving neighboring hydrogens, decide which hydrogen is responsible for each multiplet in the spectrum. Typical coupling constants between hydrogens *ortho*, *meta*, and *para* to each other are given in Table 25.3.

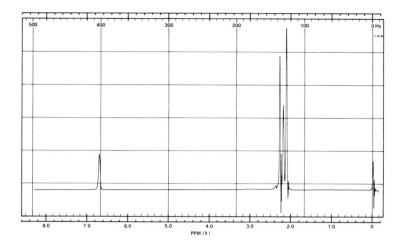

Figure 25.21. NMR spectrum of a compound of molecular formula $C_{11}H_{14}O$; CCl_4 solution.

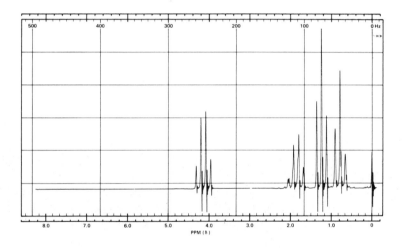

Figure 25.22. NMR spectrum of a compound of molecular formula $C_{11}H_{20}O_4$; CCl_4 solution.

References

Introductory texts:

1. R. M. Silverstein, G. C. Bassler, and T. C. Morrill, *Spectrometric Identification of Organic Compounds*, 3rd edition, Wiley, New York, 1974, pp. 159–199.

2. J. R. Dyer, *Applications of Absorption Spectroscopy of Organic Compounds*, Prentice-Hall, Englewood Cliffs, N.J., 1969, pp. 58–132.

3. A. Ault and G. O. Dudek, *An Introduction to Proton Nuclear Magnetic Resonance Spectroscopy*, Holden-Day, San Francisco, 1976.

Collections of spectra:

4. *High Resolution NMR Spectra Catalog*, Varian Associates, Palo Alto, California; Vol. 1, 1962; Vol. 2, 1963.

5. C. J. Pouchert and J. R. Campbell, *The Aldrich Library of NMR Spectra*, Aldrich Chemical Co., Inc., Milwaukee, Wisconsin; ten volumes plus index; 1974.

$$\boxed{\S 26 \text{ MASS SPECTROMETRY}}$$

26.1 THEORY OF MASS SPECTROMETRY

Mass spectrometry involves bombarding a vaporized sample of a substance with a beam of electrons and determining the relative abundance of the resulting positively charged molecular fragments. The analysis is made in terms of the mass-to-charge ratio of the various fragments, but since almost all fragments will bear a single positive charge, the mass-to-charge ratio will be numerically equal *Mass spectrum* to the mass of the fragment. The record of relative abundance versus mass-to-charge ratio that is produced from the substance by the mass spectrometer is called the *mass spectrum* of the substance.

The production of positively charged ionic fragments of various masses may be thought to occur as follows. An electron from the electron beam knocks an *Ionization* electron from a molecule to give the parent ion, or molecular ion, $M \cdot^+$ (which must now have an unpaired electron):

$$M + e^- \longrightarrow M \cdot^+ + 2e^- \qquad (26.1\text{-}1)$$

The molecular ion may fall apart very quickly (within 10^{-8} to 10^{-10} second) to give one charged and one uncharged fragment:

$$M \cdot^+ \longrightarrow A^+ + B \cdot \quad \text{and/or} \quad A \cdot + B^+ \qquad (26.1\text{-}2)$$

$$M \cdot^+ \longrightarrow C^+ + D \cdot \quad \text{and/or} \quad C \cdot + D^+, \quad \text{etc.} \qquad (26.1\text{-}3)$$

The charged fragmentation products may themselves undergo further fragmenta-*Fragmentation* tion to give smaller charged and neutral pieces. The relative abundance of positively charged fragments of various mass-to-charge ratios is characteristic of the molecules of a substance and may serve to identify the substance.

The relative abundance of the positively charged ions of various mass-to-*Acceleration* charge ratios (m/e) is determined by accelerating the ions by means of an electrostatic field. The kinetic energy acquired by each ion in this process is equal to eV, where e is the charge on the particle and V is the potential difference through which the ion has fallen:

$$\text{Kinetic energy} = eV = \tfrac{1}{2}mv^2 \qquad (26.1\text{-}4)$$

where m is the mass of the ion, v is its velocity, and "kinetic energy" is the kinetic energy gained upon acceleration in the electrostatic field. Solving for v^2 gives:

$$v^2 = \frac{2eV}{m} \qquad (26.1\text{-}5)$$

The accelerated ions then move into a magnetic field whose direction is perpendicular to their path. The action of the magnetic field is to force the ions to follow a circular path: the force of the magnetic field is a centripetal force. Since the magnitude of the force on an ion equals Hev (where H is the strength of the magnetic field, e is the charge of the ion, and v is its velocity) and the centripetal acceleration is given by v^2/r (where v is the tangential velocity and r

is the radius of the path) we can write, according to Newton's second law, the relationship:

$$Hev = m\frac{v^2}{r} \qquad (26.1\text{-}6)$$

Solving for v and squaring gives:

$$v^2 = \frac{H^2e^2r^2}{m^2} \qquad (26.1\text{-}7)$$

Substituting into this equation the value of v^2 resulting from the acceleration by the electrostatic field (Equation 26.1-5) will give

$$\frac{2eV}{m} = \frac{H^2e^2r^2}{m^2} \qquad (26.1\text{-}8)$$

and solving for m/e gives

$$\frac{m}{e} = \frac{H^2r^2}{2V} \qquad (26.1\text{-}9)$$

Thus, for constant electrostatic and magnetic fields V and H, fragments of different m/e will be separated because they will follow paths of different radii in the magnetic field. A detector that can determine the relative numbers of ions that have followed paths of different radii can thus provide the desired record of relative abundance of fragments of different m/e. An alternative often used is to position a detector so that it can continuously record the intensity of the beam of particles that follows a path of a given fixed radius and vary either the magnetic field H or the electrostatic field V in such a way that ions of increasing m/e will fall in succession upon the detector. A record of the output of the detector versus H^2 or $1/V$ is equivalent to a record of relative abundance versus m/e. Reference 1 gives a brief description of how a mass spectrometer is constructed so that it will perform these functions.

Detection

26.2 INTERPRETATION OF MASS SPECTRA

Three types of information may be obtained from a mass spectrum: the molecular weight of a substance, the molecular formula of the substance, and the molecular structure of the substance.

Molecular Weight Determination

From what has been previously said, you would expect that the mass corresponding to the largest observed m/e of the mass spectrum corresponds also to the molecular ion M^+ and therefore gives the molecular weight of the substance. This is essentially true, but with two important qualifications. The first is that if most of the initially formed molecular ions undergo fragmentation within 10^{-10} second, there may not be enough ions left to accelerate and detect; the peak corresponding to the molecular ion (the parent peak, P) may not be observable. This is not often the case, but the possibility must be kept in mind.

Parent peak

The second qualification is best illustrated by means of an example. In the mass spectrum of carbon monoxide, a peak at $m/e = 29$ appears with about 1.12 percent of the intensity of the parent peak P, corresponding to the molecular ion at $m/e = 28$. The reason for this is that carbon from natural sources contains an amount of ^{13}C equal to 1.08 percent of the amount of naturally occurring ^{12}C, and that oxygen from natural sources contains an amount of ^{17}O equal to 0.04 percent of the amount of ^{16}O. The peak at $m/e = 29$ is due to the presence of $^{13}C^{16}O$ and $^{12}C^{17}O$ isotopes of carbon monoxide, which are $1.08\% + 0.04\% = 1.12\%$ as abundant as $^{12}C^{16}O$. Thus, the molecular weight of a substance may be determined from its mass spectrum in those cases in which the molecular ion is not too unstable, by identifying the peak of highest m/e (the parent ion, P), not counting peaks due to molecules containing relatively small amounts of heavier isotopic atoms (P + 1, P + 2, etc.). The molecular weight obtained from a mass spectrum is the exact molecular weight, not an approximate one as is the case with molecular weights determined from freezing-point depression or titration, for example.

Molecular Formula Determination

The fact that substances from natural sources are composed of a certain calculable fraction of molecules containing atoms of heavier isotopes makes it possible in many cases to eliminate all but a few of the possible molecular formulas that correspond to a given molecular weight, on the basis of the intensity of the P + 1 and P + 2 peaks relative to the molecular ion peak P. For example, as previously explained, for carbon monoxide you would expect a P + 1 peak at $m/e = 29$ that would be 1.12 percent as intense as the parent peak P at $m/e = 28$. Also, since ^{18}O is 0.20 percent as abundant as ^{16}O, a P + 2 peak due to $^{16}C^{18}O$ that would be 0.2 percent as intense as P would be expected. ($^{13}C^{17}O$ is present only to the extent of $0.0108 \times 0.0004 = 0.000432\%$ of the amount of $^{12}C^{16}O$; the amount of $^{14}C^{16}O$ present is also negligible.) On the other hand, molecular nitrogen, which also has a nominal molecular weight of 28, would be expected to have a P + 1 peak at $m/e = 29$ that would be 0.76 percent as intense as P, since ^{15}N occurs 0.38 percent as abundantly as ^{14}N ($0.38\% \times 2$ atoms of N $= 0.76\%$). And the P + 2 peak at $m/e = 30$ would be expected to be $0.0038 \times 0.0038 = 0.001444\%$ as intense as P. If the experimental uncertainty is sufficiently small, the intensities of P + 1 and P + 2 relative to P will allow the assignment of CO or N_2 as the molecular formula of a sample of a gas whose parent peak appears at $m/e = 28$.

Tables of calculated intensities of P + 1 and P + 2 relative to P for many combinations of C, H, O, and N for integral values of P have been prepared (Reference 1). With the help of these tables, it is possible to reduce the number of possible molecular formulas corresponding to a particular molecular weight to a very few, as long as accurate intensities of P + 1 and P + 2 relative to P can be obtained from the mass spectrum. If the molecular ion is fairly unstable, the intensities of P, P + 1, and P + 2 may be too low to be measured with the necessary accuracy. The variations of this kind of analysis that must be made if the substance contains elements other than C, H, O, and N are described in Reference 1.

Molecular Structure Determination

Since the relative intensities of the various positive ions produced upon electron bombardment in the mass spectrometer (fragmentation patterns) are different for substances of different molecular structure, it must be possible, at least in principle, to deduce the molecular structure of a substance from its mass spectral fragmentation pattern. Comparison of the mass spectra of many substances containing different functional groups has made it possible to rationalize many

features of the fragmentation patterns in terms of the functional groups present
in the molecule. Conversely, then, it is possible to infer the presence of certain

187

SECTION 26

functional groups or structural features from the fragmentation patterns. The
subject of the correlations between molecular structure and fragmentation patterns
is very large and can only be hinted at in this brief discussion. Reference 1 is an
excellent introduction.

26.3 TIME-OF-FLIGHT MASS SPECTROMETRY

From Equation 26.1-4, you can see that since eV must be the same for all molecular fragments with a single positive charge, the heavier ions will be moving more
slowly than the lighter ones after acceleration by the electrostatic field V. Rather
than sorting the fragments according to mass by means of a magnetic field, the
time-of-flight mass spectrometer allows them to pass into a tube in which they
can drift in the absence of any magnetic or electrostatic field. If the ionizing
electron beam has been turned on for a brief moment only, a detector at the end
of the tube will receive the resulting charged fragments in order of decreasing
velocity or increasing mass. A record of detector output versus time is thus the
desired record of relative abundance versus mass. With this technique, successive
spectra may be obtained at a rate of about 1000 per second. The output of the
detector is often displayed on a cathode ray tube, and may be photographed.

26.4 HIGH-RESOLUTION MASS SPECTROMETRY

So far, it has been assumed that mass spectrometers can distinguish between ions
that differ in mass by one atomic mass unit (unit-resolution mass spectrometry).
Instruments are available, at a high price, which can resolve masses to approximately 0.001 atomic mass unit (high-resolution mass spectrometry). With this
kind of resolution, it would be possible to distinguish between $^{12}C^{16}O$ (molecular
weight = 27.9949) and $^{14}N^{14}N$ (molecular weight = 28.0062) without reference
to the relative intensities of P + 1 and P + 2. In fact, the molecular formula of
every fragment in the mass spectrum can be determined, if its mass can be
measured to about 0.001 atomic mass unit.

Problems

1. For each case, explain why the parent peak will always be either odd or even. If
   the parent peak can be either odd or even, explain how this can be so.
   a. The compound contains carbon and hydrogen.
   b. The compound contains carbon, hydrogen, and oxygen.
   c. The compound contains carbon, hydrogen, and fluorine.
   d. The compound contains carbon, hydrogen, and nitrogen.
2. Bromine is made up of almost equal parts of ^{79}Br and ^{81}Br.
   a. Taking the parent peak of CH_3Br to be 94 (or 12 + 3 + 79), estimate the
   intensity relative to the parent peak of P + 1, P + 2, etc.
   b. Taking the parent peak of CH_2Br_2 to be 172 (or 12 + 2 + 79 + 79), estimate
   the intensity relative to the parent peak of P + 1, P + 2, etc.
   c. Do the same for $CHBr_3$ and CBr_4.
   d. In what simple way might the number of bromine atoms per molecule in a
   compound containing bromine be determined from its mass spectrum?
3. Chlorine is made up of ^{35}Cl and ^{37}Cl in a ratio of about 3 to 1. Interpret the mass
   spectrum of carbon tetrachloride, shown in Figure 26.1. The parent peak of
   carbon tetrachloride fragments very rapidly.

DETERMINATION OF
PHYSICAL PROPERTIES

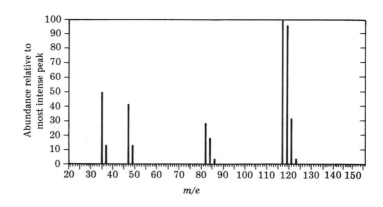

Figure 26.1. Mass spectrum of CCl₄.

4. The mass spectrum shown in Figure 26.2 is that of either methyl *p*-hydroxybenzoate or *p*-methoxybenzoic acid. Which substance can account for the spectrum? Explain.

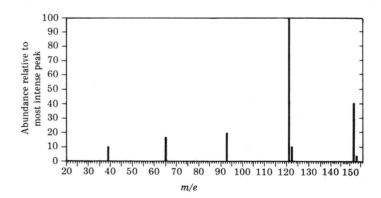

methyl *p*-hydroxybenzoate   *p*-methoxybenzoic acid

Figure 26.2.

References

1. R. M. Silverstein, G. C. Bassler, and T. C. Morrill, *Spectrometric Identification of Organic Compounds*, 3rd edition, Wiley, New York, 1974, pp. 5–71.

2. D. J. Pasto and C. R. Johnson, *Organic Structure Determination*, Prentice-Hall, Englewood Cliffs, N.J., 1969, pp. 243–294.

Determination of Chemical Properties; Qualitative Organic Analysis

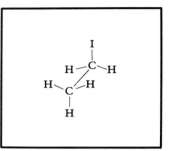

As with physical properties, chemical properties of substances can be used both for the characterization of substances and for the determination of the molecular structures of substances.

In many cases, the presence or absence of functional groups can be determined in a routine way. Since the advent of infrared spectroscopy, chemical methods for the detection of functional groups are used less often. For that reason, procedures given in the following sections have been chosen mostly to complement the information that can be obtained from a routine infrared spectrum.

If it appears that a sample may be identical with a substance of known structure, identity may often be established by preparing a solid derivative and comparing its melting point with that of the same derivative of the compound of known structure (Section 29). In most cases now, such comparisons are made between infrared and nuclear magnetic resonance spectra of substances of known structure.

A part of many organic laboratory programs is often devoted to qualitative organic analysis. *When the time comes to do an analysis of this sort, you will be given a small sample of one or more substances whose molecular structure is not known to you* (an unknown) *and you will be asked to identify or determine the structure of the compounds. While the best strategy must necessarily be different for each substance, there are several things that can be done routinely with any unknown. For liquids, one usually determines the boiling point (by distillation or by the micro method; Section 16), the density (Section 18), the index of refraction (Section 19), and the solubility behavior—and whether or not the compound is soluble in water, ether, 5% NaOH, 5% NaHCO$_3$, 5% HCl, conc. H$_2$SO$_4$, or 85% H$_3$PO$_4$ (Section 22). For solids, one normally determines the melting point (Section 17) and the solubility behavior. Other physical properties that can be determined routinely include the infrared spectrum (Section 23) and, if possible, the NMR spectrum (Section 25).*

Fusion with sodium followed by qualitative tests for nitrogen (as cyanide), sulfur (as sulfide), and the halogens (Section 27.3) often provides

Unknowns

189

important information, especially when positive indications are obtained.

Since time is always in short supply, qualitative tests for functional groups (Section 28) should be chosen with care so that the information obtained will not merely duplicate that which is already in hand. However, it is often a good idea to confirm tentative conclusions that have been based on other evidence.

If it is not possible to compare IR or NMR spectra of the unknown with those of an authentic sample it is usually necessary to prepare one or more solid derivatives (Section 29) in order to confirm an identification. Preparing derivatives is also a very good way to develop chemical judgment and laboratory technique.

§27 QUALITATIVE TESTS FOR THE ELEMENTS

A common first step in the identification of an unknown substance is to determine what elements (other than carbon, hydrogen, and possibly oxygen) are present in the sample. In this section qualitative tests are described for the detection of the presence of metals, nitrogen, sulfur, and the halogens.

27.1 IGNITION TEST; TEST FOR METALS

Procedure. Place 50–100 mg of the substance on a porcelain crucible cover and heat the cover with a flame. Heat gently at first, and then finally heat the cover to dull redness. If a metal was present in the sample, a nonvolatile residue will remain. A drop of water added to the residue will give a basic reaction with litmus.

While this test is primarily aimed at determining whether a solid is a metal salt or other metal-containing substance, other information may be obtained as well. If the substance burns with a fairly sooty flame, a high ratio of carbon to hydrogen is indicated (an aromatic rather than aliphatic compound is indicated). If the substance burns with a blue rather than a yellow flame, a high ratio of oxygen to other atoms is indicated. Certain types of compounds may explode under these conditions (see Section 1.2).

Warning

27.2 BEILSTEIN TEST; TEST FOR HALOGENS (EXCEPT FLUORINE)

Procedure. Form a small coil at one end of a 20-cm length of stiff copper wire by making a few turns around a glass rod 2 or 3 mm in diameter. Heat the coil in the oxidizing part of the flame of a Bunsen burner until it does not color the flame any longer. Allow the wire to cool (by waving it in the air) and then dip the coil into the material to be tested and heat it again in the same part of the flame. If the substance contains chlorine, bromine, or iodine, a transient green or blue-green color will be imparted to the flame; the color is due to copper that has been made volatile by combination with the halogen of the sample.

The Beilstein test is very sensitive. It should be confirmed by means of the sodium fusion test (Section 27.3) and the tests of Sections 28.11 and 28.13 to make sure that the effect was not due to the presence of a halogen-containing impurity —or to verify that the substance is not one that gives a positive test even though it contains no halogen. (Unfortunately, the test is not specific since some organic nitrogen compounds that do not contain halogen also give the effect.)

In the sodium fusion test, a substance is heated with sodium under conditions that ensure the conversion of any nitrogen to cyanide ion, any sulfur to sulfide ion, and any halogens to the corresponding halide ions. The presence or absence of these ions in the solution that results from the fusion (the *fusion solution*) indicates the presence or absence of these elements in the sample.

Sodium Fusion

Procedure.   Support a test tube 8 mm in diameter by 50 mm long by its lip in a hole in an asbestos board or sheet which is in turn supported on an iron ring. Add to the test tube about 40 mg of sodium (Note 1) and heat the test tube with a burner until the sodium has melted and the grey vapors rise about 2 cm in the tube. Adjust the rate of heating, by adjusting the height of the burner flame or the height of the iron ring, so that this condition can be maintained. Then add, in two or three portions, 50 mg (or 2 or 3 small drops) of the compound over about a 30-second interval (Note 2). After the addition is complete, turn up the burner and heat the tube as strongly as possible until the lower 1 or 2 cm is red to orange-red hot; heat the tube at this rate for about 2 minutes (Note 3). Now, holding the asbestos sheet or board with a gloved hand or tongs, lower the red-hot tube into a *Caution* small beaker (obviously no more than 50 mm high) containing about 10 ml of distilled water (Note 4). The tube will shatter, and any residual sodium will react with the water. Boil the remains of the test tube with the water and filter by gravity. The clear, almost colorless filtrate (fusion solution) will be used in the tests that follow (Note 5).

Notes

1. For precautions to be taken in handling sodium, see Section 1.7. A piece the size of half a medium pea, or a 4-mm cube, will be satisfactory.
2. Safety glasses must be worn; nitro compounds, azo compounds, and poly-halogen compounds such as carbon tetrachloride or chloroform may cause a slight explosion. Try to drop the sample straight down, without letting it hit the side of the test tube, where it may simply evaporate.
3. It is not possible to heat too strongly. Some burners are too feeble; make sure that yours can give a strong, blue, nonluminous flame.
4. Safety glasses must be worn; hold the asbestos board at arm's length; don't aim the open top of the tube at anyone.
5. If the filtrate is more than just slightly yellow, it will usually be a waste of time to continue. Better do the procedure again and heat more strongly.

Tests for the Elements: Test for Nitrogen; Test for Cyanide Ion (Test a)

Procedure.   Add 2 or 3 ml of the fusion solution to a test tube containing 100–200 mg of powdered ferrous sulfate crystals. Heat the mixture cautiously and with shaking until it boils (Note 1). Then, without cooling, add just enough dilute sulfuric acid to dissolve the gelatinous hydroxides of iron. The appearance of a blue precipitate of ferric ferrocyanide, $Fe_4[Fe(CN)_6]_3$ (Prussian Blue), indicates that cyanide ion is present and, therefore, that the original compound contained nitrogen (Note 2). If no cyanide ion is present, the solution should be a pale yellow color (Note 3).

Notes

1. It is often recommended that 2 drops of a 30% aqueous solution of potassium fluoride be added before boiling. If sulfide ion is present, a precipitate of black ferrous sulfide may appear; it will dissolve upon acidification. Ferric ions are produced by air oxidation during boiling.

2. If a precipitate is not immediately apparent, allow the mixture to stand for 15 minutes and then filter it. After washing the filter paper with a little water to remove all the colored solution, any Prussian Blue present should be visible on the paper. If there is still doubt as to whether a precipitate of Prussian Blue was formed, another sodium fusion should be carried out and the test repeated.

3. A positive test for cyanide is good evidence for the presence of nitrogen in the original sample, but the presence of nitrogen should never be ruled out solely on the basis of a negative test here for cyanide ion. You should make sure that your technique is adequate by carrying out the sodium fusion and the test for nitrogen on a sample known to contain nitrogen. Nitrogen in a high oxidation state, for example as a nitro group, is especially difficult to detect.

Test for Sulfur; Test for Sulfide Ion (Test b)

Procedure. Acidify about 2 ml of the fusion solution with dilute acetic acid and then add a few drops of approximately 1% lead acetate solution. A black precipitate of lead sulfide indicates that sulfide ion is present and, therefore, that sulfur was present in the original sample.

Alternatively, 2 or 3 drops of a freshly prepared 0.1% solution of sodium nitroprusside (Note 1) may be added to 2 ml of the fusion solution. A purple coloration indicates the presence of sulfide ion.

Note

1. This may be prepared by dissolving a tiny crystal of sodium nitroprusside ($Na_2[Fe(CN_5)NO]$) in 1 or 2 ml of water.

Test for Halogen; Test for Halide Ion (Test c)

Procedure. Acidify 2 ml of the fusion solution with dilute nitric acid (Note 1) and add a few drops of 5% aqueous silver nitrate solution. A heavy precipitate indicates the presence of chloride, bromide, or iodide ion, or any combination of the three.

If only a single halide should be present, you may be able to distinguish among the possibilities since silver chloride is white (slightly purplish due to liberation of free silver by light) and is easily soluble in dilute ammonium hydroxide, silver bromide is slightly yellow and is only slightly soluble in ammonium hydroxide, and silver iodide is definitely yellow and is insoluble in ammonium hydroxide. To determine the solubility of the precipitate in dilute ammonium hydroxide, remove the supernatant liquid by centrifugation and decantation, add 2 ml of dilute ammonium hydroxide solution, and stir the mixture to see if the solid will dissolve.

Note

1. If either cyanide or sulfide is present (tests a and b above), the acidified solution must be boiled gently for a few minutes in order to expel hydrogen cyanide and hydrogen sulfide.

Test for Iodine; Test for Iodide Ion (Test d)

Procedure. Acidify 1 or 2 ml of the fusion solution with dilute sulfuric acid (Note 1). After cooling, add 1 ml of carbon tetrachloride; and then add dropwise with good mixing a solution of chlorine water (Note 2). Iodide ion, if present, will be oxidized to free iodine under these conditions and will be extracted into the carbon tetrachloride to give a purple color (Note 3).

Notes

1. If cyanide or sulfide ion is present (tests a and b above), acidify the sample of fusion solution with dilute nitric acid, boil the mixture until its volume has been reduced by one-half in order to expel hydrogen cyanide and hydrogen sulfide, and dilute the resulting solution with an equal volume of distilled water.
2. A stabilized solution of sodium hypochlorite such as Chlorox may be used; make sure that the test solution remains acidic to litmus by adding acid as necessary. If this type of commercial bleach solution is used, a blank should be run since often a yellow-brown color is produced just from the reagent. Chlorine water may be prepared by acidifying 10% aqueous sodium hypochlorite solution with one-fifth of its volume of dilute hydrochloric acid.
3. The absence of iodide ion but the presence of bromide ion will result in the appearance of a brown color in the carbon tetrachloride layer, due to oxidation of the bromide ion to form bromine followed by extraction of the bromine into the carbon tetrachloride. If both iodide and bromide are present, the initial purple color will give way to a reddish-brown color.

Test for Bromine; Test for Bromide Ion (Test e)

In the absence of iodide ion, the conditions of Test d will result in a brown color being produced in the carbon tetrachloride layer, due to the oxidation of bromide to bromine. If iodide is present, a purple color will appear first. This will change to reddish-brown with the continued addition of chlorine water.

Test for Chlorine; Test for Chloride Ion (Test f)

If the presence of halide ion was indicated by Test c above and no color was produced in the carbon tetrachloride layer in Test d above, the halide ion present must be chloride.

If either iodide or bromide ion has been shown to be present by means of Test d, apply the following procedure.

Procedure. Acidify 1 or 2 ml of the fusion solution with glacial acetic acid, add 0.5 gram of lead dioxide, and boil the mixture gently until all iodine and bromine has been liberated and boiled off. When the mixture is allowed to stand so that the lead dioxide will settle to the bottom, the solution will be colorless when this point has been reached. Dilute the mixture with an equal volume of water, remove the excess lead dioxide either by filtration or centrifugation and decantation, and test for the presence of chloride ion in the filtrate by adding a little dilute nitric acid followed by a few drops of 5% aqueous silver nitrate solution. The formation of a white precipitate confirms the presence of chloride ion.

---

§28 QUALITATIVE CHARACTERIZATION TESTS: TESTS FOR THE FUNCTIONAL GROUPS

Before the advent of spectroscopic methods such as infrared, ultraviolet, nuclear magnetic resonance, and mass spectroscopy (Sections 23 through 26), the determination of chemical properties was of paramount importance for the identification, characterization, and determination of structure of pure substances. Many reagents or reaction conditions were found to give moderately characteristic and specific results with substances containing certain functional groups. These reagents or reaction conditions therefore serve to distinguish between substances with different functional groups, and to indicate the presence or absence of certain functional groups in substances of unknown structure.

Sections 28.1 through 28.14 will describe a number of tests that are useful for distinguishing between certain functional groups, or to indicate the presence or absence of certain functional groups. The comparison of the results of these tests on samples of substances of known and unknown structure will, in many cases, allow a tentative conclusion to be drawn about the nature of the functional group of the substance of unknown structure.

However, different substances with the same functional group will (in principle) give at least slightly different results. The question is, then, how different may the results be before you must conclude that the substances have different functional groups? Obviously there will be borderline cases. Also, certain individual substances may react atypically, to give either a "false positive test" (the unknown behaves as if a certain functional group were present, but it really isn't) or a "false negative test" (the unknown behaves as if the functional group were not there, but it really is). That is, most tests are not completely specific or characteristic. Finally, some tests are quite sensitive, and a small amount of a substance as an impurity might be sufficient to give a result that you could interpret as a "positive test."

The uncertainty in the interpretation of these tests can be decreased by testing the behavior of compounds of known structure under the same conditions. The substances of known structure should be either substances that are said to give typical or normal results, or, better, substances that are very similar in structure (especially if more than one functional group is present) to the suspected structure of the unknown substance.

When recording the results of one of these characterization tests, you must be careful to record *your observations* (a yellow-orange crystalline precipitate was formed within thirty seconds after adding two drops of the liquid of unknown structure; the mixture was not heated; no further changes occurred upon standing; similar behavior but slower than with two drops of acetone) and *not your tentative conclusion* (positive test for a ketone). It is impossible to reinterpret a conclusion; only data—observations—can be reinterpreted. If your conclusions are inconsistent, there is no way to reinterpret them, and you have wasted your time; data can be continually reinterpreted until the conclusions are consistent.

There are many hundreds of qualitative tests that can be used to characterize unknown substances and to distinguish between different functional groups. The procedures presented in the following sections can therefore be seen to be only a very small sample of such tests. They have been chosen because they involve some of the most familiar functional groups, and because the information they give complements the information that can be obtained from infrared and nuclear magnetic resonance spectra. Table 28.1 is a summary and index for the fourteen tests presented.

Table 28.1. Summary of classification tests

Test	Section	Application
Ammonia	28.1	Ammonium salts; primary amides; nitriles
Baeyer's test: *see* Potassium permanganate test		
Benzenesulfonyl chloride (Hinsberg's test)	28.2	Amines (distinguishes among primary, secondary, or tertiary)
Bromine in carbon tetrachloride	28.3	Alkenes; alkynes
Chromic anhydride	28.4	Alcohols (distinguishes primary and secondary from tertiary)
2,4-Dinitrophenylhydrazine	28.5	Aldehydes and ketones
Ferric chloride	28.6	Phenols and enols
Ferric hydroxamate test	28.7	Acid halides and anhydrides; esters; amides; nitriles
Hinsberg's test: *see* Benzenesulfonyl chloride		
Hydrochloric acid/zinc chloride test (Lucas's test)	28.8	Alcohols (distinguishes among primary, secondary, or tertiary)
Iodoform test	28.9	Methyl ketones; secondary methyl carbinols
Lucas's test: *see* Hydrochloric acid/ zinc chloride test		
Potassium permanganate test (Baeyer's test)	28.10	Alkenes; alkynes
Silver nitrate	28.11	Halides
Sodium hydroxide test	28.12	Esters; lactones; anhydrides
Sodium iodide	28.13	Halides
Tollens' test	28.14	Aldehydes

The references for each test are given at the end of this section. These references provide more information concerning the scope and limitations of certain procedures, variations in procedures, and exceptional behavior of certain compounds or types of compounds. You should consult these references in order to minimize the chances of being misled by the results of the tests.

28.1 DETECTION OF AMMONIA FROM AMMONIUM SALTS, PRIMARY AMIDES, AND NITRILES

Ammonium salts, primary amides, and nitriles that are relatively easily hydrolyzed will liberate ammonia under the conditions of the procedure that follows.

Materials Required.

20% aqueous sodium hydroxide solution

10% aqueous copper sulfate solution

Procedure. Place 50 mg of the substance in a small test tube and add 2 ml of 20% sodium hydroxide solution. Fasten a small circle of filter paper tightly over the mouth of the test tube and wet it with a couple of drops of the copper sulfate solution. Heat the contents of the test tube to boiling for about one minute, and look to see if the filter paper has turned blue. The blue color is due to the formation of the copper-ammonia complex ion from ammonia either liberated from the ammonium salt or formed by hydrolysis of the primary amide or nitrile. (Ref. 1, p. 164; Ref. 2, pp. 109, 123; Ref. 3, pp. 404, 410, 798, 805.)

28.2 BENZENESULFONYL CHLORIDE (HINSBERG'S TEST)

Useful for distinguishing among primary, secondary, and tertiary amines.

Materials Required.

10% aqueous sodium hydroxide solution

Benzenesulfonyl chloride

benzenesulfonyl
chloride

Procedure. Add 5 ml of 10% sodium hydroxide to 0.2 ml of the liquid amine or 0.2 gram of the solid amine in a test tube, and then add 0.4 ml benzenesulfonyl chloride. Stopper the test tube and occasionally shake it vigorously over a period of 5 or 10 minutes, cooling the tube in water if it heats up a lot. By this time, all the benzenesulfonyl chloride should have reacted, either with the amine to form the sulfonamide or with the basic solution to form the water-soluble sodium benzenesulfonate; complete reaction is indicated by the disappearance of the distinctive odor of benzenesulfonyl chloride. Now make sure that the mixture is basic (if it is not, add 10% sodium hydroxide to make it strongly basic) and observe whether any insoluble material is present. Remove any insoluble material, by filtration if it is a solid and by decanting if it is a liquid.

Test the solubility of the insoluble material that was removed, both in water and in dilute hydrochloric acid.

Acidify the filtrate (or the basic solution that was separated by decantation, or the clear reaction mixture) and attempt to promote crystallization by scratching and by cooling. Benzenesulfonic acid is soluble in water.

Primary amines Primary amines normally give a sulfonamide that is soluble in the basic reaction mixture; no insoluble material should be present at the end of the first part of the test. Acidification of the clear reaction mixture should then result in the precipitation or crystallization of the water-insoluble sulfonamide of the amine. However, the sodium salts of the sulfonamides of some primary amines—for example, cyclohexyl- through cyclodecylamine and certain high-molecular-weight amines—are not very soluble in 10% sodium hydroxide solution. If the unknown is one of these compounds, a precipitate will be present at the end of the first part of the test, but it will be found to be soluble in water.

Secondary Secondary amines normally give a sulfonamide that is insoluble in the basic reaction mixture; a solid residue will be present at the end of the first part of the test. This residue will be insoluble in both water and dilute hydrochloric acid. The filtrate should give no precipitate upon acidification.

Tertiary Tertiary amines normally give no reaction with benzenesulfonyl chloride; the amine itself should be present at the unchanged liquid or solid at the end of the first part of the test. The liquid or solid residue will be soluble in dilute hydrochloric acid.

It should be apparent that water-soluble tertiary amines and secondary amines with a carboxyl group on another part of the molecule will not show the typical behavior previously outlined. (Ref. 1, p. 119; Ref. 3, p. 650. See also Fanta and Wang, *J. Chem. Educ.* **41**, 280 (1964). For an alternative method, see Ritter, *J. Chem. Educ.* **29**, 506 (1952).)

28.3 BROMINE IN CARBON TETRACHLORIDE

This test is useful for indicating the presence of many olefinic or acetylenic functional groups. It should be used in conjunction with a similar test involving aqueous potassium permanganate solution (Section 28.10).

Materials Required.

Bromine in carbon tetrachloride: a solution of 2 ml in 100 ml carbon tetrachloride

Carbon tetrachloride

Procedure. Dissolve 0.2 ml of a liquid or 0.1 gram of a solid in 2 ml of carbon tetrachloride, and add the bromine in carbon tetrachloride solution drop by drop with shaking. The presence of an olefinic or acetylenic linkage in the sample will cause more than 2 or 3 drops of the bromine solution to be required before the characteristic orange-brown color of bromine will persist for one minute. The presence of electron-withdrawing groups such as phenyl or substituted phenyl, carboxyl, or halogen will reduce the rate of addition and in some cases completely inhibit the reaction. The presence of phenols, enols, amines, aldehydes, or ketones can result in the consumption (as indicated by loss of color) of large amounts of bromine, accompanied by *the evolution of hydrogen bromide*, which may be detected by exhaling cautiously over the top of the test tube and noting the fog that is produced in the moist air. The test should be carried out in diffuse light, since light can catalyze a free-radical substitution reaction that will consume bromine and produce hydrogen bromide. (Ref. 1, p. 121; Ref. 2, p. 111; Ref. 3, p. 1058.)

28.4 CHROMIC ANHYDRIDE

Used to distinguish primary and secondary alcohols from tertiary alcohols.

Materials Required.

Reagent-grade acetone

Chromic anhydride reagent: prepared by pouring slowly and with good stirring a suspension of 25 grams chromic anhydride (CrO_3) in 25 ml conc. sulfuric acid into 75 ml of water and allowing the deep orange-red solution to cool to room temperature.

Procedure. Dissolve 1 drop of a liquid or 15–30 mg of a solid in 1 ml of reagent-grade acetone and then add 1 drop of the chromic anhydride reagent. Primary and secondary alcohols produce an opaque blue-green suspension within two seconds; tertiary alcohols give no visible reaction and the solution remains orange in color. Aldehydes give a positive test, but ketones, alkenes, acetylenes, amines, and ethers give negative tests in two seconds. Enols may give a positive test, and phenols cause the solution to turn a much darker color than the blue-green produced by primary and secondary alcohols. (See Bordwell and Wellman, *J. Chem. Educ.* **39**, 308 (1962). Ref. 1, p. 125; Ref. 2, p. 121. For a means of distinguishing among primary, secondary, and tertiary alcohols according to ease of oxidation by potassium permanganate in acetic acid, see Ritter, *J. Chem. Educ.* **30**, 395 (1953).)

28.5 2,4-DINITROPHENYLHYDRAZINE

Useful for identifying aldehydes and ketones.

HN—NH$_2$
NO$_2$

NO$_2$
2,4-dinitro-
phenylhydrazine

Materials Required.

2,4-Dinitrophenylhydrazine reagent: prepared by dissolving 3 grams of 2,4-dinitrophenylhydrazine in 15 ml conc. sulfuric acid and adding this solution, with good stirring, to a mixture of 20 ml of water and 70 ml of 95% ethanol. After thorough mixing, the reagent should be filtered.
95% ethanol

Procedure. Add a solution of 1 or 2 drops of the liquid, or 25–50 mg of the solid, in 2 ml of 95% ethanol to 3 ml of the 2,4-dinitrophenylhydrazine reagent. Mix the solution well and allow the mixture to stand for 15 minutes.

Most aldehydes and ketones give a solid 2,4-dinitrophenylhydrazone under these conditions. The derivative is yellow if the carbonyl group is not conjugated with a double bond or an aromatic ring. The 2,4-dinitrophenylhydrazones of most conjugated aldehydes and ketones are orange-red or red. The fact that the product is non-yellow should be interpreted with caution, as the color might be due to an impurity. The most obvious impurity is 2,4-dinitrophenylhydrazine itself, which is orange-red. (Ref. 1, p. 126; Ref. 2, p. 136; Ref. 3, p. 1060.)

28.6 FERRIC CHLORIDE SOLUTION

Useful for the recognition of phenols and enols.

Materials Required.

Procedure a: 2.5% aqueous ferric chloride solution; Ethanol
Procedure b: Ferric chloride in chloroform; prepared by dissolving 1 gram ferric chloride in 100 ml chloroform; Pyridine

Procedure a. Dissolve 1 drop or 30–50 mg of the compound in 1 or 2 ml of water or a mixture of ethanol and water, and add up to 3 drops of the aqueous ferric chloride solution. Note color changes or formation of a precipitate; the color may not be permanent. Most phenols produce red, blue, purple, or green colorations; enols usually produce tan, red, or red-violet colorations. However, many phenols do not give a color.

Procedure b. Dissolve (or suspend) 1 drop of a liquid or 30 mg of a solid in 1 ml chloroform, and add 2 drops of the solution of ferric chloride in chloroform; then add 1 drop of pyridine and note color changes. This test appears to be more sensitive than the test described in Procedure a, and some phenols that do not give a color under the conditions of Procedure a will do so under these conditions. (Ref. 1, p. 127; Ref. 2, p. 147. Proc. a: Wesp and Brode, *J. Am. Chem. Soc.* **56**, 1037 (1934). Proc. b: Soloway and Wilen, *Anal. Chem.* **24**, 979 (1952).)

28.7 FERRIC HYDROXAMATE TEST

Useful for the identification of acid halides and anhydrides; esters; amides; and nitriles.

Test for Acid Halides and Anhydrides (Test a)

Materials Required.
    1.0 M hydroxylamine hydrochloride in 95% ethanol
    6 M hydrochloric acid
    2 M hydrochloric acid
    10% aqueous ferric chloride solution

Procedure.   Add 30–40 mg of the solid, or 1 drop of the liquid, to 0.5 ml of the hydroxylamine hydrochloride solution. Add 2 drops of the 6 M hydrochloric acid solution, warm the mixture slightly for 2 minutes, and then heat it to boiling for a few seconds. After cooling the solution, add 1 drop of the 10% ferric chloride solution.

    A reddish-blue or bluish-red color will result if the original substance was an acid halide or anhydride. If the color is more red than blue, adjust the pH to 2 or 3 by dropwise addition of 2 M hydrochloric acid.

$$H \overset{\displaystyle H}{\underset{\displaystyle H}{\overset{|}{\underset{|}{-N}}}} - OH \quad Cl^-$$

hydroxylamine
hydrochloride

Tests for Esters of Carboxylic Acids (Test b)

Materials Required.
    1.0 M hydroxylamine hydrochloride in 95% ethanol
    2 M potassium hydroxide in methanol
    2 M hydrochloric acid
    10% aqueous ferric chloride solution

Procedure.   Add 30–40 mg of the solid, or 1 drop of the liquid, to 0.5 ml of the hydroxylamine hydrochloride solution. Now add the 2 M potassium hydroxide solution dropwise until the mixture is basic to litmus, and then add 4 more drops. Heat the mixture just to boiling, cool it to room temperature, and add dropwise (with good mixing) 2 M hydrochloric acid until the pH is approximately 3. Add 1 drop of the ferric chloride solution and note the color.

    Esters of carboxylic acids, including lactones and polyesters, give definite magenta colors of varying degrees of intensity. Acid chlorides and anhydrides, and trihalo compounds such as benzotrichloride and chloroform give a positive test. Formic acid is the only acid that produces a color (red). Carbonic acid esters, sulfonic acid esters, urethanes, chloroformates, and esters of inorganic esters all give a negative result.

Test for Amides and Nitriles (Test c)

Materials Required.
    1 M hydroxylamine hydrochloride in propylene glycol
    1 M potassium hydroxide in propylene glycol
    Propylene glycol
    5% aqueous ferric chloride

Procedure.   Add to 2 ml of the solution of hydroxylamine hydrochloride in propylene glycol a solution of 1 drop of the liquid, or 30–40 mg of the solid, dissolved in a minimum amount of propylene glycol. To this add 1 ml of the potassium hydroxide solution, and then boil the mixture for 2 minutes. Cool the solution to room temperature, and then add 0.5–1.0 ml of the ferric chloride solution. A red-to-violet color is a positive test; yellow colors should be interpreted as negative tests; brown colors or precipitates are indeterminate.

$$CH_3 - \overset{\displaystyle OH}{\underset{\displaystyle H}{\overset{|}{\underset{|}{C}}}} - CH_2OH$$

propylene
glycol

You should realize that esters, anhydrides, and acid halides will be converted to hydroxamic acids under these conditions as well as under the milder conditions of Tests a and b. Before a positive result in this test can be interpreted as indicating the presence of an amide or nitrile functionality, the possibility that the substance might be an ester, anhydride, or acid halide must be ruled out. (Ref. 1, p. 135; Ref. 2, pp. 119, 140, 144; Ref. 3, p. 1062. See also Davidson, *J. Chem. Educ.* **17**, 81 (1940).)

28.8 HYDROCHLORIC ACID/ZINC CHLORIDE TEST (LUCAS'S TEST)

Useful for distinguishing among lower-molecular-weight primary, secondary, and tertiary alcohols. Since the test requires that the alcohol initially be in solution, it is limited in its application to monohydroxylic alcohols more soluble than *n*-hexanol and to certain polyfunctional molecules.

Materials Required.
Lucas's reagent: prepared by dissolving 136 grams (1 mole) of zinc chloride in 89 ml (105 grams; 1 mole) conc. hydrochloric acid with cooling in ice.

Procedure. Add 2 ml of the reagent to 0.2 ml of the alcohol in a test tube; swirl to dissolve. Note the time required to form the insoluble alkyl chloride, which appears as a layer or emulsion.

The substitution of hydroxyl by chloride apparently takes place by an S_N1 mechanism, since tertiary alcohols, as well as allylic alcohols and benzyl alcohol, react very quickly. Secondary alcohols give indications of the formation of a second phase within 2 or 3 minutes, and primary alcohols react very much more slowly to form the insoluble alkyl chloride.

Tertiary alcohols will give a similar but slower reaction with concentrated hydrochloric acid alone. (Ref. 1, p. 131; Ref. 2, p. 121; Ref. 3, p. 261. See also Lucas, *J. Am. Chem. Soc.* **52**, 802 (1930).)

28.9 IODOFORM TEST

Useful for the identification of methyl ketones and secondary methyl carbinols.

iodoform

Materials Required.
Iodine/potassium iodide reagent: prepared by adding 200 g potassium iodide and 100 g iodine to 800 ml distilled water, stirring until solution is complete
Dioxane
10% aqueous sodium hydroxide solution

Procedure. Dissolve 4 drops of a liquid or 100 mg of a solid in 5 ml of dioxane (use 1 ml of water in place of the dioxane if the compound is soluble in water to this extent). Add 1 ml of the sodium hydroxide solution and then the iodine/potassium iodide solution with shaking until the definite dark color of iodine persists. If less than 2 ml of the iodine/potassium iodide solution was consumed, place the test tube in a beaker of water at 60°C. If the dark color of iodine now disappears, continue to add the iodine/potassium iodide solution until the dark color that represents an excess of iodine is not discharged by heating at 60°C for 2 minutes. Now add 10% sodium hydroxide solution dropwise until the dark iodine color is gone, remove the tube from the heating bath, add 15 ml of water, and allow the mixture to cool to room temperature. Collect by suction filtration any solid that is formed and determine its melting point. Iodoform melts at 119–121°C. If the iodoform is reddish, dissolve it in 3 or 4 ml of dioxane, add 1 ml of 10% sodium hydroxide solution, and shake the mixture until the reddish color gives way to the lemon-yellow color of iodoform. Slowly dilute the mixture with water and collect the precipitated iodoform by suction filtration.

Methyl ketones (and acetaldehyde) and methyl carbinols including ethanol (compounds that can be oxidized to methyl ketones by the reagent) give iodoform under these conditions; acetic acid does not. Compounds that can react with the reagent to generate one of these functional groups will also give iodoform; conversely, it is possible that the functionality that might be expected to result in the formation of iodoform can be removed by hydrolysis before iodoform formation is complete (see Fuson and Bull). (Ref. 1, p. 137; Ref. 2, p. 112; Ref. 3, p. 1068. See also Fuson and Bull, *Chem. Revs.* **15**, 275 (1934).)

28.10 AQUEOUS POTASSIUM PERMANGANATE SOLUTION (BAEYER'S TEST)

This test is useful for indicating the presence of most olefinic or acetylenic functional groups.

As a test for unsaturation, it should be used in conjunction with a similar test involving a solution of bromine in carbon tetrachloride (Section 28.3).

Materials Required.
2% aqueous potassium permanganate solution
Reagent-grade acetone (free of alcohol)

Procedure. Dissolve 0.2 ml of a liquid or 0.1 gram of a solid in 2 ml of water or acetone. Add the potassium permanganate solution drop by drop with good shaking of the mixture. If more than one drop of the permanganate solution is consumed, as shown by the loss of the characteristic purple color, the presence of an olefin or acetylene or other functional group that can be oxidized by permanganate under these conditions is indicated. Such other groups include phenols and aryl amines, most aldehydes (but not benzaldehyde or formaldehyde) and

formate esters (which contain the $-\overset{\overset{\text{O}}{\|}}{\text{C}}-\text{H}$ group), primary and secondary alcohols (see, however, Swinehart), mercaptans and sulfides, and thiophenols. Aryl-substituted alkenes are oxidized by permanganate under these conditions.

Certain carefully purified alkenes are not oxidized under the conditions of this test (acetone solvent), but can be oxidized if ethanol is used instead; ethanol does not react with potassium permanganate within 5 minutes at room temperature.

In this test you must take care not to be misled by the limited reaction of impurities. (Ref. 1, p. 149; Ref. 2, p. 112; Ref. 3, p. 1058; see also Swinehart, *J. Chem. Educ.* **41**, 392 (1964).)

28.11 ALCOHOLIC SILVER NITRATE SOLUTION

This test is useful for classifying compounds known to contain chlorine, bromine, or iodine. It should be used in conjunction with a similar test using sodium iodide in acetone (Section 28.13).

The rate of reaction of the halogen in this test gives an indication of its structural environment. Reference 1 includes a discussion of the relationship between structure and reactivity for organic halides. Table 28.2 summarizes the results that can be expected for the most common functional groups.

Materials Required.
2% silver nitrate in 95% ethanol
5% aqueous nitric acid

Procedure. Add 1 drop of the liquid (or 30–40 mg of the solid) to 2 ml of the silver nitrate solution at room temperature. If no precipitate is formed upon standing at room temperature for 5 minutes, heat the solution to boiling for 30 seconds and note whether a precipitate is formed under these conditions. If a precipitate is formed, note the color and determine whether it dissolves upon addition of two drops of the nitric acid solution. Silver chloride is white, silver bromide is pale yellow, and silver iodide is yellow; the silver halides are insoluble in dilute nitric acid, but silver salts of organic acids will dissolve. (Ref. 1, p. 152; Ref. 2, p. 122; Ref. 3, p. 1059.)

Table 28.2. Alcoholic silver nitrate test

Water-soluble compounds that give an immediate precipitate at room temperature with *aqueous* silver nitrate:

Salts of amines and halogen acids
Low-molecular-weight acid halides (their water solubility results in their rapid hydrolysis to give halide ion)

Water-insoluble compounds that give an immediate precipitate at room temperature with *alcoholic* silver nitrate:

Acid halides
Tertiary alkyl halides
Allylic halides
Alkyl iodides
α-Haloethers
1,2-Dibromides

Water-insoluble compounds that react slowly or not at all at room temperature, bu give a precipitate readily at higher temperatures with *alcoholic* silver nitrate:

Primary and secondary alkyl chlorides
Geminal dibromides
Activated aryl halides

Water-insoluble compounds that do not give a precipitate at higher temperatures with *alcoholic* silver nitrate:

Unactivated aryl halides
Vinyl halides
Chloroform, carbon tetrachloride

28.12 SODIUM HYDROXIDE TEST

Useful for identifying esters, lactones, and anhydrides.

Materials Required.
0.1 *M* sodium hydroxide in 95% ethanol
Phenolphthalein indicator solution, in ethanol
95% ethanol

Procedure. Dissolve 0.1 gram of the substance in 3 ml ethanol. Add 3 drops of the phenolphthalein indicator solution and then add, dropwise, sufficient 0.1 M sodium hydroxide solution to turn the solution pink. Warm the mixture in a beaker of water at 40°C. Disappearance of the pink color indicates consumption of base.

Under these conditions, esters, lactones, and anhydrides are hydrolyzed, resulting in the consumption of base; additionally, some alkyl halides, amides, and nitriles will undergo solvolysis or hydrolysis and will likewise result in the use of base. In order to determine that the base was not consumed by an impurity, the cycle of adding base to turn the indicator pink and heating at 40°C to decolorize the solution should be repeated several times. (Ref. 1, p. 164.)

28.13 SODIUM IODIDE IN ACETONE

This test is useful for classifying compounds known to contain bromine or chlorine. It should be used in conjunction with a similar test using a solution of silver nitrate in 95% ethanol (Section 28.11).

Materials Required.

Acetone, reagent grade

Sodium iodide in acetone reagent: prepared by dissolving 15 g of sodium iodide in 100 ml reagent-grade acetone. The solution is colorless at first, but develops a pale lemon-yellow color on standing. When the color becomes definitely red-brown, the solution should be discarded.

Procedure. Add 2 drops of the liquid, or a solution of 100 mg of the solid dissolved in a minimum amount of acetone, to 1 ml of the sodium iodide in acetone solution. After mixing, allow the solution to stand at room temperature for 3 minutes. Note whether a precipitate has formed (sodium chloride or sodium bromide) and whether the solution has become reddish-brown (liberation of iodine). If no change has occurred, heat the solution in a beaker of water at 50°C for 6 minutes and note any relevant changes.

This test depends upon two facts: (1) sodium chloride and sodium bromide, which can be formed in an S_N2-type displacement of the halide by iodide, are insoluble in acetone, and (2) iodine can be formed by the reaction of iodide with 1,2-dichloro and 1,2-dibromo compounds.

Table 28.3 summarizes the results that can be expected for some of the more common functional groups. (Ref. 1, pp. 169, 152; Ref. 3, p. 1059.)

28.14 TOLLENS' REAGENT: SILVER–AMMONIA COMPLEX ION

Useful for distinguishing aldehydes from ketones and other carbonyl compounds.

Materials Required.

5% aqueous silver nitrate solution

10% aqueous sodium hydroxide solution

2 M aqueous ammonium hydroxide solution

Procedure. Clean a test tube thoroughly, preferably by boiling in it a 10% solution of sodium hydroxide and then discarding the sodium hydroxide solution and rinsing the test tube with distilled water. To the clean tube add 2 ml of the silver nitrate solution and 1 drop of the sodium hydroxide solution. Add the

Table 28.3. Sodium iodide in acetone test

Formation of a precipitate within 3 minutes at 25°C:

  Primary alkyl bromides
  Benzylic and allylic chlorides and bromides
  α-Halo ketones, esters, amides, and nitriles
  Acid chlorides and bromides

Formation of a precipitate within 6 minutes at 50°C:

  Primary and secondary alkyl chlorides
  Secondary and tertiary alkyl bromides
  Cyclopentyl chloride
  Benzal chloride, benzotrichloride, bromoform, 1,1,2,2-
    tetrabromoethane

Unreactive under the conditions specified:

  Tertiary alkyl chlorides
  Cyclohexyl chloride and bromide
  Vinyl halides; aryl halides
  Chloroform, carbon tetrachloride, trichloroacetic acid

ammonium hydroxide solution drop by drop with good shaking until the dark precipitated silver oxide just dissolves.

Add 1 drop of the liquid or 30–50 mg of the solid to be tested, shake the tube to mix, and allow it to stand at room temperature for 20 minutes. If nothing happens, heat the tube in a beaker of water at 35°C for five minutes.

Aldehydes and other substances that can be oxidized by means of the silver-ammonia complex ion will reduce the silver ion to metallic silver, which will precipitate as a "mirror" on the test tube if it is sufficiently clean, or as a black colloidal suspension if the test tube is not clean.

The solution should not be allowed to stand after the test is completed, but should be discarded down the drain immediately, as the highly explosive silver fulminate may be formed on standing. Rinse the test tube with dilute nitric acid.

Substances that can be oxidized by means of the silver–ammonia complex ion include most aldehydes, "reducing sugars," hydroxylamines, acyloins, amino-phenols, and polyhydroxyphenols. (Ref. 1, p. 173; Ref. 2, p. 138; Ref. 3, pp. 330, 1060.)

References

1. R. L. Shriner, R. C. Fuson, and D. Y. Curtin, *The Systematic Identification of Organic Compounds*, 5th edition, Wiley, New York, 1965.
2. N. D. Cheronis and J. B. Entrikin, *Identification of Organic Compounds*, Interscience, New York, 1963.
3. A. I. Vogel, *A Textbook of Practical Organic Chemistry*, 3rd edition, Wiley, New York, 1957.

§29 CHARACTERIZATION THROUGH FORMATION OF DERIVATIVES

Sections 29.1 through 29.31 describe a number of procedures whereby substances containing certain functional groups may be converted into reasonably well-behaved crystalline products (derivatives) through reaction at the functional group. As in the qualitative characterization tests of Section 28, the fact that the substance of unknown structure gives a solid product is consistent with the presence in its molecules of the suspected functional group, but the formation (or nonformation) of a solid product is by no means proof of the presence (or absence) of the suspected functional group. The reasons are the same as those discussed for the qualitative characterization tests: the derivative-forming reactions are not completely specific or characteristic (some substances may react atypically), and a small amount of an impurity, possibly water, may be sufficient to result in a solid product.

If a solid product is obtained, however, it can be purified—usually by recrystallization until successive recrystallizations do not raise the melting point significantly—and the melting point determined. Comparing the melting point of the derivative with the melting points recorded in the chemical literature for the corresponding derivatives of substances of known structure can often serve to eliminate all but one or two possibilities for the structure of the substance of unknown structure. A hypothetical case will show how this is done, and some of the problems and pitfalls that must be considered and avoided.

Suppose the substance of unknown structure is a liquid that distilled between 96.5 and 98°C at 740 Torr. On the basis of other chemical and physical properties, you deduced that the liquid must be an alcohol. When you treated a sample with 3,5-dinitrobenzoyl chloride according to the procedure of Section 29.1, you obtained a solid that melted after two recrystallizations at 74.5–75.5°C. You presumed it to be the 3,5-dinitrobenzoate ester of the unknown alcohol.

At this point, alcohols of higher and lower boiling points have already been tentatively eliminated, because their boiling points are believed to be well outside the limits of the experimental error of the observed boiling range of the unknown. Shriner, Fuson, and Curtin (see References, end of this section) offer the following additional information:

| Compound | b.p. | m.p. of Derivative | | |
		α-Naphthylurethan	Phenylurethan	3,5-Dinitrobenzoate
3-Butene-2-ol	96	—	—	—
Allyl alcohol	97	109	70	48
n-Propyl alcohol	97	80	51	74
sec-Butyl alcohol	99	97	65	75
tert-Amyl alcohol	102	71	42	117

From these data, it appears that tert-amyl alcohol and allyl alcohol may also be eliminated, since the melting points reported for their 3,5-dinitrobenzoate esters are well outside the range of experimental error of the melting point observed for the 3,5-dinitrobenzoate of the unknown. If you knew the melting point of the 3,5-dinitrobenzoate of 3-butene-2-ol, you might be able to eliminate this possibility as well. Now you can see that it might have been smarter to try to make the α-naphthylurethan of the unknown, since it might serve to distinguish among all possibilities.

Looking in Cheronis and Entrikin (see References) to try to find melting points for the derivatives of 3-butene-2-ol, you locate the following information:

$H_2C{=}CH{-}CH_2OH$
allyl alcohol

| Compound | b.p. | m.p. of Derivative | | |
		α-Naphthylurethan	Phenylurethan	3,5-Dinitrobenzoate
3-Butene-2-ol	—	—	—	—
Allyl alcohol	97.1	108	—	49
n-Propyl alcohol	97.15	105	—	74
sec-Butyl alcohol	99.5	97	—	76
tert-Amyl alcohol	102.35	75	—	116

You did not find what you wanted, but you did discover a remarkable disagreement between the values reported for the melting point of the α-naphthylurethan of *n*-propyl alcohol by Shriner, Fuson, and Curtin, and by Cheronis and Entrikin. The disagreement between the other values reported for boiling points and melting points is typical and is an indication of the uncertainty of these values.

Looking in Vogel (see References) for melting points for derivatives of 3-butene-2-ol and to get another opinion on the melting point of the α-naphthylurethan of *n*-propyl alcohol, you find the following information:

| | | *m.p. of Derivative* | | | |
Compound	b.p.	α-Naphthyl-urethan	Phenyl-urethan	3,5-Dinitro-benzoate	Hydrogen 3-nitrophthalate
3-Butene-2-ol	—	—	—	—	—
Allyl alcohol	97	109	70	50	124
n-Propyl alcohol	97	80	57	75	145
sec-Butyl alcohol	99.5	98	64	76	131
tert-Amyl alcohol	102	72	42	118	—

You did not find out anything here, either, about the melting points of derivatives of 3-butene-2-ol, but you did discover that Vogel and Shriner, Fuson, and Curtin agree on the melting point of the α-naphthylurethan of *n*-propyl alcohol. However, this indicates no more than that these authors are reporting a value from the same source in the original literature, and that Cheronis and Entrikin are reporting a value from a different source (assuming no typographical or transcription errors). You also see that Vogel and Shriner, Fuson, and Curtin report significantly different values for the melting point of the phenylurethan of *n*-propyl alcohol.

One more easy place to look for melting points of derivatives is the Chemical Rubber Company's *Handbook of Tables for Organic Compound Identification, 3rd edition* which contains the following information:

| | | *m.p. of Derivative* | | | |
Compound	b.p.	α-Naphthyl-urethan	Phenyl-urethan	3,5-Dinitro-benzoate	Hydrogen 3-nitrophthalate
3-Butene-2-ol	94–6	—	—	—	43–4
Allyl alcohol	97.1	108	70	49–50	124
n-Propyl alcohol	97.1	80; 76	57	74	145.5
sec-Butyl alcohol	99.5	97	64.5	76	131
tert-Amyl alcohol	102.3	72	42	116	—

From this research, you see that if the compound is an alcohol and the solid formed from the unknown by treatment of a sample by the procedure of Section 28.1 is the corresponding 3,5-dinitrobenzoate ester, the boiling point of the unknown liquid and the melting point of the derivative serves to rule out all possibilities except *n*-propyl alcohol, *sec*-butyl alcohol, 3-butene-2-ol, and other alcohols of which we are not aware that boil near 96–98°C. You also see that the derivative that should have been made was not the 3,5-dinitrobenzoate but the hydrogen 3-nitrophthalate (by the procedure of Section 28.2), which would have allowed the elimination of at least two of the remaining possibilities. The moral of your search is that a few minutes of library research can save hours of laboratory work.

The next step would be to prepare the hydrogen 3-nitrophthalate of the unknown in order to reduce the number of possibilities for its structure. The α-naphthylurethan should not be prepared for at least two reasons: (1) on the basis

of the present information, it would not allow elimination of 3-butene-2-ol, and
(2) there is uncertainty about the melting point of this derivative of *n*-propanol.

Of course, the number of possibilities may be reduced by consideration of
other physical and chemical properties. For example, the following values for
density and index of refraction are reported in the *Handbook of Tables for Organic
Compound Identification.*

Compound	d_4^{20}	n_D^{20}
3-Butene-2-ol	—	—
n-Propyl alcohol	0.80359	1.38499
sec-Butyl alcohol	0.80692	1.39495^{25}

It should be possible to distinguish between *n*-propyl alcohol and *sec*-butyl
alcohol on the basis of the index of refraction, but of course 3-butene-2-ol cannot
be ruled out by either the density or the index of refraction.

Assume now that you have prepared the hydrogen 3-nitrophthalate and found
that it has a melting point of 144–145°C. All possibilities except *n*-propyl alcohol
and alcohols of which we are not aware that boil near 96–98°C have now been
eliminated. Of course, you would have to be very unlucky indeed to have an
unknown sample that had the same boiling point as *n*-propyl alcohol and whose
derivatives melted at the same temperature as those of *n*-propyl alcohol, but yet
was not *n*-propyl alcohol.

The identity of the liquid of unknown structure and *n*-propyl alcohol could
be established in several ways if a sample of *n*-propyl alcohol were available. For
example, the "identity" of the infrared or mass spectra would be completely
acceptable as evidence of the identity of the two substances. The comparison of
spectra could also be made by way of published spectral data for *n*-propyl
alcohol. Either the hydrogen 3-nitrophthalate or 3,5-dinitrobenzoate of *n*-propyl
alcohol could be prepared, and the identity of the melting points and non-
depression of the mixed melting point (or identity of the infrared spectra of the
two derivatives) could establish the identity of the unknown liquid as *n*-propyl
alcohol.

Finally, of course, it must be admitted that the NMR spectrum would immediately establish whether or not the unknown liquid was *n*-propyl alcohol, or,
for that matter, any of the other four alcohols that were considered. The NMR
spectrum would also reveal whether the unknown substance had a structure different from that of any of the five possibilities considered.

Table 29.1 is a summary and index for the 31 procedures for the formation
of derivatives that are presented in the following sections. The references at the
end of each procedure give more information concerning the scope and limitations
of the procedures, possible procedural variations, and exceptional behavior of
certain compounds or types of compounds; references that are given in abbreviated form are listed in full at the end of this section.

The book by Shriner, Fuson, and Curtin and the Cheronis and Entrikin
book give many references to the original literature in which the procedures for
preparations of derivatives are described. If you think that you know the identity
of the substance for which you are attempting to prepare a derivative, you will
often save time by looking up in these references the exact procedure used with
that compound. You may find that the substance is best converted to the derivative by a modification of the general procedure, or that the derivative is formed in
poor yield (by an experienced chemist, which might mean that you should consider another derivative), or that the product is best isolated in a certain way or
recrystallized from a certain solvent, or that it has some interesting or characteristic property by which it may be recognized. This kind of information can sometimes save hours of lab time.

Table 29.1. Procedures for the formation of derivatives

Type of Compound	Type of Derivative	Section
Alcohol	Benzoate, p-nitrobenzoate, and 3,5-dinitrobenzoate esters	29.1
Alcohol	Hydrogen 3-nitrophthalate ester	29.2
Alcohol	Plenylurethan; α-naphthylurethan	29.3
Aldehyde	Methone derivative	29.4
Aldehyde	2,4-Dinitrophenylhydrazone	29.5
Aldehyde	Semicarbazone	29.6
Aldehyde	Oxime	29.7
Amide, unsubstituted	Hydrolysis	29.8
Amide, unsubstituted	9-Acylamidoxanthene	29.9
Amide, N-substituted	Hydrolysis	29.10
Amine, primary and secondary	N-Substituted acetamide	29.11
Amine, primary and secondary	N-Substituted benzamide	29.12
Amine, primary and secondary	N-Substituted p-toluenesulfonamide	29.13
Amine, primary and secondary	Phenylthiourea; α-naphthylthiourea	29.14
Amine, tertiary	Picrate	29.15
Amine, tertiary	Methiodide; p-toluenesulfonate	29.16
Carboxylic acid	Unsubstituted amide	29.17
Carboxylic acid	Anilide; p-toluidide; p-bromoanilide	29.18
Carboxylic acid	Phenacyl ester; p-substituted phenacyl esters	29.19
Carboxylic acid	p-Nitrobenzyl ester	29.20
Carboxylic acid anhydride	Anilide; p-toluidide; p-bromoanilide	29.18
Carboxylic acid ester	N-Benzylamide	29.21
Carboxylic acid ester	3,5-Dinitrobenzoate	29.22
Carboxylic acid ester	Hydrolysis	29.23
Carboxylic acid halide	Anilide; p-toluidide; p-bromoanilide	29.18
Ether, aromatic	Bromination product	29.24
Ether, aromatic	Picrate	29.15
Halide, aliphatic	S-Alkylthiuronium picrate	29.25
Halide, aromatic	o-Aroylbenzoic acid	29.26
Halide, aromatic	Oxidation	29.27
Halide	Anilide; p-toluidide; α-naphthalide	29.28
Hydrocarbon, aromatic	o-Aroylbenzoic acid	29.26
Hydrocarbon, aromatic	Oxidation	29.27
Hydrocarbon, aromatic	2,4,7-Trinitrofluorenone adduct	29.29
Hydrocarbon, aromatic	Picrate	29.15
Ketone	2,4-Dinitrophenylhydrazone	29.5
Ketone	Semicarbazone	29.6
Ketone	Oxime	29.7
Nitrile	Hydrolysis	29.8
Phenol	α-Naphthylurethan	29.3
Phenol	Bromination product	29.30
Phenol	Aryloxyacetic acid	29.31

These two very useful books also give literature references to many more procedures for preparation of derivatives, both for substances with the functional groups for which procedures are given in this book and for compounds with other types of functional groups.

29.1 BENZOATES, *p*-NITROBENZOATES, AND   **209**
3,5-DINITROBENZOATES OF ALCOHOLS
         SECTION 29

$$\text{Ar—C(=O)—Cl} + \text{R—O—H} \xrightarrow{\text{pyridine}} \text{Ar—C(=O)—O—R} + \text{"HCl"}$$

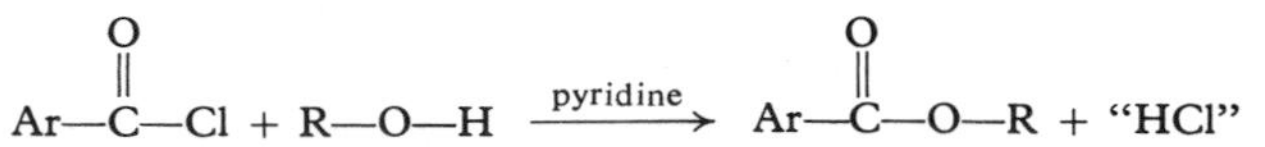

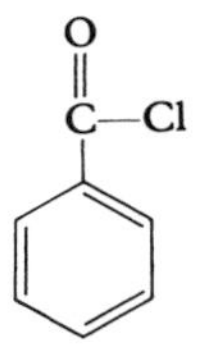

benzoyl
chloride

Procedure. Treat a solution of 0.5 gram of the alcohol in 3 ml of pyridine with about 2 grams of the appropriate acid chloride (benzoyl chloride, *p*-nitrobenzoyl chloride, or 3,5-dinitrobenzoyl chloride) with cooling in an ice bath. Heat the resulting mixture on the steam bath with exclusion of moisture for 10 minutes if the alcohol is primary or secondary, and for 30 minutes if it is tertiary (Note 1). Pour the mixture with stirring into 10 ml of ice water and cautiously acidify it with conc. hydrochloric acid. Thoroughly triturate the residue, which frequently is an oil, with 5 ml of 5% sodium carbonate solution. Finally, collect the solid by suction filtration and recrystallize it from aqueous alcohol, or methanol, or ethanol, or acetone/petroleum ether. (Ref. 1, pp. 246, 247; Ref. 2, p. 249; Ref. 3, p. 262.)

Note

1. Tertiary alcohols are esterified poorly under these conditions.

29.2 HYDROGEN 3-NITROPHTHALATES OF ALCOHOLS

Derivative for primary and secondary alcohols (Note 1).

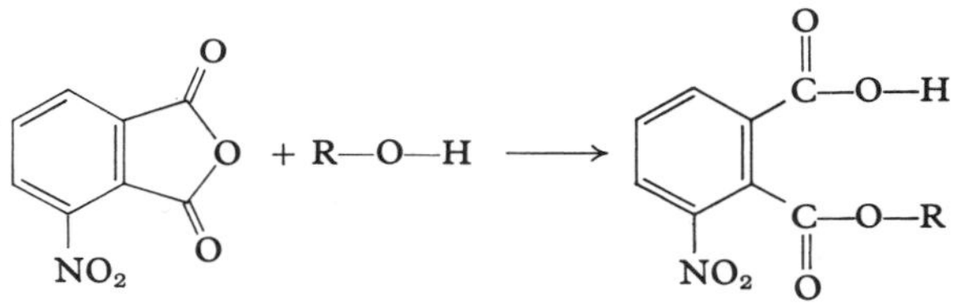

Procedure. Heat for 2 hours on the steam bath a mixture of 0.3 ml of the alcohol, 0.3 gram of 3-nitrophthalic anhydride, and 0.5 ml of pyridine. Pour the mixture onto ice, acidify with conc. hydrochloric acid, and isolate the solid ester either by filtration or by extraction with benzene or chloroform. Recover the acid phthalate by extraction into dilute sodium hydroxide followed by acidification of the extract and suction filtration. Recrystallize from water, alcohol/water, or toluene. (Ref. 1, p. 248.)

Note

1. Tertiary alcohols undergo elimination under these conditions.

29.3 PHENYL- AND α-NAPHTHYLURETHANS

Derivatives for primary and secondary alcohols (Note 1) and phenols.

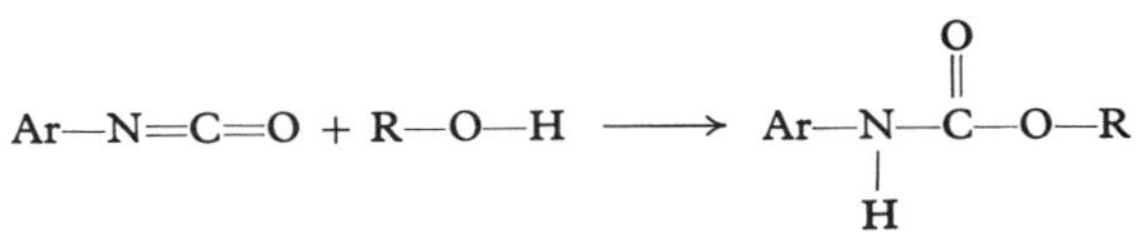

$$\text{Ar—N=C=O} + \text{R—O—H} \longrightarrow \text{Ar—N(H)—C(=O)—O—R}$$

Procedure. Add a solution of 0.5 gram of the alcohol (Note 2) in 5 ml of ligroin (b.p. 80–100°C) to a solution of 0.5 gram of phenylisocyanate or α-naphthylisocyanate (Note 3) in 10 ml of the same solvent. Heat the mixture for 1–3 hours on the steam bath, filter it while hot (Note 2), and allow the filtrate to cool. Collect the product by suction filtration, and recrystallize it from ligroin or carbon tetrachloride. (Ref. 1, p. 246; Ref. 2, p. 252; Ref. 3, pp. 264, 683.)

Notes

1. The urethans of tertiary alcohols are difficult to prepare by this method.
2. The alcohol must be dry; water leads to the formation of the very insoluble diphenylurea or di-α-naphthylurea. The hot filtration will remove most of this if it is formed, but some will crystallize with the product.
3. If a phenol is used, the α-naphthylurethan is the preferred derivative and the reaction should be catalyzed by the addition of a few drops of pyridine.

29.4 METHONE DERIVATIVES OF ALDEHYDES

Derivative for low-molecular-weight aldehydes.

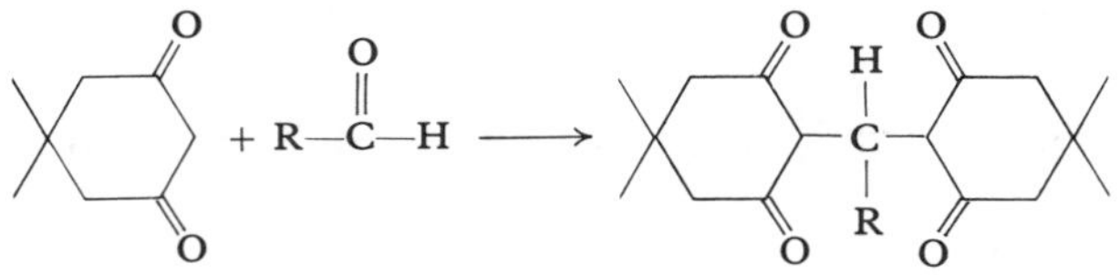

Procedure. Dissolve 300 mg of the reagent (Dimedon; dimethyldihydroresorcinol; 5,5-dimethyl-1,3-cyclohexanedione—Section 67.2; 2.1 mmole) in 4 ml of 50% aqueous ethanol. Add to this 1 mmole of the aldehyde and boil the mixture for about 30 seconds. Allow it to cool and stand for crystallization for at least 4 hours. Collect the product by suction filtration and recrystallize from a minimum amount of methanol and water. (Ref. 1, p. 254; Ref. 2, p. 262; Ref. 3, p. 332.)

29.5 2,4-DINITROPHENYLHYDRAZONES

Derivative for aldehydes and ketones.

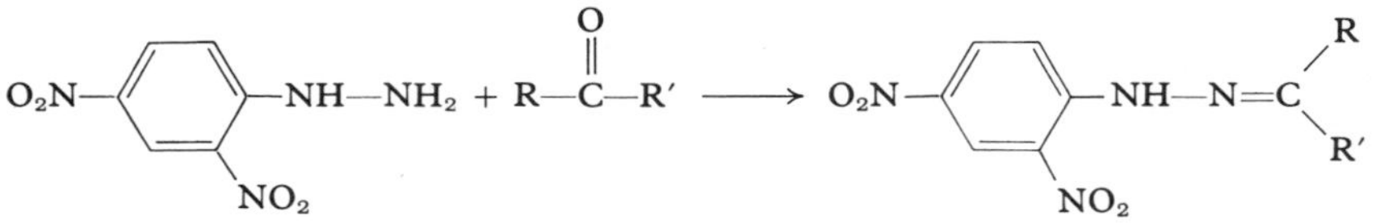

Preparation of a Solution of 2,4-Dinitrophenylhydrazine in 30% Perchloric Acid. Dissolve 1.2 grams (6 mmole) of 2,4-dinitrophenylhydrazine in a mixture of 16 ml of 60% perchloric acid plus 34 ml of water at room temperature.

Preparation of the Derivative. Take 4 ml of the perchloric acid solution of 2,4-dinitrophenylhydrazine and dilute it with 8 ml of water; stir well. Add quickly to this well-stirred mixture 0.5 mmole of the carbonyl compound as a 10–20% solution in ethanol. Collect the resulting 2,4-dinitrophenylhydrazone by suction filtration, and recrystallize it from ethanol, ethyl acetate, dioxane, dioxane/water, or ethanol/water. (Ref. 1, p. 253; Ref. 2, p. 260; Ref. 3, p. 344.)

Derivative for aldehydes and ketones.

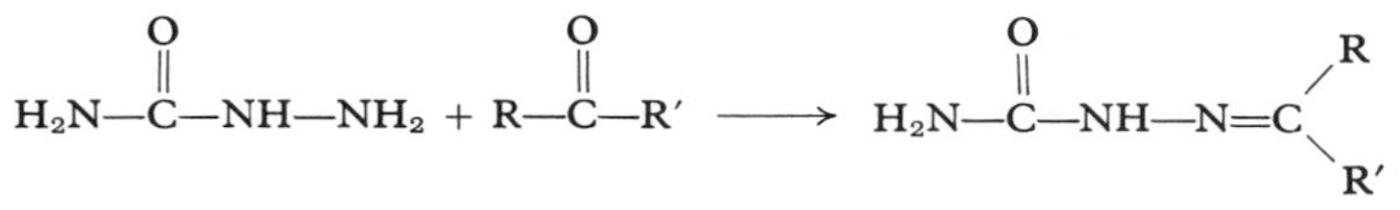

$$H_2N—\overset{\overset{\displaystyle O}{\|}}{C}—NH—NH_2 + R—\overset{\overset{\displaystyle O}{\|}}{C}—R' \longrightarrow H_2N—\overset{\overset{\displaystyle O}{\|}}{C}—NH—N{=}C\begin{smallmatrix}R\\ \\R'\end{smallmatrix}$$

Preparation of an Alcoholic Solution of Semicarbazide Acetate.   Grind 1 gram of semicarbazide hydrochloride with 1 gram of anhydrous sodium acetate in a mortar. Transfer the mixture to a flask, boil it with 10 ml of absolute ethanol, and filter the suspension while hot.

Preparation of the Derivative.   Add to the freshly prepared solution of semi-carbazide acetate about 0.2 gram of the carbonyl compound and heat the mixture under reflux on the steam bath for 30–60 minutes. Dilute the hot solution with water to incipient cloudiness, and allow it to cool slowly to room temperature. Collect the semicarbazone by suction filtration and recrystallize it from aqueous alcohol or alcohol. (Ref. 1, p. 253; Ref. 2, pp. 262, 320; Ref. 3, p. 344.)

29.7 OXIMES

Derivative for higher-molecular-weight aldehydes and ketones.

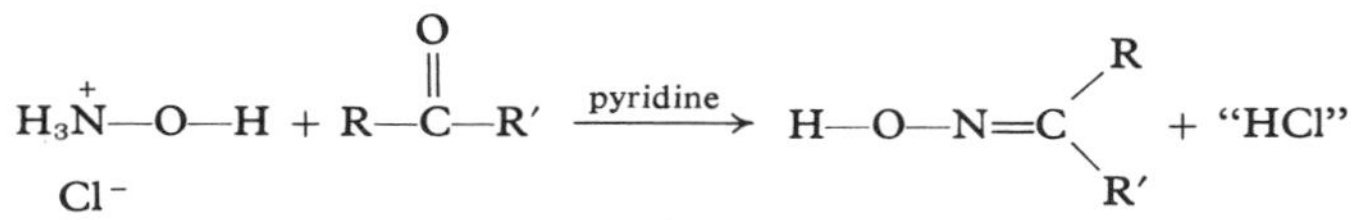

$$\overset{+}{H_3N}—O—H \atop Cl^- \quad + R—\overset{\overset{\displaystyle O}{\|}}{C}—R' \xrightarrow{\text{pyridine}} H—O—N{=}C\begin{smallmatrix}R\\ \\R'\end{smallmatrix} + \text{``HCl''}$$

Procedure.   Heat for 2 hours under reflux on the steam bath a mixture of 0.5 gram of the carbonyl compound, 0.5 gram of hydroxylamine hydrochloride, 3 ml of pyridine, and 3 ml of absolute ethanol. Remove the solvent by evaporation (by heating the mixture while drawing a current of air over it or with a rotary evaporator; Section 36) and recrystallize the residue from methanol or methanol/water. (Ref. 1, p. 289; Ref. 2, p. 319; Ref. 3, p. 721.)

29.8 CARBOXYLIC ACIDS BY HYDROLYSIS OF PRIMARY AMIDES AND NITRILES

Derivative for primary amides and nitriles.

$$R—\overset{\overset{\displaystyle O}{\|}}{C}—\underset{\underset{\displaystyle H}{|}}{N}—H + H_2O \longrightarrow R—\overset{\overset{\displaystyle O}{\|}}{C}—OH + NH_3$$

$$R—C{\equiv}N + 2H_2O \longrightarrow R—\overset{\overset{\displaystyle O}{\|}}{C}—OH + NH_3$$

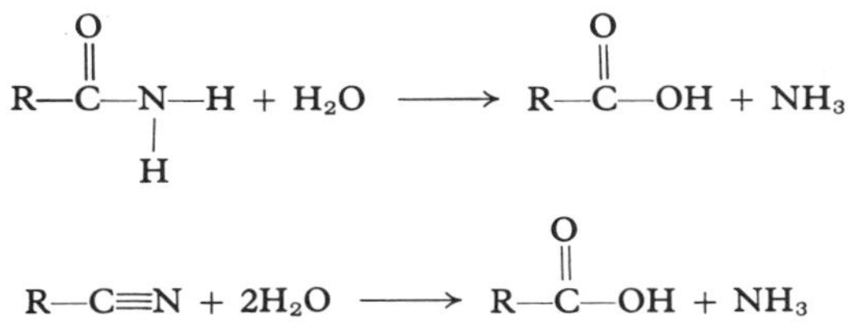

Basic Hydrolysis

Procedure a (*easily hydrolyzed amides or nitriles*).   Boil under reflux 10 mmole of the substance with 0.8 gram (20 mmoles) of sodium hydroxide in 3 ml of water until ammonia evolution can be detected no longer (4–10 hours; Notes 1 and 2).

Procedure b (*more difficultly hydrolyzed amides or nitriles*). Boil under reflux 10 mmole of the substance with 1.2 grams (20 mmoles) of potassium hydroxide in 4 ml of mono-, di-, or triethyleneglycol until ammonia evolution has ceased (about 5 hours; Note 1). At this time dilute the mixture with 10 ml of water.

Workup for Both Procedures a and b. Acidify the aqueous solution with 20% sulfuric acid. Collect the precipitated acid by suction filtration, wash it with water, and recrystallize it from water or aqueous methanol.

If the acid is a liquid or if it is relatively soluble in water, it should be converted to a phenacyl ester by the procedure of Section 29.19.

Acidic Hydrolysis: For Nitriles That Resist Basic Hydrolysis

Procedure. Heat in an oil bath at 160°C for 30 minutes, with stirring and using a reflux condenser, a mixture of 5 ml of 75% sulfuric acid, 200 mg of sodium chloride, and 10 mmole of the nitrile. Raise the temperature of the bath to 190°C and continue to heat and stir for another 30 minutes. Cool the mixture to room temperature and pour it onto 20 grams of ice. Collect the precipitate by suction filtration and dissolve it in 5 ml of 10% sodium hydroxide solution. Any insoluble amide should be removed by suction filtration and, if it is a major product, recrystallized from water or aqueous methanol. Acidify the filtrate to obtain the acid. Collect it by suction filtration and recrystallize from water, benzene, acetone/water, or alcohol/water. (Ref. 1, p. 291; Ref. 2, p. 321; Ref. 3, pp. 404, 410, 798, 805.)

Notes

1. Ammonia may be detected by the method described in Section 28.1.
2. If the amide or nitrile solidifies in the condenser, 0.5 ml of ethanol can be added to help to redissolve it; this should be removed by distillation at the end of the heating period.

29.9 9-ACYLAMIDOXANTHENES FROM AMIDES

Derivative for unsubstituted amides.

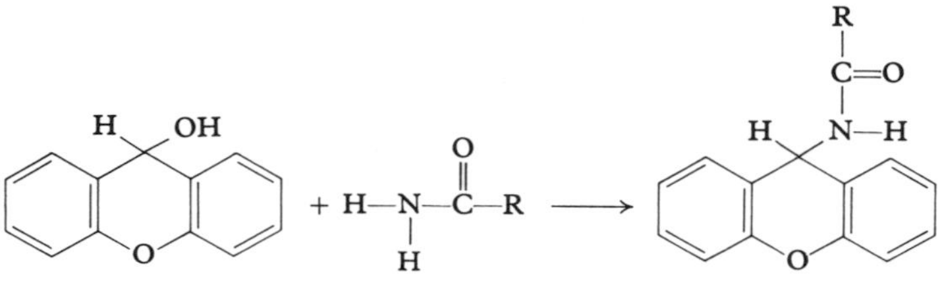

Procedure. Dissolve 2 mmoles (400 mg) of xanthydrol in 5 ml of glacial acetic acid (Note 1). Add to this 1–1.5 mmole of the unsubstituted amide (Note 2), heat the mixture in a beaker of water at 85°C for 20–30 minutes, and then allow it to cool for crystallization. Collect the product by suction filtration and recrystallize it from 2:1 dioxane:water or 2:1 ethanol:water. (Ref. 1, p. 256; Ref. 2, p. 269; Ref. 3, p. 405.)

1. If the solution is not clear, allow it to stand for a few minutes and decant or filter the clear liquid from the insoluble material.

2. If the amide is not soluble in glacial acetic acid, it may be added to the reaction mixture as a solution in 2 ml of ethanol; if this is done, add 1 ml of water to the mixture after heating and before cooling.

29.10 HYDROLYSIS OF N-SUBSTITUTED AMIDES

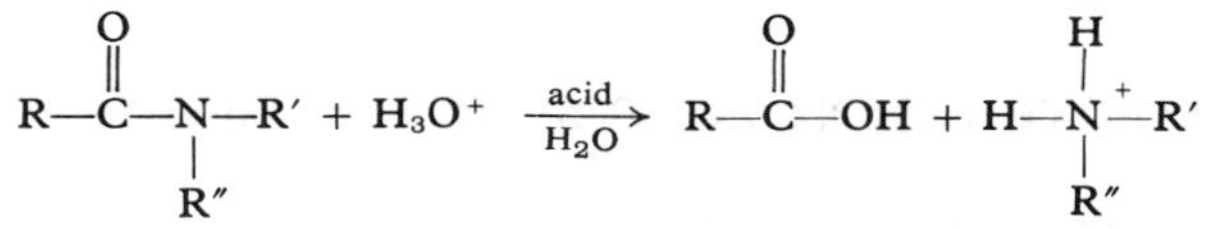

$$
\underset{\overset{|}{R''}}{R-\overset{\overset{\text{O}}{\|}}{C}-\overset{}{N}-R'} + H_3O^+ \xrightarrow[\text{H}_2\text{O}]{\text{acid}} R-\overset{\overset{\text{O}}{\|}}{C}-OH + H-\overset{\overset{\text{H}}{|}}{\underset{\overset{|}{R''}}{N}}{}^+{-}R'
$$

Procedure a. Boil under reflux for 2 hours a mixture of 0.5–1 gram of the amide and 10–20 ml of conc. hydrochloric acid. Cool the mixture and collect any solid by suction filtration. The solid may be the acid corresponding to the amide, the hydrochloride salt of the amine corresponding to the amide, or unchanged amide. If the solid is soluble in water, it is probably the salt of the amine. If it is insoluble in water but soluble in 5% aqueous sodium bicarbonate, it is probably the acid. The melting point as well as solubility properties will indicate unchanged starting material.

The amine may be recovered by making the solution alkaline (use care in neutralizing a concentrated acid solution) and then either filtering with suction or extracting with ether, depending upon the nature of the amine. It may then be purified by recrystallization and/or converted to a derivative by procedures suitable for primary and secondary amines.

The acid may be isolated from the solution by reacidification and extraction (if it were not soluble, it would have separated from the solution originally) and converted to a derivative.

Procedure b (*for amides that resist Procedure a*) (*Note 1*). Boil under reflux for 30 minutes to an hour a mixture of 1 gram of the amide in 10 ml of 70% sulfuric acid (Note 2). Cautiously pour 10 ml of water down the condenser and allow the solution to cool. Remove any solid by suction filtration and continue as in Procedure a. (Ref. 1, p. 255; Ref. 2, p. 267; Ref. 3, p. 801.)

Notes

1. Benzanilide and related compounds, for example.
2. Prepared by pouring 8 ml of conc. sulfuric acid onto 6 grams of ice.

29.11 SUBSTITUTED ACETAMIDES FROM AMINES

Derivative for water-insoluble primary and secondary amines.

$$
\underset{\overset{|}{R'}}{R-\overset{\cdot\cdot}{N}-H} + CH_3-\overset{\overset{\text{O}}{\|}}{C}-O-\overset{\overset{\text{O}}{\|}}{C}-CH_3 \longrightarrow \underset{\overset{|}{R'}}{R-\overset{\cdot\cdot}{N}-\overset{\overset{\text{O}}{\|}}{C}-CH_3} + CH_3-\overset{\overset{\text{O}}{\|}}{C}-O-H
$$

Procedure. Dissolve 0.5 gram of the amine in 25 ml of 5% hydrochloric acid. Add 5% sodium hydroxide solution in small portions until the mixture becomes cloudy (due to precipitation of the amine). Redissolve the precipitate by adding a little more 5% hydrochloric acid. Add about 10 grams of ice and then 5 ml of acetic anhydride. Next add with good stirring and all in one portion a previously prepared solution of 5 grams of sodium acetate trihydrate in 5 ml of water. Cool the mixture in an ice bath and allow it to stand (possibly overnight) for crystallization. (Compare the preparation of acetanilide, Section 60.1).

Collect the product by suction filtration and wash it well with water. Recrystallize from ethanol or ethanol/water or, if the crude product has been thoroughly dried, from benzene/cyclohexane. (Ref. 1, p. 259; Ref. 3, pp. 576, 652.)

29.12 SUBSTITUTED BENZAMIDES FROM AMINES

Derivative for primary and secondary amines.

$$R-\overset{..}{\underset{R'}{N}}-H + Cl-\overset{O}{\overset{||}{C}}-\phi \xrightarrow{\text{pyridine}} R-\overset{..}{\underset{R'}{N}}-\overset{O}{\overset{||}{C}}-\phi + \text{``HCl''}$$

pyridine

Procedure. Add dropwise 0.5 ml of benzoyl chloride to a solution of 0.5 gram of the amine in 5 ml of dry pyridine and 10 ml of dry benzene. Heat the resulting mixture in a beaker of hot water at 60–70°C for 30 minutes. Pour the mixture into 100 ml of water, separate the benzene layer, extract the aqueous layer with one 10-ml portion of benzene, and combine this with the benzene layer. Wash the combined benzene layers with water followed by 5% sodium carbonate solution, dry over anhydrous magnesium sulfate, and, after removing the magnesium sulfate by filtration, concentrate to a volume of 3 or 4 ml. Stir about 20 ml of hexane into the concentrated benzene solution in order to precipitate the derivative, collect the product by suction filtration, and wash it with hexane. Recrystallize the benzamide from cyclohexane/hexane or cyclohexane/ethyl acetate. (Ref. 1, p. 260; Ref. 2, p. 272; Ref. 3, p. 652.)

29.13 *p*-TOLUENESULFONAMIDES FROM AMINES

Derivative for primary and secondary amines (see Note 1).

$$R-\overset{..}{\underset{R'}{N}}-H + Cl-\overset{O}{\underset{O}{\overset{||}{\underset{||}{S}}}}-\langle\rangle-CH_3 \longrightarrow R-\overset{..}{\underset{R'}{N}}-\overset{O}{\underset{O}{\overset{||}{\underset{||}{S}}}}-\langle\rangle-CH_3 + HCl$$

Procedure. Boil for 30 minutes under reflux a mixture of 0.5 g of the amine, 1–1.5 g (5–8 mmoles) of *p*-toluenesulfonyl chloride (Note 2), and 3 ml of pyridine. Pour the reaction mixture into 5 ml of cold water and stir until the product crystallizes. Collect the sulfonamide by suction filtration, wash it well with water, and recrystallize it from alcohol or aqueous alcohol. (Ref. 1, p. 261; Ref. 2, p. 274; Ref. 3, p. 653.)

Notes

1. Compare the Hinsberg test, Section 28.2.
2. An equivalent amount of benzenesulfonyl chloride may be used; the *p*-toluenesulfonamides are said to be more satisfactory derivatives.

Derivatives for primary and second amines.

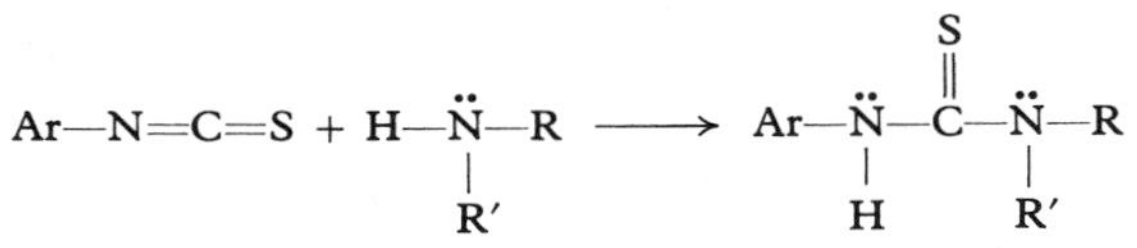

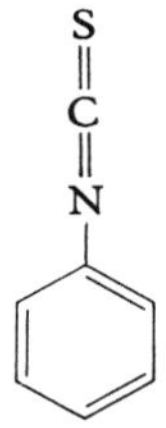

$$Ar-N=C=S + H-\overset{\cdot\cdot}{N}-R \longrightarrow Ar-\overset{\cdot\cdot}{\underset{H}{N}}-\overset{\overset{\displaystyle S}{\|}}{C}-\overset{\cdot\cdot}{\underset{R'}{N}}-R$$

Procedure. Dissolve 0.2 gram of the amine in 5 ml of ethanol (Note 1) and add to this a solution of 0.2 g (0.18 ml; 1.5 mmole) of phenylisothiocyanate or 0.28 g (1.5 mmole of α-naphthylisothiocyanate in 5 ml of ethanol. If no reaction takes place at room temperature, heat the mixture for one or two minutes. If no crystals separate upon cooling and scratching (as with aromatic amines), reheat the mixture for another 10 minutes and then cool again (Note 2). Collect the product by suction filtration and recrystallize it from ethanol. (Ref. 1, p. 261; Ref. 2, p. 275; Ref. 3, p. 422.)

phenyl-
isothiocycanate

Notes

1. These reagents are not reactive toward water (compare the analogous isocyanates, Section 28.3), and so an aqueous solution of the amine may be used if it is difficult to obtain the pure amine.

2. If no product can be obtained with an aromatic amine, the reaction can be tried without a solvent with heating for 10 minutes. The product can be isolated at the end of the reaction by the addition of 50% aqueous ethanol.

29.15 PICRATES

Derivative for tertiary amines, aromatic ethers, and aromatic hydrocarbons (Note 1).

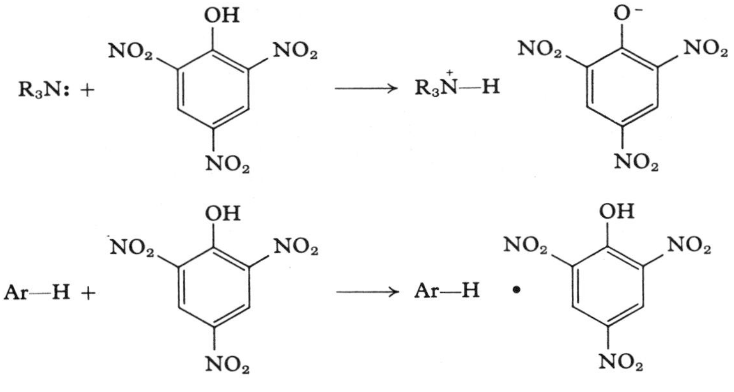

$$R_3N: + \quad \text{(picric acid)} \longrightarrow R_3\overset{+}{N}-H \quad \text{(picrate anion)}$$

$$Ar-H + \quad \text{(picric acid)} \longrightarrow Ar-H \bullet \text{(picric acid)}$$

Procedure a (*for tertiary amines; also for aromatic ethers and hydrocarbons that form relatively stable picrates*). Dissolve 2 mmoles of the substance in 5 ml of 95% ethanol, and add to this solution a solution of 2.2 mmoles of picric acid (500 mg) in 10 ml of 95% ethanol. Heat the mixture to boiling on the steam bath and allow it to cool slowly for crystallization. Collect the product by suction filtration and recrystallize it from ethanol (Note 2).

Procedure b (*for substances that form relatively unstable picrates*). Dissolve 5 mmoles of the substance in about 5 ml of boiling chloroform and add a solution of 5 mmoles (1.15 grams) of picric acid in 3 ml of boiling chloroform. Swirl the resulting mixture and allow it to cool for crystallization. Collect the product by suction filtration and recrystallize it from a minimum amount of chloroform (Notes 2 and 3). (For amine picrates see Ref. 1, p. 263; Ref. 2, p. 277; Ref. 3, p. 422. For ether and hydrocarbon picrates, see Ref. 1, p. 277; Ref. 2, pp. 298, 315; Ref. 3, pp. 518, 672.)

Notes

1. Certain other aromatic compounds such as primary and secondary amines and halides will form picrates.

2. The picrates of some substances cannot be recrystallized because they dissociate to too large a degree in solution.

3. The m.p. should be determined immediately because some picrates decompose.

29.16 QUATERNARY AMMONIUM SALTS: METHIODIDES AND *p*-TOLUENESULFONATES

Derivatives for tertiary amines.

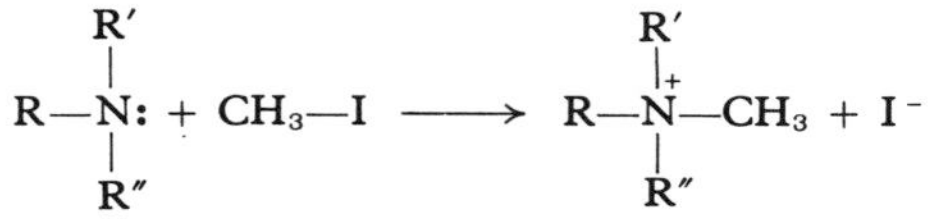

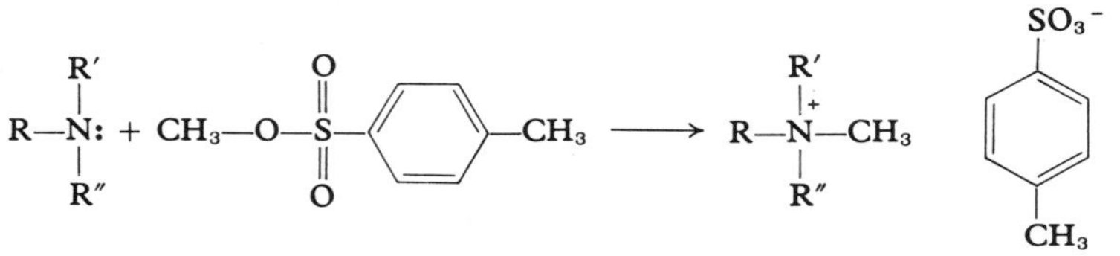

Procedure. Dissolve 0.5 gram of the tertiary amine in twice its volume of nitromethane, acetonitrile, or alcohol (Note 1); do the same for 1 gram of the quaternizing agents (methyl iodide or methyl *p*-toluenesulfonate). Combine the solutions, allow the mixture to stand at room temperature for an hour, and then heat it for 30 minutes on the steam bath. If the quaternary salt crystallizes out, it should be collected by suction filtration; if not, the solvent should be removed under vacuum (Section 36). The crude product or residue should be recrystallized from ethyl acetate/ethanol. (Ref. 1, p. 262; Ref. 2, p. 277; Ref. 3, p. 660.)

Note

1. The solvents named are listed in decreasing order of suitability for the reaction.

29.17 CARBOXYLIC ACID AMIDES

A procedure for deriving amides from carboxylic acids. Lower-molecular-weight amides, which are relatively soluble in water, are difficult to isolate by this procedure.

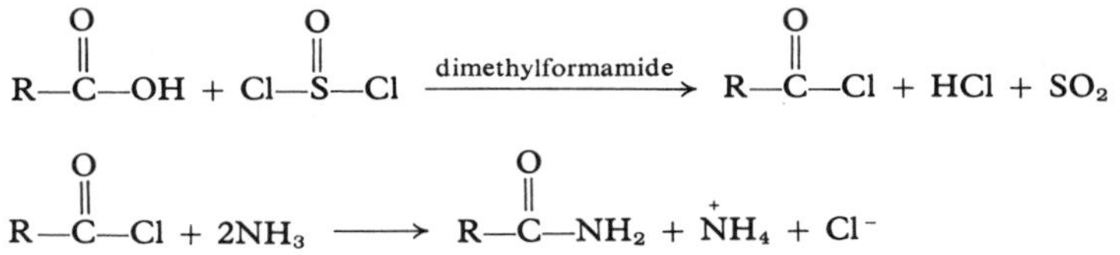

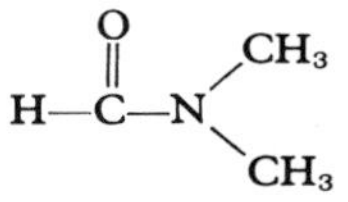

$$R-\overset{O}{\underset{||}{C}}-OH + Cl-\overset{O}{\underset{||}{S}}-Cl \xrightarrow{\text{dimethylformamide}} R-\overset{O}{\underset{||}{C}}-Cl + HCl + SO_2$$

$$R-\overset{O}{\underset{||}{C}}-Cl + 2NH_3 \longrightarrow R-\overset{O}{\underset{||}{C}}-NH_2 + \overset{+}{N}H_4 + Cl^-$$

$$H-\overset{O}{\underset{||}{C}}-N\overset{CH_3}{\underset{CH_3}{}}$$
dimethylformamide

Procedure. Boil under reflux on the steam bath, using a calcium chloride drying tube in the condenser, a mixture of 1 gram of the carboxylic acid, 5 ml of thionyl chloride, and 1 drop of dimethylformamide for 15–30 minutes. Pour the mixture *cautiously* into 15 ml of ice-cold conc. ammonium hydroxide. Collect the crude amide by suction filtration and recrystallize it from water or aqueous alcohol. (Ref. 1, p. 235; Ref. 3, p. 361.)

29.18 ANILIDES, *p*-TOLUIDIDES, AND *p*-BROMOANILIDES OF CARBOXYLIC ACIDS

Derivatives for carboxylic acids, acid halides, and anhydrides.

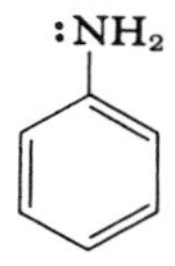

$$R-\overset{O}{\underset{||}{C}}-OH + Cl-\overset{O}{\underset{||}{S}}-Cl \xrightarrow{\text{dimethylformamide}} R-\overset{O}{\underset{||}{C}}-Cl + HCl + SO_2$$

$$R-\overset{O}{\underset{||}{C}}-X + 2Ar-\ddot{N}H_2 \longrightarrow R-\overset{O}{\underset{||}{C}}-\underset{\underset{H}{|}}{\overset{..}{N}}-Ar + Ar-\overset{+}{N}H_3 + X^-$$

aniline

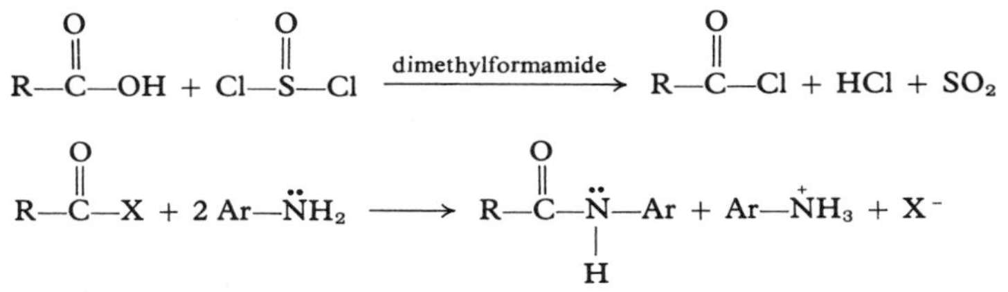

Preparation of the Acid Chloride. Heat for 30 minutes 1 mmole of the acid with 0.8 ml (1.3 g; 11 mmole) of thionyl chloride, and one drop of dimethylformamide on the steam bath under reflux, using a calcium chloride drying tube on the condenser.

Treatment of the Acid Halide or Anhydride with the Aromatic Amine. Add to the acid chloride just prepared, or to 1 mmole of the acid halide or anhydride, dissolved in dry benzene if it is a solid, a solution of 5 mmoles of the aromatic amine (aniline, *p*-toluidine, or *p*-bromoaniline) dissolved in 20 ml of benzene. Heat the mixture under reflux on the steam bath for about 15 minutes. Cool the mixture to room temperature, transfer it to a separatory funnel, and extract it with 2 ml of water, 5 ml of 5% hydrochloric acid, 5 ml of 5% sodium hydroxide, and 2 ml of water. Evaporate the benzene solution to dryness and recrystallize the residual amide from aqueous alcohol, methanol, or ethanol. (Ref. 1, p. 236; Ref. 2, p. 238; Ref. 3, p. 361.)

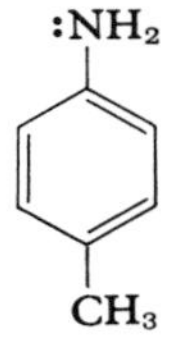

p-toluidine

29.19 PHENACYL AND SUBSTITUTED PHENACYL ESTERS OF CARBOXYLIC ACIDS

Derivative for carboxylic acids.

$$Ar-\overset{O}{\underset{||}{C}}-CH_2-Br + R-\overset{O}{\underset{||}{C}}-O^- \longrightarrow Ar-\overset{O}{\underset{||}{C}}-CH_2-O-\overset{O}{\underset{||}{C}}-R + Br^-$$

Procedure a (*for pure acids*).   Dissolve 100 mg (1 mmole) triethylamine in 2 ml of dry acetone and neutralize the solution by the addition of the carboxylic acid. To this solution, add a solution of 0.5 mmole of the phenacyl bromide (Note 1) (phenacyl bromide, *p*-chlorophenacyl bromide, *p*-bromophenacyl bromide, or *p*-phenylphenacyl bromide) in 3 ml of dry acetone. Allow the mixture to stand for 3 hours at room temperature (a precipitate of triethyl ammonium bromide will form after a short time); then dilute it with 10 ml of water and collect the precipitate by suction filtration. After thoroughly washing the crude phenacyl ester with 5% sodium bicarbonate solution and finally with water, recrystallize it from aqueous alcohol.

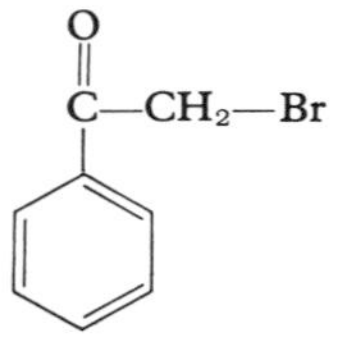

Procedure b (*for salts or aqueous solutions of the acid*).   Dissolve 150 mg of the salt in 2 ml of water and add 2 drops of 5% hydrochloric acid; *or* add 5% hydrochloric acid to 2 ml of a basic aqueous or aqueous alcoholic solution of the acid containing about 100 mg of the acid until it is neutral to litmus, and then add 2 more drops of 5% hydrochloric acid. To this slightly acidic solution of the sodium salt of the acid (Note 2), add a solution of 200 mg of the phenacyl bromide (Notes 1 and 3) in ethanol and heat the mixture under reflux—(1 hour for a monocarboxylic acid; 2 hours for a dicarboxylic acid; 3 hours for a tricarboxylic acid. Occasionally a solid separates from solution during the reflux period; it should be brought into solution by the addition of a few milliliters of ethanol. At the end of the heating period, allow the mixture to cool to room temperature and collect the crude phenacyl ester by suction filtration. It should be washed and recrystallized as in Procedure a. (Ref. 1, p. 235; Ref. 2, p. 243; Ref. 3, p. 362.)

Notes

1. The phenacyl halides are lachrymatory and tend to irritate the skin. They should be treated with respect.
2. The reaction mixture should not be alkaline; alkali causes the hydrolysis of the phenacyl bromide to the phenacyl alcohol.
3. *p*-Bromophenacyl bromide is converted to the relatively insoluble *p*-bromophenacyl chloride (m.p. 117°C) by chloride ion. If the mixture contains relatively large amounts of chloride ion, another phenacyl halide should be used.

29.20 *p*-NITROBENZYL ESTERS OF CARBOXYLIC ACIDS

Derivative for carboxylic acids.

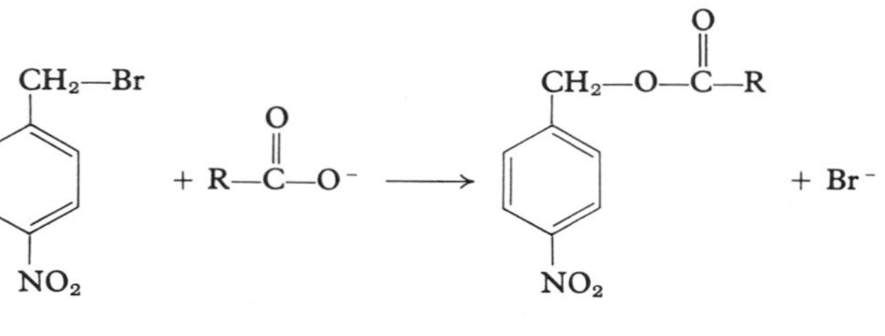

Procedure.   The procedure for the preparation of phenacyl esters (Section 29.19) may be used with the following exceptions:

· *Procedure a*: *p*-Nitrobenzyl bromide is used in place of the phenacyl bromide; ethanol is used as solvent in place of acetone.

- *Procedures a and b:* If *p*-nitrobenzyl chloride is substituted for *p*-nitrobenzyl
bromide, 10 mg of sodium iodide should be added.

The benzyl halides are lachrymatory and tend to irritate the skin; they should be treated with respect. (Ref. 1, p. 235; Ref. 2, p. 244; Ref. 3, p. 362.)

29.21 N-BENZYLAMIDES FROM ESTERS

Derivate for esters.

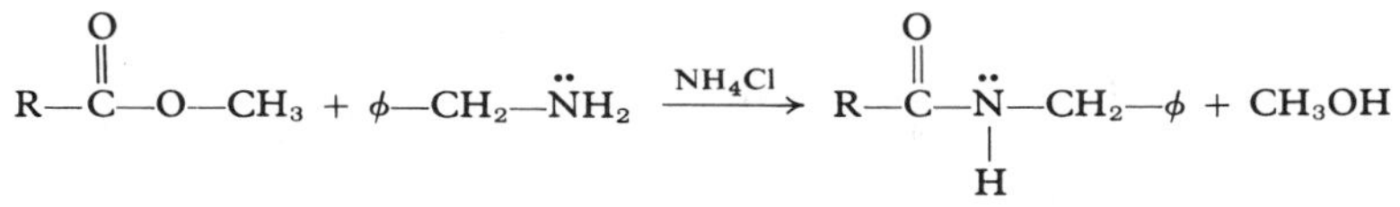

$$R-\overset{\overset{\textstyle O}{\|}}{C}-O-CH_3 + \phi-CH_2-\overset{..}{N}H_2 \xrightarrow{NH_4Cl} R-\overset{\overset{\textstyle O}{\|}}{C}-\underset{\underset{\textstyle H}{|}}{\overset{..}{N}}-CH_2-\phi + CH_3OH$$

Procedure. Heat under reflux for about 1 hour a mixture of 0.5 g of the methyl or ethyl ester (Note 1), 1.5 ml of benzylamine, and 50 mg of ammonium chloride. Cool the mixture and stir it with a little water and finally a little dilute hydrochloric acid (Note 2). Collect the solid by suction filtration and recrystallize it from alcohol/water or acetone/water. (Ref. 1, p. 272; Ref. 2, p. 290; Ref. 3, p. 394.)

Notes

1. Higher esters must first be transesterified to the methyl ester by heating 0.6–1 gram under reflux for 30 minutes in 10 ml of anhydrous methanol in which 0.1 g of sodium or sodium methoxide has been dissolved. The residue obtained by evaporation of the excess methanol can then be used in the procedure described.

2. Avoid an excess of hydrochloric acid, as it may dissolve the product as well.

29.22 3,5-DINITROBENZOATES FROM ESTERS

Derivative for the alcohol fragment of an ester (Note 1).

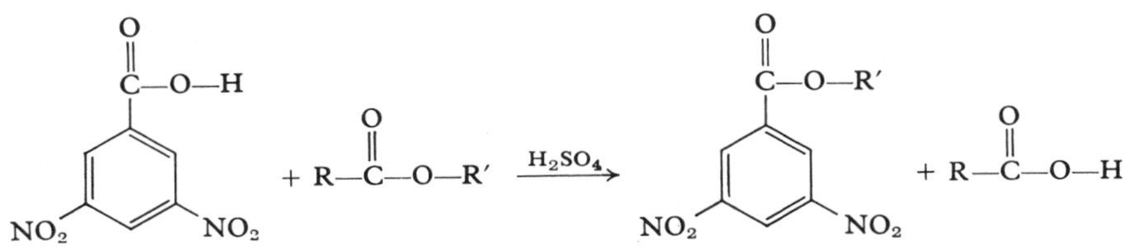

Procedure. Mix 0.5 gram of the ester with 0.5 gram of powdered 3,5-dinitrobenzoic acid, add a small drop of conc. sulfuric acid, and heat the mixture. Heat under reflux if the ester boils below 150°C, and at 150°C in an oil bath if it boils above 150°C; heat until the 3,5-dinitrobenzoic acid dissolves and then for an additional 30 minutes. Cool the mixture, dissolve it in 30 ml of ether, and extract it twice with 5% sodium bicarbonate solution to remove 3,5-dinitrobenzoic acid and sulfuric acid (*caution:* foaming from CO_2 evolution). Finally, after washing the ether extract with water, evaporate it to dryness and recrystallize the residue by dissolving it in a minimum of hot ethanol, adding water to incipient cloudiness, and allowing the solution to cool. It may be necessary to induce crystallization by scratching. (Ref. 1, p. 273; Ref. 2, p. 292; Ref. 3, p. 393.)

Note

1. This procedure is not satisfactory for esters of alcohols that are unstable to conc. sulfuric acid, such as tertiary alcohols or certain unsaturated alcohols. Higher-molecular-weight esters react slowly or not at all.

29.23 HYDROLYSIS OF ESTERS

$$\underset{\substack{\| \\ \text{R—C—O—R'}}}{\overset{O}{}} + HO^- \xrightarrow{H_2O} \underset{\substack{\| \\ \text{R—C—O}^-}}{\overset{O}{}} + H—O—R'$$

Procedure. Boil under reflux a mixture of 1 gram of the ester and 10 ml of 1 M sodium hydroxide until a clear solution has been obtained (Note 1).

Some of the solution may be used to identify the acid corresponding to the ester by means of Procedure b in Section 29.19 (Note 2). The remainder of the solution may be saturated with potassium carbonate to salt out the alcohol, which can then be extracted and identified by one of the procedures described in Sections 29.1–29.3. (Ref. 2, p. 287; Ref. 3, p. 391.)

Notes

1. If the alcohol corresponding to the ester is not soluble in water, it will be present as an oil after hydrolysis.
2. If the acid is relatively insoluble in water, it may be isolated by acidification of the hydrolysate and collected by suction filtration.

29.24 BROMINATION OF AROMATIC ETHERS

Derivative for aromatic ethers.

$$\text{Ar—H} + \text{Br}_2 \longrightarrow \text{Ar—Br} + \text{HBr}$$

Procedure. Add 1 mmole of the aromatic ether to the calculated amount of a 1% (by volume) solution of bromine in acetic acid (Note 1) that has been cooled in an ice bath; continue cooling during the addition. After a few more minutes, remove the mixture from the ice bath and allow it to stand at room temperature for 15 minutes. Isolate the product by diluting the mixture with water and collecting the precipitate by suction filtration. Remove traces of bromine from the product by washing it with dilute sodium bisulfite solution and then water. It may be recrystallized from alcohol or alcohol/water. (Ref. 1, p. 276; Ref. 2, p. 297.)

Note

1. A solution of 1 ml of bromine in 100 ml of acetic acid will contain 1 millimole of bromine per 5.2 ml of solution. Depending upon whether a mono-, di-, or tribromo derivative is expected, a 10–20% excess over 1, 2, or 3 millimoles of bromine should be used.

Derivative for primary and secondary alkyl chlorides, bromides, and iodides.

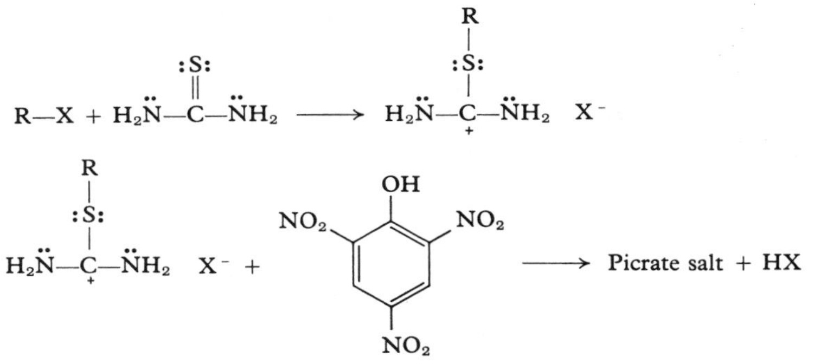

Procedure (Reference 4).   Place in a 10-ml standard-taper Erlenmeyer flask 5 ml of ethylene glycol, 300 mg of thiourea, 6 drops of the alkyl halide, and two boiling stones. Fit the flask with a reflux condenser and clamp the flask in position in an oil bath that has been preheated to 117°C. After the flask has been held at this temperature for 30 minutes (Note 1), add down the condenser 1 ml of a saturated solution of picric acid in ethanol (Note 2). Keep the flask in the oil bath for another 15 minutes; then remove it, cool it to room temperature, and then add 5 ml of water. Finally, cool the mixture in an ice bath for 15 minutes, with occasional swirling. Collect the S-alkylthiuronium picrate by suction filtration. The derivative may be recrystallized from methanol. (Ref. 1, p. 280; Ref. 2, p. 302; Ref. 3, p. 291; Ref. 4 H. M. Crosby and J. B. Entrikin, *J. Chem. Educ.* **41**, 360 (1964).)

$$HO—CH_2CH_2—OH$$
ethylene glycol

Notes

1. If decomposition occurs (odor of mercaptan; red color), repeat the experiment with the oil bath held at 65°C.
2. The solubility of picric acid is reported to be 1 gram per 12 ml of ethanol.

29.26 *o*-AROYLBENZOIC ACIDS FROM AROMATIC HYDROCARBONS

Derivative for aromatic hydrocarbons (Note 1).

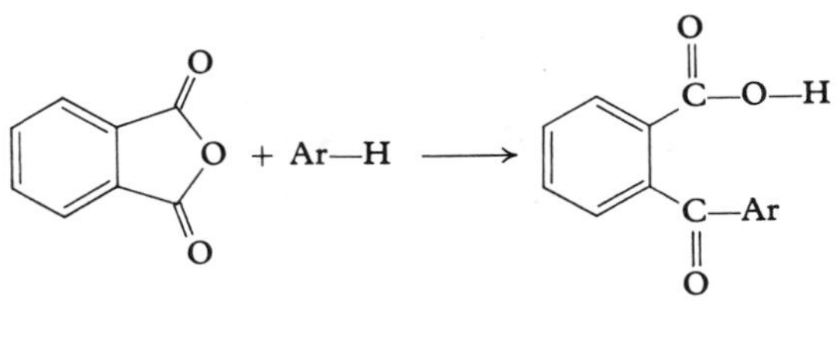

Procedure.   Add 2.5 grams of powdered anhydrous aluminum chloride to a mixture of 0.5 gram of the hydrocarbon and 0.6 gram of phthalic anhydride in 2 or 3 ml of dry dichloromethane while the mixture is cooled in an ice bath. After the initial reaction has subsided, either allow the mixture to stand at room temperature or heat it under reflux, until the evolution of hydrogen chloride has occurred (about half an hour); if heating has been used, allow the mixture to cool. Then add 5 grams of ice and 5 ml of conc. hydrochloric acid. When the addition product has been hydrolyzed, collect the product by suction filtration and wash it well with water (Note 2). The aroylbenzoic acid should be purified by dissolving it with heating in 5 ml of conc. sodium carbonate solution, boiling the solution with activated carbon for 5 minutes, filtering, and acidifying to a pH of 3 by the addition of 20% hydrochloric acid. The precipitated acid should then be collected by suction filtration, washed with water, and recrystallized from aqueous alcohol or (after drying) from toluene/petroleum ether. (Ref. 1, p. 285; Ref. 2, p. 519.)

Notes

1. Certain aryl halides may be characterized in this manner also.
2. If the product does not crystallize immediately, it should be allowed to stand overnight.

29.27 AROMATIC ACIDS BY OXIDATION BY PERMANGANATE

Derivative for aromatic hydrocarbons and aryl bromides and chlorides.

$$\text{Ar-R} \xrightarrow{\text{KMnO}_4} \text{Ar-}\overset{\displaystyle O}{\overset{\|}{C}}\text{-O-H}$$

Procedure. Add 1 gram of the hydrocarbon to a solution of 3 grams of potassium permanganate and 1 gram of sodium carbonate in 75 ml of water and heat the mixture under reflux until the permanganate color has disappeared (in 15 minutes to 4 hours; Note 1). Cool the solution to room temperature and acidify it cautiously by the addition of 50% sulfuric acid (Note 2). Remove the manganese dioxide by the addition of a concentrated solution of sodium bisulfite (with good stirring and possibly heating on the steam bath) and, after cooling the mixture thoroughly in an ice bath, collect the acid by suction filtration. It may be recrystallized from water. (Ref. 1, p. 285; Ref. 2, pp. 315, 326; Ref. 3, p. 520.)

Notes

1. Potassium permanganate can be detected by dipping a stirring rod into the mixture and touching it to a piece of filter paper; a pink color in the ring around the dark spot of manganese dioxide indicates the presence of permanganate.
2. 50% sulfuric acid may be prepared by cautiously pouring 5 ml conc. sulfuric acid over 10 grams ice.

29.28 ANILIDES, p-TOLUIDIDES, AND α-NAPHTHALIDES FROM ALKYL HALIDES

Derivatives for alkyl halides and aryl bromides and iodides.

$$\text{R-X} + \text{Mg} \xrightarrow{\text{dry ether}} \text{R-Mg-X}$$

$$\text{R-Mg-X} + \text{Ar-N=C=O} \longrightarrow \text{Ar-}\underset{\displaystyle H}{\overset{\displaystyle }{N}}\text{-}\overset{\displaystyle O}{\overset{\|}{C}}\text{-R}$$

Preparation of the Grignard Reagent. Add a solution of 1 ml of the halide in 5 ml of absolute ether to 0.4 gram of magnesium turnings and a crystal of iodine in a dry flask.

Preparation of the Derivative. To the freshly prepared Grignard reagent, add a solution of 0.5 ml of the isocyanate (phenyl-, p-tolyl-, or α-naphthylisocyanate) in 10 ml of absolute ether. After swirling the mixture and allowing it to stand for 10 minutes, add 20 grams of crushed ice and 1 ml of conc. hydrochloric acid. After hydrolysis of the solid, separate the ether layer; then, after drying over anhydrous magnesium sulfate, evaporate the ether to give the derivative as a solid residue. The residual amide may be recrystallized from methanol, ethanol, or aqueous alcohol. (Ref. 1, p. 279; Ref. 3, p. 290.)

29.29 2,4,7-TRINITROFLUORENONE ADDUCTS OF AROMATIC HYDROCARBONS

Derivative for aromatic hydrocarbons.

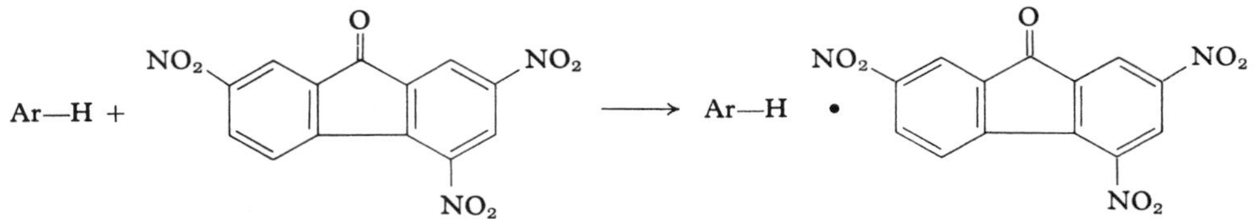

Procedure. Dissolve 100 mg of 2,4,7-trinitrofluorenone (0.3 mmole) in a mixture of 10 ml of absolute methanol or ethanol and 2 ml of benzene. To this hot solution add a solution of an equivalent amount (0.3 mmole) of the aromatic hydrocarbon in the minimum amount of 2:1 alcohol:benzene. Heat the mixture for 30 seconds and then allow it to cool for crystallization. Collect the product by suction filtration. It may be recrystallized from absolute ethanol or ethanol/benzene. (Ref. 2, p. 315.)

29.30 BROMINATION OF PHENOLS

Derivative for phenols.

$$Ar-O-H + Br_2 \longrightarrow \text{Brominated phenol} + HBr$$

Procedure. Prepare a solution of 7.5 grams of potassium bromide and 5 grams (1.6 ml; 31 mmoles) of bromine in 50 ml of water. Add this solution, dropwise, to a well-stirred solution of 0.5 gram of the phenol dissolved in water, dioxane, ethanol, or acetone until a weak yellow color persists. Precipitate the brominated phenol by stirring in 25 ml of cold water and collect the crude product by suction filtration. Wash the product with dilute sodium bisulfite solution to remove free bromine and then recrystallize it from ethanol or aqueous alcohol. (Ref. 1, p. 298.)

29.31 ARYLOXYACETIC ACIDS FROM PHENOLS

Derivative for phenols.

Procedure. Add 1.25 grams of monochloroacetic acid to a mixture of 1 gram of the phenol and 4 ml of 10 M sodium hydroxide solution. Add 1 or 2 ml of water as necessary to give a homogeneous solution. Heat the mixture for 1 hour on the steam bath, cool it to room temperature, dilute it with 10 or 15 ml of water, and acidify it to a pH of 3 with dilute hydrochloric acid. Extract the product with 50 ml of ether, wash the extract with 10 ml of water, and then extract the product from the ethereal solution by means of 25 ml of 5% aqueous sodium carbonate solution. Precipitate the product by acidifying the sodium carbonate solution with dilute hydrochloric acid (CO_2 evolution) and collect it by suction filtration. It may then be recrystallized from water. (Ref. 1, p. 298; Ref. 2, p. 331; Ref. 3, p. 682.)

References

1. R. L. Shriner, R. C. Fuson, and D. Y. Curtin, *The Systematic Identification of Organic Compounds*, 5th edition, Wiley, New York, 1965.
2. N. D. Cheronis and J. B. Entrikin, *Identification of Organic Compounds*, Interscience, New York, 1963.
3. A. I. Vogel, *A Textbook of Practical Organic Chemistry*, 3rd edition, Wiley, New York, 1957.
4. Z. Rappoport, *Handbook of Tables for Organic Compound Identification*, 3rd edition, The Chemical Rubber Co., Cleveland, Ohio, 1967.

Apparatus and Techniques for Chemical Reactions

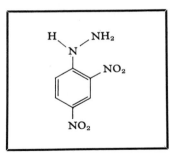

Effecting a chemical reaction in which a substance of one structure is converted to a substance of a different structure typically involves (1) combining the materials that are to react, usually with a solvent and/or a catalyst, (2) maintaining the mixture at a certain temperature for a certain length of time, with stirring if necessary, and then (3) isolating the desired product by means of the separation procedures described in Sections 7–15. The following sections describe some of the ways in which these various phases of the operation may be carried out on the usual laboratory scale, as well as a number of variations of this basic procedure. Industrial-scale and micro-scale reactions often involve considerably different procedures.

§30 ASSEMBLING THE APPARATUS

The apparatus in which a reaction is to be carried out may be as simple as a single test tube or flask. Often it is made up of several glass components, such as a boiling flask and condenser, plus a drying tube to exclude moisture, a separatory funnel by which a liquid reagent or solution can be added, a trap to adsorb harmful gases that may be evolved, or any of a number of other parts. Figures 30.1 and 30.2 illustrate two of the assemblies that are most often used for carrying out reactions.

There is no doubt that the nicest and most convenient components from which the apparatus can be assembled are those that can be connected by means of ground-glass joints. The joints should normally be put together *without* the use of a lubricant. A lubricant may be used on ground-glass joints for either of two reasons: (1) to make them vacuum-tight, and (2) to keep them from sticking together. If the joints are well ground, and most of them are, lubrication will be needed for vacuum tightness only in vacuum distillation at oil pump pressures (less than about 5 Torr). If the apparatus is taken apart *immediately* after use, a lubricant will be needed to prevent sticking only if strong basic solutions (NaOH, KOH) have been involved in the experimental procedure—as, for example, in the steam distillation of aniline (Section 59). If a lubricant must be

No grease except...

225

employed, use only a very small amount spread out in a thin, even layer on the inner member. Avoid silicone-based lubricants, as they are practically impossible to remove and will prevent the glass from ever being wet by water again.

If the glass components do not have ground-glass joints, they must be joined by means of corks or rubber stoppers: one member is placed through a hole in the cork or stopper, and this is then plugged into the other.

As illustrated in Figures 30.1 and 30.2, the weight of the apparatus should be supported by clamps so that torque is minimized.

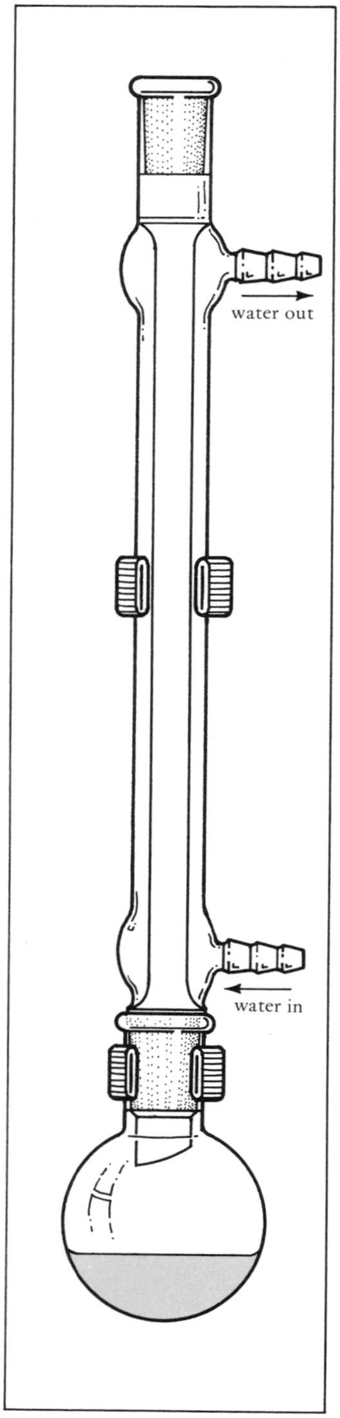

§31 TEMPERATURE CONTROL

For two reasons, a reaction should be carried out at a definite, constant temperature. The first reason is that *the rate of a reaction increases sharply with increasing temperature, and a reaction will proceed at a convenient or desirable rate only within a relatively narrow temperature range.* The rate increases with temperature according to the relationship

$$\text{Rate} \propto e^{-\Delta G^{\ddagger}/RT}$$

where $\Delta G^{\ddagger}$ is the free energy of activation of the reaction, R is the gas constant, and T is the absolute temperature. The second reason is that *the rates of competing or side reactions increase faster with a given increase in temperature than the rate of the desired reaction.* This is because the free energies of activation of the side reactions are greater than the free energy of activation of the desired reaction. Thus, at higher temperatures, the relative amount of side products will increase. Therefore, the choice of the optimum temperature at which to carry out a reaction always represents a compromise between a desire to carry out the reaction as fast as possible and a desire to minimize side reactions. The exponential dependence of rate upon temperature makes it necessary to control the temperature over a narrow range.

One of the easiest ways to maintain a reaction mixture at an approximately constant temperature is to heat it until it boils, using a condenser to condense the vapor and return the condensate to the flask (to heat under reflux). As long as the composition of the mixture does not change, the boiling point will remain constant. Since the composition must change as the reaction progresses, an excess of one of the reactants may be used as a solvent so that the boiling point will change less during the course of the reaction. Sometimes the solvent is chosen so that the mixture will boil at the desired temperature. This is the usual procedure if the reactants are solids.

The reaction mixture may be boiled by heating the flask with a flame, an oil or steam bath, or a heating mantle. These methods of heating are described in Section 32.1.

Control of Highly Exothermic Reactions

A highly exothermic reaction presents an additional problem since, as the rate of a reaction increases, the rate of heat evolution will increase. If the rate of heat evolution is greater than the rate at which heat can be removed (by condensation of the vapors of the boiling mixture and all other cooling methods), the reaction will proceed at an increasingly greater rate as the temperature of the mixture rises until it boils over or blows out the condenser.

One way of controlling a highly exothermic reaction is to add one of the reactants to the mixture at such a rate that the reaction can be maintained at the

Figure 30.1. Apparatus for carrying out a reaction under reflux.

water out

water in

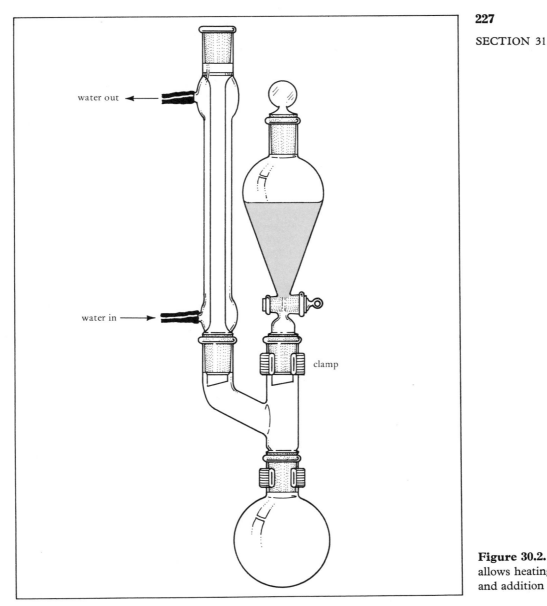

water out

water in

clamp

Figure 30.2. Apparatus that allows heating under reflux and addition of liquid reagent.

desired temperature (the boiling point or below) by the heat evolved in the reaction. The heat of the reaction may be the only heat source, or, if the reaction is only moderately exothermic or if the temperature is relatively high, an external source of heat can be used as well. More often it is necessary to remove heat by external cooling. This will always be necessary if the temperature is to be kept below the boiling point of the mixture. In these cases, a cooling bath must be used, and the rate of addition of the reactant will be determined by how efficiently heat can be removed from the mixture. Different types of cooling baths are described in Section 32.2.

By rate of addition of reactant

Warning: The danger of this method of controlling exothermic reactions is that the reactant may unintentionally be added faster than it reacts. It is particularly easy to add too much reactant at the beginning, when the temperature may be too low for the reaction to proceed at a reasonable rate. If too much reactant accumulates in the mixture, the reaction may accelerate out of control when it finally starts. The only way to be sure the reaction is under way and under control

Warning

is to determine that the temperature rises when the reactant is added and starts to fall when the addition is stopped or slowed down.

The second method of controlling an exothermic reaction is to run the reaction in small batches, or on a small scale, in a large flask so that a large surface area is available for efficient cooling. When the temperature of the reaction starts to climb, the flask can be plunged into a cold bath and swirled for a short time; remove it when the temperature rise has been checked. The use of a relatively large amount of solvent (and a still larger flask so that the mixture can be cooled efficiently in a cold bath) can serve to make the reaction more easily controlled both by decreasing the rate (of bimolecular reactions) by dilution and by increasing the total heat capacity of the mixture. This second method requires a little more judgment and skill on the part of the experimenter, and is dangerous if large amounts are used.

From this discussion of exothermic reactions, it should be obvious that new or unknown reactions should be carried out on a small scale until their exothermicity has been determined. Another, less obvious consequence is that, as a reaction is scaled up, more attention must be paid to the problem of heat transfer. If the volume of the reaction (and thus of the apparatus) is increased by a factor of 8, the surface area of the apparatus (through which the heat must flow) is increased only by a factor of $8^{2/3} = 4$. Whereas a cooling bath might not have been needed on the small-scale reaction, it might be required for the large-scale reaction.

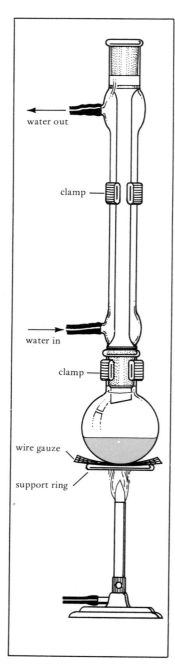

Figure 32.1. Heating with a burner flame. Rate of heating is determined by the size of the flame and the height of the flask above the flame. The wire gauze moderates and spreads the heat of the flame.

§32 METHODS OF HEATING AND COOLING

The typical chemical reaction is not carried out at room temperature. The next two sections describe methods by which reaction mixtures may be heated or cooled.

32.1 HEATING

Heating with a Flame

One very common method of heating is with a Bunsen burner flame. The height of the flame and the distance of the flask from the flame control the rate of heat input (see Figure 32.1).

The advantages of using a burner as a source of heat are that it can be set up very quickly and easily, and that the mixture can be brought up to temperature very rapidly. However, a burner has three important disadvantages. The first is that it must *not* be left unattended. It might go out, creating a gas leak that could lead to an explosion; or the height of the flame might change, resulting in either too little heat to maintain boiling or so much heat that the mixture boils over and catches fire. The second disadvantage is that if the flask should crack or break, or if the reaction should boil over because it suddenly became more exothermic or for some other reason, the burner could also set the mixture on fire. The third is that a burner flame is an exceedingly uneven source of heat. The glass at the bottom of the flask will be very much hotter than the rest of the flask, which may result in some undesired reactions. If there is a solid in the reaction mixture, boiling may be uneven with lots of bumping. However, if relatively small amounts are involved, if the heating period is short, and if the mixture will be watched, heating with a flame will often be most satisfactory.

Three methods of electrical heating are very often used. In the first, an electrically heated jacket (*heating mantle*) is placed around the lower half of the flask (sometimes a second part is used around the upper half also). Figure 32.2 illustrates the use of such a mantle. The rate of heating is controlled by the voltage input to the mantle from the variable voltage transformer. A similar device, the Thermowell heater, does not require the use of a variable voltage transformer. The rate of heating of the Thermowell device is set by means of a proportional controller, which determines the fraction of the time that current is allowed to flow through the heater.

Heating mantle

The advantages of this method are that the apparatus can more safely be left unattended (overnight, for example); it is a very constant and very easily controlled source of heat; and it is not nearly as likely to result in ignition of flammable materials as is a flame. There are several disadvantages, however. One is that the glass of the flask is likely to be at a considerably higher temperature than the contents of the flask. As in the case of heating with a flame, this can cause local superheating and possible decomposition, as well as uneven boiling. The temperature of the mantle must be determined by means of the thermocouple, and this is somewhat inconvenient. As a result most people use a heating mantle without bothering to determine its temperature. Also, the level of the liquid in the flask should not fall below the top of the mantle. If it does, the glass between the top of the liquid and the top of the mantle will become very much hotter than the rest of the flask. Material that splashes onto this part of the flask will be superheated. For this reason, heating with a mantle is not really satisfactory for a distillation in which the pot will end up nearly empty. Another disadvantage is that a heating mantle takes quite a long time to warm up and reach thermal equilibrium, perhaps an hour or more. Also, if you wish to turn off the heat in a hurry, it is not enough to turn off the electricity, because the mantle is hotter than the flask and it has a high heat capacity. The mantle itself must be removed. For this reason, you should never set up an apparatus using a heating mantle with the mantle resting on the desk top. An additional disadvantage is that the mantle must approximately fit the flask, and a different one may be required for each size of flask. You can buy at least a dozen burners for the price of a mantle and transformer. However, for a reaction involving a large volume of material, a heating mantle is more satisfactory than any other apparatus. A heating mantle will also be most useful in a situation where you will have to run the same reaction a number of times, and a relatively high temperature and/or long reaction time is required.

A second method of electrical heating uses an immersion heater, supplied by a variable autotransformer, in a bath of a liquid heat-transfer medium. The immersion heater can be of the commercially available Calrod type, or can be easily made from a power resistor, a piece of lamp cord, and a plug (see Figure 32.3). The bath liquid can be a mixture of hydrocarbons (mineral oil), a silicone oil, or a polyethylene glycol fraction that freezes just below room temperature (Carbowax 600). The latter is generally most useful since it is cheap, can be used at temperatures up to about 175–180°C without appreciable decomposition, and has the very great advantage of being completely water soluble, which makes cleaning up very much easier. The disadvantage of polyethylene glycol is that it is somewhat hygroscopic and, if exposed for a long time to high humidity, it can boil below the desired temperature; the water will boil out with continued heating.

The advantages of an electrically heated oil bath are that the temperature of the bath (and thus of the contents of the flask) may be determined by a thermometer in the bath; no superheating and local decomposition are possible; and, in the case of the power resistor used as heating element, the heat capacity of the heater

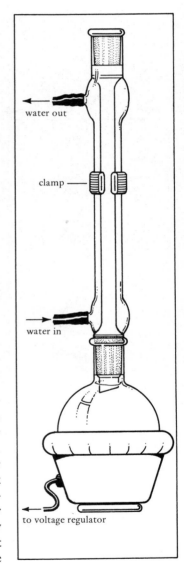

water out

clamp

water in

to voltage regulator

Figure 32.2. Use of a heating mantle. The mantle should be supported on an iron ring or other removable support.

is so small that the rate of heating can be changed almost instantly by adjusting the voltage from the variable transformer. The electrically heated oil bath has the advantage over the heating mantle that it can be brought up to temperature relatively quickly, the rate of heating is more quickly and easily adjusted, the contents of the flask are more visible, and a magnetic stirrer can easily be used (Section 33) with it. The disadvantages of the oil bath are that it is a nuisance to store, it starts to smoke at elevated temperatures (about 150°C for mineral oil, acrid smell; about 180° for polyethylene glycol, burnt sugar smell), and it presents the danger of spilling hot oil if the container is tipped or broken during use. For small flasks (up to about 500 ml), at temperatures below 150–180°C, and for short reaction times, an electrically heated oil bath is most useful. For the constant and exact heat control required for a distillation, the electrically heated oil bath is unsurpassed. Of course, the immersion heater requires the use of an expensive variable autotransformer, just as the heating mantle does.

A third method of electrical heating makes use of a hotplate. The hotplate, which may also contain a magnetic stirrer (Section 33), may be used either to heat a flask directly or to heat an oil bath in which the flask is supported. Since the surface of the hotplate will be much hotter than the contents of the flask, local superheating is possible, just as with heating with a flame. Indirect heating via an oil bath avoids the problem of superheating and generally provides a much more even and constant source of heat for the flask. The temperature of the bath can be determined with a thermometer.

Electric hotplate

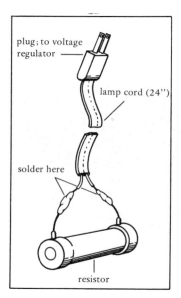

Figure 32.3. An inexpensive immersion heater, made from 125-ohm, 5-watt resistor (Ohmite "Brown Devil"), lamp cord, and plug.

Steam Bath Heating

A steam bath, illustrated in Figure 32.4, is often used to heat solutions that boil below about 90°C, or to heat a mixture to approximately 100°C. Most laboratories are supplied with steam from a central boiler and in this case use of the steam bath eliminates the hazards of a flame. The steam bath heats the contents of a flask relatively quickly, but it has the disadvantage of being good for only one temperature, approximately 100°C.

Hot Water Heating

A useful device for heating at temperatures below 100°C is a hot water bath—simply a beaker or steam bath full of hot water. In some cases, hot water from the faucet may be hot enough; sometimes it may be necessary to heat (and occasionally reheat) the water in the beaker or steam bath with a flame. A steam bath or hot water bath should be used with flammable substances whenever possible.

32.2 COOLING

Cold water; freezing mixture

Depending upon the temperature of the flask to be cooled, a cooling bath may be simply a beaker of water or a specially prepared cold bath. Air cooling is very inefficient compared to a bath of a cold liquid, since a liquid has a much higher heat capacity. For cooling below room temperature and down to about 0°C, a mixture of ice and water may be used (Figure 32.5). It is much more satisfactory for the ice to be in small pieces than as ice cubes. For cooling to a few degrees below zero, or for more efficient cooling to 0°C, a freezing mixture may be used. This is made by thoroughly mixing 1 part sodium chloride and 3 parts of snow or finely chopped ice (by weight). Ice/salt baths, however, are fairly short-lived and inefficient.

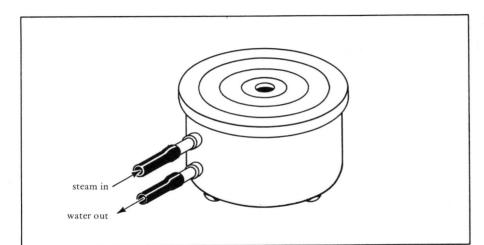

Figure 32.4. A steam bath. Rings can be removed to accommodate flasks of different sizes.

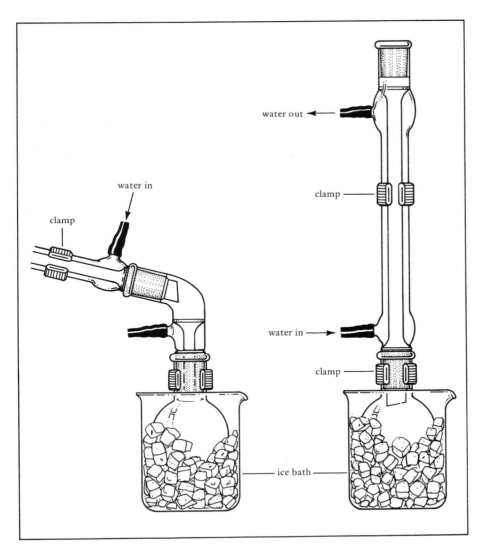

Figure 32.5. Use of an ice bath.

Dry Ice bath

For efficient cooling for extended periods, a Dry Ice acetone bath may be used. This is prepared by adding lumps of Dry Ice (do not handle with bare hands) to acetone until the desired temperature is reached. The temperature can be maintained at any given level by adding Dry Ice as needed. Do not add the Dry Ice too rapidly, or the carbon dioxide evolved will cause the bath to overflow. The ultimate temperature that can be reached with such a bath is about $-78°C$; it must be measured with an alcohol or toluene thermometer, as mercury freezes at $-40°C$.

A crystallizing dish is the most conveniently shaped container for hot or cold baths, since it has a flat bottom and straight sides, is wider than it is tall, and comes in several sizes. Don't forget that it is glass and can be broken. The life of a low-temperature bath may be greatly extended by using a Dewar flask (like a Thermos bottle) as a container, or by making the bath in a beaker that is nested inside a large beaker with a folded towel in between for insulation.

Efficient cooling

For most efficient cooling, the reaction mixture or solution should be in a relatively large flask to give a large area of cold surface, and it should be swirled or stirred continuously.

---

§33 STIRRING

Usually a reaction mixture must be mixed or stirred. This is done to mix two relatively insoluble phases, or to mix a reagent as it is gradually added, or to promote efficient heat transfer to or from a cooling or heating bath and to obtain a constant temperature throughout the mixture. If the reaction time is short, swirling by hand may be acceptable. If the mixture is homogeneous and the reaction is neither very endothermic nor very exothermic, so that the heat transfer does not have to be very efficient, occasional swirling in the heating or cooling bath may be adequate. However, if constant mixing or stirring is required for more than a short time, a mechanical means of stirring is almost a necessity. The boiling action of a vigorously boiling mixture can often provide adequate mixing, but boiling will be smoother and superheating can be minimized if the mixture is stirred.

Swirling by hand

Mechanical stirring

Two methods of stirring are commonly used. The first makes use of a paddle or wire agitator, which extends on a shaft into the mixture from above the reaction vessel. The shaft is usually aligned by means of one or two bearings clamped in place or by means of a glass sleeve with a ground joint that fits into the center ground-glass-jointed neck of the flask. The vigor of the stirring is determined by the speed of the motor, which is governed by a rheostat. Figure 37.1 illustrates this type of stirring apparatus.

Magnetic stirring

The second method makes use of a glass- or Teflon-covered bar magnet that is placed in the reaction flask. The flask is positioned over a motor, which turns a magnet fastened to its shaft (magnetic stirrer); as the motor-driven magnet turns, the magnet inside the flask turns and stirs the mixture. Simultaneous heating and stirring are often carried out by use of a combination magnetic stirrer-hotplate. If the flask is heated by the hot plate via an oil bath, the magnetic stirrer can be used to stir both the bath and the mixture. Figure 33.1 shows another arrangement in which a magnetic stirrer is used to stir both an electrically heated oil bath and the contents of a reaction flask. The magnet in the bath should be larger than the one in the flask.

Direct mechanical stirring has an advantage over magnetic stirring in that it can provide the greater torque necessary to stir suspended solids or viscous mixtures or very large volumes. The mechanical stirrer can also be used when

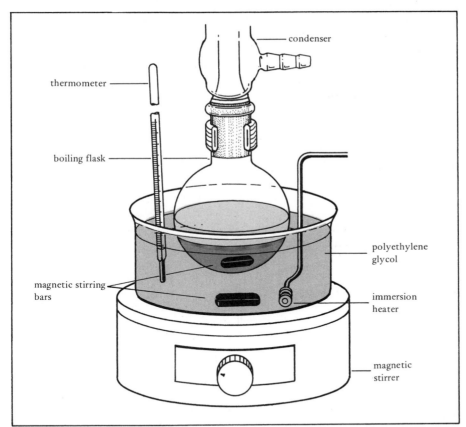

thermometer

condenser

boiling flask

polyethylene glycol

magnetic stirring bars

immersion heater

magnetic stirrer

Figure 33.1. Magnetic stirrer arranged to stir electrically heated oil bath and reaction flask.

heating with a flame or with a heating mantle. On the other hand, a magnetic stirrer is much easier to set up, and is generally much more convenient if the mixture needs only to be stirred and not heated or cooled at the same time. It is most useful when used in conjunction with an electrically heated oil bath, as the stirring motor can stir both the bath and the reaction mixture. This combination is especially suited for heating and stirring during a distillation, particularly a vacuum distillation. Use of a small magnetic stirrer in the pot will essentially eliminate superheating and bumping. Magnetic stirring is somewhat more convenient when the apparatus is not at atmospheric pressure, or when an inert atmosphere is being used, as magnetic stirring does not require a flask opening. Both types of stirring can be used when a heating or cooling bath is being employed, but with the magnetic stirrer, the flask must be positioned near the bottom of the bath, and the bath container must be made of nonferrous material.

Small air or water turbine-driven magnetic stirring motors are also available. These are useful for stirring "under water" in a constant-temperature bath.

§34 ADDITION OF REAGENTS

Many chemical reactions are carried out by combining the reagents (and solvent and catalyst, if necessary) all at once and keeping the mixture for a length of time at a certain temperature. In some cases, however, it is necessary to add one or more reagents gradually over a period of time. This would be done, for example, if the reaction is strongly exothermic and the rate can be easily controlled only by adding one reagent a little at a time. Or it would be done if one reagent might tend to undergo an undesirable bimolecular reaction with itself: if this reagent is added slowly to an excess of the other materials to keep its concentration low, the rate of its bimolecular reaction with itself will be decreased relative to its bimolecular reaction with another substance. A gaseous reagent must usually be added slowly because of limited solubility; after the solution is saturated, it can be added no faster than it can be consumed.

34.1 ADDITION OF SOLIDS

If a solid must be added gradually, it is usually added as a solution. If it is necessary or desirable to add it as a solid, portions can often be dropped right into the flask or down the condenser; sometimes it will be necessary to poke a length of glass rod down the condenser to dislodge material that has stuck to the walls. If larger amounts need to be added, or if air or water vapor must be excluded from the reaction, an Erlenmeyer flask containing the solid can be attached to a neck

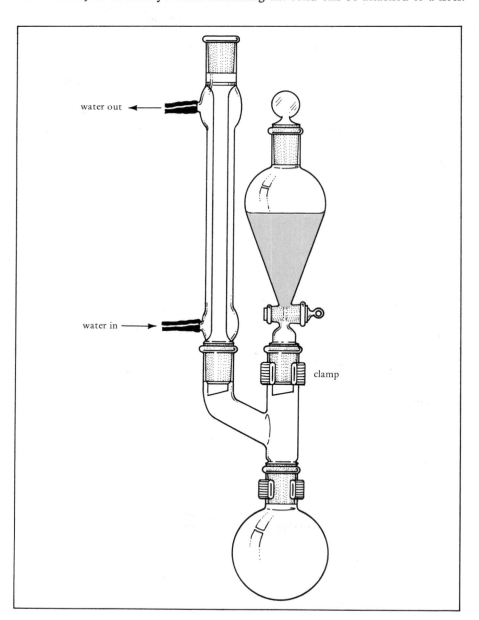

Figure 34.1. Reaction apparatus that allows addition of a liquid reagent or solution.

of the flask with a length of wide, thin-walled tubing (Gooch tubing). Portions can be added by shaking the solid in through the tubing. When the Erlenmeyer flask is hanging down between additions, the tubing is held shut, preventing the vapors from the reaction from getting into the flask and caking the solid.

34.2 ADDITION OF LIQUIDS AND SOLUTIONS

Liquids or solutions may be added by means of a separatory funnel, as shown in Figure 34.1. If it is necessary to add a liquid to a closed system, as when an inert atmosphere is being used, a pressure-equalizing addition funnel is required (Figure 37.1). The rate of addition is controlled by means of the stopcock. If the addition must be made very slowly over a long period of time, funnels with stopcocks are not very satisfactory since the flow rate is variable, increasing as the stopcock grease is dissolved away (unless a Teflon stopcock is used), and decreasing as dust or other bits of solid collect at the opening. In these cases an addition funnel may be used that has, instead of a stopcock, a narrow capillary tube fitted with a tungsten wire that may be moved in and out of the capillary (a Hershberg funnel). The flow rate is determined by the length of wire extending into the capillary.

34.3 ADDITION OF GASES

A gas may be added to a reaction mixture through a tube that dips below the surface of the liquid. If the gas is only slightly soluble, it must be passed in only as quickly as it is consumed, in order not to waste it. The solubility of a gas decreases with increasing temperature, and stirring promotes more rapid solution. Hopefully, the gas will not be so soluble or reactive that the solution is sucked up the tube, and it will not react to produce a solid that clogs the addition tube.

If a gas is very soluble or very reactive, the addition tube should be open-ended and of wide bore so that the solution cannot be drawn up into it very far before the level falls below the end of the tube. Alternatively, the gas can be led just to the surface of the mixture in the vortex formed by the stirrer.

If the product of the reaction with the gas is a solid that tends to clog the tube, as in the treatment of a Grignard reagent with carbon dioxide, a solid rod may be incorporated into the wide-bore addition tube, as shown in Figure 34.2, so that the solid can be dislodged from time to time.

Measuring Gases

If the weight of the gas added is large enough compared to the weight of the cylinder, the amount added can be determined by the weight loss of the cylinder. A lecture bottle is a good source of gas, from this point of view. In some cases, it may be possible to pass the gas in at a steady rate. If so, the use of a flowmeter in the line will allow the amount of gas added to be determined from the length of time of addition. If the gas is not too low-boiling, it can be passed into a tube cooled in a Dry Ice/acetone bath and measured by volume (if the density of the liquid at the bath temperature can be estimated). When the desired volume has been condensed, the gas can be passed into the reaction mixture as it vaporizes. Section 76.1 describes this procedure for a chlorination reaction. Of course, if an excess of an inexpensive gas is not objectionable, the amount added need be estimated only very roughly.

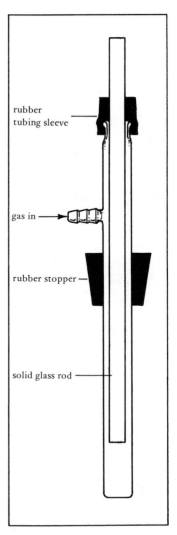

Figure 34.2. A type of gas-addition tube.

rubber tubing sleeve

gas in →

rubber stopper

solid glass rod

APPARATUS AND
TECHNIQUES FOR
CHEMICAL REACTIONS

Use the hood

Gas trap

Reactions in which a poisonous or irritating gas is formed are best carried out in a ventilated enclosure, or hood. If hood space is limited, it is possible, when the gas is readily soluble in water and not too poisonous, to set up or modify the apparatus in such a manner that the poisonous or irritating gas can be prevented from escaping into the atmosphere (to trap the gas). One way to do this is to lead the gas from the apparatus to a beaker of water, or an appropriate aqueous solution, where it will dissolve rather than escape into the room, as illustrated in Figure 35.1. This method is satisfactory if the rate of gas evolution is not too great and if relatively small amounts are involved. The capacity and efficiency of this system can be increased by using a dilute sodium hydroxide solution (for acidic gases

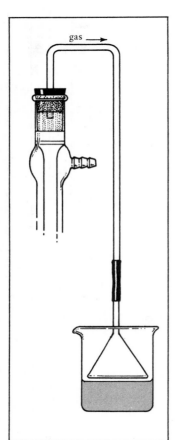

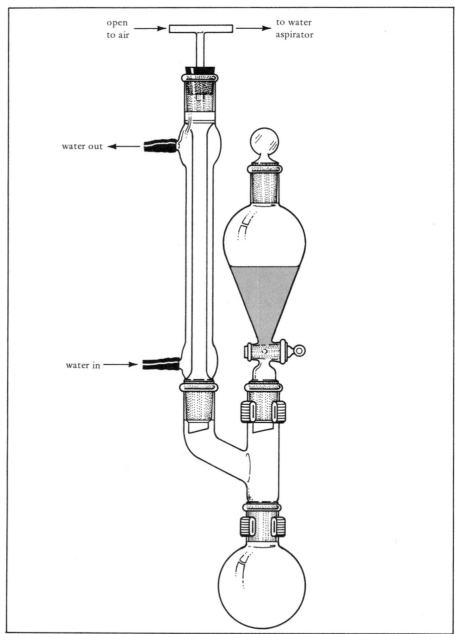

Figure 35.1. A way of absorbing soluble noxious gases. The funnel should not touch the solution in the beaker.

Figure 35.2. A method of control of noxious water-soluble gases. A vacuum adapter can be used in place of the cork and *T* tube.

such as HCl or HBr) or a dilute solution of sulfuric acid (for basic gases such as ammonia) in the beaker.

Another way to dispose of a water-soluble gas involves an *aspirator*; an arrangement is illustrated in Figure 35.2. While it is slightly easier to set up, it is less reliable than the first method and uses a lot of water.

Of course, gases that are not soluble in water, such as carbon monoxide, cannot be controlled by either of these methods.

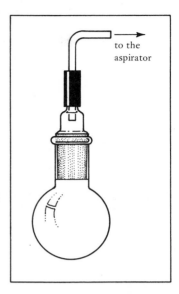

Figure 36.1. Removal of solvent under vacuum. Heat the flask in a beaker of hot water or on the steam bath.

---

§36 CONCENTRATION; EVAPORATION

It is often necessary to either concentrate a solution or completely remove the solvent, in order to leave a relatively nonvolatile solid or liquid residue. The most common situations involve the recovery of material after an extraction, evaporation of chromatographic fractions, or concentration of the filtrate after recrystallization.

Distillation

Possibly the most often used method of concentration or evaporation is the removal of solvent by distillation. If the source of heat is a steam bath or an oil bath, this method is entirely satisfactory for either partial or complete removal of the solvent.

The complete removal of solvent by distillation usually involves two steps. First, the solution is boiled at a bath temperature 25–50°C above the boiling point of the solvent until distillation has slowed almost to a complete stop. The second step involves *heating the residue under vacuum* (usually at the same temperature at which the bulk of the solvent was removed) until the remainder of the solvent is gone, as illustrated in Figure 36.1. Evaporation may be judged to be complete when the residue shows no appreciable loss in weight between successive periods of heating under vacuum. If only a small amount of a volatile solvent is involved, it may be removed by distillation using an aspirator arrangement like that illustrated in Figure 36.2.

While you may concentrate a solution by distillation using a flame or heating mantle as a source of heat, you should not attempt to completely evaporate a solution by heating by either of these methods, as you will most likely superheat and begin to decompose the residue when it becomes viscous or when solid separates after the solution becomes more concentrated. If a large amount of solvent is to be evaporated to give only a small residue, it may be appropriate to fit a small flask with a separatory funnel so that solution can be added as the distillation progresses. In this way, the small residue will end up in a small flask, and mechanical losses can be minimized. Or the bulk of the solvent can be removed using a large flask and the remainder of the solution can be transferred (with rinsing) to a small flask for completion of the evaporation.

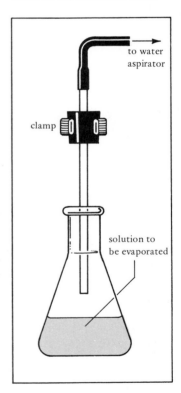

Figure 36.2. A way to concentrate a solution rapidly. The flask may be heated on the steam bath.

Other Techniques

If large amounts of a volatile solvent must be evaporated, it may be convenient to place the solution in a large shallow dish (evaporating dish) and keep the dish in the hood with the draft on until evaporation is complete. This method is slower, but it requires no attention.

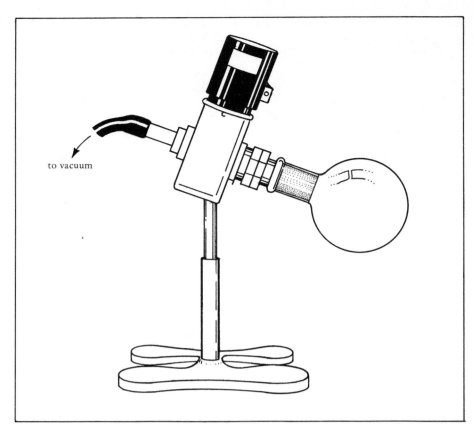

Figure 36.3. A rotary evaporator. The flask may be heated with warm water or steam.

It is possible to evaporate solutions very rapidly by means of a *rotary evaporator*, a device that can apply an aspirator vacuum to a flask while rotating it in a heating bath (see Figure 36.3). Rotation of the flask minimizes superheating and bumping by keeping the contents well mixed, and provides a large surface area for evaporation by constantly rewetting the walls of the flask.

§37 USE OF AN INERT ATMOSPHERE

Certain reactions, for instance the Grignard reaction, are better carried out in the absence of oxygen or water vapor because these substances either undergo or catalyze undesired side reactions. If the acceptable levels of oxygen and water vapor are not too low, simply providing for flushing out the air of the reaction flask with dry nitrogen and maintenance of a slight positive pressure of nitrogen will be adequate. Figure 37.1 illustrates the essential features of an arrangement *Keeps out water and* to do this: the tank of nitrogen connected to the apparatus through a wash bottle *oxygen* for removal of traces of water, and the bubbler in the exit line to maintain the slight positive pressure and to indicate the flow rate. The apparatus is flushed initially at a relatively high flow rate, and then maintained at a slight positive pressure throughout the reaction at a minimum flow rate.

If the acceptable levels of water or oxygen are rather low, usually because their action is catalytic, much more elaborate procedures must be used; these might include working on a vacuum line, recrystallization and redistillation of the reactants and solvents under nitrogen, and heating and pumping out of the reaction vessels.

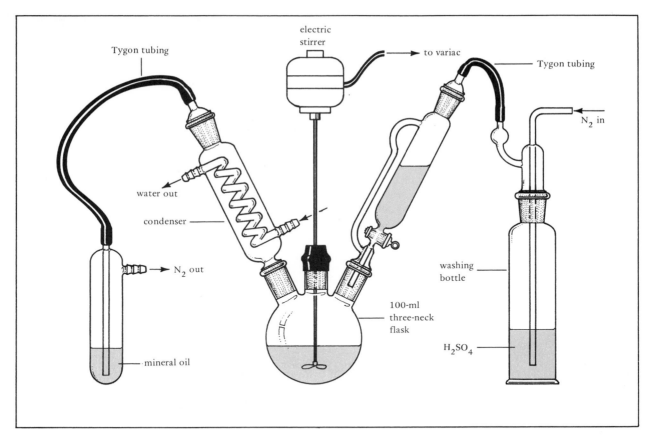

Figure 37.1. An apparatus for carrying out a reaction in an atmosphere of nitrogen. If magnetic stirring is used rather than the mechanical stirring shown, a three-neck flask is not needed. Instead, a single-neck flask fitted with a Claisen adapter may be used. The condenser will go in one neck of the adapter, and the pressure-equalizing addition funnel in the other.

§38 WORKING UP THE REACTION; ISOLATION OF THE PRODUCT

After the reaction has taken place, the product must be separated from the reaction mixture by one or more of the separation procedures described in Sections 7–15. Although each reaction mixture presents a unique problem in separation, most reactions may be worked up (the product isolated) more or less according to one of a few general schemes.

If the product is a solid, it may crystallize or separate spontaneously from the reaction mixture. In this case, it need only be collected by a suction filtration and washed with the appropriate solvents. Depending upon the degree of purity required and the nature of any impurities, it may be further purified by recrystallization or sublimation or some other method, perhaps chromatography. *Suction filtration*

If the product is a liquid, it may occasionally be isolated simply by fractional distillation of the reaction mixture. The initial product fraction may then be further purified, if necessary, by refractionation or some other method; gas chromatography may be found to be effective. *Fractional distillation*

In most cases, the workup will involve adding water (or water and ice) to the reaction mixture, or vice versa. If the product is a water-insoluble solid, it may then be collected by suction filtration, washed with appropriate solvents, dried, and recrystallized. If the product is a liquid—or often even if it is a solid—an *Aqueous workup* organic solvent such as ether or dichloromethane can be added to dissolve it. The resulting ethereal or dichloromethane solution is first separated from the aqueous phase, using the separatory funnel (Section 13.3). The solution of the neutral product in the organic solvent is then washed with portions of either aqueous base (such as dilute sodium hydroxide or dilute sodium bicarbonate) or aqueous acid (such as dilute hydrochloric acid or dilute sulfuric acid). If the organic solution is washed with both acid and base, the washings must be done in succession. Usually, the order of washing with both acid and base is not important. Washing with base will remove acidic impurities, and washing with acid will eliminate basic contaminants, as described in Section 13.2. A final wash with water will then remove traces of the previous aqueous solution. The organic solution of the neutral compound is then dried over a suitable drying agent (Section 15.2), filtered by gravity to remove the hydrated drying agent (Section 7.1), and finally evaporated to remove the solvent if the product is a solid, or fractionally distilled if the product is a liquid. These operations are summarized in the flow diagram in Figure 38.1.

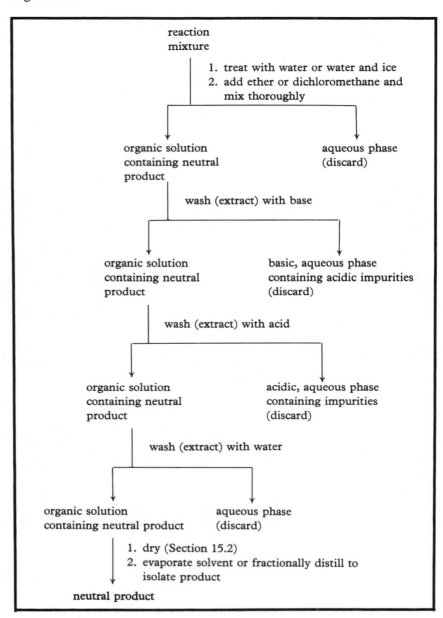

Figure 38.1. Flow diagram for isolation and purification of a neutral product.

If the product is a water-insoluble acid, a typical aqueous workup will involve extraction from the organic solvent into a basic aqueous solution, such as dilute sodium hydroxide. After the organic phase has been removed by using the separatory funnel (Section 13.3), the basic aqueous solution of the salt of the acid is acidified and the free acid is then extracted into a fresh portion of the organic solvent. After the organic solution of the free acid has been separated from the aqueous layer, the organic solution is washed with a small portion of water (to remove traces of the previous aqueous solution) and dried over a suitable drying agent (Section 15.2). The hydrated drying agent is then removed by gravity filtration and the organic solution is evaporated to remove the solvent if the product is a solid, or fractionally distilled if the product is a liquid. This procedure is summarized in the flow diagram of Figure 38.2.

If the product to be purified is a water-insoluble base, an aqueous workup may involve extraction of the base from the organic solvent into aqueous acid, basification of the aqueous solution, and reextraction by a fresh portion of the organic solvent. After the organic solution is dried, the product can be isolated

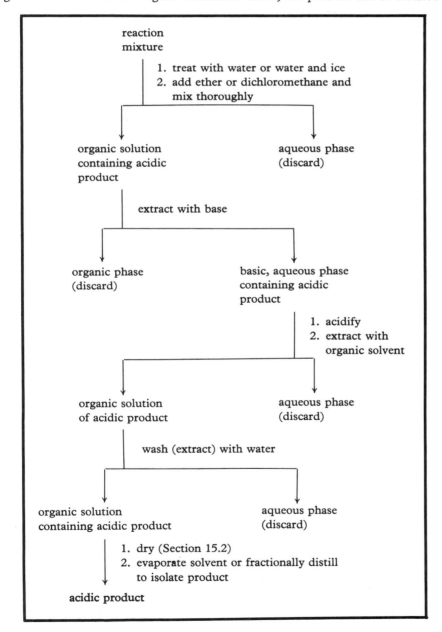

Figure 38.2. Flow diagram for isolation and purification of an acidic product.

either by evaporation or fractional distillation. These operations are summarized in the flow diagram of Figure 38.3.

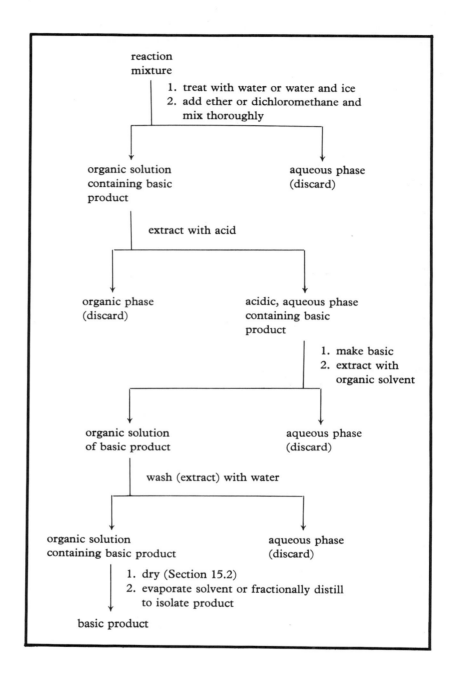

Figure 38.3. Flow diagram for isolation and purification of a basic product.

Question

1. In preparing N,N-diethyl-*m*-toluamide ("Off") from *m*-toluic acid and diethylamine, described in Section 60.3, the toluene solution of the crude product is extracted first with dilute sodium hydroxide solution and then with dilute hydrochloric acid solution. Explain how these extractions remove unreacted starting materials.

EXPERIMENTS

Separations

Transformations

*Synthetic Sequences: Synthesis Experiments
Involving a Sequence of Reactions*

Projects

Four types of experiments are presented in the four main sections of Part II: separations, one-step transformations, synthetic sequences, and synthesis projects.

The *separations* are mostly normal applications of the techniques described in Sections 7 through 15 in which the compound to be isolated and purified is already present as a component of a mixture.

The *transformations* generally involve a one-step synthesis of the compound followed by isolation and purification. Sometimes more than one procedure is given and the idea is to choose between them or to combine the best parts of each.

The *synthetic sequences* present syntheses that must be carried out by a series of reactions. In most cases the crude product of the preceding reactions can be used without purification.

The *synthesis projects* are suggestions for the preparation of a number of unusual compounds. For each project, procedures from the original chemical literature are presented. Some are rather brief and others give quite detailed directions. In all cases, however, they appear to be the best the literature has to offer and should serve as the starting point for the development of a procedure for the preparation of the compound. The procedures presented in the projects seem reasonable to the author, but *they have not been tested*.

Sometimes it will take more than one laboratory period to complete an experiment. Or there may be some other reason for setting an experiment aside for a while. With experience it is possible to recognize good stopping points. For example, one of the best is when a solution is being set aside to stand for crystallization. Other good times to stop are when a mixture is ready to be distilled or when a solution is being kept over a drying agent. Sometimes a good time to stop is at the end of a heating period. In the following procedures some of the possible stopping points are indicated by the symbol ■. All the experiments have been designed for use in a three-hour laboratory period. Some can be done easily in less time, but a few will require more than three hours for many workers. The estimates of the time required, which are given at the end of each procedure, are of the time needed by students who are taking their first course in organic chemistry. The estimates do not include the time needed for any optional recrystallizations.

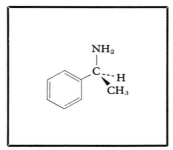

Separations

The separation procedures described in this section include the isolation of cholesterol from gallstones (Section 39), lactose from dry milk (Section 40), acetylsalicylic acid from aspirin tablets (Section 41), caffeine from tea or NoDoz (Section 42), clove oil from cloves (Section 43), eugenol from clove oil (Section 44), and (R)-(+)-limonene from grapefruit or orange peel (Section 45). The two remaining experiments involve the isolation of enantiomeric molecules. In one, the R and S forms of carvone can be isolated from oil of caraway and oil of spearmint. An interesting property of these enantiomers is that they have different odors. In the final experiment, the R and S forms of α-phenylethylamine are separated from the racemic mixture by use of the chiral reagent (R),(R)-(+)-tartaric acid.

---

§39 ISOLATION OF CHOLESTEROL FROM GALLSTONES

In this experiment cholesterol is isolated from human gallstones by a simple crystallization process. On the average, the cholesterol content of human gallstones is about 75 percent.

The infrared spectrum of cholesterol is shown in Figure 39.1.

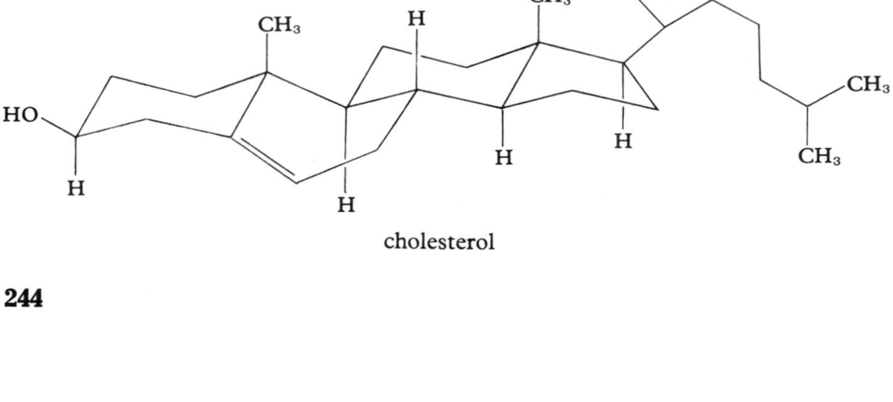

cholesterol

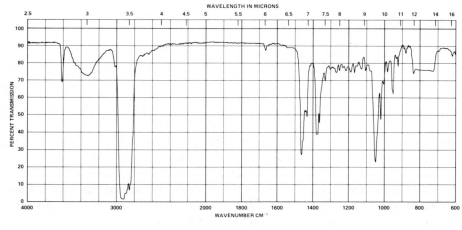

Figure 39.1. Infrared spectrum of cholesterol; CCl$_4$ solution.

Procedure (Reference 1).   Place a 2-gram sample of crushed gallstones in a 50-ml Erlenmeyer flask. To this, add 10 ml of dioxane and heat the mixture on the steam bath with occasional swirling for about 5 minutes. During this time most of the solid in the flask should dissolve. While the mixture is still hot, remove the brown residue of bile pigments by gravity filtration, collecting the filtrate in a 50-ml Erlenmeyer flask. Dilute the filtrate with 10 ml of methanol, add some decolorizing carbon, and warm the mixture on the steam bath for a minute or two. While the mixture is still hot, remove the carbon by gravity filtration and collect the filtrate in a 50-ml Erlenmeyer flask. Reheat the filtrate on the steam bath, add 1 ml of hot water, and swirl to dissolve any precipitated solid. (If it does not all dissolve, add a little methanol and heat with swirling until it does.) Allow the solution to come to room temperature, during which time cholesterol will separate as colorless plates. ■* Collect the crystals of cholesterol by suction filtration, and wash them with a few milliliters of cold methanol. Leave the vacuum on for a few minutes to suck the crystals as dry as possible. Cholesterol may be recrystallized from methanol, using 35 ml per gram. Pure cholesterol melts at 149°C.

Time: less than 3 hours.

Cholesteryl benzoate, the ester between cholesterol and benzoic acid, is converted to a "liquid crystal" when heated to a temperature above 147°C but below 180°C. Section 56.2 tells how to prepare cholesteryl benzoate and how to observe the liquid crystal phenomenon.

Reference

1. L. F. Fieser, *Organic Experiments*, Heath, Boston, 1964, p. 70.

§40 ISOLATION OF LACTOSE FROM POWDERED MILK

In this experiment lactose is recovered from the whey that remains after precipitation and removal of the milk proteins. Lactose is a disaccharide that, upon hydrolysis, yields a molecule of glucose and a molecule of galactose. Dry milk is approximately one-half lactose.

* The symbol ■ indicates a stopping point in the procedure.

OH
CH₂OH   H   H   H   H
                                    OH
H   O   HO   OH
                          O
HO   CH₂OH   O
        OH   H   H
H   H   H   H

lactose (β anomer)

Procedure. Place 25 grams of nonfat dry milk powder in a 250-ml beaker (Note 1). Add 75 ml of warm water and stir to mix. Adjust the temperature of the mixture to between 40 and 50°C by heating or cooling. Add about 10 ml of 10% acetic acid solution (Note 2) and stir the mixture to coagulate the casein. Precipitation can be judged to be complete when the liquid changes from milky to clear. Remove the precipitated casein by filtering the mixture by gravity through cheesecloth (Note 3). Collect the filtrate in a 250-ml beaker. Add about 2 grams of calcium carbonate powder to the filtrate, stir it well, and boil the suspension for about 10 minutes (Note 4). Add to the hot mixutre about as much decolorizing carbon as would cover a nickel, stir the mixture thoroughly, and filter it by suction through a layer of wet filter aid on a Buchner funnel (Note 5). ■ Transfer the filtrate to a 250-ml beaker and concentrate it to a volume of about 30 ml by boiling over a low flame with a wire gauze between the flame and the beaker (Note 6). When the volume has been reduced to 30 ml, turn off the burner and add 125 ml of 95% ethanol and about the same amount of decolorizing carbon as used before. Stir the mixture well and filter it through a layer of wet filter aid on a Buchner funnel (Note 5). Allow the clear filtrate to stand for crystallization for at least 24 hours in a stoppered Erlenmeyer flask. ■ Collect the crystals of lactose by suction filtration. They may be washed with a small amount of 95% ethanol. Yield: between 2.5 and 4.5 grams.

Notes

1. The powdered milk is most easily measured by volume using a beaker; 25 grams is about 100 ml.

2. Ten percent acetic acid is prepared by diluting 10 ml of glacial acetic acid to 100 ml. More than 10 ml of 10% acetic acid seems to be required if the dry milk is not fresh.

3. Use two pieces of cheesecloth about 12 in. square. Lay them over a conical funnel large enough to contain all the curds. First decant as much as possible of the liquid from the coagulated casein into the funnel and then carefully transfer the wet protein mass onto the cheesecloth. After most of the liquid has drained through, a bit more can be recovered by wrapping the cheesecloth around the curds and squeezing.

4. The mixture will foam occasionally.

5. Add about 5 grams of filter aid (we have used Celite) to about 25 ml of water in a small beaker. Wet the filter paper on the Buchner funnel with water, fit the funnel to the suction flask, and apply a gentle suction. Swirl the mixture of filter aid and water to suspend the solid and pour the suspension all at once into the Buchner funnel. When the water has been drawn through the funnel, a damp pad of filter aid should be in place on top of the filter paper. Remove the funnel momentarily and pour the water out of the filter

flask. Be careful not to disturb the layer of filter aid when pouring the mixture to be filtered into the funnel.

6. Some bumping and foaming occur.

Time: 3 hours plus $\frac{1}{2}$ hour to isolate the product after crystallization is complete.

Reference

1. J. Cason and H. Rapoport, *Laboratory Text in Organic Chemistry*, Prentice-Hall, New York, 1950, p. 113.

---

§41 ISOLATION OF ACETYLSALICYLIC ACID FROM ASPIRIN TABLETS

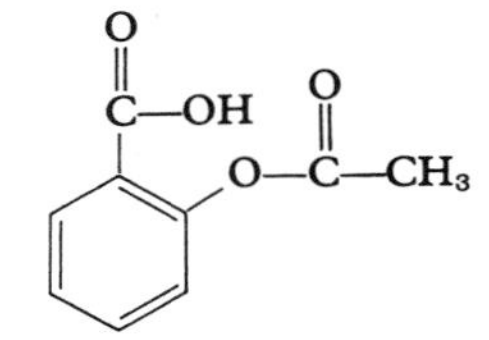

---

acetylsalicylic acid
(aspirin)

Aspirin tablets consist of acetylsalicylic acid plus a binder that helps to keep the tablet from breaking up. In this experiment the acetylsalicylic acid is recovered by a crystallization process. A procedure for the preparation of acetylsalicylic acid from salicylic acid is described in Section 56.5. The infrared and NMR spectra of acetylsalicylic acid are shown in Figures 56.8 and 56.9.

Procedure. Place ten 5-grain aspirin tablets (3.24 grams acetylsalicylic acid) in a 125-ml Erlenmeyer flask. Add to the flask 10 ml of 95% ethanol and heat the mixture to boiling on the steam bath. Swirl the hot mixture until the tablets disintegrate. The acetylsalicylic acid will dissolve, leaving a white residue (the binder). Filter the hot solution by gravity to remove the insoluble material and add to the filtrate 50 ml of cool water. Acetylsalicylic acid will begin to crystallize almost immediately. Allow crystallization to take place for about 10 minutes at room temperature and then for about 5 minutes in an ice bath. Collect the acetylsalicylic acid by suction filtration and wash the product with a little cold water. Recovery: about 2.4 grams (75%).

Time: about 2 hours.

---

§42 ISOLATION OF CAFFEINE FROM TEA AND NODOZ

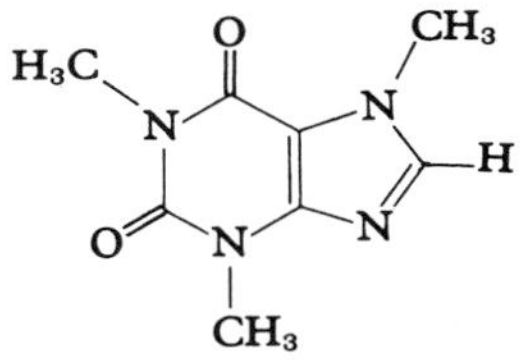

---

caffeine

In these experiments caffeine is isolated from tea or NoDoz tablets by extraction into water followed by extraction into chloroform, evaporation, and recrystallization from ethanol. It may also be purified by sublimation (Section 12).

Caffeine is described as a cardiac, respiratory, and psychic stimulant, and as a diuretic. It is reported to have a melting range of 235–236°C. The infrared spectrum is shown in Figure 42.1 and the NMR spectrum in Figure 42.2.

Caffeine from Tea

Procedure (Reference 1). Place 125 ml of water in a 500-ml Erlenmeyer flask. Add 12.5 grams of tea and 12.5 grams of powdered calcium carbonate (Note 1). Boil the contents of the flask with constant stirring for about 20 minutes. At this time, filter the hot mixture with suction, and press out the liquid from the tea leaves with a large cork. Transfer the filtrate to a 250-ml Erlenmeyer flask and

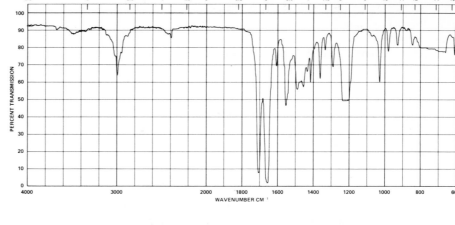

Figure 42.1. Infrared spectrum of caffeine; CHCl₃ solution.

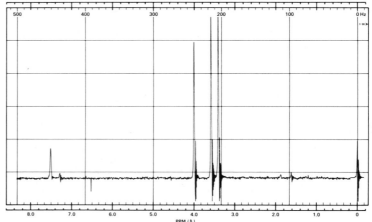

Figure 42.2. NMR spectrum of caffeine; CDCl₃ solution.

add 100 ml of chloroform. Swirl or gently stir the two layers together for about 10 minutes (Note 2) and then separate them by means of a separatory funnel. Distill the chloroform layer until only about 10 ml remains, transfer the remaining solution to a tared (weighed) 25-ml Erlenmeyer flask (filtering if necessary), and evaporate to dryness on the steam bath. Recrystallize the solid residue of caffeine from 95% ethanol, using 5 ml per gram (Note 3).

Caffeine from NoDoz

Procedure. Place 100 ml of water in a 250-ml Erlenmeyer flask. Add two NoDoz tablets that have been crushed and weighed. Heat the suspension to boiling to ensure solution of the caffeine in the NoDoz tablets; the binder will remain in suspension. Cool the mixture, add 100 ml of chloroform, and gently swirl or stir the two layers together for about 10 minutes (Note 2). Continue as in the isolation of caffeine from tea.

Notes

1. R. H. Mitchell, W. A. Scott, and P. R. West suggest using 12.5 grams of tea *bags* and 9 grams of *sodium* carbonate [*J. Chem. Educ.* **51**, 69 (1974)]. This variation makes it possible to omit the suction filtration.

2. More vigorous extraction procedures often result in very troublesome emulsions. Gentle stirring of the mixture with a magnetic stirrer is quite satisfactory. The NoDoz extraction is less likely to form an emulsion.

3. One recrystallization from ethanol is sufficient and yields caffeine in the form of small needles. A reasonable expected recovery of recrystallized caffeine is about 25–50% of the total present, depending upon the thoroughness of the extraction. The greenish color that is present in the case of the tea extraction can be completely removed by careful washing of the crystals with ethanol.

Time: about 3 hours.

Reference

1. G. K. Helmkamp and H. W. Johnson, Jr., *Selected Experiments in Organic Chemistry*, 2nd edition, W. H. Freeman, San Francisco, 1968, p. 157.

§43 ISOLATION OF CLOVE OIL FROM CLOVES

In this experiment, oil of cloves is isolated from cloves by steam distillation followed by extraction of the distillate. The extraction solution may be further treated to yield oil of cloves, or it may be used for the isolation of eugenol, as described in the procedure of Section 44.

Procedure (Reference 1). Place 75 grams of cloves (Note 1) and 250 ml of water in a 500-ml boiling flask. Clamp the flask so that it can be heated with a burner flame, and then fit it with a Claisen adapter. Fit the center neck of the Claisen adapter with a separatory funnel, and the side neck with a distillation adapter carrying a thermometer and a condenser set for downward distillation. (The arrangement is similar to that shown in Figure 11.3, except that the steam inlet tube is replaced by a separatory funnel.) Distill the mixture rapidly by heating the flask with a burner until no more oil can be observed to form in the condensate (Note 2). ■

Extract the oil of cloves from the distillate with 50 ml of dichloromethane or petroleum ether, batchwise if necessary, as described in Section 13.3.

The oil of cloves may be isolated by drying the extract over anhydrous magnesium sulfate (Section 15) and removing the dichloromethane or petroleum ether by distillation on the steam bath (Section 36).

Notes

1. Whole cloves are much more satisfactory than ground cloves.
2. Steam distillation for 1 hour will remove most of the oil of cloves.

Time: 3 hours.

Reference

1. Information Division, Unilever, Ltd., Unilever House, Blackfriars, London.

eugenol

§44 ISOLATION OF EUGENOL FROM CLOVE OIL

In this experiment, eugenol is isolated from oil of cloves by extraction with base. Eugenol has the characteristic odor of cloves. Its infrared spectrum is shown in Figure 44.1.

Procedure (Reference 1). Dissolve 5 ml of oil of cloves in 50 ml of carbon tetrachloride and extract the solution with three 50-ml portions of 5% aqueous potassium hydroxide solution. Combine the basic extracts and acidify them to litmus with 5% aqueous hydrochloric acid; about 100 ml will be needed.

Extract the free eugenol from the aqueous mixture with 50 ml of carbon tetrachloride, dry the organic extract over anhydrous magnesium sulfate, and remove the carbon tetrachloride to give a residue of eugenol by distillation using a steam bath.

Time: about 2 hours.

eugenol

Figure 44.1. Infrared spectrum of eugenol; thin film.

Reference

1. Information Division, Unilever Ltd., Unilever House, Blackfriars, London.

§45 ISOLATION OF (R)-(+)-LIMONENE FROM GRAPEFRUIT OR ORANGE PEEL

The major constituent of the steam-volatile oil of grapefruit or orange peel is (R)-(+)-limonene:

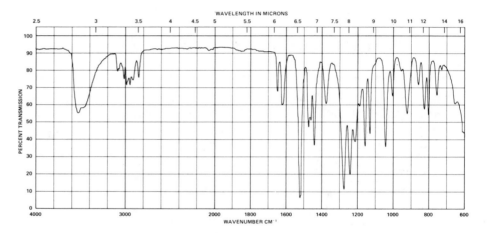

(R)-(+)-limonene

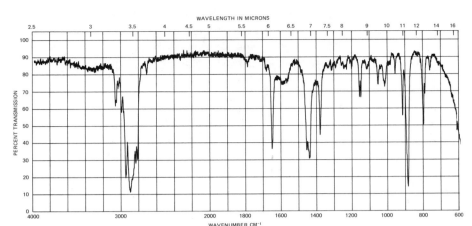

(R)-(+)-limonene

Figure 45.1. Infrared spectrum of (R)-(+)-limonene; thin film.

It can be isolated as a material of about 97% purity by a simple steam distillation of citrus fruit peels. It is responsible for the characteristic smell of citrus peel. The infrared spectrum of a commercial sample of (R)-(+)-limonene is shown in Figure 45.1.

Procedure. Obtain the peel from a grapefruit or from two oranges. Cut the peel into pieces about 5-10 mm square, and put the peel into a 500 ml boiling flask along with about 250 ml of water. Clamp the flask so that it can be heated with a burner flame, and then fit it with a disillation adapter carrying a condenser set for downward distillation. Distill the mixture rapidly by heating the flask with a burner until you have collected about 50 ml of distillate.

Extract the limonene from the distillate with 20 ml of either pentane or dichloromethane, separate the extract by means of a separatory funnel, and dry the extract over anhydrous magnesium sulfate for a few minutes (Note 1). Isolate the limonene by filtering the extract by gravity from the magnesium sulfate into a tared (previously weighed) flask and removing the solvent by distillation on the steam bath. Yield: between $\frac{1}{2}$ and $1\frac{1}{2}$ ml. Physical properties reported for purified (R)-(+)-limonene include b.p.^{763} = 175.5–176°C, d_4^{21} = 0.8403, n_D^{21} = 1.4743, and $[\alpha]_D^{19.5}$ = +124.2°.

Note

1. As an alternative, you can collect the distillate in two 20 × 150 mm test tubes. The limonene will separate to form a clear layer above the water and most of the limonene can be easily removed using a medicine dropper.

Time: less than 3 hours.

Reference

1. F. H. Greenberg, *J. Chem. Educ.* **45**, 537 (1968).

---

§46 ISOLATION OF (R)-(−)- OR (S)-(+)-CARVONE FROM OIL OF SPEARMINT OR OIL OF CARAWAY

Enantiomeric molecules have identical properties, with two exceptions. Enantiomers rotate the plane of polarization of plane-polarized light in opposite directions, and they interact differently with chiral molecules. The latter difference is like the difference in the ways that enantiomeric members of a pair of gloves will interact with a right hand. Enzymes, which are chiral molecules present in every living cell, often appear to react exclusively with one member of a pair of enantiomers. It is therefore not surprising that enantiomers have often been reported to differ in taste or smell. One of the cases in which the evidence is very strong (Reference 1) involves the enantiomers of carvone. (R)-(−)-carvone, isolable from oil of spearmint, is said to smell like spearmint, while (S)-(+)-carvone, isolable from oil of caraway, is said to smell like caraway. About nine people out of ten can distinguish the enantiomers by smell (Reference 2).

carvone

(R)-(−)-carvone (spearmint)
$[\alpha]_D^{20} = -62.5°$

(S)-(+)-carvone (caraway)
$[\alpha]_D^{20} = +61.2°$

The experiment in this section involves the isolation of the enantiomeric forms of carvone, one from each oil. The isolation may be effected by distillation at atmospheric pressure, distillation under reduced pressure, or column chromatography on silica gel. As you can see from Figures 46.1 and 46.2, the infrared spectra of the enantiomeric forms are identical.

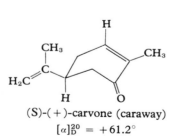

(R)-(−)-carvone (spearmint)
$[\alpha]_D^{20} = -62.5$

Figure 46.1. Infrared spectrum of (R)-(−)-carvone; thin film.

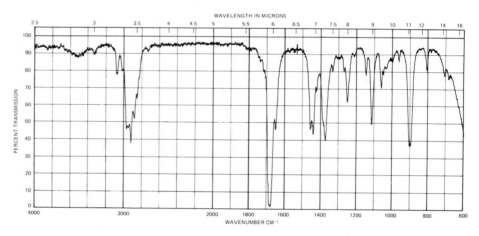

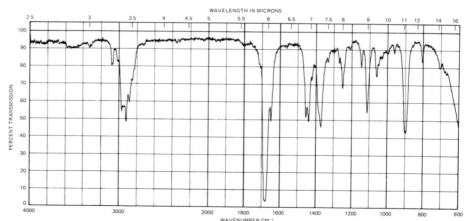

(S)-(+)-carvone (caraway)
$[\alpha]_D^{20} = +61.2°$

Figure 46.2. Infrared spectrum of (S)-(+)-carvone; thin film.

The lower-boiling and more easily eluted components of the oils are mostly isomers of molecular formula $C_{10}H_{16}$:

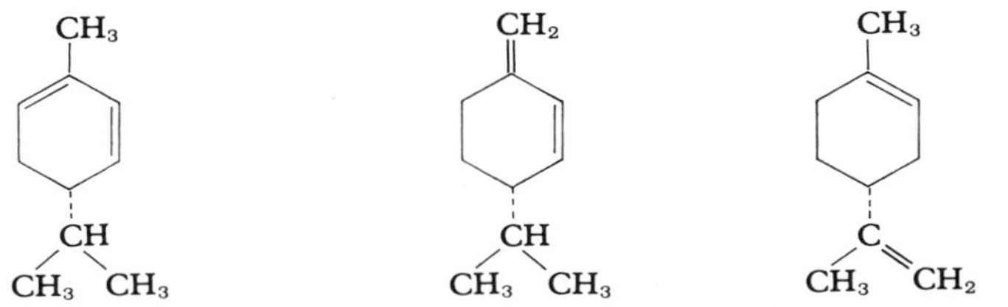

(S)-(+)-α-phellandrene   (S)-(+)-β-phellandrene   (R)-(+)-limonene

Distillation at Atmospheric Pressure

Procedure. Put 10 ml of one of the oils and one or two boiling stones in a 25-ml round-bottom flask; place a Claisen adapter in the neck of the flask (Note 1). Clamp the flask in position on a wire gauze supported by an iron ring. Put a stopper in the center neck of the Claisen adapter. If you want you may poke a small amount of stainless steel sponge into the curved part of the Claisen adapter in order to increase the efficiency of the separation (Section 9.4). Then place a distillation adapter in the side neck of the Claisen adapter. Fit a thermometer in the top neck of the distilling adapter and connect the side neck directly to a vacuum adapter. The apparatus should look like the one shown in Figure 10.2 except that no condenser or vacuum connection is used. Heat the flask with a burner until the oil boils and continue to heat it strongly until material begins to distill. Then adjust the heat so that the distillate accumulates at a rate of 1 drop every 3–5 seconds (Note 2). Collect the distillate in at least three fractions: $C_{10}H_{16}$ isomers, intermediate fraction, and carvone fraction.

Atmospheric-pressure boiling points for carvone and the isomers of molecular formula $C_{10}H_{16}$ have been reported as follows. Carvone: various values between 224 and 231°C; limonene: 176°C; α-phellandrene: 176°C; β-phellandrene: 172°C.

Notes

1. By omitting the Claisen adapter and using a 10-ml flask, you can distill as little as 5 ml of oil. If the distillation is done carefully, the separation is quite good.
2. In order to drive the carvone over, it will probably be necessary to insulate the apparatus by wrapping the adapters with a single layer of aluminum foil. We did not experience the difficulty with foaming reported in Reference 3.

Time: less than 3 hours.

Distillation under Reduced Pressure

Procedure (Reference 3). Place 10 ml of one of the oils in a 25-ml round-bottom flask. Assemble the apparatus as shown in Figure 10.2, but use a fraction cutter (Figure 10.5) or a cow (Figure 10.3) in place of the simple vacuum adapter. Connect the fraction cutter or cow to a source of vacuum (Section 10.3) and a manometer (Section 10.4). It is preferable to use a magnetically stirred oil bath for heating, as shown in Figure 33.1. If magnetic stirring is available, place a very small magnetic stirring bar in the flask instead of boiling stones. If the flask must be heated with a flame, add a few boiling stones and clamp the flask in position on a wire gauze supported by an iron ring. Reduce the pressure of the system by turning on the water or oil pump, and allow the pressure to stabilize at 15–30 Torr. It is vital to the success of the distillation that the pressure of the system does not fluctuate. Heat the flask until the oil starts to boil (Note 1); continue heating until distillation begins. At this time, adjust the rate of heating so that the distillate accumulates at the rate of 1 drop every 3–5 seconds (Sections 9.6 and 10.5). Collect the distillate in at least three fractions: $C_{10}H_{16}$ isomers, an intermediate fraction, and the carvone fraction.

The expected boiling points will have to be estimated from the reported reduced-pressure boiling points (Section 10.1): Carvone: 115°C (20 Torr), 100°C (10 Torr), 91°C (6 Torr); limonene: 64°C (15 Torr); α-phellandrene: 61°C (11 Torr); β-phellandrene: 57°C (11 Torr).

Note

1. The air-saturated oil will de-gas under reduced pressure long before it starts to boil.

Time: less than 3 hours.

Column Chromatography

Procedure (Reference 4). Prepare a chromatographic column as described in Section 14.2, using a 22-mm-diameter tube and 25 grams of chromatographic-grade silica gel. Fit the bottom of the tube with a short length of plastic tubing with an adjustable clamp to control the rate of flow. Add the silica gel as a slurry in petroleum ether. After draining the petroleum ether from the column so that the level of the liquid is even with the top of the sand above the silica gel, add 2 ml of one of the oils. Use a little petroleum ether to assist in transferring the oil and to rinse down the sides of the tube.

For elution of the column use four portions of solvent:

1. 25–30 ml of petroleum ether
2. 50–60 ml of 1:9 dichloromethane: petroleum ether
3. 25 ml of 1:4 dichloromethane: petroleum ether
4. 100–130 ml of 1:1 dichloromethane: petroleum ether

Collect the eluant in four fractions whose volumes correspond to the volumes of the four portions of solvent added. The second fraction should contain $C_{10}H_{16}$ hydrocarbons, the third should be an intermediate fraction, and the fourth should contain carvone. ■ Recover the material in each fraction by distilling off the solvent, using the steam bath for heating. The last traces of solvent may be removed by heating under vacuum (Section 36).

Time: 3 hours.

Question

1. The dextrorotatory isomers of limonene and the two phellandrenes are illustrated. However, the phellandrene isomers are assigned the S configuration, while the limonene isomer is assigned the R configuration. Explain.

1. G. F. Russell and J. I. Hills, *Science* **172**, 1043 (1971).
2. L. Friedman and J. G. Miller, *Science* **172**, 1044 (1971).
3. S. L. Murov and M. Pickering, *J. Chem. Educ.* **50**, 74 (1973).
4. R. H. Mitchell and P. R. West, *J. Chem. Educ.* **51**, 274 (1974).

---

§47 RESOLUTION OF (R),(S)-α-PHENYLETHYL-AMINE BY MEANS OF (R),(R)-(+)-TARTARIC ACID

In this experiment the enantiomeric forms of α-phenylethylamine react with the chiral reagent (R),(R)-(+)-tartaric acid to give diastereomeric salts that differ greatly in their solubility in methanol. The enantiomeric reactants have the same structure but different configurations, being mirror images of one another; they have identical physical and chemical properties, except for the direction of rotation of plane-polarized light and the interaction with chiral molecules. The diastereomeric products have the same structure but different configurations; they are not mirror images of one another. The (R)-(+)-amine (R),(R)-(+)-tartrate is quite soluble in methanol, whereas its diastereomeric companion (S)-(−)-amine (R),(R)-(+)-tartrate is relatively insoluble in methanol.

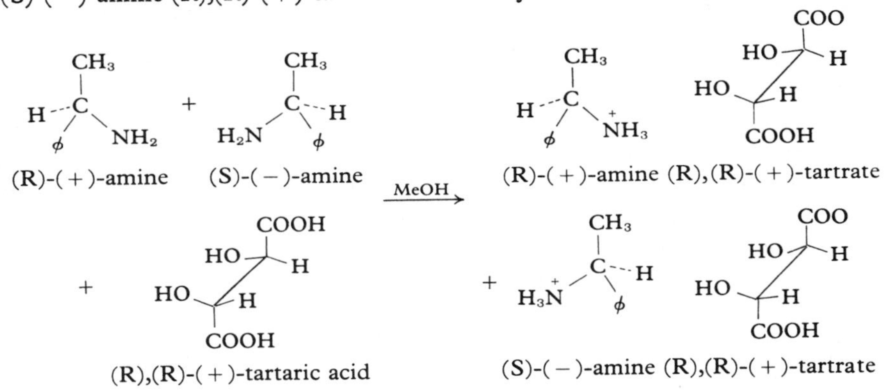

When a hot methanolic solution of the racemic amine and an equivalent amount of (R),(R)-(+)-tartaric acid is allowed to cool, the less soluble of the diastereomeric salts crystallizes out. Treatment of this salt with excess sodium hydroxide converts it to the free (S)-(−)-amine and disodium tartrate.

When the methanolic solution remaining after crystallization of the (S)-(−)-amine (R),(R)-(+)-tartrate is evaporated, the residue consists of the (R)-(+)-amine (R),(R)-(+)-tartrate, which ideally did not crystallize out at all, and that part of the (S)-(−)-amine (R),(R)-(+)-tartrate that remained in solution after crystallization. The amine recovered from this residue by treatment with sodium hydroxide will therefore contain more (R)-(+)-amine than (S)-(−)-amine.

Part of this excess (R)-(+)-amine can be isolated by treating a hot ethanolic solution of this amine mixture with an amount of sulfuric acid in methanol just great enough to convert the excess R-(+)-amine to its neutral sulfate salt:

$$\text{(R)-(+)-amine} + \text{(S)-(−)-amine} + H_2SO_4 \xrightarrow{\text{ethanol}}$$
$$x + y\ mole \qquad\qquad x\ mole \qquad\qquad \tfrac{1}{2}y\ mole$$

$$[\text{(R)-(+)-amine}^+]_2SO_4^= + \text{(R)-(+)-amine} + \text{(S)-(−)-amine}$$
$$\tfrac{1}{2}y\ mole \qquad\qquad\qquad x\ mole \qquad\qquad x\ mole$$
$$crystallizes\ out$$

(The experimental procedure for this step is described in detail under (R)-(+)-α-phenylethylamine.) The (R)-(+)-amine is then recovered from its neutral sulfate salt by treatment with excess sodium hydroxide.

The infrared spectrum of (R),(S)-α-phenylethylamine is shown in Figure 47.1. In contrast to the enantiomeric forms of carvone (Section 46), the enantiomeric forms of α-phenylethylamine appear to smell the same.

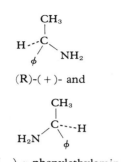

(R)-(+)- and

(S)-(−)-α-phenylethylamine

Figure 47.1. Infrared spectrum of (R),(S)-α-phenylethylamine; thin film.

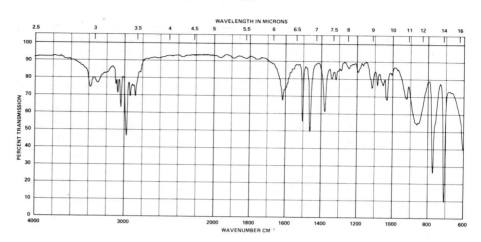

(S)-(−)-α-Phenylethylamine

Procedure. Add 31.25 grams of *d*-tartaric acid (0.208 mole) to 450 ml of methanol in a 1-liter Erlenmeyer flask, and heat the mixture almost to boiling. To the hot solution, add cautiously 25 grams of *d,l*-α-phenylethylamine (26.6 ml; 0.206 mole); too rapid an addition will cause the mixture to boil over. Since crystallization occurs slowly, the solution must be allowed to stand at room temperature for about 24 hours. ■ The (−)-amine-(+)-hydrogen tartrate separates as prismatic crystals (Note 1). Collect the product by suction filtration, and wash it with a little methanol (Note 2). Yield: 18.1 grams (65%).

A second crop (3.8 grams) may be obtained by concentrating the combined mother liquor and washings to 225 ml, and allowing crystallization to proceed at room temperature for another 24 hours. ■

Place the product (21.9 grams) in a 250-ml Erlenmayer flask and add about 90 ml of water; the amine tartrate salt will partially dissolve. Add to this mixture 12.5 ml of 50% sodium hydroxide solution (0.24 mole) in order to convert the amine salt to the free base. Then add about 50 ml of ethyl ether and swirl the mixture in order to extract the free amine into the ether. Separate the ethereal extract using a separatory funnel (Section 13.3) and then dry the extract over anhydrous magnesium sulfate for about 10 minutes (Section 15.2). ■ Remove the drying agent from the ethereal extract by gravity filtration (Section 7.1); collect the filtrate in a 100-ml round-bottom flask. Add a boiling stone to the flask and, heating with the steam bath, remove most of the ether by distillation. Transfer the residue to a 25-ml round-bottom flask, using a little ether for rinsing. Add a boiling stone to the flask and set up an apparatus for distillation like that shown in Figure 9.3. Remove the last of the ether by heating the flask strongly with the steam bath. ■ Now, being sure you are in a part of the lab that is free of ether, distill the residue to obtain (S)-(−)-α-phenylethylamine by heating the flask with a flame (Note 3). Yield: 6.9 grams (55%); b.p., 184–186°C; d_4^{25}, 0.953; $[\alpha]_D^{27}$, −38.2° (lit.: $[\alpha]_D^{22}$, −40.3°).

Time: 3 hours plus $\frac{1}{2}$ hour of a previous laboratory period to prepare the initial solution.

Procedure. Allow the methanolic solution remaining from the isolation of the (−)-amine-(+)-hydrogen tartrate to evaporate to dryness (this is most easily done by leaving the solution in an evaporating dish in the hood overnight ■), and recover the remainder of the amine by treating the residual salt with sodium hydroxide, extracting with ether, and distilling as previously described. Determine the specific rotation of the recovered amine.

The (+)-amine is isolated by treating a hot ethanolic solution of the recovered amine with an amount of sulfuric acid in ethanol slightly greater than that necessary to convert the excess (+)-amine to the neutral sulfate salt. The resulting crystals of (+)-α-phenylethylamine sulfate yield the (+)-amine (Note 4).

Dissolve 12.5 grams (0.103 mole) recovered amine ($[\alpha]_D^{27}$; $+24.5°$) in 88 ml of 95% ethanol. (This solution contains 0.062 mole excess (+)-amine, based on a $[\alpha]_D^{22}$ of $+40.7°$ for the pure (+)-amine, and 0.041 mole racemic amine.) Heat the solution to boiling and add to it a solution of conc. sulfuric acid (3.2 grams of 98% H_2SO_4; 0.032 mole H_2SO_4) in 180 ml of 95% ethanol. ■ After allowing the mixture to cool slowly to room temperature, collect the crystalline (+)-amine sulfate by suction filtration and wash it thoroughly with ethanol. Yield: 7.8 grams (74%).

Isolate the free (+)-amine as described (treatment with sodium hydroxide, extraction, and distillation), using 40 ml of water and 5 ml of 50% sodium hydroxide for each 10 grams of the sulfate salt. Yield: 4.4 grams (59%); b.p., 184–186°C; $[\alpha]_D^{27}$, $+38.3°$ (lit.: $[\alpha]_D^{22}$, $+40.67°$).

Notes

1. Sometimes a substance separates in the form of very fine needles. In this event, warm the mixture to dissolve the crystals, and allow it to cool again. The needle-like crystals dissolve much more rapidly than the prismatic ones. If possible, seed the solution with prismatic crystals. Workup of the needle-like crystals gives α-phenylethylamine of a $[\alpha]_D^{25}$ of -19 to $-21°$.

2. As an alternative, the mother liquor may be decanted from the crystals and the crystals washed by adding a little methanol, swirling, and decanting again. If no second crop is to be taken, the experiment can be continued by adding the water and the base to these crystals in the original flask.

3. Sometimes there is a problem with foaming. The conditions that cause foaming are not known.

4. You will have to calculate the amount of conc. sulfuric acid that you will need as the amount required depends both upon the yield of the recovered amine and upon its optical purity. The following directions illustrate one possibility.

Time: 3 hours to recover the amine enriched in the (R)-(+)-isomer and to prepare the solution. Two hours to isolate the (+)-isomer.

Questions

1. What would be the result of the experiment if (S),(S)-(−)-tartaric acid were used in place of (R),(R)-(+)-tartaric acid?
2. If 0.100 mole "recovered amine" had $[\alpha]_D^{25} = +20.35°$:
   a. How many moles of (+)-amine and how many moles of (−)-amine would there be in the sample?

b. How many moles of excess (+)-amine and how many moles of racemic amine would there be in the sample?

c. How much sulfuric acid (both moles and grams) would be needed to convert the excess (+)-amine to its neutral sulfate salt?

3. a. Are the (+)-amine sulfate and (−)-amine sulfate salts of α-phenylethylamine enantiomeric or diastereomeric?

b. Are their solubilities different in 95% ethanol?

c. Why is it possible to isolate the (+)-amine in this experiment?

d. What may be the possible consequences if more sulfuric acid is used than the amount necessary to convert the excess (+)-amine to its neutral sulfate salt?

4. What might be the molecular composition of the needle-like crystals?

5. What experiments might be done in order to determine the conditions for foaming (see Note 3)?

6. How would you expect the infrared spectrum of either separate enantiomer of α-phenylethylamine to compare with that of the racemic mixture (shown in Figure 47.1)?

References

1. A. Ault, *J. Chem. Educ.* **42**, 269 (1965).

2. A. Ault, *Organic Syntheses*, Vol. 49, Wiley, New York, 1969, p. 93.

Transformations

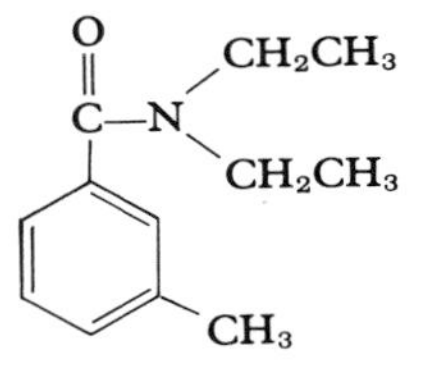

$$O$$

(structure of N,N-diethyl-m-toluamide, showing a benzene ring with a CH₃ substituent, and a C(=O)–N(CH₂CH₃)₂ group)

The experiments in the following sections are one-step preparations of various compounds. They are grouped by type of reaction, and, in general, reactions typical of aliphatic compounds are presented before the reactions of aromatic compounds.

Often, several examples of each kind of reaction are set forth. By comparing them you can see what features are essential to that kind of reaction and which are incidental to the particular compounds involved. Amounts of materials have been given in moles as well as the units normally used to measure them, so that comparisons between procedures can be made more easily.

In some cases, more than one procedure is given. Sometimes a choice will be determined by the starting materials, time, or equipment available. However, the most interesting choices will be those that must be made between variations of the same basic procedure. In these cases, experience, insight, and understanding will allow an intelligent decision as to what procedure or combination of procedures provides the best chance of giving the best result (yield, purity) with the least effort.

---

§48 ISOMERIZATIONS

The two experiments described in this section are concerned with changing one substance into a second that has the same molecular formula as the first. The first experiment involves a very subtle change: both the molecular formula and the structural formula of the starting material and the product are the same; only the configurations, or geometries, are different. In the second experiment, one structural isomer of molecular formula $C_{10}H_{16}$ is converted into another, more stable, structural isomer. The third experiment is a problem involving the determination of a mechanism of a *cis-trans* isomerization: the conversion of maleic acid to fumaric acid.

259

48.1 *cis*-1,2-DIBENZOYLETHYLENE FROM *trans*-1,2-DIBENZOYLETHYLENE

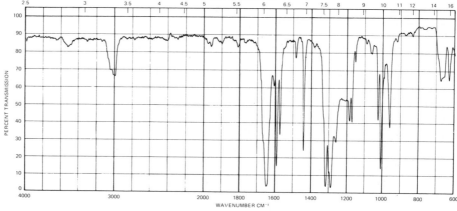

trans-1,2-dibenzoylethylene
m.p. 113–115°C; yellow

cis-1,2-dibenzoylethylene
m.p. 137–138°C; colorless

In this experiment the colorless *cis* isomer can be prepared from the yellow *trans* isomer if the hot solution of the *trans* isomer is allowed to stand in the sun during recrystallization. If the solution is not placed in the sun, the *trans* isomer recrystallizes without change.

A detailed procedure for the preparation of *trans*-1,2-dibenzoylethylene is given in Reference 1.

Recrystallization of *trans*-1,2-Dibenzoylethylene

Procedure (Reference 2).   Place 3 grams of *trans*-1,2-dibenzoylethylene in a 125-ml Erlenmeyer flask and add 100 ml of 95% ethanol. Heat the mixture almost to boiling on the steam bath and swirl the flask to dissolve the solid. Remove the flask from the heat and add as much decolorizing carbon as would cover a nickel. Filter the hot solution by gravity and collect the filtrate in another 125-ml Erlenmeyer flask. Reheat the filtrate if necessary to dissolve any crystals that may have formed during the filtration. Set the flask aside to cool slowly. ■ After several hours the long, canary-yellow needles of the *trans* isomer may be collected by suction filtration, using a small amount of ethanol for rinsing and washing. Recovery: 2.2 grams (73%). Melting point: 113–115°C (Reference 2).

Time: 1 hour plus $\frac{1}{2}$ hour to collect the product. This does not include the time required to stand for crystallization.

The infrared and proton NMR spectra of *trans*-1,2-dibenzoylethylene are shown in Figures 48.1 and 48.2. In the NMR spectrum, the singlet due to the resonance of the olefinic protons appears at $\delta = 8.0$.

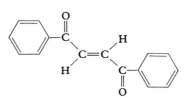

trans-1,2-dibenzoylethylene

Figure 48.1. Infrared spectrum of *trans*-1,2-dibenzoylethylene; CHCl$_3$ solution.

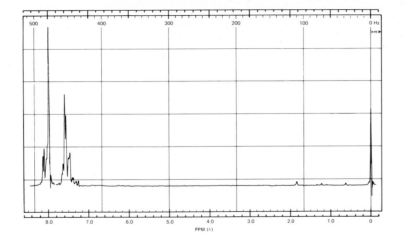

trans-1,2-dibenzoylethylene

Figure 48.2. NMR spectrum of *trans*-1,2-dibenzoylethylene; CDCl₃ solution.

Formation of *cis*-1,2-Dibenzoylethylene

Procedure (Reference 2).   Prepare a solution of 3 grams of *trans*-1,2-dibenzoyl-ethylene in 100 ml of 95% ethanol as described in the previous procedure. After removing the decolorizing carbon by gravity filtration and reheating to dissolve any crystals that may have formed during the filtration, place the loosely stoppered flask in bright sunlight (Note 1). ■ After several hours, very fine, colorless needles start to form (Note 2). When the flask appears to be about two-thirds full of crystals they may be collected by suction filtration and washed with a little ethanol. Yield: 2.2 grams (73%). Melting point: 137–138°C (Reference 2).

Notes

1. A 275-watt sunlamp may be used instead of the sun (References 1 and 2).
2. If yellow needles of the *trans* isomer form, they will redissolve and give way to the colorless needles of the *cis* isomer on further irradiation.

Time: 1 hour plus ½ hour to collect the product. This does not include the time required for irradiation and crystallization.

The infrared and proton NMR spectra of *cis*-1,2-dibenzoylethylene are shown in Figures 48.3 and 48.4. In the NMR spectrum, the singlet due to the resonance of the olefinic protons appears at δ = 7.1.

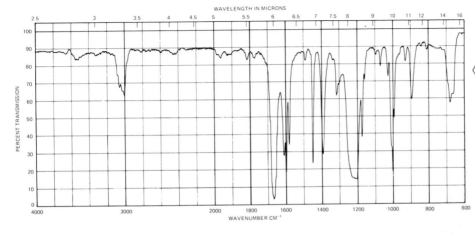

cis-1,2-dibenzoylethylene

Figure 48.3. Infrared spectrum of *cis*-1,2-dibenzoylethylene; CHCl₃ solution.

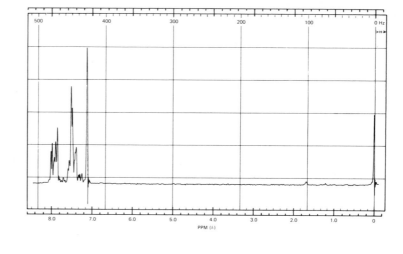

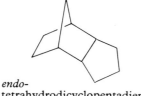

cis-1,2-dibenzoylethylene

Figure 48.4. NMR spectrum of *cis*-1,2-dibenzoylethylene; $CDCl_3$ solution.

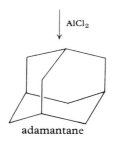

endo-tetrahydrodicyclopentadiene

↓ $AlCl_2$

adamantane

Nice crystals

References

1. D. J. Pasto, J. A. Duncan, and E. F. Silversmith, *J. Chem. Educ.* **51**, 227 (1974).
2. E. F. Silversmith and F. C. Dunson, *J. Chem. Educ.* **50**, 568 (1973).

48.2 ADAMANTANE FROM *endo*-TETRAHYDRODICYCLOPENTADIENE VIA THE THIOUREA CLATHRATE

In this experiment, adamantane is formed by the aluminum chloride catalyzed rearrangement of *endo*-tetrahydrodicyclopentadiene. Because of its symmetry and freedom from angle strain, adamantane is the structural isomer of molecular formula $C_{10}H_{16}$, which would be expected to predominate under equilibrating conditions. The adamantane is recovered from solution in hexane by treatment with a methanolic solution of thiourea to form a crystalline inclusion complex in which the molecules of adamantane occupy channels in the crystal lattice of thiourea. When the inclusion complex is shaken with a mixture of water and ether, it breaks down. The thiourea dissolves in the water and the adamantane in the ether, from which it is recovered by evaporation.

The infrared spectrum of adamantane is shown in Figure 48.5.

adamantane

Figure 48.5. Infrared spectrum of adamantane; CCl_4 solution.

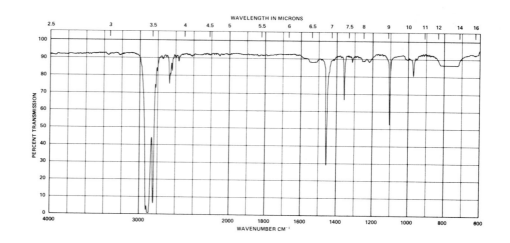

Procedure (Reference 1).   Weigh out 4 grams of aluminum chloride and store it until ready for use in a stoppered sample vial. Place 10.0 grams of *endo*-tetra-hydrodicyclopentadiene in a 50-ml 19/22 Erlenmeyer flask, and fit the flask with a well-greased inner joint that will serve as an air condenser. Clamp the flask so that it can be heated by means of a small Bunsen burner flame. Heat the flask gently to melt the tetrahydrodicyclopentadiene. When it has melted, add about one-fourth of the aluminum chloride down the air condenser and suspend a thermometer in the mixture so that the bulb dips into the liquid as far as possible but does not touch the bottom of the flask. Adjust the flame of the burner so that the mixture is heated just barely to boiling (180–185°C) and is maintained in that condition. Add the remaining aluminum chloride down the condenser in three or four portions at about 5-minute intervals, and continue the heating for a total of one hour. During this time, aluminum chloride will sublime into the condenser tube. It should be pushed down into the reaction mixture from time to time with a spatula.

At the end of the heating period, remove the thermometer and allow the black mixture to cool for an hour. ■ During this time, prepare a solution of 10 grams of thiourea in 150 ml of methanol (Note 1).

At the end of the cooling period, extract the product from the flask by adding hexane, swirling the mixture, and decanting, being careful not to pour out any of the black, tarry material; use a total of 70–80 ml of hexane in four or five portions. Add 1 gram of chromatography-grade alumina to the combined hexane extracts and swirl occasionally for about a minute. Filter the extract by gravity into the methanolic solution of thiourea, using 10–20 ml of hexane to rinse the flask and filter paper. Stir the two layers together for 2 or 3 minutes to permit the beauti-fully crystalline inclusion complex to complete its formation (Note 2). ■ Collect the complex by suction filtration and wash it with about 20 ml of hexane. The yield, after drying to constant weight, is 6–7 grams.

Transfer the crystals (which need not be dry) to a 125-ml separatory funnel and add 80 ml of water and 40 ml of ether. Shake the funnel vigorously for about 5 minutes, releasing the pressure occasionally, and then allow the layers to separate. If any solid remains, draw off as much of the lower layer as is con-venient, add 50 ml more water, and shake vigorously until all the solid has disappeared (Note 3). Dry the ethereal extract over anhydrous magnesium sulfate and evaporate it to dryness in a tared (weighed) 50-ml Erlenmeyer flask. The last of the ether can be removed by keeping the flask on its side for about 20 minutes. The yield is 1.45–1.6 grams (14.5–16%) of material with a melting range of 258–265°C (sealed capillary).

The product may be recrystallized from isopropyl alcohol, using 13 ml per gram, with a recovery of about 60%, to give an overall yield of 0.93 gram (9.3%) of adamantane with a melting point of 268–270°C (sealed capillary).

$$\overset{\text{S}}{\underset{\text{thiourea}}{\text{H}_2\text{N}-\overset{\|}{\text{C}}-\text{NH}_2}}$$

Notes

1. The process of dissolution can be greatly hastened by stirring with a magnetic stirrer or by heating on the steam bath.
2. Saturation of a similar solution with hexane does not result in crystal formation.
3. If a larger separatory funnel is available, all the water may be used at once.

Time: 4 hours.

Reference

1. A. Ault and R. Kopet, *J. Chem. Educ.* **46**, 612 (1969).

When maleic acid is warmed with hydrochloric acid, fumaric acid is formed:

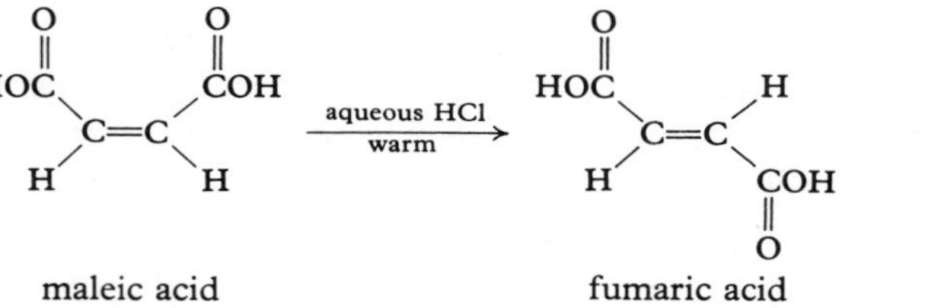

maleic acid
m.p. 139–140°C

fumaric acid
m.p. 300–302°C

Some of the mechanisms that can be conceived of for this reaction are listed here:

A thermal mechanism:

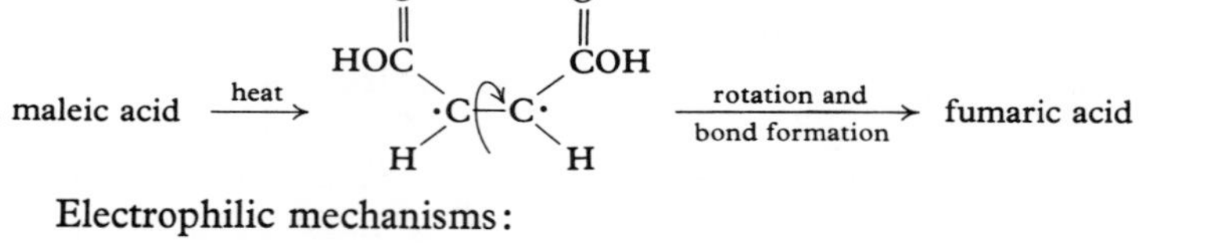

1. maleic acid $\xrightarrow{\text{heat}}$ [intermediate] $\xrightarrow[\text{bond formation}]{\text{rotation and}}$ fumaric acid

Electrophilic mechanisms:

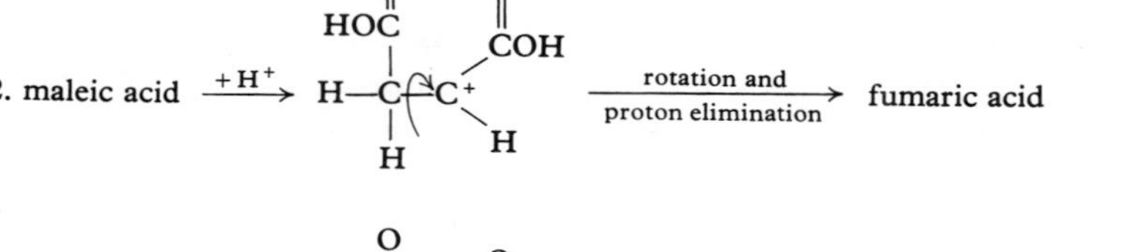

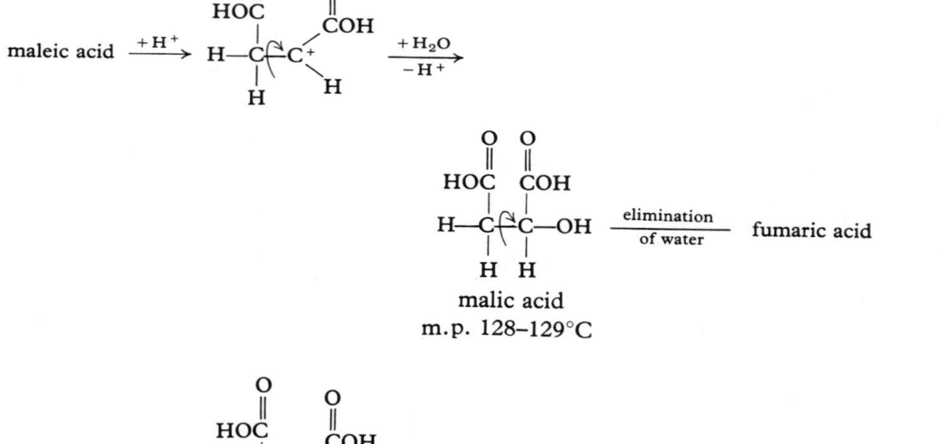

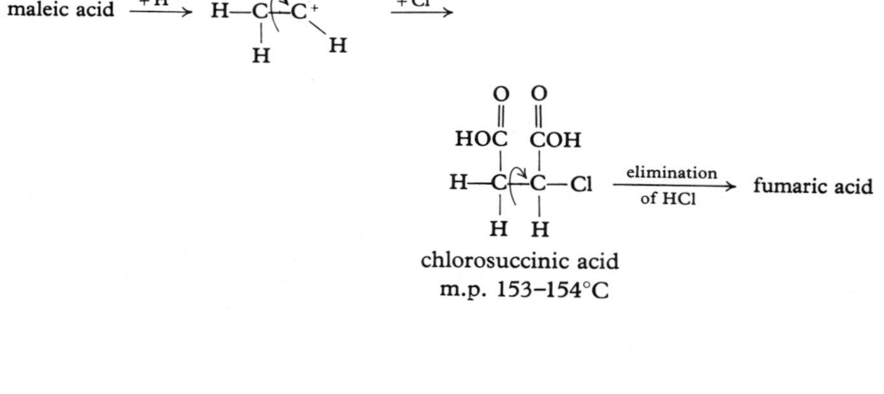

2. maleic acid $\xrightarrow{+\text{H}^+}$ [intermediate] $\xrightarrow[\text{proton elimination}]{\text{rotation and}}$ fumaric acid

3. maleic acid $\xrightarrow{+\text{H}^+}$ [intermediate] $\xrightarrow[-\text{H}^+]{+\text{H}_2\text{O}}$

malic acid
m.p. 128–129°C

[malic acid] $\xrightarrow[\text{of water}]{\text{elimination}}$ fumaric acid

4. maleic acid $\xrightarrow{+\text{H}^+}$ [intermediate] $\xrightarrow{+\text{Cl}^-}$

chlorosuccinic acid
m.p. 153–154°C

[chlorosuccinic acid] $\xrightarrow[\text{of HCl}]{\text{elimination}}$ fumaric acid

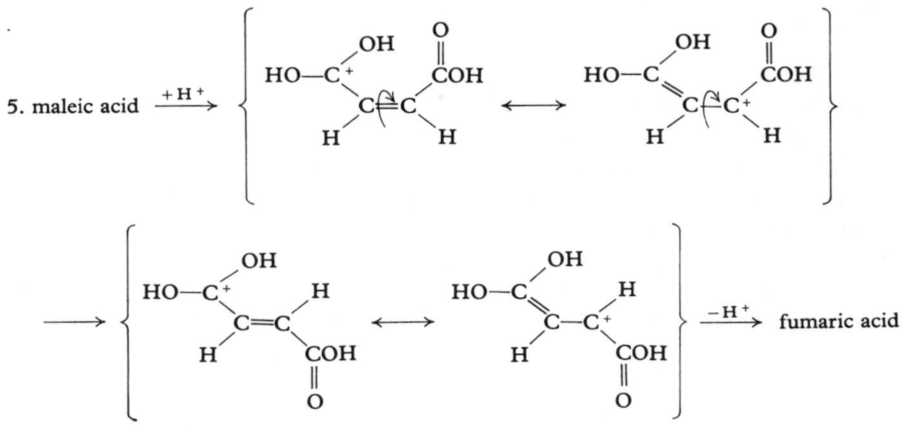

5. maleic acid $\xrightarrow{+H^+}$

$$\left\{ \text{HO—C}^+ \text{(OH) } \text{C=C } \text{(H)(H) COH(O)} \longleftrightarrow \text{HO—C(OH) } \text{C=C}^+ \text{(H)(H) COH(O)} \right\}$$

$$\longrightarrow \left\{ \text{HO—C}^+\text{(OH) } \text{C=C } \text{(H)(H) COH(O)} \longleftrightarrow \text{HO—C(OH) } \text{C—C}^+ \text{(H)(H) COH(O)} \right\} \xrightarrow{-H^+} \text{fumaric acid}$$

A nucleophilic mechanism:

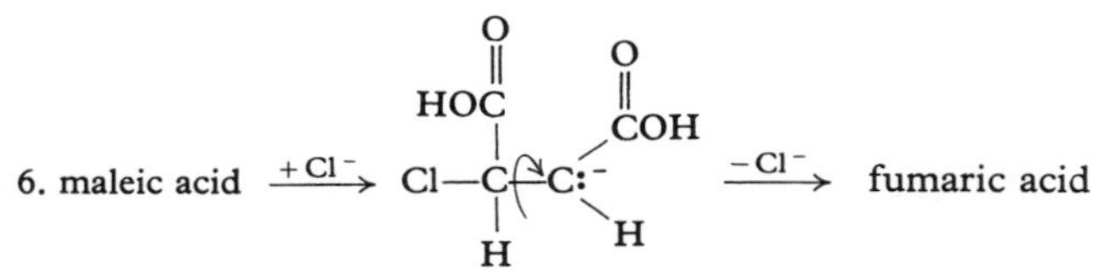

6. maleic acid $\xrightarrow{+Cl^-}$ $\text{HOC(O) } \text{Cl—C(H) C:}^-\text{(H) COH(O)}$ $\xrightarrow{-Cl^-}$ fumaric acid

The results of the six experiments that are described next will allow you to rule out some or all of these mechanisms.

Procedure. Label six medium-sized test tubes and then add to the tubes first the solid and then the solution as prescribed here:

Tube		Solution (Note 1)
1.	1 gram maleic acid	3 ml hydrochloric acid
2.	1 gram maleic acid	3 ml sulfuric acid
3.	1 gram maleic acid	3 ml water
	1 gram ammonium chloride	
4.	1 gram maleic acid	3 ml sulfuric acid
	1 gram ammonium chloride	
5.	1 gram malic acid	3 ml hydrochloric acid
6.	1 gram chlorosuccinic acid	3 ml hydrochloric acid

Place the six tubes in either a steam bath or a beaker of boiling water. Swirl each tube in order to dissolve the solid material. Once the solids are dissolved no further swirling is needed. Heat the tubes for 15 minutes and record in your notebook any changes that you observe. If by the end of the 15 minute heating period any solid has formed in a test tube, cool the contents of the tube and collect the solid by suction filtration. Wash the product with one milliliter of water, allow it to dry, weigh it, and identify it by determining its melting point (Notes 2 and 3).

Notes

1. Make up the acid solutions by pouring cautiously two volumes of the concentrated acid into one volume of water.

2. The melting points of the acids are: maleic, 139–140°C; fumaric, 300–302°C; malic (racemic mixture), 128–129°C; chlorosuccinic (racemic mixture), 153–154°C.

3. If a sample has not melted at a temperature of 160°C, it can be assumed to be fumaric acid.

Questions

1. through **6**.
   a. Describe briefly each of the six mechanisms previously presented.
   b. For each mechanism, which tubes should give rise to fumaric acid?

7. Which, if any, of the six mechanisms can account for your results?

8. If none of the mechanisms can account for your results, propose one or more mechanisms which are consistent with them.

9. When the reaction is run with DCl in D_2O and the fumaric acid that is formed is recrystallized from ordinary water, it is found that the product contains no deuterium. What mechanisms can be ruled out on the basis of this observation?

Reference

1. This experiment is based on the work of J. S. Meek, University of Colorado, Boulder.

---

§49 ALKENES FROM ALCOHOLS

Alkenes can often be formed from alcohols when the alcohol is heated with a strong acid whose anion is a weak oxidizing agent and a poor nucleophile. The transformation begins as follows:

$$H-\overset{|}{\underset{|}{C}}-\overset{|}{\underset{|}{C}}-OH + HA \underset{rapid}{\overset{}{\rightleftharpoons}} H-\overset{|}{\underset{|}{C}}-\overset{|}{\underset{|}{C}}-\overset{+}{O}H_2 + A^-$$

Then Path 1 is taken:

$$H-\overset{|}{\underset{|}{C}}-\overset{|}{\underset{|}{C}}-\overset{+}{O}H_2 \xrightarrow{slow} H-\overset{|}{\underset{|}{C}}-\overset{|}{C}{}^+ + H_2O \longrightarrow {\underset{\diagdown}{\overset{\diagup}{C}}}{=}{\underset{\diagdown}{\overset{\diagup}{C}}} + ''H^+''$$

or Path 2:

$$H-\overset{|}{\underset{|}{C}}-\overset{|}{\underset{|}{C}}-\overset{+}{O}H_2 \xrightarrow{slow} {\underset{\diagdown}{\overset{\diagup}{C}}}{=}{\underset{\diagdown}{\overset{\diagup}{C}}} + H_2O + ''H^+''$$

Path 1 is more likely to represent the reaction of tertiary alcohols or other alcohols which can form a relatively stable carbonium ion. Path 2 is more likely to represent the action of primary alcohols.

Substitution versus elimination

If A^- is a good nucleophile (for example, Br^- or Cl^-), it may either compete successfully for the carbonium ion of Path 1 or displace H_2O from the conjugate acid of the alcohol in Path 2 to form the substitution product:

$$H-\overset{|}{\underset{|}{C}}-\overset{|}{C}{}^+ + A^- \longrightarrow H-\overset{|}{\underset{|}{C}}-\overset{|}{\underset{|}{C}}-A$$

$$H-\overset{|}{\underset{|}{C}}-\overset{|}{\underset{|}{C}}-\overset{+}{O}H_2 + A^- \longrightarrow H-\overset{|}{\underset{|}{C}}-\overset{|}{\underset{|}{C}}-A + H_2O$$

Examples of reactions in which substitution predominates are given in Section 50 (cyclohexyl bromide from cyclohexanol), and Section 55 (alkyl halides from alcohols).

Oxidation

If A^- is a good oxidizing agent, the alcohol (or products derived from it) may undergo oxidation. In the experiment of Section 53, where sodium dichromate and sulfuric acid are present, oxidation of the secondary alcohol to the ketone is the predominant reaction. Sulfuric acid alone causes some oxidation when it is used to effect the dehydration of cyclohexanol.

Procedure (Reference 1).   Introduce 20.0 g (21 ml; 0.20 mole) of cyclohexanol, 5 ml of 85% phosphoric acid (Note 1), and a boiling stone into a 50-ml boiling flask; swirl to mix the layers. Attach the flask to a fractionating column (Note 2) fitted with a distilling adapter and condenser, and distill until the volume of the residue has been reduced to 5–10 ml. To reduce loss by evaporation of the product, cool the receiver in an ice bath.

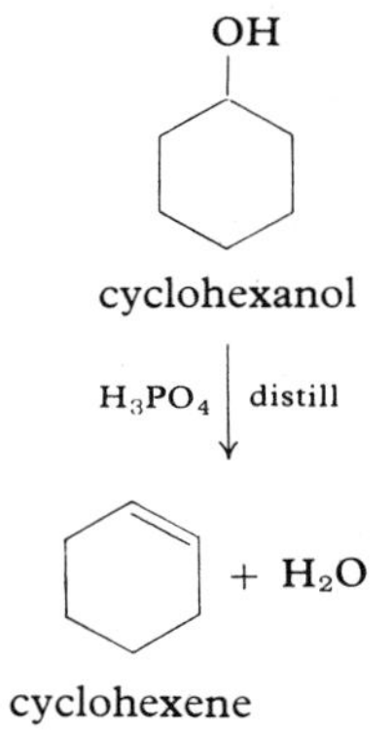

Let the boiling flask cool a little and then pour 20 ml of xylene (Note 3) down through the fractionating column into the boiling flask. Note the size of the upper layer, and continue distillation until the upper layer has been reduced by about one-half.

Pour the total distillate into a small separatory funnel, wash it with about 20 ml of water, and, after separating it from the water and decanting into a small Erlenmeyer flask, dry it over anhydrous magnesium sulfate. ■

While waiting for the distillate to dry, clean and dry the boiling flask, fractionating column, and adapters. After filtering the dried distillate into the boiling flask, reassemble the apparatus and fractionally distill the product, taking precautions against losses by evaporation. A typical yield is about 13 grams.

Gas chromatography (Note 4) and infrared spectroscopy are recommended as methods of determining the purity of your product with respect to starting material and chaser solvent.

The infrared spectrum of the product of this transformation is shown in Figure 49.1.

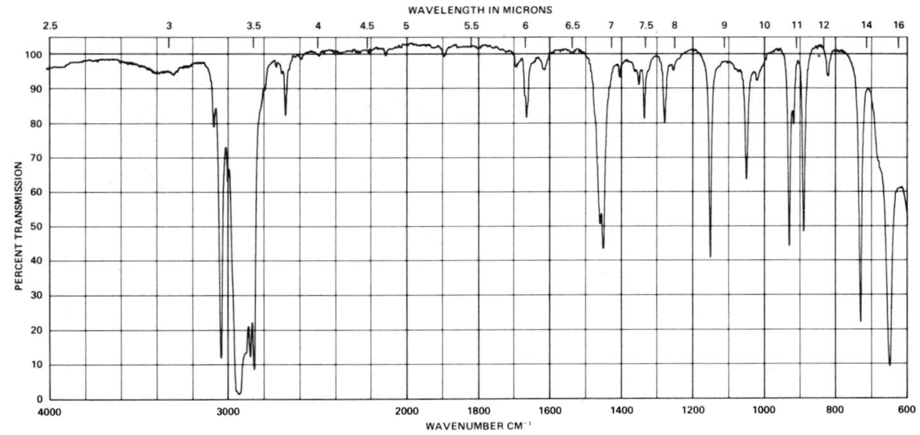

OH

cyclohexanol

H_3PO_4 | distill

+ H_2O

cyclohexene

Analyze your product

WAVELENGTH IN MICRONS

PERCENT TRANSMISSION

WAVENUMBER CM⁻¹

cyclohexene

Figure 49.1. Infrared spectrum of cyclohexene; thin film.

Notes

1. Sulfuric acid may be used in place of phosphoric acid, but it causes some oxidation of the organic materials.

2. A Vigreux column or a tube packed loosely with stainless steel sponge is suitable.

3. Xylene functions as a "chaser solvent." That is, since it boils considerably higher than the compound of interest (in this case, cyclohexene), it serves to displace the product from the boiling flask and distilling column in the first distillation, and thus reduces losses due to holdup. It is fairly easy to contaminate the product with chaser solvent in the final distillation unless it is done slowly, especially toward the end, and with even heating.

4. Suitable conditions for gas chromatographic analysis are: column, Apiezon L; column temperature, 180°C; helium pressure, 25 lb; flow rate, 120 ml/min. Under these conditions, the analysis takes 2 minutes.

Time: 3 hours.

Questions

1. Why do you suppose that the dehydration of cyclohexanol to cyclohexene has been the traditional experiment to illustrate the acid-catalyzed dehydration of an alcohol to an alkene?

2. Suppose 2-methylcyclohexanol were subjected to the conditions of this experiment instead of cyclohexanol.
   a. What alkenes would you expect to be formed?
   b. In what relative amounts would you expect them to be formed?

3. When 3,3-dimethyl-2-butanol is boiled with 85% H_3PO_4 according to the procedure of this section, an elimination reaction takes place. If a carbonium ion mechanism is involved and the initial carbonium ion undergoes elimination before rearrangement, the product should contain only 3,3-dimethyl-1-butene:

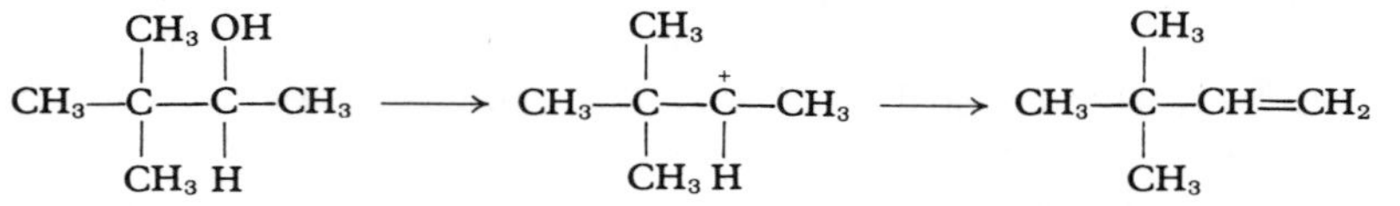

3,3-dimethyl-1-butene

If the initial carbonium ion should rearrange by methyl migration, two other alkenes could be formed:

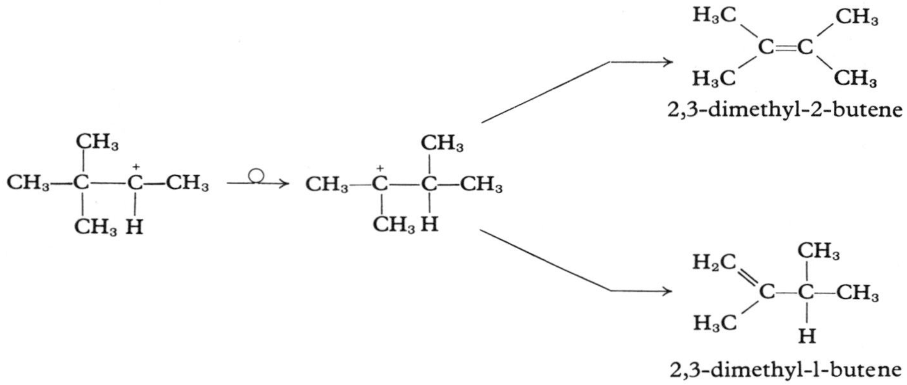

2,3-dimethyl-2-butene

2,3-dimethyl-1-butene

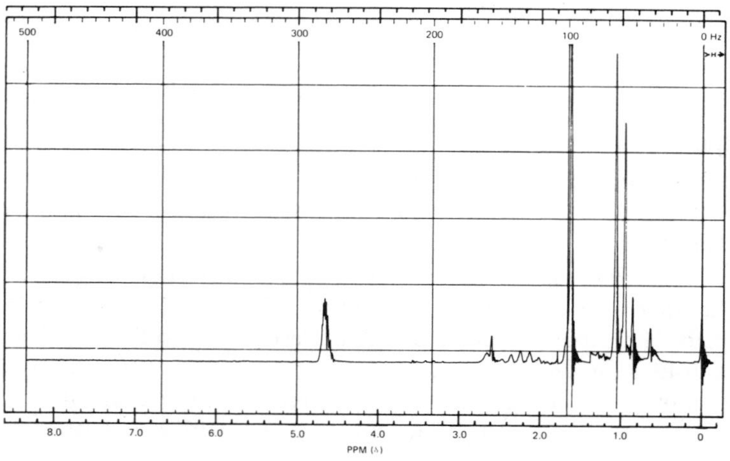

Figure 49.2. NMR spectrum of a mixture of alkenes produced by boiling 3,3-dimethyl-2-butanol with 85% H_3PO_4. The peak at $\delta = 1.6$ was recorded at one-tenth the amplitude used to record the rest of the spectrum.

Figure 49.2 shows the proton NMR spectrum of the product mixture obtained in this elimination reaction. The integral of the multiplet at $\delta = 4.6$ is 4.2 units, and the integral of the remainder of the spectrum is 124 units. From these data estimate the relative amounts of the three products formed and interpret the results in terms of the relative stability of the products and the mechanism of the reaction.

Reference

1. L. F. Fieser, *Experiments in Organic Chemistry*, 3rd edition, revised, Heath, Boston, 1955, p. 62.

49.2 DEHYDRATION OF 2-METHYLCYCLOHEXANOL

Procedure. In the *Journal of Chemical Education* **44**, 620 (1967), Taber and Champion reported that 2-methylcyclohexanol can be dehydrated under conditions that are essentially the same as those that serve to convert cyclohexanol to cyclohexene in the preceding experiment (20 grams of compound boiled with 5 ml of 85% phosphoric acid). They did not suggest the use of a chaser solvent, and said that the rate of heating should be such that the temperature of the distillate does not rise much above 95°C, since higher temperatures will result in distillation of too much unreacted alcohol. After distillation has become very slow, they suggest that the distillate be worked up by washing it with water, 10% sodium bicarbonate, and finally with water again. It should then be dried over a little anhydrous calcium chloride and filtered.

The composition of the product is then determined by gas chromatography (Sections 14.7 and 14.8). The authors suggest the use of a silicone grease column at 75°C.

As an alternative to gas chromatographic analysis, they say that one can determine the refractive index of the mixture and estimate the composition by assuming a linear relationship between molar concentration and refractive index.

Time: 3 hours.

2-methylcyclohexanol

H_3PO_4 | distill

alkenes

Questions

1. What are the most likely products of this reaction?
2. **a.** How many stereoisomers can there be of 2-methylcyclohexanol?
   b. To what extent would you expect that these isomers could be separated by fractional distillation or gas chromatography? Explain.
   c. Which isomers would be expected to give exactly the same results in this experiment if each could be tested separately? Explain.
3. In the gas chromatographic analysis of the product of this reaction:
   a. How do you determine what peak corresponds to which compound?
   b. Does it matter whether you assume that the ratio of peak areas corresponds to the weight ratio or the mole ratio of the alkenes?
   c. How can one demonstrate that the ratio of peak areas is a valid measure of product composition?
4. In the analysis of the product by index of refraction:
   a. How can one determine the validity of the assumption of a linear relationship between molar concentration and index of refraction?
   b. Would you expect a linear relationship between weight concentration and index of refraction?
   c. What assumption has been made about the number of components in the product mixture?

5. How can you tell whether the relative amounts of alkenes found in the product of this experiment are determined by their relative rates of formation or by their relative stability?

OH

cyclohexanol

NaBr | H₂SO₄
↓

Br

cyclohexyl bromide

---

§50 CYCLOHEXYL BROMIDE FROM CYCLOHEXANOL

A general reaction which serves to convert an alcohol to the corresponding alkyl chloride or bromide involves treatment of the alcohol with a solution that is strongly acidic and contains a high concentration of halide ion. A number of reagents have been used, including a solution containing conc. sulfuric acid and sodium bromide (as in this experiment and the experiment of Section 55.1, *n*-butyl bromide from *n*-butyl alcohol), and conc. hydrochloric acid (Section 55.2, *tert*-butyl chloride from *tert*-butyl alcohol).

The course of the reaction is usually represented by the following equations:

$$H—\underset{|}{\overset{|}{C}}—\underset{|}{\overset{|}{C}}—OH + HX \underset{\xrightarrow{\hspace{1cm}}}{\overset{rapid}{\xleftarrow{\hspace{1cm}}}} H—\underset{|}{\overset{|}{C}}—\underset{|}{\overset{|}{C}}—\overset{+}{O}H_2$$

Then Path 1:

$$H—\underset{|}{\overset{|}{C}}—\underset{|}{\overset{|}{C}}—\overset{+}{O}H_2 \xrightarrow{slow} H—\underset{|}{\overset{|}{C}}—\overset{|}{C}^+ + H_2O \xrightarrow{+X^-} H—\underset{|}{\overset{|}{C}}—\underset{|}{\overset{|}{C}}—X$$

or Path 2:

$$H—\underset{|}{\overset{|}{C}}—\underset{|}{\overset{|}{C}}—\overset{+}{O}H_2 + X^- \xrightarrow{slow} H—\underset{|}{\overset{|}{C}}—\underset{|}{\overset{|}{C}}—X + H_2O$$

Path 1 represents the reaction of tertiary alcohols or other alcohols that can form a relatively stable carbonium ion, whereas path 2 represents the bimolecular substitution reaction to be expected of primary alcohols. Presumably, the attack of the halide ion as a nucleophile on either the carbonium ion in Path 1 or the conjugate acid of the alcohol in Path 2 is sufficiently fast to compete successfully with elimination to form the alkene.

The infrared spectrum of cyclohexyl bromide is shown in Figure 50.1.

Procedure. Place a magnetic stirring bar in a 100-ml boiling flask, and add to the flask 20 ml of water, 20 g (0.19 mole) of sodium bromide, and 10 g (10.5 ml; 0.10 mole) of cyclohexanol. Fit the flask with a Claisen adapter and put the

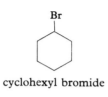

Br

cyclohexyl bromide

Figure 50.1. Infrared spectrum of cyclohexyl bromide; thin film.

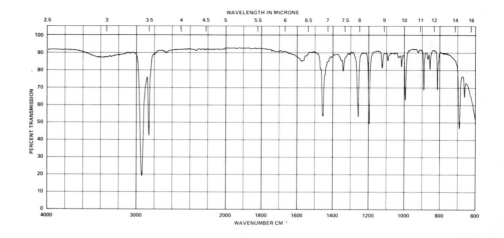

bottom joint of the separatory funnel in the center joint of the adapter. Add 20 ml (37 g; 0.37 mole) of conc. sulfuric acid to the separatory funnel.

SECTION 51

Support the flask over the magnetic stirring motor, and adjust the speed of the motor so that the contents of the flask are stirred quite vigorously. While the contents of the flask are being stirred, add the sulfuric acid dropwise but otherwise as rapidly as possible; the addition should take about 2 minutes.

After completing the addition of the acid, let the mixture stir for 5 more minutes. At this time, pour the hot (Note 1) mixture into the separatory funnel and, after allowing sufficient time for separation of the layers, draw off the aqueous phase. To the cyclohexyl bromide, which remains in the separatory funnel, add a solution of about 5 grams potassium carbonate in about 50 ml of water. Mix the layers thoroughly, and, after separating the layers, dry the crude cyclohexyl bromide over a mixture of anhydrous potassium carbonate and anhydrous magnesium sulfate for about 10 minutes.

Filter the crude product from the drying agent into a 25-ml boiling flask, add a pinch of anhydrous potassium carbonate, and distill (it is not necessary to use a condenser, Figure 9.17). Collect as product the material boiling between 155 and 165°C. Yield: about 9 grams.

Note

1. Be very careful when working with the hot reaction mixture, as it contains sulfuric and hydrobromic acids; it will cause painful burns if it is spilled on your hands, and it will burn holes in your clothes.

Time: 3 hours.

Questions

1. It was assumed in the discussion of the mechanism of this reaction that cyclohexyl bromide is not formed via elimination to give cyclohexene followed by addition of hydrogen bromide to the cyclohexene.
a. Is there any evidence to support this assumption?
b. What kinds of experiments might be done to test this assumption?

§51 CYCLOHEXANOL FROM CYCLOHEXENE

Cyclohexanol will form cyclohexene on being heated with conc. sulfuric acid or 85% phosphoric acid (Section 49.1). The reverse reaction may be carried out as described in the procedure that follows.

Procedure (Reference 1). Cautiously add 7.0 ml (0.126 mole) of conc. sulfuric acid to 3.4 ml (3.4 g; 0.19 mole) of water in a 50-ml ground-glass-stoppered Erlenmeyer flask. Cool the solution to room temperature. Add 10.1 ml (8.2 g; 0.10 mole) of cyclohexene. Stopper the flask and shake to mix; the mixture should be shaken or stirred until a clear homogeneous solution is formed (Note 1 ■). At this point, pour the mixture into a 250-ml boiling flask and rinse the Erlenmeyer flask with a total of about 120 ml of water, adding the rinsings to the boiling flask. Distill the mixture until 50 or 60 ml has been collected. Saturate the distillate with sodium chloride and separate the cyclohexanol by extracting with ether. Dry the ethereal extract over anhydrous potassium carbonate, filter, and distill the ether on the steam bath. Distill the residue and collect the fraction boiling between 155–162°C as cyclohexanol.

cyclohexene

H_2SO_4

H_2O

OH

cyclohexanol

1. Some time can be saved by allowing the mixture to stand between laboratory periods.

Time: 3 hours.

The infrared spectrum of cyclohexanol is shown in Figure 51.1.

cyclohexanol

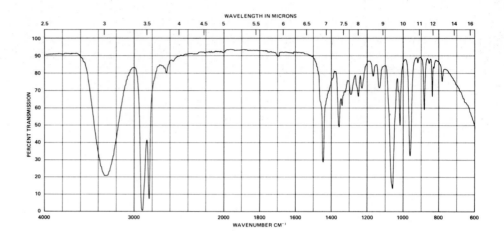

Figure 51.1. Infrared spectrum of cyclohexanol; thin film.

Questions

1. Write balanced equations for the reactions that take place in this experiment.

2. At which point in the experiment does each reaction take place?

3. Compare this procedure with that for the conversion of cyclohexanol to cyclohexene (Section 49.1). Explain why the reactions go in opposite directions under the different conditions.

Reference

1. G. H. Coleman, S. Wawzonek, and R. E. Buckles, *Laboratory Manual of Organic Chemistry*, 2nd edition, Prentice-Hall, Englewood Cliffs, N.J., 1962, p. 52.

---

§52 ADDITION OF DICHLOROCARBENE TO CYCLOHEXENE BY PHASE TRANSFER CATALYSIS

One method for the formation of a cyclopropane ring is through the addition of a carbene to a carbon–carbon double bond. This experiment involves the addition of dichlorocarbene to cyclohexene to give 2,2-dichlorobicyclo[4.1.0]heptane, or dichloronorcarane:

cyclohexene   +   :C ⟶   dichloronorcarane

Although carbenes are uncharged the valence shell of the carbon atom is electron deficient since it contains only six electrons. For this reason carbenes act as powerful electrophiles. In order to avoid reaction with water, a nucleophile, carbene reactions are usually run under anhydrous conditions. In this procedure dichlorocarbene is generated in a water-free environment in an unusual way. The reaction mixture consists of two phases: an aqueous phase containing sodium hydroxide and a catalytic amount of tetra-*n*-butylammonium bromide, and an organic phase containing chloroform and cyclohexene. Apparently tetra-*n*-butylammonium ions and hydroxide ions migrate into the organic phase as ion pairs where the hydroxide ion removes a proton from a chloroform molecule to give the trichloromethyl anion:

$$HO^- + H-\underset{\underset{Cl}{|}}{\overset{\overset{Cl}{|}}{C}}-Cl \longrightarrow H_2O + {:}\underset{\underset{Cl}{|}}{\overset{\overset{Cl}{|}}{C}}-Cl$$

The trichloromethyl anion then eliminates chloride ion in an alpha elimination to form dichlorocarbene:

$$^-{:}\underset{\underset{Cl}{|}}{\overset{\overset{Cl}{|}}{C}}-Cl \longrightarrow {:}\underset{\underset{Cl}{|}}{\overset{\overset{Cl}{|}}{C}} + Cl^-$$

The dichlorocarbene, formed in the organic layer where the water concentration is very low and the cyclohexene concentration very high, will usually react to give dichloronorcarane. The chloride ion formed and the tetra-*n*-butylammonium ion then migrate as an ion pair into the water layer where the ammonium cation can again pick up an hydroxide ion to take into the organic layer. The catalytic role of the tetra-*n*-butylammonium cation can be summarized in this way:

aqueous layer

$$(n\text{-Bu})_4 N^+ OH^- \xleftarrow[\text{anion exchange}]{\text{NaOH}} (n\text{-Bu})_4 N^+ Cl^-$$

$$(n\text{-Bu})_4 N^+ OH^- \xrightarrow[\substack{\text{carbene generation} \\ \text{and reaction with} \\ \text{cyclohexene}}]{\text{CHCl}_3} (n\text{-Bu})_4 N^+ Cl^-$$

organic layer

The catalytic function of the tetra-*n*-butylammonium salt in this reaction has been called *phase transfer catalysis*. Other quaternary ammonium salts have the same effect, and no reaction takes place in the absence of the salt.

The infrared spectrum of the product, dichloronorcarane, is shown in Figure 52.1.

Procedure. Put a magnetic stirring bar in a 250-ml round-bottom boiling flask. Place a steam bath on a magnetic stirrer and clamp the flask in position in the rings of the steam bath so that the contents of the flask can be heated by steam and magnetically stirred at the same time. Add to the flask 19 ml of chloroform (27.5 grams; 230 mmole), 7.7 ml of cyclohexene (6.2 grams; 75 mmole), and 250 mg (0.78 mmole) of tetra-*n*-butylammonium bromide. Fit the flask with a reflux condenser and then pour in one portion down the condenser into the flask a mixture of 30 ml of 50% sodium hydroxide (570 mmole) and 22 ml of water. Gently heat and vigorously stir the resulting mixture for 30 minutes (Note 1). At the end of the heating period add ice in order to cool the mixture to room temperature. Then add to the flask 50 ml of ether and vigorously stir the resulting mixture for three minutes (Note 2). Then transfer the mixture to a

dichloronorcarane

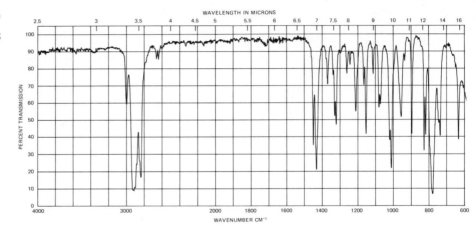

WAVELENGTH IN MICRONS

PERCENT TRANSMISSION

WAVENUMBER CM⁻¹

Figure 52.1. Infrared spectrum of dichloronorcarane; thin film.

separatory funnel, allow the layers to separate, and draw off the lower, aqueous, layer, returning it to the original 250 ml boiling flask. Save the ether extract in a 125-ml Erlenmeyer flask. Add 25 ml of ether to the aqueous layer in the boiling flask and again vigorously stir the mixture for three minutes. Again, using the separatory funnel, separate the layers and then dry the combined ether extracts over anhydrous magnesium sulfate. Filter the dried ethereal solution by gravity into a round-bottom boiling flask and remove the ether by distillation on the steam bath. Distill the residue at atmospheric pressure using a flame. Collect the fraction boiling between 192 and 197°C as dichloronorcarane. Yield: about 7 grams about 55%).

Notes

1. Heat the mixture just to boiling so that an occasional drop falls from the reflux condenser.
2. The yield is critically dependent on the thoroughness of the extraction.

Time: 3 hours.

References

1. A. Ault and B. Wright, *J. Chem. Educ.* **53**, 486 (1976).
2. M. Makosza and M. Wawrzyniewicz, *Tetrahedron Lett.* 4659 (1969).
3. G. C. Joshi, N. Singh, and L. M. Pande, *Tetrahedron Lett.* 1461 (1972).
4. K. Isagawa, Y. Kimura, and S. Kwon, *J. Org. Chem.* **39**, 3171 (1974).

§53 CYCLOHEXANONE FROM CYCLOHEXANOL

One of the common methods for the preparation of ketones is the oxidation of a secondary alcohol by means of a sulfuric acid solution of a chromate or dichromate salt:

$$3 \: R-\underset{\underset{H}{|}}{\overset{\overset{OH}{|}}{C}}-R + H_2Cr_2O_7 + 6\,H^+ \longrightarrow 3 \: R-\overset{\overset{O}{\|}}{C}-R + 2\,Cr^{+++} + 7\,H_2O$$

The following experiment is an example of this method. For an infrared spectrum of the product, see Figure 53.1.

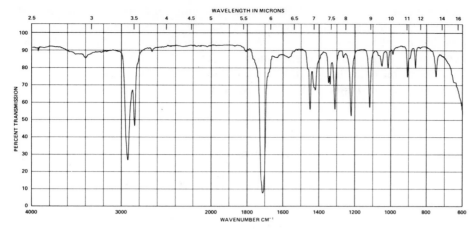

cyclohexanone

Figure 53.1. Infrared spectrum of cyclohexanone; thin film.

OH

cyclohexanol

| oxidation

O

cyclohexanone

Make sure the reaction starts

Procedure a (Reference 1). Cautiously add 20 ml (37 g; 0.37 mole) of conc. sulfuric acid to 60 grams of crushed ice. After mixing thoroughly, pour the solution into a 250-ml boiling flask and add a stirring bar. Fit the flask with a Claisen adapter, and put the bottom joint of a separatory funnel in the outer neck of the adapter. Add 20 g (21.0 ml; 0.20 mole) of cyclohexanol to the flask and suspend a thermometer in the center neck of the adapter so that the bulb is entirely in the liquid, but not in the way of the stirring bar. Don't make a closed system! Place a solution of 21 g (0.070 mole) of sodium dichromate dihydrate in 10 ml of water in the separatory funnel. While the mixture is being stirred, add about 1 ml of the dichromate solution from the separatory funnel. The temperature should increase, and the color should change from orange to green (Note 1). Add the remainder of the dichromate solution at such a rate that the temperature remains between 25 and 35°C. After the addition is complete (30–40 minutes), continue stirring until the temperature spontaneously falls one or two degrees. At this time, add about 1 gram of oxalic acid in order to reduce any excess dichromate.

Now remove the thermometer, and, after rinsing the separatory funnel with water, add 100 ml of water to it and place it in the center neck of the Claisen adapter. Fit the other neck of the Claisen adapter with a distilling adapter and condenser. Distill the reaction mixture until about 70–90 ml has been collected, running in water from the separatory funnel to replace the material removed by distillation (See Section 11). ∎

"Salting out"

Saturate the distillate with sodium chloride (about 15–18 grams), and ex- extract it with ether. Dry the ethereal extract over anhydrous magnesium sulfate, filter and distill. Collect as cyclohexanone the portion boiling between 150 and 155°C.

Note

1. This color change must occur before more dichromate solution is added. See Section 31.

Procedure b (Reference 2). Dissolve 20 g (0.067 mole) sodium dichromate dihydrate in 35 ml of glacial acetic acid in a 125-ml Erlenmeyer flask by swirling the mixture on the steam bath. Then cool the solution in an ice bath to 15°C. Prepare a mixture of 20 g (21 ml; 0.20 mole) of cyclohexanol in 13 ml of glacial acetic acid in a second 125-ml Erlenmeyer flask, and cool it to 15°C as well. When the temperature of both solutions has been adjusted to 15°C, pour the dichromate solution into the cyclohexanol solution. Remove the mixture from the ice bath, but be prepared to put it back for a moment in order to keep the temperature of the reaction mixture from exceeding 65°C. By means of intermittent brief swirling in the ice bath, keep the temperature between 60 and 65°C until the temperature begins to drop spontaneously (about 25–30 minutes). After 5–10 minutes more, pour the reaction mixture into a 250-ml boiling flask, add 75 ml of water, and work up as in Procedure **a**. The ethereal extract should be washed with saturated sodium chloride solution.

Question

1. Upon what basis did you choose between Procedures **a** and **b**?

References

1. G. K. Helmkamp and H. W. Johnson, Jr., *Selected Experiments in Organic Chemistry*, 2nd edition, W. H. Freeman, San Francisco, 1968, p. 108.
2. L. F. Fieser, *Organic Experiments*, Heath, Boston, 1964, p. 106.

§54 REDUCTION OF KETONES TO SECONDARY ALCOHOLS

Sodium borohydride is an efficient and specific reducing agent for aldehydes, ketones, and acid chlorides, converting them to the corresponding alcohols. Other reducible groups such as carbon–carbon double or triple bonds, nitro or cyano groups, and other carbonyl-containing functional groups such as esters or lactones and amides are not reduced under conditions whereby aldehydes and ketones may be reduced. Sodium borohydride is relatively soluble in water, methanol, and ethanol. In these solvents, in the presence of base, it hydrolyzes to form hydrogen very slowly; acidification, however, leads to rapid evolution of hydrogen. The reaction appears to proceed by hydride transfer from the borohydride ion to the carbonyl group:

$$4 \underset{R}{\overset{R}{C}}{=}O + Na^+BH_4^- \longrightarrow [(H{-}\underset{R}{\overset{R}{C}}{-}O)_4{-}B]^-Na^+$$

The borate ester may then be hydrolyzed by addition of water:

$$[(H{-}\underset{R}{\overset{R}{C}}{-}O)_4{-}B]^-Na^+ + 2\,H_2O \longrightarrow 4\,H{-}\underset{R}{\overset{R}{C}}{-}O{-}H + Na^+BO_2^-$$

54.1 CYCLOHEXANOL FROM CYCLOHEXANONE

Procedure. Dissolve 9.8 g (10.3 ml; 0.10 mole) cyclohexanone in 25 ml of methanol. To this solution, add a solution of about 0.5 g sodium methoxide and 1.0 g (0.026 mole) sodium borohydride in 25 ml of methanol. After about 5 minutes, pour the mixture into 100 ml of ice water and add 10 ml of dilute hydrochloric acid. Extract the aqueous mixture with 50 ml of ether, and wash the ethereal extract with three 15-ml portions of water. After drying the extract

cyclohexanone

reduction

cyclohexanol

over anhydrous magnesium sulfate, remove the drying agent by gravity filtration and isolate the product by distillation. Collect as cyclohexanol the portion boiling between 155 and 165°C.

Time: less than 3 hours.

The IR spectra of cyclohexanol and cyclohexanone are presented in Figures 51.1 and 53.1 (pages 272 and 275).

54.2 *cis-* AND *trans*-4-*tert*-BUTYLCYCLOHEXANOL FROM 4-*tert*-BUTYLCYCLOHEXANONE

Reduction of 4-*tert*-butylcyclohexanone under the conditions of the preceding experiment leads to a mixture of the *cis-* and *trans*-4-*tert*-butylcyclohexanols:

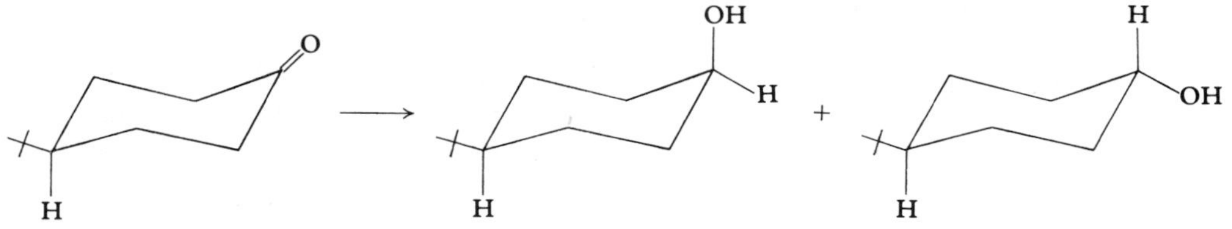

The ratio of *cis* to *trans* isomers may be determined by gas chromatographic analysis of the dried ethereal extract on a 20% Carbowax(polyethylene glycol)-on-firebrick column at 150°C; the *cis* isomer has the shorter retention time (1).

Time: less than 3 hours.

Questions

1. How do you determine that the *cis* isomer has the shorter retention time?
2. What assumption is made when taking the peak area ratio as the isomer ratio?
3. How do you explain the observed isomer ratio?

Reference

1. *Organic Syntheses*, Vol. 47, p. 18.

---

§55 ALKYL HALIDES FROM ALCOHOLS

In Sections 49 and 50, it was explained how treatment of an alcohol with a strong acid in the presence of chloride or bromide ion could serve to convert the alcohol to the corresponding alkyl halide. The three experiments in this section illustrate this type of reaction.

55.1 *n*-BUTYL BROMIDE FROM *n*-BUTYL ALCOHOL

n-Butyl bromide is prepared by boiling *n*-butyl alcohol with a mixture of sodium bromide, water, and conc. sulfuric acid. It is likely that the reaction proceeds by way of a nucleophilic displacement of water by bromide ion from the conjugate acid of the alcohol:

$$CH_3CH_2CH_2CH_2{-}OH + "H^+" \underset{\xleftarrow{\hspace{1cm}}}{\overset{rapid}{\xrightarrow{\hspace{1cm}}}} CH_3CH_2CH_2CH_2{-}\overset{+}{O}H_2$$

$$CH_3CH_2CH_2CH_2{-}OH_2 + Br^- \xrightarrow{slow} CH_3CH_2CH_2CH_2{-}Br + H_2O$$

The infrared spectrum of *n*-butyl bromide is shown in Figure 55.1.

Procedure. Add to a 250-ml boiling flask 27.0 g (0.26 mole) sodium bromide, 30 ml of water, and 20 ml (16.2 g; 0.22 mole) of *n*-butyl alcohol. Cool the mixture in an ice bath, and add, with swirling and cooling, 23 ml (42.3 g; 0.44 mole) of conc. sulfuric acid. Fit the flask with a condenser and heat the mixture under reflux for one-half hour (Notes 1 and 2). At the end of the heating period, distill the mixture until no more water-insoluble material comes over (see Section 11). Pour the distillate into a separatory funnel and add 20 ml of water. Add a pinch of sodium bisulfite to remove any free bromine and shake the funnel to mix the contents. Remove the lower organic layer, and wash it successively with a 20-ml portion of ice-cold sulfuric acid, 20 ml of water, and 20 ml of 10% aqueous sodium carbonate solution (Note 3). After drying the *n*-butyl bromide over anhydrous calcium chloride, distill it and collect the fraction boiling between 99 and 103°C.

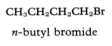

CH₃CH₂CH₂CH₂Br

n-butyl bromide

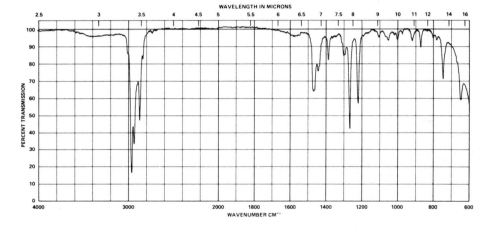

Figure 55.1. Infrared spectrum of *n*-butyl bromide; thin film.

Notes

1. A longer heating time increases the yield only slightly.
2. Some hydrogen bromide is evolved. Usually the amount is not objectionable, but it can be trapped by absorption in water if desired (Section 35).
3. Except for the wash with conc. sulfuric acid, the organic layer will be the bottom layer. In these cases, the bottom layer will have to be removed, the separatory funnel rinsed with water, and the organic layer added again to the funnel. It is easy to lose material in these transfers.

Time: 3 hours.

55.2 *tert*-BUTYL CHLORIDE FROM *tert*-BUTYL ALCOHOL

tert-Butyl chloride is prepared by treating *tert*-butyl alcohol with conc. hydrochloric acid. It appears that, in this case, the *tert*-butyl carbonium ion can be formed at a convenient rate near room temperature and without additional Lewis acid catalysis:

$$
\underset{\underset{CH_3}{|}}{\overset{\overset{CH_3}{|}}{CH_3-C-OH}} + HCl \underset{\xrightarrow{\hspace{1cm}}}{\overset{rapid}{\longleftarrow}} \underset{\underset{CH_3}{|}}{\overset{\overset{CH_3}{|}}{CH_3-\overset{+}{C}-OH_2}} + Cl^-
$$

$$
\underset{\underset{CH_3}{|}}{\overset{\overset{CH_3}{|}}{CH_3-\overset{+}{C}-OH_2}} \xrightarrow{slow} \underset{\underset{CH_3}{|}}{\overset{\overset{CH_3}{|}}{CH_3-\overset{+}{C}}} \xrightarrow[fast]{+\,Cl^-} \underset{\underset{CH_3}{|}}{\overset{\overset{CH_3}{|}}{CH_3-C-Cl}}
$$

The infrared and NMR spectra of *tert*-butyl chloride are shown in Figures 55.2 and 55.3.

Procedure.   Place 60 ml (71.5 g; 0.72 mole) conc. hydrochloric acid, which has been cooled in an ice bath, in a 125-ml separatory funnel. To this, add 20 ml (15.8 g; 0.21 mole) *tert*-butyl alcohol. Shake the mixture occasionally during twenty minutes, relieving the internal pressure by inverting the funnel and cautiously opening the stopcock. Allow the mixture to stand until the layers have separated cleanly, and then separate and discard the hydrochloric acid layer. Wash the product with 10 ml water and then with 10 ml of aqueous sodium

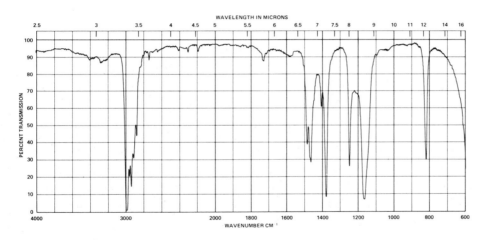

$$
\underset{\underset{CH_3}{|}}{\overset{\overset{CH_3}{|}}{CH_3-C-Cl}}
$$
tert-butyl chloride

Figure 55.2. Infrared spectrum of *tert*-butyl chloride; thin film.

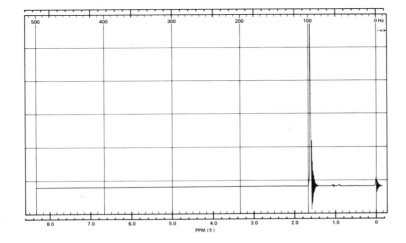

$$
\underset{\underset{CH_3}{|}}{\overset{\overset{CH_3}{|}}{CH_3-C-Cl}}
$$
tert-butyl chloride

Figure 55.3. NMR spectrum of *tert*-butyl chloride; neat.

bicarbonate solution. Dry the product over anhydrous calcium chloride and distill, collecting the fraction boiling between 49 and 52°C as *tert*-butyl chloride.

Time: less than 3 hours.

55.3 COMPETITIVE NUCLEOPHILIC SUBSTITUTION OF BUTYL ALCOHOLS BY BROMIDE AND CHLORIDE ION

The relative reactivities of two species can sometimes be determined by allowing them to compete for a limited amount of a second reagent. In this experiment, bromide ion and chloride ion compete to form the corresponding alkyl halides from one of the isomeric butyl alcohols. The relative nucleophilicity of the two ions is inferred from the relative amounts of the two alkyl halides formed.

Procedure (Reference 1). Cautiously pour 50 ml of conc. sulfuric acid over 60 grams of ice. Pour the resulting solution into a 250-ml boiling flask. To this, add 13.5 g (0.25 mole) ammonium chloride and 24.5 g (0.25 mole) ammonium bromide. Fit the flask with a reflux condenser and bring the solids into solution by swirling and warming on the steam bath. Now add 0.20 mole of one of the isomeric butyl alcohols (Note 1) and boil the mixture very gently for 15–30 minutes, depending upon your estimate of the reactivity of the alcohol. Cool the reaction mixture in an ice bath, and transfer it to a separatory funnel. Save the organic layer and wash it twice with 20-ml portions of cold conc. sulfuric acid, once with 50 ml of water, and once with aqueous sodium bicarbonate solution. Dry the organic phase over anhydrous calcium chloride.

Determine the ratio of alkyl bromide to alkyl chloride by one of the following methods:

1. Determination of the index of refraction (Section 19), assuming a linear relation between the refractive index and molar composition of the sample (Note 2).
2. Determination of the density (Section 18), assuming a linear relation between the density and molar composition of the sample (Note 2).
3. Vapor-phase chromatography (Sections 14.7 and 14.8), assuming that the ratio of the peak areas equals the molar ratio (Note 3).

Notes

1. Normal, secondary, and tertiary butyl alcohol appear to work satisfactorily in this experiment. In the case of isobutyl alcohol, the index of refraction of the product was greater than that of isobutyl bromide.
2. The following values have been reported for the density and the index of refraction of isomeric butyl chlorides and bromides. (The data are from A. I. Vogel, *Practical Organic Chemistry*, 3rd edition, Wiley, New York, 1957, pp. 293 and 294. The values in parentheses were estimated from the other values.)

		n-Butyl	*i*-Butyl	*s*-Butyl	*t*-Butyl
d_4^{20}	Cl	0.886	0.881	0.874	0.846
	Br	1.274	1.253	1.256	(1.226)
n_D^{20}	Cl	1.402	1.398	1.397	1.386
	Br	1.440	1.435	1.437	(1.424)

3. Satisfactory conditions for vapor-phase chromatographic analysis of the product mixtures are $6' \times \frac{1}{4}''$ Carbowax 6000 column; column temperature, 80°C; detector temperature, 162°C; helium pressure, 7 lb; flow rate, 30 ml/min; sample size, 1 microliter. Under these conditions, the longest retention time, that of n-butyl bromide, is 11 minutes.

Time: 3 hours.

Questions

1. What effect would the following procedural variations have on the numerical value of the observed ratio, compared to what the value would be if the variation had not occurred?
   a. The mixture was boiled so vigorously that some of the product escaped from the condenser.
   b. Some of the product evaporated during work-up.
   c. The product contained substances other than the two alkyl halides. (Consider each method of analysis.)
   d. The alcohol was added before the ammonium salts had completely dissolved.

2. In interpreting the results, it will be assumed that the observed ratio of R—Br to R—Cl is the result of the relative rates of formation of the two compounds, and not of their relative stability. How could one show experimentally whether or not this is a valid assumption?

3. One possible result is that the ratios would be the same for each alcohol. How would this result be interpreted?

4. Another result might be that the ratio of R—Br to R—Cl would be different for each alcohol. What would be a reasonable variation in this ratio with the structure of the alcohol?

5. Assuming that class results are available from the reaction of normal, secondary, and tertiary butyl alcohols, interpret them in terms of the presently accepted mechanisms for this type of reaction.

6. What difference would it make if the initial concentrations of ammonium chloride and ammonium bromide were not equal? How would you take this into account?

7. What effect would the following factors have on the numerical value of the observed product ratio, compared to what the value would be otherwise? Assume an observed ratio of 2.0.
   a. 0.20 mole of each ammonium salt was used.
   b. 0.10 mole of alcohol was used.
   c. The reaction was not complete (some alcohol remained unreacted).
   d. Some of the alcohol was converted to the alkene.

8. Answer Question 7 assuming an observed ratio of 1.0.

9. Answer Question 7 assuming an observed ratio of 0.5.

10. What would be the result in this experiment if 0.50 mole of alcohol were used?

11. It should now be apparent that taking the observed ratio of R—Br to R—Cl as a measure of the relative nucleophilicity of bromide and chloride ions in this reaction is an approximation. In fact, in an experiment using 0.25 mole each of bromide and chloride and 0.20 mole of alcohol, a relative nucleophilicity of 2.80 is necessary to account for an observed ratio of 2.0. If only 0.10 mole of alcohol had been used in an experiment that was otherwise the same, a relative nucleophilicity of 2.23 would account for an observed ratio of 2.0. Derive an exact expression, verify the examples given, and show under what conditions the exact expression reduces to the approximation.

Reference

1. G. K. Helmkamp and H. W. Johnson, *Selected Experiments in Organic Chemistry*, 2nd edition, Freeman, San Francisco, 1968, p. 59.

A very common method of preparing esters of carboxylic acids is to heat together a mixture of the acid and the alcohol until equilibrium has been established; a small amount of a mineral acid is usually added to speed attainment of equilibrium:

$$\underset{\substack{\displaystyle \text{O} \\ \displaystyle \|}}{\text{R—C—O—H}} + \text{R'—O—H} \xrightleftharpoons{H^+} \underset{\substack{\displaystyle \text{O} \\ \displaystyle \|}}{\text{R—C—O—R'}} + \text{H—O—H}$$

The yield of ester can be increased by using an excess of one of the reactants (the cheaper or more easily available one), or by removing from the reaction mixture one or both of the products.

56.1 ISOAMYL ACETATE, A COMPONENT OF THE ALARM PHEROMONE OF THE HONEY BEE

Isoamyl acetate, sometimes called pear oil or banana oil, can be prepared as previously described by heating together acetic acid and isoamyl alcohol in the presence of concentrated sulfuric acid. An excess of acetic acid is used to shift

$$\underset{\substack{\displaystyle \text{O} \\ \displaystyle \|}}{\text{CH}_3\text{—C—OH}} + \text{HO—CH}_2\text{CH}_2\text{CH}\overset{\text{CH}_3}{\underset{\text{CH}_3}{\diagdown}} \xrightleftharpoons{H^+} \underset{\substack{\displaystyle \text{O} \\ \displaystyle \|}}{\text{CH}_3\text{—C—O—CH}_2\text{CH}_2\text{CH}}\overset{\text{CH}_3}{\underset{\text{CH}_3}{\diagdown}} + \text{H}_2\text{O}$$

the equilibrium toward product formation, since the acid is a little less expensive and is easier to remove from the product.

Isoamyl acetate has been shown to be one of the active compounds in the alarm pheromone of the honey bee (1). Cotton balls containing freshly excised honey bee stings and placed near the hive entrance were seen to be more frequently stung than control balls. Gas chromatographic and infrared analysis of 700 stings indicated that approximately one microgram of isoamyl acetate can be isolated from each sting. Other substances in addition to isoamyl acetate must be involved, however. While cotton balls treated with isoamyl acetate do alert and agitate the guard bees, other balls containing an equivalent number of stings incite the bees to sting as well.

The infrared and proton NMR spectra of isoamyl acetate are shown in Figures 56.1 and 56.2.

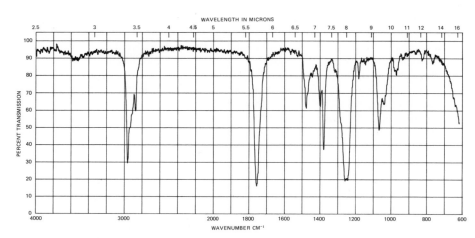

$$\underset{\substack{\displaystyle \text{O} \\ \displaystyle \|}}{\text{CH}_3\text{—C—O—CH}_2\text{CH}_2\text{CH}}\overset{\text{CH}_3}{\underset{\text{CH}_3}{\diagdown}}$$

iso-amyl acetate

Figure 56.1. Infrared spectrum of isoamyl acetate; thin film.

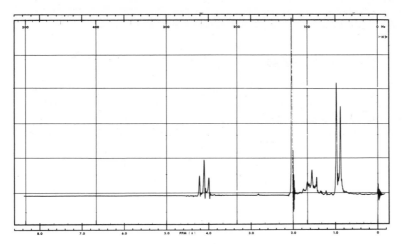

iso-amyl acetate

Figure 56.2. NMR spectrum of isoamyl acetate; CCl$_4$ solution.

Procedure. Place in a 100-ml round-bottom boiling flask 10.9 ml of isoamyl alcohol (3-methyl-l-butanol; 8.8 grams; 0.1 mole), 23 ml of glacial acetic acid (24 grams; 0.4 mole), and 3 ml of conc. sulfuric acid (5.5 grams; 0.054 mole). Add a few boiling chips, fit the flask with a condenser, and boil the mixture gently, using a flame, for one to two hours (Note 1). At the end of the heating period cool the contents of the flask by adding 50 grams of ice (Notes 2 and 3). Transfer the reaction mixture to a separatory funnel. Add to the separatory funnel 40 ml of diethyl ether, using some of the ether to rinse the boiling flask. Thoroughly mix the contents of the separatory funnel, allow the layers to separate, and draw off the lower, aqueous, layer. Wash the ether layer with a 50 ml portion of cold water and then extract the ether layer with a solution of 3 grams of sodium carbonate dissolved in 50 ml of water, using this solution in two 25-ml portions (Note 4). Dry the ether extract over anhydrous magnesium sulfate, remove the drying agent by suction filtration, and strip off the ether by distillation on the steam bath. Distill the residue, using a flame, to isolate the isoamyl acetate. The boiling point of isoamyl acetate is reported to be 142°C at 756 Torr. Yield (Note 1): 8 to 9 grams (60 to 70%).

Notes

1. A slightly higher yield is obtained with the longer reaction time.
2. Fifty grams of ice is approximately the amount of crushed ice that can be contained in a 100-ml beaker.
3. The point is to cool the mixture well below the boiling point of the solvent that will be used for extraction. In this case the solvent will be diethyl ether, b.p. 35°C, and the mixture should be cooled to below 25°C. See Section 13.3.
4. Take care not to build up excessive pressure in the separatory funnel due to the carbon dioxide that will be produced. See Section 13.3.

Time: 3 to 4 hours.

Question

1. How is the excess acetic acid removed from the product mixture?

Reference

1. R. Boch, D. A. Shearer, and B. C. Stone, *Nature* **195**, 1018 (1962).

56.2 CHOLESTERYL BENZOATE FROM CHOLESTEROL; LIQUID CRYSTALS

In 1888 an Austrian botanist prepared cholesteryl benzoate and noticed that at about 150°C it "melted" to give a cloudy liquid that became clear only on further heating to above 180°C. Since this first observation of the phenomenon, which has come to be called the formation of a liquid crystal, many other compounds have been found that behave in a similar way. The liquid crystal phase differs from the ordinary liquid phase in this way: although it takes the shape of its container, its molecules are still ordered to a certain extent and so the liquid still exhibits some of the properties of the crystalline state. In the case of cholesteryl benzoate the molecules are generally associated in layers with their long axes parallel to one another and the layers are stacked in a spiral arrangement. This is the so-called "twisted nematic" form, which was earlier called the "cholesteric" form. All substances that give the twisted nematic form of liquid crystal are made up of chiral molecules; it is presumably the chirality of the molecules that leads to the large-scale spiral organization. This spiral organization makes it possible for certain cholesterol derivatives, though colorless in themselves, to selectively scatter light into different colors. As cholesteryl benzoate enters and leaves the liquid crystal phase, reddish purple colors can be seen if the sample is strongly illuminated from the side. An easy way to see this is described below. Reference 1 discusses the phenomenon of liquid crystals in more detail.

It is often desirable to obtain only one member of a pair of enantiomers. One way to do this is by resolving the racemic mixture, as in the resolution of α-phenylethylamine (Section 47). Another is to start the synthesis with only one member of a pair of enantiomers rather than starting with the racemic mixture, as in the synthesis of (R)-(−)-carvone from (R)-(+)-limonene (Section 77.5). However it is done, *something* chiral must be involved. A particularly attractive concept is the use of a chiral catalyst, and much research is being done to find ways to form or consume exclusively only one member of a pair of enantiomers in a process that makes use of a chiral catalyst. One recent experiment of this general nature was to heat phenyl ethyl malonic acid in the liquid crystal phase of cholesteryl benzoate in order to form 2-phenylbutyric acid by thermal decarboxylation (2):

phenyl ethyl malonic acid → liquid crystal phase of cholesteryl benzoate → (R)-(−)-2-phenylbutyric acid 59% + (S)-(+)-2-phenylbutyric acid 41%

Apparently the two carboxyl groups, which reside in enantiomeric environments, react at slightly different rates in the chiral reaction medium, although there is disagreement about the relative importance in this experiment of the chirality of the individual molecules and the chirality of the large scale spiral structure (3).

Although the application of this concept to synthetic organic chemistry is relatively recent, enzymes have been doing similar things for millions of years. For example, the alcohol dehydrogenases appear to be able to react exclusively with either the R or S proton in the alpha-methylene group of ethanol:

acetaldehyde that contains H_S or H_R exclusively

One class of alcohol dehydrogenases removes exclusively H_R, and the other removes exclusively H_S (4).

In this experiment the ester is prepared by treating the alcohol with an acid chloride in pyridine solution:

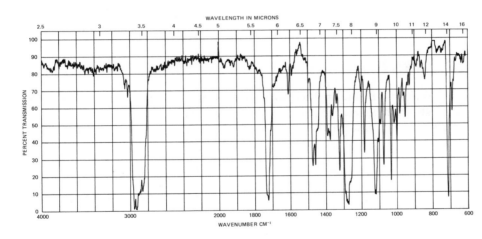

The infrared spectrum of cholesteryl benzoate is shown in Figure 56.3.

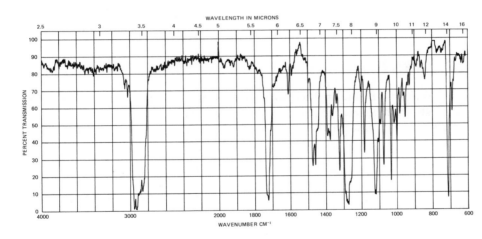

WAVELENGTH IN MICRONS

PERCENT TRANSMISSION

WAVENUMBER CM⁻¹

Figure 56.3. Infrared spectrum of cholesteryl benzoate; CCl_4 solution.

Procedure. Place 1.0 gram (2.6 mmole) of cholesterol in a 50 ml Erlenmeyer flask. To this add 3 ml of pyridine and swirl the mixture to dissolve the cholesterol. Then add 0.4 ml (0.48 gram; 3.5 mmole) of benzoyl chloride (Note 1) and heat the resulting mixture on the steam bath for about 10 minutes. At the end of the heating period cool the mixture somewhat by swirling the flask in a beaker of cold water. Then dilute the mixture with 15 ml of methanol and collect the solid cholesteryl benzoate by suction filtration using a little methanol to rinse the flask and to wash the crystals. Recrystallize the cholesteryl benzoate by heating it in an Erlenmeyer flask with 20 ml of ethyl acetate until it has dissolved, filtering the hot solution by gravity, allowing the filtrate to cool to room temperature, and collecting the crystals by suction filtration (Note 2). Yield: from 0.6 to 0.8 grams (45 to 65%).

TRANSFORMATIONS

1. Benzoyl chloride is a lachrymator. It should be stored in the hood and you should take your flask to the hood and work there while making the addition. A graduated pipet should be supplied to use to transfer the benzoyl chloride.

2. The recovery may be improved somewhat by cooling the mixture in an .ice bath before collecting the crystals.

The phenomenon of the formation of the liquid crystal phase of cholesteryl benzoate can be easily seen by placing 100–200 milligrams of the compound on the end of a microscope slide and heating the sample by holding the slide with a pair of tongs above a small burner flame. The solid will turn first to a cloudy liquid and then with further heating to a clear melt. On cooling, the cloudy liquid will first appear, and then it will change to a hard, crystalline solid. With strong lighting from the side, purple and red colors will be seen as the sample changes phases on both heating and cooling. The more cautious the heating, the better you can see the changes. You can repeat the heating and cooling many times with the same sample.

Time: 3 hours.

References

1. L. Verbit, *J. Chem. Educ.* **49**, 37 (1972).
2. L. Verbit, T. R. Halbert, and R. B. Patterson, *J. Org. Chem.* **40**, 1649 (1975).
3. W. H. Pirkle and P. L. Rinaldi, *J. Am. Chem. Soc.* **99**, 3510 (1977).
4. W. L. Alworth, *Stereochemistry and Its Application in Biochemistry*, Wiley-Interscience, 1972, p. 77.

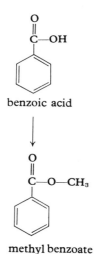

benzoic acid

methyl benzoate

56.3 METHYL BENZOATE

In this section, two related methods are presented for the preparation of methyl benzoate from benzoic acid and methanol. The first makes use of conc. sulfuric acid as catalyst, while the second uses the Lewis acid boron trifluoride:

$$\phi-\overset{\overset{\displaystyle O}{\|}}{C}-O-H + CH_3-O-H \underset{\xrightarrow{acid}}{\rightleftharpoons} \phi-\overset{\overset{\displaystyle O}{\|}}{C}-O-CH_3 + H_2O$$

The second method, that of Procedure b, seems to be suitable for the preparation of the methyl esters of a large number of aromatic carboxylic acids.

The infrared and NMR spectra of the product methyl benzoate are shown in Figures 56.4 and 56.5.

Catalysis by Conc. H₂SO₄ (Procedure a)

Procedure. Add to a 100-ml boiling flask 12.2 g (0.1 mole) of benzoic acid, 25 ml (19.7 g; 0.62 mole) of methanol, and 3 ml conc. sulfuric acid (Note 1). Fit the flask with a reflux condenser and boil the mixture for about 45 minutes. Cool the mixture to room temperature and pour it into a separatory funnel that contains 50 ml of cold water. Rinse the flask with 25 ml of ether and pour this into the separatory funnel. Mix the contents of the separatory funnel, allow them to settle, and then draw off the aqueous layer. Wash the ethereal layer with 25 ml of water and then 25 ml of 5% aqueous sodium carbonate solution (Note 2). Dry the ethereal extract over anhydrous magnesium sulfate, filter, and

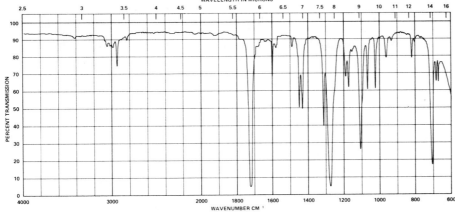

methyl benzoate

Figure 56.4. Infrared spectrum of methyl benzoate; thin film.

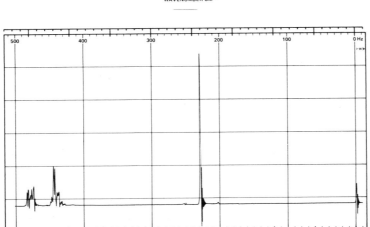

methyl benzoate

Figure 56.5. NMR spectrum of methyl benzoate; CCl$_4$ solution.

remove the ether by distillation on the steam bath. Complete the distillation with a flame, collecting as methyl benzoate the material boiling above 190°C. Yield: about 70%.

Notes

1. Add the sulfuric acid cautiously, allowing it to run down the wall of the flask. Swirl to mix.
2. Add the sodium carbonate solution in portions; swirl to mix; watch for foaming.

Time: 3 hours.

Catalysis by BF$_3$ (Procedure b)

Procedure (Reference 1). Add to a 100-ml boiling flask 2.44 g (0.020 mole) of benzoic acid, 22 ml of anhydrous methanol (0.54 mole), and 4.4 ml of the boron trifluoride/methanol complex (0.040 mole of BF$_3 \cdot$2MeOH). Fit the flask with a condenser and boil the mixture for one hour. Cool the mixture to room temperature and pour it into a saturated solution of sodium bicarbonate (Note 1). Isolate the ester by extraction with ether and recover it as described in Procedure a. Yield: about 96%.

1. The amount of sodium bicarbonate required is not specified. If all the boron trifluoride is hydrolyzed to give hydrogen fluoride, at least 0.12 mole of base will be required for neutralization.

Time: 3 hours.

Questions

1. Write a mechanism for the BF$_3$-catalyzed reaction of Procedure b.
2. Is it possible that the BF$_3$ is doing more than just catalyzing the reaction that takes place in Procedure b? If so, what else might it be doing?

Reference

1. G. Hallas, *J. Chem. Soc.* **1965**, 5770.

56.4 METHYL SALICYLATE; OIL OF WINTERGREEN

salicylic acid → methyl salicylate

Procedure b of Section 56.3 is said to give a 91% yield of methyl salicylate from salicylic acid with a reflux time of three hours.

Methyl salicylate is widely used as a perfume and flavor ingredient; it smells like wintergreen. Its infrared and NMR spectra are shown in Figures 56.6 and 56.7.

56.5 ACETYLSALICYLIC ACID; ASPIRIN

In the preceding experiment, salicylic acid participates in an esterification reaction through its carboxyl group. In this experiment, it takes part through the phenolic hydroxyl group.

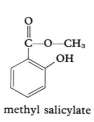

methyl salicylate

Figure 56.6. Infrared spectrum of methyl salicylate; thin film.

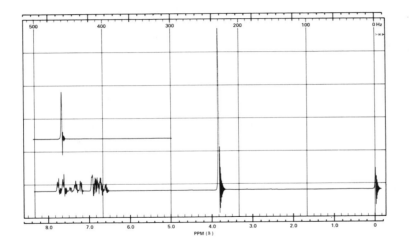

methyl salicylate

Figure 56.7. NMR spectrum of methyl salicylate; CCl$_4$ solution.

The industrial method of production of acetylsalicylic acid is similar to the method given in this section. In 1976, industrial production in the United States was in the amount of about 29 million pounds, or about 1.3 trillion one-gram doses.

Procedure. Add to a 125-ml Erlenmeyer flask 2.0 grams (14.5 mmole) of salicylic acid and 4 ml (4.3 grams; 42 mole) of acetic anhydride. To this mixture add 5 drops of 85% phosphoric acid and swirl to mix. Fit the flask with a reflux condenser and heat the mixture on the steam bath for about 5 minutes. Without cooling the mixture add 2 ml of water in one portion down the condenser. The excess acetic anhydride will hydrolyse and the contents of the flask will come to a boil. When the vigorous reaction has ended add 40 ml of cold water, cool the mixture to room temperature, stir and rub the mixture with a stirring rod if necessary to induce crystallization, and finally allow the mixture to stand in the ice bath to complete crystallization. Collect the product by suction filtration and wash it with a little water. The product may be recrystallized from water.

Time: 2 hours.

Figures 56.8 and 56.9, on page 290, show the infrared and NMR spectra of acetylsalicylic acid.

salicylic acid

acetylsalicylic acid

56.6 α- or β-D-GLUCOSE PENTAACETATE FROM D-GLUCOSE

When D-glucose is heated with acetic anhydride and zinc chloride, α-D-glucose pentaacetate is formed.

α-D-glucose pentaacetate

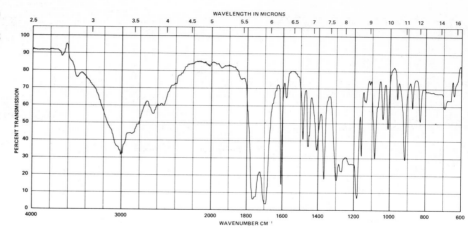

Figure 56.8. Infrared spectrum of acetylsalicylic acid; CHCl₃ solution.

acetylsalicylic acid

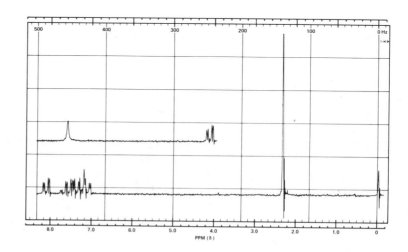

acetylsalicylic acid

Figure 56.9. NMR spectrum of acetylsalicylic acid; CDCl₃ solution.

If sodium acetate is used in place of zinc chloride, the product is β-D-glucose pentaacetate.

β-D-glucose pentaacetate

The two products differ only in the configuration at C-1, the anomeric carbon.* The scheme of Figure 56.10 is proposed to account for the different results of these two procedures.

In both experiments β-D-glucose pentaacetate is the initial product, being formed more rapidly than α-D-glucose pentaacetate (kinetic control). However,

---

* Since these diastereoisomers differ in configuration at only a single chiral center, they can be called *epimers*. Epimers of this type, created by formation of a new chiral center upon cyclization of an aldose, can also be called *anomers*.

when zinc chloride is the catalyst, the initially formed β-D-glucose pentaacetate is rapidly isomerized to the more stable α isomer (thermodynamic control). Apparently, sodium acetate is not an effective catalyst for this isomerization.

It may be surprising that the α anomer is the more stable since in this isomer the anomeric acetoxyl group has the axial configuration. However, as required by this scheme, when β-D-glucose pentaacetate is treated with acetic anhydride in the presence of zinc chloride, the α isomer can be isolated from the reaction mixture in good yield.

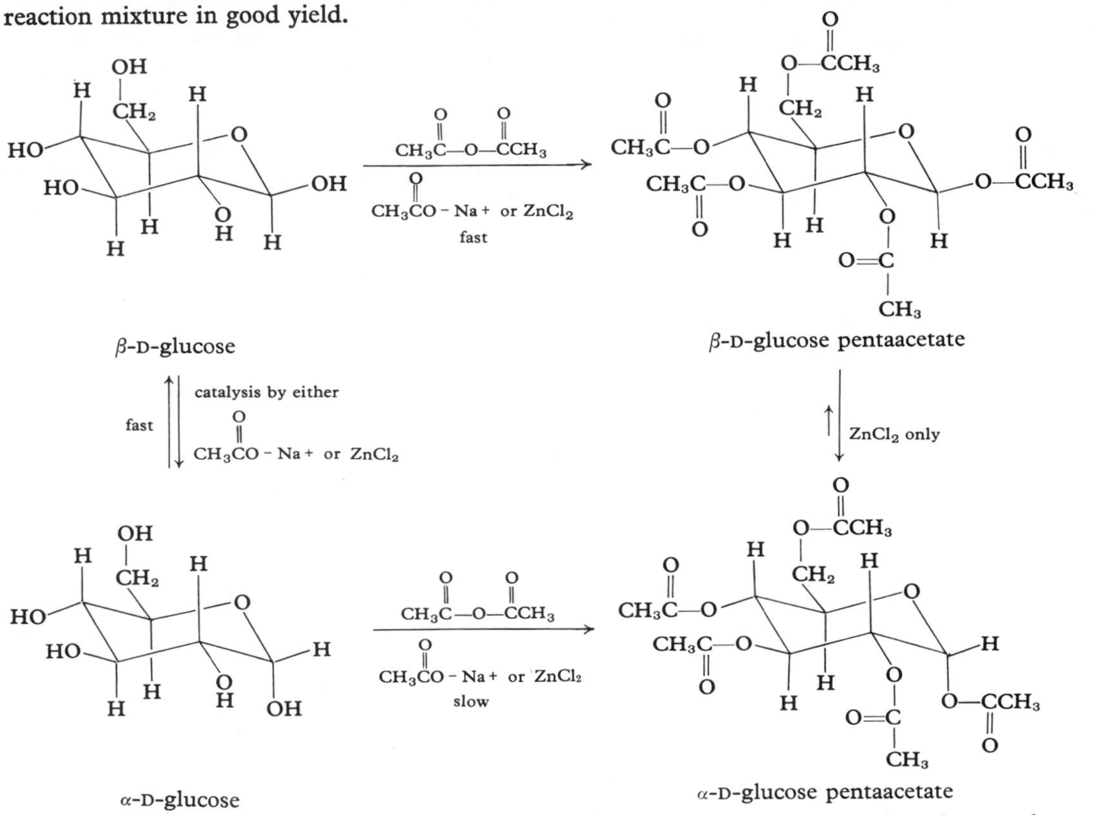

Figure 56.10. Scheme for the formation of α- or β-D-glucose pentaacetate from D-glucose.

α-D-Glucose Pentaacetate

Procedure. Place 0.5 g anhydrous zinc chloride (3.7 mmole), 12 ml acetic anhydride (13.0 g; 127 mmole), and 2.0 g anhydrous D-glucose (11.1 mmole) in a 50-ml round-bottom boiling flask. Add a boiling stone, fit the flask with a condenser and heat the flask cautiously with a burner flame until the contents start to boil. Remove the flask from the heat until the brief exothermic reaction is over, and then, by means of further heating with the flame, gently boil the mixture for about two more minutes. While it is still hot, pour the solution in a thin stream with good stirring into about 200 ml of a mixture of water and ice. Stir the resulting suspension until the oil has solidified. Break up any lumps and collect the solid by suction filtration. Recrystallize the crude α-D-glucose pentaacetate from 10 ml of methanol. The melting point of purified α-D-glucose pentaacetate is reported to be 112–113°C. Yield: about 1.9 grams (about 50%).

Time: 2 hours.

Procedure. The procedure is the same as that just given for the preparation of α-D-glucose pentaacetate except that 1.2 g anhydrous sodium acetate (14.7 mmole) is used in place of the anhydrous zinc chloride. The melting point of purified β-D-glucose pentaacetate is reported to be 132–134°C. Yield: about 2.4 grams (about 62%).

Time: 2 hours.

Questions

1. The scheme proposed to account for these results specifies explicitly that the equilibrium between the α and the β isomers of D-glucose pentaacetate favors the α isomer. Why (or under what conditions) is the position of the equilibrium between the α and the β forms of D-glucose irrelevant?

2. An alternative explanation for the results of these two reactions is that when zinc chloride is used as catalyst the α isomer is formed more rapidly, but when sodium acetate is the catalyst the β isomer is formed more rapidly. Explain how this interpretation is or is not consistent with the experimental facts.

3. Why is it reasonable that the —OH of the anomeric carbon atom of β-D-glucose is acetylated more rapidly than that of α-D-glucose?

4. Why is it reasonable that β-D-glucose pentaacetate can be converted to the α isomer by zinc chloride catalysis but not by sodium acetate catalysis?

§57 THE GRIGNARD REACTION

One important synthetic method for the formation of a carbon–carbon bond is the Grignard reaction. The first step of this three-step reaction is to prepare the Grignard reagent by allowing an ethereal solution of an alkyl halide to react with magnesium:

$$R—X + Mg \xrightarrow{\text{dry ether}} R—Mg—X$$

Dry equipment and reagents

The reaction is strongly inhibited by traces of water and therefore dry reagents and equipment are essential (Note 1). Although there is disagreement as to the molecular structure of the active Grignard reagent, it behaves as though it were a source of $R:^-$. Thus, it is a very good nucleophile and a very strong base.

The second step is to treat the ethereal solution of the Grignard reagent with an aldehyde, ketone, or ester. The three equations illustrate the mode of reaction of the reagent with each of these three types of compound:

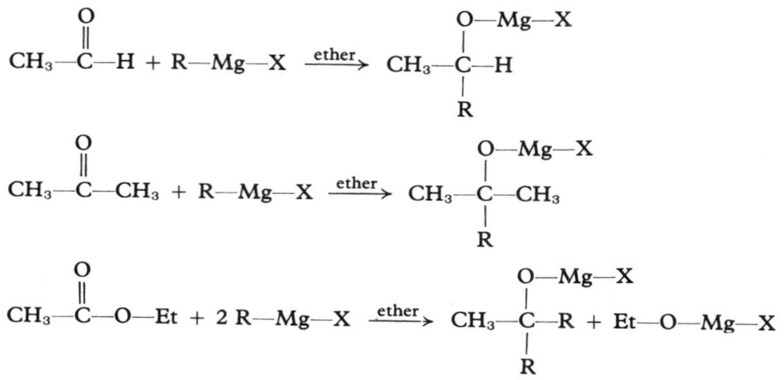

$$CH_3—\overset{\overset{\displaystyle O}{\|}}{C}—H + R—Mg—X \xrightarrow{\text{ether}} CH_3—\overset{\overset{\displaystyle O—Mg—X}{|}}{\underset{\underset{\displaystyle R}{|}}{C}}—H$$

$$CH_3—\overset{\overset{\displaystyle O}{\|}}{C}—CH_3 + R—Mg—X \xrightarrow{\text{ether}} CH_3—\overset{\overset{\displaystyle O—Mg—X}{|}}{\underset{\underset{\displaystyle R}{|}}{C}}—CH_3$$

$$CH_3—\overset{\overset{\displaystyle O}{\|}}{C}—O—Et + 2\,R—Mg—X \xrightarrow{\text{ether}} CH_3—\overset{\overset{\displaystyle O—Mg—X}{|}}{\underset{\underset{\displaystyle R}{|}}{C}}—R + Et—O—Mg—X$$

The third step involves the hydrolysis of the halomagnesium complex with aqueous acid and isolation of the product alcohol.

In this experiment you are given a general procedure for the preparation of an aliphatic alcohol by the Grignard synthesis.

For your synthesis, choose a combination of alkyl halide and aldehyde, ketone, or ester from the following lists. Your product should have more than five carbon atoms so that it will not be too soluble in water. Higher-boiling alcohols, especially those that can form a tertiary carbonium ion, may have to be distilled under reduced pressure in order to prevent elimination to give an alkene.

- *Alkyl halides:* methyl iodide, ethyl iodide, ethyl bromide, *n*-propyl bromide, *n*-butyl bromide.
- *Aldehydes:* acetaldehyde, propionaldehyde, *n*-butyraldehyde, isobutyraldehyde (Note 2).
- *Ketones:* acetone, 2-butanone, 2-pentanone, 3-pentanone, cyclopentanone, cyclohexanone.
- *Esters:* ethyl acetate, ethyl propionate.

Preparation of the Grignard Reagent

Procedure a. Place 4.0 g (0.16 mole) of magnesium in a dry 250-ml boiling flask (Note 3). Fit the flask with a Claisen adapter, and fit an addition funnel in the center neck of the adapter and a condenser with a drying tube in the side neck of the adapter (Note 4). Add to the addition funnel a mixture of 0.16 mole of alkyl halide and 60 ml of absolute ether. Allow 8–10 ml of the solution to run into the flask with the magnesium and wait until the reaction has started as indicated by the boiling of the reaction mixture (Note 5). Then add the remainder of the mixture of alkyl halide and ether at such a rate that the reaction mixture boils gently from the heat of the reaction (Note 6). After all of the alkyl halide/ether mixture has been added, allow the reaction mixture to stand for about 20 minutes with occasional swirling.

Procedure b. Place 4.0 g (0.16 mole) of magnesium in a dry 250-ml boiling flask (Note 3). Fit the flask with a Claisen adapter, and fit an addition funnel in the center neck of the adapter and a condenser with a drying tube in the side neck of the adapter (Note 4). Add to the flask 50 ml of anhydrous ether and 0.04 mole of alkyl halide. The reaction should start immediately (Note 5). When the reaction has moderated, add via the addition funnel another 0.04-mole portion of the alkyl halide, and repeat with two more 0.04-mole portions. When the reaction has moderated from the last addition of alkyl halide, gently reflux the mixture on the steam bath for about 15 minutes.

Treatment of the Grignard Reagent with Aldehyde, Ketone, or Ester

Procedure. Cool the flask containing the Grignard reagent with an ice bath. With swirling or stirring and cooling, add dropwise from the addition funnel a solution of 0.15 mole of the aldehyde or ketone (or 0.075 mole of the ester) in 20 ml of dry ether. The reaction is often quite exothermic and the addition should therefore be made with caution. After all the solution has been added, remove the ice bath and allow the mixture to stand at room temperature for about 20 minutes. ■

Procedure.   Pour the reaction mixture slowly, with stirring, onto a mixture of chipped ice and dilute sulfuric acid (prepared by adding 6 ml of conc. sulfuric acid to about 50 grams of ice and then adding about 50 ml of water). Any addition complex remaining in the reaction flask may be hydrolyzed by pouring the aqueous acid/ether mixture back in and swirling. After separating the ether layer, extract the water layer with two 25-ml portions of ordinary solvent ether. Combine all the ether solutions, wash with aqueous sodium bisulfite (if necessary; Note 5) and aqueous sodium bicarbonate, and dry over anhydrous magnesium sulfate. Isolate the product alcohol by evaporation of the ether on the steam bath and distillation of the residue.

Careful ⟵

Notes

1. Ether and alkyl halide can be dried satisfactorily over anhydrous magnesium sulfate.

2. Unless fresh aldehydes are used, they must be purified by distillation.

3. If the humidity is not high, the flask need not be dried by any special method. If desired, the flask may be dried by heating it with a flame, fitting it with a calcium chloride drying tube, and then allowing it to cool to room temperature.

4. The apparatus should look like that shown in Figure 34.2 except that a drying tube will be in the top of the condenser.

5. In case the reaction does not start within a few minutes, the following may be tried: (a) With a clean, dry stirring rod (not fire-polished) crush two or three pieces of the magnesium under the surface of the solution in order to break the magnesium and expose fresh surfaces. If this is successful, little bubbles will appear where the magnesium has been crushed, and the reaction mixture will become slightly cloudy. (b) Add a tiny crystal of iodine. If this is done, the ethereal solution of the *final* reaction product should be treated with a solution of sodium bisulfite in order to remove the iodine. (c) Warm the mixture on the steam bath and see if boiling will continue when it has been removed from the steam bath. (d) Add about 1 ml of someone else's successfully formed Grignard reagent. (e) Start over, taking more care to see that the apparatus and reagents are dry.

6. Have a beaker or pan of cold water available in case the reaction mixture boils so vigorously that ether starts to escape from the condenser. If this happens, the reaction can be moderated by immersing the flask in the cold water.

Time: 3 hours to the end of the second step; 3 hours to work up the reaction mixture and isolate the product.

Questions

1. What combinations of alkyl halide and aldehyde or ketone could be used to prepare 2-butanol by a Grignard synthesis?

2. What combinations of alkyl halide and aldehyde or ketone could be used to prepare 2-methyl-2-butanol by a Grignard synthesis?

3. **a.** What would be the final product of a Grignard synthesis involving treatment of methyl propionate with two equivalents of ethyl magnesium iodide?
   b. What product would ethyl propionate give when treated as in **a**?

Triphenylcarbinol can be prepared by treating the Grignard reagent formed from bromobenzene and magnesium with methyl benzoate:

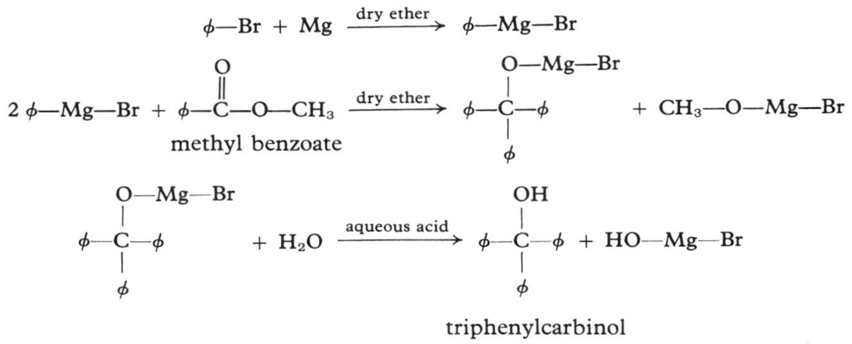

$$\phi\text{—Br} + \text{Mg} \xrightarrow{\text{dry ether}} \phi\text{—Mg—Br}$$

$$2\,\phi\text{—Mg—Br} + \phi\text{—}\overset{\displaystyle O}{\overset{\|}{C}}\text{—O—CH}_3 \xrightarrow{\text{dry ether}} \phi\text{—}\overset{\displaystyle O\text{—Mg—Br}}{\underset{\displaystyle \phi}{\overset{|}{\underset{|}{C}}}}\text{—}\phi + \text{CH}_3\text{—O—Mg—Br}$$

methyl benzoate

$$\phi\text{—}\overset{\displaystyle O\text{—Mg—Br}}{\underset{\displaystyle \phi}{\overset{|}{\underset{|}{C}}}}\text{—}\phi + \text{H}_2\text{O} \xrightarrow{\text{aqueous acid}} \phi\text{—}\overset{\displaystyle OH}{\underset{\displaystyle \phi}{\overset{|}{\underset{|}{C}}}}\text{—}\phi + \text{HO—Mg—Br}$$

triphenylcarbinol

Presumably the reaction takes place by formation of benzophenone as an intermediate, which then adds a second molecule of the Grignard reagent:

$$\phi\text{—Mg—Br} + \phi\text{—}\overset{\displaystyle O}{\overset{\|}{C}}\text{—O—CH}_3 \longrightarrow \phi\text{—}\overset{\displaystyle O\text{—Mg—Br}}{\underset{\displaystyle \phi}{\overset{|}{\underset{|}{C}}}}\text{—O—CH}_3$$

$$\phi\text{—}\overset{\displaystyle O\text{—Mg—Br}}{\underset{\displaystyle \phi}{\overset{|}{\underset{|}{C}}}}\text{—O—CH}_3 \longrightarrow \phi\text{—}\overset{\displaystyle O}{\overset{\|}{C}}\text{—}\phi + \text{CH}_3\text{—O—Mg—Br}$$

benzophenone

$$\phi\text{—}\overset{\displaystyle O}{\overset{\|}{C}}\text{—}\phi + \phi\text{—Mg—Br} \longrightarrow \phi\text{—}\overset{\displaystyle O\text{—Mg—Br}}{\underset{\displaystyle \phi}{\overset{|}{\underset{|}{C}}}}\text{—}\phi$$

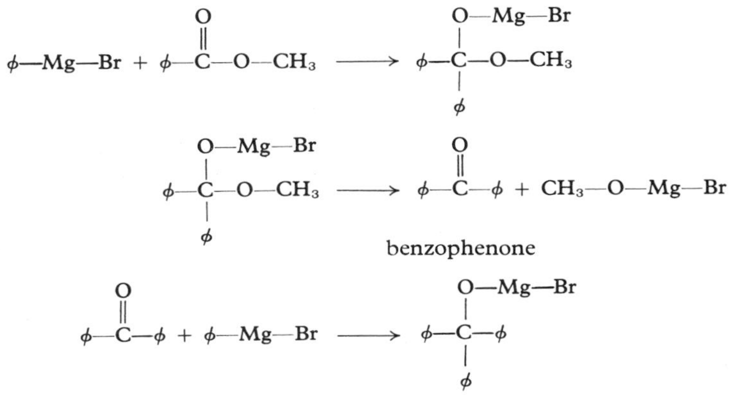

Preparation of the Grignard Reagent

Procedure. Prepare an ethereal solution of phenylmagnesium bromide according to procedure b of Section 57.1 using 2.0 grams (0.08 mole) of magnesium, 8.4 ml (12.5 grams; 0.08 mole) of bromobenzene, and 25 ml of anhydrous ether. At first, add about one quarter of the bromobenzene and then, after the initial reaction has moderated, add the remaining three quarters of the bromobenzene dropwise from the addition funnel at such a rate that vigorous boiling is maintained but not so fast that ether escapes from the top of the condenser.

Reaction with Methyl Benzoate

Procedure. Cool the ethereal solution of phenylmagnesium bromide by swirling or stirring it in a pan of cold water. Then, while continuing to swirl or stir the mixture, add from the addition funnel a solution of 4.9 ml (5.4 grams; 0.04 mole) of methyl benzoate in 15 ml of anhydrous ether. Add the solution of methyl benzoate at such a rate that the reaction mixture boils gently. When the addition is complete, boil the mixture under reflux on the steam bath for about 15 minutes.

Procedure. Carry out the hydrolysis as described in the procedure in Section 57.1, but extract the water layer with only one 25-ml portion of solvent ether. Dry the combined ether solutions of the product over anhydrous magnesium sulfate and filter the solution into a 100 ml boiling flask. Add to the flask 25 ml of hexane and concentrate the solution by distilling off the majority of the ether on the steam bath. Condense and collect the ether, and put the recovered ether in the container reserved for it. Continue the distillation until crystals start to appear. At this time, remove the flask from the distillation apparatus and set it aside to cool to room temperature. Complete the crystallization by cooling the flask in an ice bath and collect the produce by suction filtration, using a small amount of cold hexane to rinse the flask and to wash the crystals.

Questions

1. Explain how triphenycarbinol could be prepared from bromobenzene and benzophenone? How would the procedure given in this section have to be modified if benzophenone were used?

2. Explain how triphenylcarbinol could be prepared from bromobenzene and ethyl benzoate? How would the procedure given in this section have to be modified if ethyl benzoate were used?

3. **a.** What would be the product of reaction of methyl benzoate with two equivalents of ethyl magnesium bromide?

   b. Show how this substance could be prepared by two other Grignard syntheses.

Reference

A good general reference on the Grignard reagent and its uses is:

1. M. S. Kharasch and O. Reinmuth, *Grignard Reactions of Nonmetallic Substances*, Prentice-Hall, Englewood Cliffs, N.J., 1964.

---

§58 ELECTROPHILIC AROMATIC SUBSTITUTION REACTIONS OF BENZENE

Most electrophilic substitution reactions of benzene and of substituted benzenes are thought to take place according to the following general mechanism:

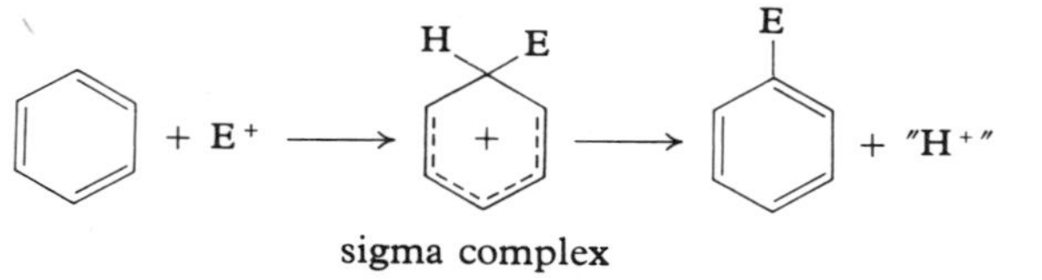

$$\text{(58.1)}$$

sigma complex

In some cases, the experimental evidence is interpreted best by proposing that E^+, the electrophile, is completely formed as a cation before it reacts with the aromatic compound, as implied by Equation 58.1. In other cases, the data are best interpreted in terms of a simultaneous attack of the aromatic system on E—X and loss of the leaving group X^-:

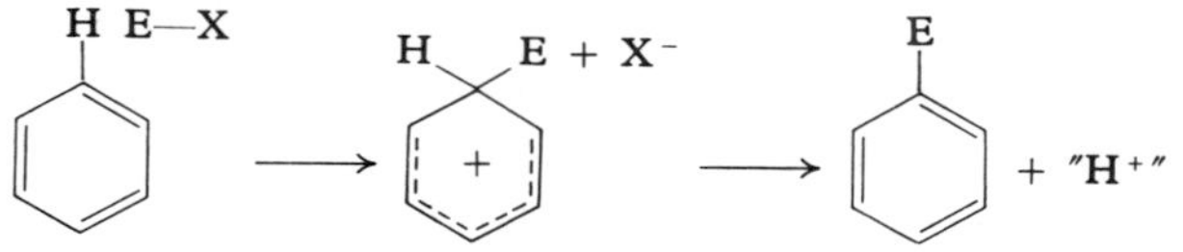

$$\text{(58.2)}$$

For convenience, however, electrophilic aromatic substitution reactions are often interpreted in terms of attack by previously formed E^+, with the mental reservation that this may not always be the best representation of the reaction.

A complete discussion of the variety of mechanistic possibilities for electrophilic aromatic substitution reactions would have to include the possibility of formation of pi complexes before and after the sigma complex. In some cases, the formation of the first pi complex appears to be the rate-determining step.

58.1 NITROBENZENE FROM BENZENE

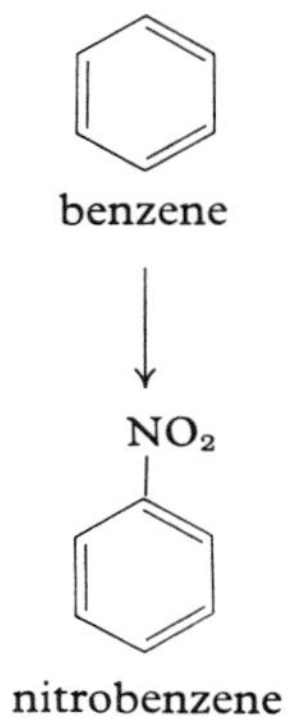

The nitrating agent in most nitration reactions of aromatic compounds appears to be $O=\overset{+}{N}=O$, the nitronium ion. The source of nitronium ion may be a nitronium salt, such as nitronium fluoroborate, $NO_2^+BF_4^-$, or, as in this experiment, the ion may be derived from nitric acid by the action of sulfuric acid:

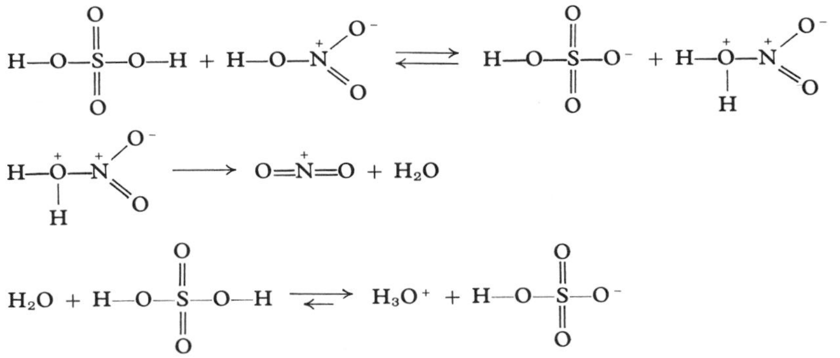

The infrared and NMR spectra of nitrobenzene are shown in Figures 58.1 and 58.2.

Procedure. Place 5 ml water and 25 ml (46 g; 0.45 mole) conc. sulfuric acid in a 125-ml Erlenmeyer flask; cool the mixture. To this, add 15 ml (0.23 mole) conc. nitric acid, and again cool the mixture. Now add 17.7 ml (15.6 g; 0.20 mole) benzene, and, after inserting a thermometer, swirl the mixture (Note 1). By constant swirling and a very brief cooling in cold water, keep the temperature of the mixture close to, but not above, 60°C, and not below 55°C. Within 10 minutes, the reaction will slow and the temperature will begin to drop. At this time, heat the mixture occasionally on the steam bath in order to maintain the

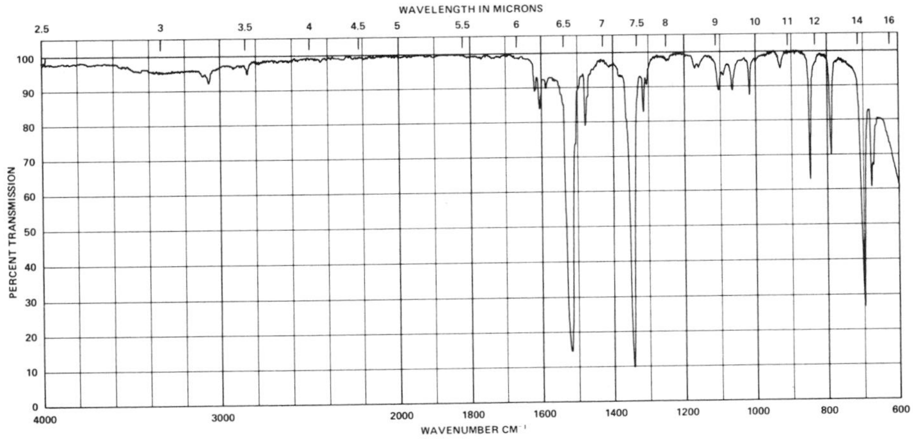

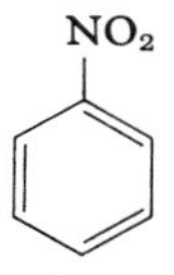

Figure 58.1. Infrared spectrum of nitrobenzene; thin film.

NO₂

nitrobenzene

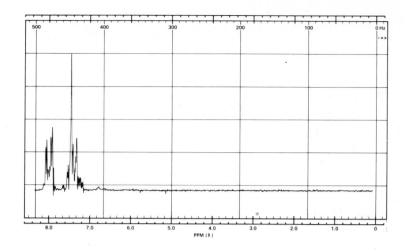

Figure 58.2. NMR spectrum of nitrobenzene; CCl₄ solution.

temperature near 60°C for another 10 minutes. Then cool in ice to room temperature, add 75 ml water, and cool to room temperature again. Extract the mixture with 30 ml ether and wash the extract with 10% sodium hydroxide until the basic aqueous phase no longer becomes bright yellow. Shake the ethereal extract with saturated salt solution, dry it over anhydrous sodium sulfate, and distill. Collect as nitrobenzene the distillate that boils between 205 and 207°C (Note 2).

Notes

1. The benzene and acid layers are only very slightly soluble in one another. When the benzene dissolves in the acid layer, it reacts. Swirling speeds the process of solution.
2. Do not distill to dryness; any residue containing *m*-dinitrobenzene might decompose explosively.

Time: 3 hours.

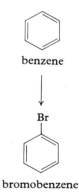

benzene

Br

bromobenzene

58.2 BROMOBENZENE FROM BENZENE

The electrophilic agent in many electrophilic aromatic bromination reactions is elemental bromine; bromide ion is displaced from the bromine molecule as the electrophilic attack is made upon the aromatic ring (see Equation 58.2). When the reaction is catalyzed by ferric bromide, the effective electrophile can be considered to be formed in the following way:

$$
\begin{array}{ccc}
& \overset{\displaystyle Br}{\underset{\displaystyle Br}{|}} & \\
Br{-}Br + Fe{-}Br & \longrightarrow & \overset{\displaystyle Br}{Br{-}\overset{+}{Br}{-}\underset{\displaystyle Br}{\overset{|}{\underset{|}{Fe}}}{-}Br}
\end{array}
$$

$$
\overset{\displaystyle Br}{Br{-}\overset{+}{Br}{-}\underset{\displaystyle Br}{\overset{|}{\underset{|}{Fe}}}{-}Br} \longrightarrow Br^{+} + \overset{\displaystyle Br}{Br{-}\underset{\displaystyle Br}{\overset{|}{\underset{|}{Fe}}}{-}Br}
$$

A catalytic amount of ferric bromide apparently may be generated in the reaction mixture from bromine and elemental iron in the form of an iron tack or iron filings.

The infrared spectrum of bromobenzene is shown in Figure 58.3.

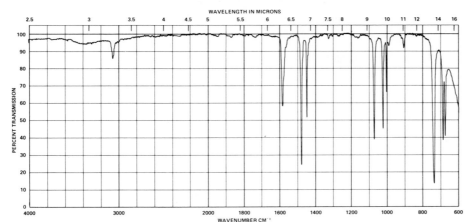

Figure 58.3. Infrared spectrum of bromobenzene; thin film.

Procedure. Fit a 50-ml boiling flask with a reflux condenser and add to the flask 10.0 ml (8.79 g; 0.113 mole) of benzene, 2.82 ml (8.8 g; 0.055 mole) of bromine (*caution:* see Note 1), and one iron tack (Note 2). Warm the flask with a beaker of water at 50–55°C (Note 3). In about 15 minutes, after the spontaneous reaction has started to subside, remove the water bath and heat the reaction mixture to boiling. Within 10 minutes the bromine vapors above the liquid in the flask will have disappeared. At this time cool the mixture, add 25 ml of ether, and extract the solution with two 5-ml portions of 10% aqueous sodium hydroxide and one 10-ml portion of water. Dry the ethereal solution over anhydrous magnesium sulfate for a few minutes, filter it, and distill it, collecting the fraction boiling between 140 and 160°C. Gas chromatographic analysis shows that the product contains about 4% benzene.

Notes

1. Add the bromine in the hood from a buret with a Teflon stopcock. For bromine burns, wash instantly with water and rub in glycerine (Section 1.7).
2. A $\frac{3}{8}$-inch blued-cut upholsterer's tack works well.
3. Since hydrogen bromide will be evolved, either the reaction must be carried out in the hood or some provision must be made to absorb the fumes.

   Time: 3 hours.

58.3 *tert*-BUTYLBENZENE FROM BENZENE

In the Friedel-Crafts alkylation of an aromatic compound, the alkylating agent, which may be considered to be the electrophile R^+, can be formed from the corresponding alkyl chloride and aluminum chloride. Alternatively, it can be generated by the action of aluminum chloride on the alcohol R—OH. The alternative procedures here make use of these two methods.

The infrared spectrum of the product *tert*-butylbenzene is shown in Figure 58.4.

benzene

$$CH_3$$
$$CH_3—C—CH_3$$

tert-butylbenzene

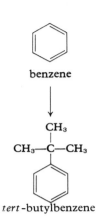

benzene

↓

tert-butylbenzene

Figure 58.4. Infrared spectrum of *tert*-butylbenzene; thin film.

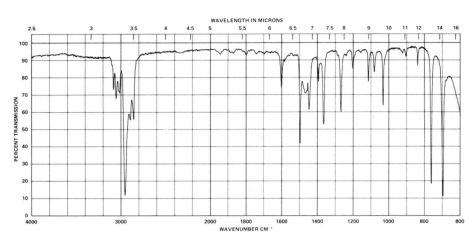

Alkylation with *tert*-Butyl Chloride (Procedure a)

Procedure (Reference 1). In a 250-ml Erlenmeyer flask place 60 ml (53 g; 0.68 mole) of dry, thiophene-free benzene and 16.4 ml (13.0 g; 0.150 mole) of *tert*-butyl chloride. Cool the flask in an ice bath and prepare a trap to be connected to the flask to absorb the hydrogen chloride that will be evolved (Section 35). Weigh out 2 grams (0.015 mole) of aluminum chloride into a vial, which can be kept stoppered, and add this in three portions separated by intervals of about five minutes to the solution of *tert*-butyl chloride in benzene. After each addition, reconnect the trap and swirl the mixture in the ice bath. After gas evolution has slowed down following the addition of the final portion of aluminum chloride, add cautiously to the flask about 60 grams of crushed ice followed by 30 ml of water. ■

Wash the benzene layer with dilute sodium bicarbonate solution, and finally with water. Dry the benzene solution over anhydrous calcium chloride. Fractionally distill the dried liquid, collecting as *tert*-butylbenzene the portion boiling between 165 and 172°C.

Alkylation with *tert*-Butyl Alcohol (Procedure b)

Procedure (Reference 2). Place 10 g (0.075 mole) of aluminum chloride and 60 ml (53 g; 0.68 mole) of benzene in a 100-ml boiling flask. Add a magnetic stirring bar and fit the flask with a Claisen adapter. Fit a separatory funnel in the middle neck of the adapter, and a condenser in the side neck. Provide for the absorption of the hydrogen chloride, which will be evolved during the reaction (Section 35). Place 14 ml (11.1 grams; 0.150 mole) of *tert*-butyl alcohol in the separatory funnel and add it dropwise to the vigorously stirred mixture in the boiling flask. Take about 15–20 minutes to add the alcohol. If the flask feels warm to your hand, cool it momentarily with a bath of cold water. Allow the mixture to stir at room temperature for one-half hour after the addition of the alcohol is complete. At this time, pour the contents of the flask cautiously and with stirring into a mixture of 50 ml conc. hydrochloric acid and 50 grams of ice. Separate the benzene layer, wash it with water, and work it up as described in the preceding experiment.

Time: less than 3 hours.

1. Write equations showing how the alkylating agent, the *tert*-butyl carbonium ion, is formed from the following:
   a. *tert*-butyl chloride and aluminum chloride.
   b. *tert*-butyl alcohol and aluminum chloride.

References

1. J. R. Mohrig and D. C. Neckers, *Laboratory Experiments in Organic Chemistry*, Reinhold, New York, 1968, p. 19.
2. C. B. Kremer and S. H. Wilen, *J. Chem. Educ.* **38**, 306 (1961).

58.4 *p*-DI-*tert*-BUTYLBENZENE FROM BENZENE

If in the reaction involving the alkylation of benzene with *tert*-butyl chloride in the presence of aluminum chloride (described in Section 58.3, Procedure a) 1.66 mole of *tert*-butyl chloride is used per mole of benzene rather than 0.22 mole of halide per mole of benzene, *p*-di-*tert*-butylbenzene can be isolated in good yield. This product should produce infrared and NMR spectra similar to those of Figures 58.5 and 58.6.

More tert-butyl chloride gives disubstitution

Procedure (Reference 1). Place 20 ml (17 g; 0.184 mole) of *tert*-butyl chloride and 10 ml (8.8 g; 0.11 mole) of benzene in a 125-ml Erlenmeyer flask. Cool the flask in an ice bath and prepare a trap to be connected to the flask to absorb the hydrogen chloride that will be evolved (Section 35). Weigh out 1.0 g (0.075 mole) of aluminum chloride in a stoppered vial and add about one-quarter of this to the solution of *tert*-butyl chloride in benzene. Connect the trap and swirl the mixture in the ice bath. After a short time, hydrogen chloride evolution will begin. After the initial reaction has subsided, add the remainder of the aluminum chloride in three portions at about 2-minute intervals, reconnecting the trap and swirling after each addition. When a white precipitate begins to appear, remove the flask from the ice bath and allow it to stand at room temperature for about 5 minutes. Now add cautiously a mixture of about 25 grams of ice and 25 ml of water. Extract the product with ether, and after transferring the ethereal extract to a separatory funnel, wash it with water and then with saturated salt solution. Dry the extract over anhydrous magnesium sulfate, and remove the ether by

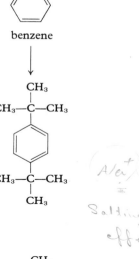

p-di-*tert*-butylbenzene

Figure 58.5. Infrared spectrum of *p*-di-*tert*-butylbenzene; CCl₄ solution.

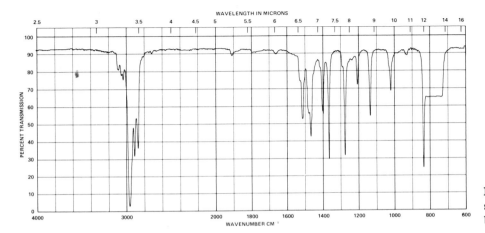

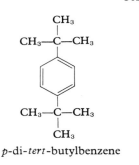

CH₃—C—CH₃ (CH₃ above)

CH₃—C—CH₃ / CH₃

p-di-*tert*-butylbenzene

Figure 58.6. NMR spectrum of *p*-di-*tert*-butylbenzene; CCl₄ solution.

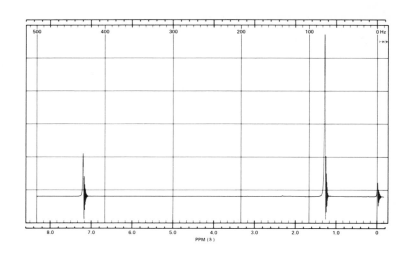

distillation on the steam bath. The last traces of ether may be removed under water-pump vacuum with heating on the steam bath. Cool the residue to induce crystallization. ■ Recrystallize the product from 20 ml of methanol. If the solution is allowed to come to room temperature slowly, beautiful needles or plates will be formed. After eventual cooling in an ice bath, collect the crystals by suction filtration and wash them with a little cold methanol.

Time: 3 hours.

Reference

1. L. F. Fieser, *Organic Experiments*, Heath, Boston, 1964, p. 184.

nitrobenzene

Sn | HCl

:NH₂

aniline

Steam distillation

§59 ANILINE FROM NITROBENZENE

Aromatic nitro compounds may be reduced to the corresponding aromatic amine by many reagents. In this experiment, aniline is produced from nitrobenzene by means of tin and concentrated hydrochloric acid. The distillate obtained in the course of the procedure, a mixture of aniline and water, may be used directly for the preparation of acetanilide (Section 60.1); assume that the mixture contains 0.09 mole of aniline.

The infrared spectrum of aniline is shown in Figure 59.1.

Procedure. Place 25 g (0.212 mole) of granulated tin and 10.3 ml (12.3 g; 0.10 mole) of nitrobenzene in a 250-ml boiling flask. Make ready an ice bath, and add to the tin and nitrobenzene mixture 55 ml (0.66 mole) of conc. hydrochloric acid. Insert a thermometer and swirl the mixture. Keep the temperature of the reaction between 55 and 60°C by swirling and occasional immersion of the flask in the ice bath. After 15 minutes, fit the flask with a condenser and heat the reaction mixture on the steam bath with frequent swirling until the condensate in the condenser shows the absence of oily drops of unreacted nitrobenzene (about 15 minutes). Cool the flask in an ice bath and add slowly with swirling and cooling 50 ml of 50% aqueous sodium hydroxide (0.95 mole), followed by 75 ml of water. Fit the flask with a Claisen adapter (Note 1), and fit the center neck with a separatory funnel and the side neck with an adapter and condenser set for downward distillation (Note 2). Steam distill the aniline from the flask by heating it with a flame, adding water by means of the separatory funnel to keep the volume constant. After the distillate begins to come over clear, continue to distill; distill about 50 ml more. ■

Aniline may be isolated from the distillate by salting it out (using 20 g sodium chloride per 100 ml distillate), extraction with ether or dichloromethane, drying over sodium sulfate, and distilling, collecting the fraction boiling between 180 and 185°C as aniline.

Notes

1. The joints involving the Claisen adapter must be well greased, and the apparatus must be disassembled immediately after the distillation is stopped. Otherwise, the joints will freeze tight (Section 30).
2. The apparatus should look like that shown in Figure 11.3 except that the separatory funnel takes the place of the steam addition tube in the center neck of the Claisen adapter.

Time: a long 3 hours if the aniline is isolated and distilled.

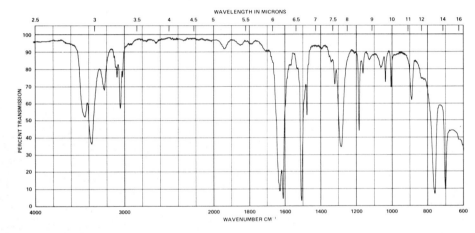

Figure 59.1. Infrared spectrum of aniline; thin film.

---

§60 PREPARATION OF AMIDES

Amines can be converted to the corresponding amides of acetic acid in several ways. In the first experiment of this section, three procedures are given for the acetylation of aniline to form acetanilide. The second experiment presents one method for the preparation of p-ethoxyacetanilide, or phenacetin. The third experiment illustrates the preparation of an amide via the acid chloride by the synthesis of the insect repellent "Off."

60.1 ACETANILIDE FROM ANILINE

The acetanilide produced by any of the three procedures that follow should have an infrared spectrum like that shown in Figure 60.1.

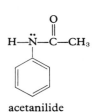

acetanilide

Figure 60.1. Infrared spectrum of acetanilide; CHCl₃ solution.

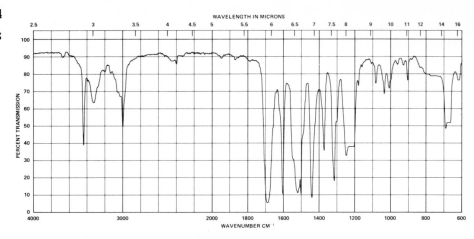

Acetylation with Acetic Acid (Procedure a)

Procedure. Place 5.0 g (4.9 ml; 0.054 mole) aniline, 21 g (20.0 ml; 0.35 mole) acetic acid, and a pinch of zinc dust (Note 1) in a 125-ml boiling flask. Fit the flask with an air condenser, and heat the mixture under reflux for about four hours. Pour the hot mixture (Note 2) in a thin stream into 200 ml of cold water. ■ After cooling the mixture in an ice bath for about 10 minutes, collect the product by suction filtration, and wash it with cold water (Note 3).

Acetylation with Acetic Anhydride (Procedure b)

Procedure. Place 5.0 g (4.9 ml; 0.054 mole) aniline, 5.4 g (5.0 ml; 0.053 mole) acetic anhydride, and a pinch of zinc dust (Note 1) in a 125-ml boiling flask. Fit the flask with a condenser and boil the mixture gently for 30 minutes. Pour the hot mixture in a thin stream into 200 ml of cold water. ■ After cooling the mixture in an ice bath for about 10 minutes, collect the product by suction filtration and wash it with cold water (Note 3).

Acetylation with Acetic Anhydride and Sodium Acetate in Water (Procedure c)

Procedure. Dissolve 5.0 g (4.9 ml; 0.054 mole) aniline in 135 ml of water and 4.5 ml (0.054 mole) of conc. hydrochloric acid. If the solution is highly colored, treat it with decolorizing carbon and filter. Prepare a solution of 5.3 g (0.065 mole) anhydrous sodium acetate in 30 ml of water. To the solution of aniline hydrochloride, add 6.6 g (6.2 ml; 0.065 mole) acetic anhydride and, as soon as this has been brought into solution by swirling or stirring, add the solution of sodium acetate. Cool the mixture in an ice bath for about 10 minutes, collect the product by suction filtration, and wash it with cold water (Note 3).

Notes

1. The zinc reduces the colored impurities in the aniline and helps to prevent its oxidation during the reaction.
2. If the mixture is allowed to cool, it will set to a solid cake.
3. Acetanilide may be recrystallized from water.

1. Justify your choice of procedure.
2. Why is a slight excess of acetic anhydride called for in Procedure c?
3. Why is a large excess of acetylating agent called for in Procedure a as compared to b and c?
4. What is the function of sodium acetate in Procedure c?
5. Could sodium acetate trihydrate be used in Procedure c? If so, what changes, if any, would be necessary in the experiment?

60.2 *p*-ETHOXYACETANILIDE FROM *p*-PHENETIDINE

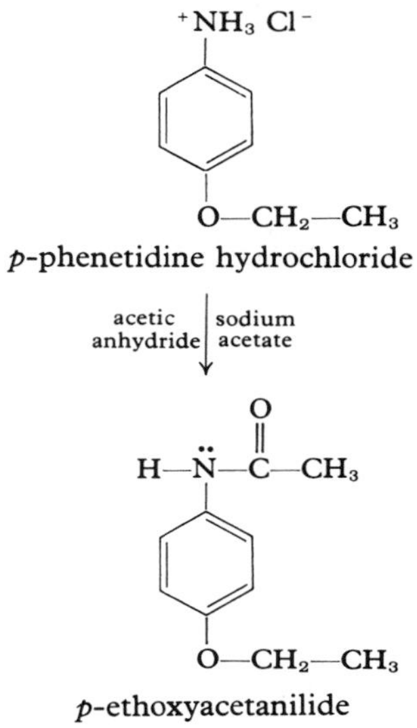

p-Ethoxyacetanilide, or phenacetin, is described by the *Merck Index* as an analgesic and antipyretic (pain reliever and fever reducer). It is a component of "APC" tablets, along with aspirin (Section 41) and caffeine (Section 42). Phenacetin can be prepared by treatment of *p*-phenetidine hydrochloride with sodium acetate and acetic anhydride.

Figures 60.2 and 60.3 show the infrared and NMR spectra of *p*-ethoxyacetanilide.

Procedure. Add 150 ml of water and 7.4 grams (7.0 ml; 0.054 mole) of *p*-phenetidine to a 250-ml Erlenmeyer flask. Dissolve the amine by adding 4.5 ml (0.054 mole) of conc. hydrochloric acid. If the solution is very dark, treat it with decolorizing carbon, and filter. Prepare a solution of 5.3 grams (0.065 mole) of anhydrous sodium acetate in 30 ml of water. To the solution of *p*-phenetidine hydrochloride add 6.6 grams (6.2 ml; 0.065 mole) of acetic anhydride. Swirl the mixture for a few seconds and then add the solution of sodium acetate. Phenacetin forms and precipitates immediately. Collect the product by suction filtration and wash it with cold water. Recrystallize the damp material from about 35 ml of 95% ethanol. On slow cooling phenacetin will crystallize as long spars. Yield after recrystallization: 7.5 grams (78%); m.p.: 137–138°C.

Time: 2 hours.

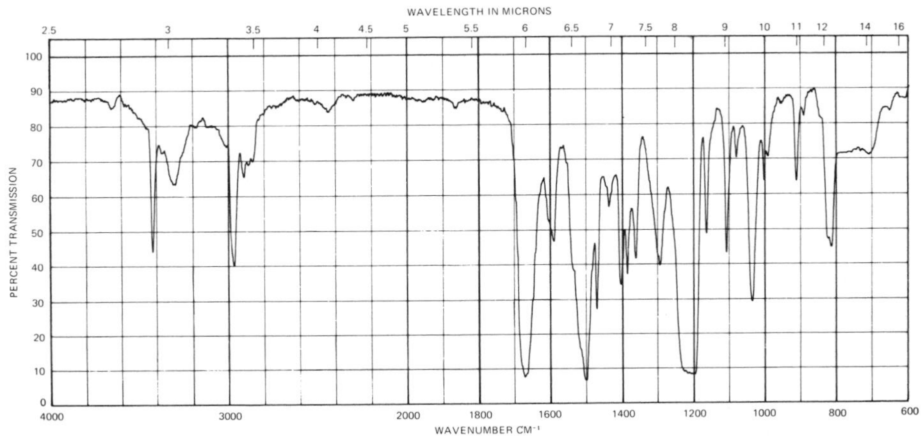

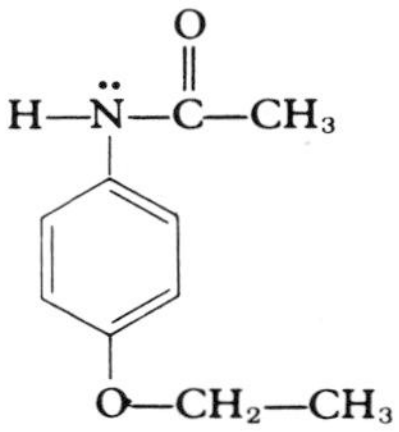

p-phenetidine hydrochloride

p-ethoxyacetanilide

p-ethoxyacetanilide

Figure 60.2. Infrared spectrum of *p*-ethoxyacetanilide; CHCl₃ solution.

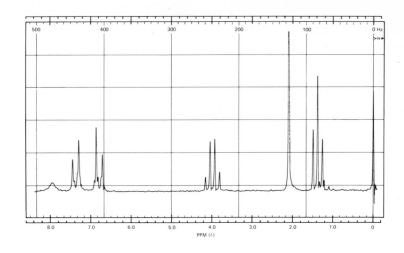

p-ethoxyacetanilide

Figure 60.3. NMR spectrum of *p*-ethoxyacetanilide; CDCl₃ solution.

60.3 N,N-DIETHYL-*m*-TOLUAMIDE FROM *m*-TOLUIC ACID; "OFF"

One of the more effective insect repellents is the N,N-diethylamide of *m*-toluic acid. It can be prepared by converting *m*-toluic acid to the acid chloride with thionyl chloride and then treatment of the product with diethylamine:

The amide is an ingredient of a number of commercial insect repellents, but it is probably most familiar as the active ingredient of "Off." Its infrared spectrum is shown in Figure 60.4.

Procedure. Place 8.4 grams (60 mmole) of *m*-toluic acid in a dry 250 ml boiling flask and add to this 9.0 ml (14.5 grams; 122 mmole) of thionyl chloride (Note 1). Fit the flask with a reflux condenser and heat the mixture on the steam bath for about 20 minutes. Since SO₂ and HCl will be evolved during the heating period you must either work in the hood or make provision for the absorption of these gases (Section 35).

N,N-diethyl-*m*-toluamide

Figure 60.4. Infrared spectrum of N,N-diethyl-*m*-toluamide; thin film.

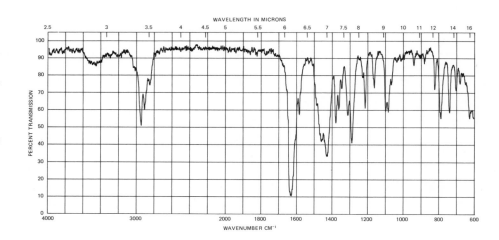

At the end of the heating period remove the condenser and add to the flask 40 ml of toluene. Add a magnetic stirring bar to the flask and clamp the flask in position over a magnetic stirring motor. Fit the flask with a Claisen adapter and place in the center neck a dropping funnel containing a solution of 20 ml (14.2 grams; 196 mmole) of diethylamine in 25 ml of toluene. Add the diethylamine solution dropwise while stirring the contents of the flask; the addition should take from 30 to 40 minutes.

When the addition of the diethylamine solution is complete add to the flask 25 ml of cold water. Continue stirring in order to thoroughly mix the contents of the flask. Transfer the dark reaction mixture to a separatory funnel and draw off the lower, aqueous, layer (Note 2). Wash the toluene solution again with a second 25-ml portion of water. Now wash the toluene solution first with a solution prepared by diluting 5 ml of 50% sodium hydroxide with water to 50 ml (use this solution in two 25-ml portions) and then with a solution prepared by diluting 10 ml of conc. hydrochloric acid with water to 50 ml (use this solution also in two 25-ml portions). Finally wash the toluene solution with two more 25-ml portions of water.

Transfer the brown toluene solution of N,N-diethyl-*m*-toluamide to a 125 ml Erlenmeyer flask and dry it for 5 to 10 minutes over anhydrous magnesium sulfate. Filter the dried solution into a 100 ml boiling flask and remove the toluene by distillation. Since the boiling point of the product is quite high it should be distilled under reduced pressure (Section 10) (Note 3). Although it is colorless when pure the product is usually obtained as a light brown oil. The boiling point has been reported to be 160° at 20 Torr and 173° at 24 Torr. Yield: about 7 grams (60%).

Notes

1. Thionyl chloride is a lachrymator. You should work in the hood while you measure it out and add it to your flask.
2. It may be impossible to distinguish the layers. If so, draw off 25 ml of the liquid.
3. The product may also be purified by chromatography on alumina (2).

Time: 4 hours.

References

1. E. T. McCabe, W. F. Barthels, S. I. Gertler, and S. A. Hall, *J. Org. Chem.* **19**, 493 (1954).
2. B. J.-S. Wang, *J. Chem. Educ.* **51**, 631 (1974).

§61 ELECTROPHILIC SUBSTITUTION REACTIONS OF BENZENE DERIVATIVES

The reactivity toward electrophilic substitution of a monosubstituted benzene may be either greater or less than that of benzene itself; the substituent is said to be either activating or deactivating. The new substituent may enter either the *ortho* and *para* positions, or the *meta* position; the original substituent is said to be either *ortho-para* directing, or *meta* directing. Substituents of the general type —A=O have always been found to be deactivating and *meta* directing; examples are the nitro group, —NO₂, and the carbonyl group, —C=O. Substituents of the general type —A: have, with the exception of bromine and chlorine, been found to be activating and *ortho-para* directing. Bromine and chlorine have been found to be *ortho-para* directing but slightly deactivating, presumably due to greater electron withdrawal through an inductive mechanism than electron release by a resonance mechanism.

The conditions of these electrophilic aromatic substitution reactions should be compared with those involving benzene itself (Section 58).

methyl benzoate

methyl *m*-nitrobenzoate

61.1 METHYL *m*-NITROBENZOATE FROM METHYL BENZOATE

The infrared spectrum of the product methyl *m*-nitrobenzoate is shown in Figure 61.1.

Procedure. Place 14.5 ml (25 g; 0.250 mole) of conc. sulfuric acid in a 125-ml Erlenmeyer flask, cool it to 0°C in an ice bath, and then add 6.3 ml (6.8 g; 0.050 mole) of methyl benzoate with swirling. Prepare a mixture of 5 ml (9 g; 0.090 mole) of conc. sulfuric acid and 5 ml (7 g; 0.078 mole) of conc. nitric acid, and cool it in an ice bath. Add the cold acid mixture dropwise over a period of about 5 minutes to the methyl benzoate solution, which is constantly swirled in the ice bath. Allow the resulting mixture to stand at room temperature for an additional 10 minutes with occasional swirling and then pour it with stirring over about 50 grams of crushed ice. Collect the resulting solid by suction filtration and wash it thoroughly with water to remove the acids. Finally, wash the product with two 5-ml portions of ice-cold methanol. The product may be recrystallized from a small amount of methanol.

Time: 2 hours.

methyl *m*-nitrobenzoate

Figure 61.1. Infrared spectrum of methyl *m*-nitrobenzoate; CCl₄ solution.

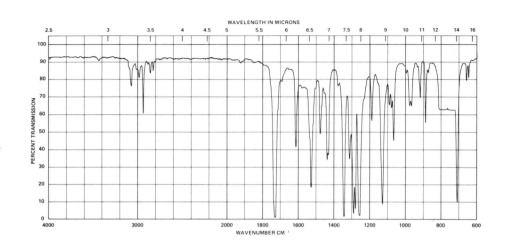

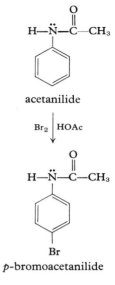

Figure 61.2 shows the infrared spectrum of *p*-bromoacetanilide.

Procedure. Dissolve 6.7 g (0.050 mole) acetanilide in 25 ml of glacial acetic acid in a 250-ml flask, and add slowly, with stirring, a solution of 8.1 g (2.60 ml; 0.051 mole) of bromine in 5 ml of glacial acetic acid (*caution:* see Note 1). Stir the mixture for 2 or 3 minutes more, and then add slowly with stirring 200 ml of water. Add enough concentrated sodium bisulfite solution to discharge the yellow color, and then collect the product by suction filtration, washing it well with water and sucking it as dry as possible. Yield: 10.2 g (96%). The product may be recrystallized from ethanol, using 3.3 ml per gram, with a recovery of 90%; m.p., 168–169°C.

Note

1. See Section 1.7 for precautions to be observed when working with bromine.

Time: 2 hours.

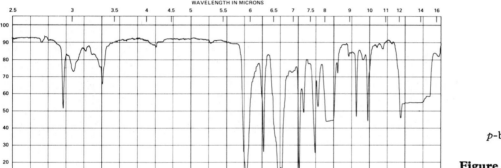

Figure 61.2. Infrared spectrum of *p*-bromoacetanilide; CHCl₃ solution.

61.3 2,4-DINITROBROMOBENZENE FROM BROMOBENZENE

The 2,4-dinitrobromobenzene produced should have infrared and NMR spectra like those of Figures 61.3 and 61.4.

Procedure. Heat a mixture of 15 ml (27 g; 0.27 mole) of conc. sulfuric acid and 5 ml (7 g; 0.08 mole) of conc. nitric acid in a 50-ml Erlenmeyer flask to 85–90°C. Add 2.0 ml (3.0 g; 0.019 mole) of bromobenzene in three or four portions during a minute, and swirl the mixture well after each addition. The temperature will rise to 130–135°C. Allow the mixture to stand with occasional swirling for 5 minutes, then cool it nearly to room temperature and pour it over about 100 g of ice. Stir the resulting mixture until the product has solidified, crush the lumps, and collect the crude product by suction filtration.

Recrystallize the product by dissolving it in 15 ml of hot 95% ethanol and allowing the solution to cool (Note 1). After crystals have started to form, cool the solution in a cold-water bath and then in an ice bath. Crystallization is complete in about 5 minutes. Collect the product by suction filtration and wash with a small amount of cold ethanol.

Note

1. The product usually separates as an oil, but vigorous swirling of the mixture when the oil appears will promote crystallization.

Time: 2 hours.

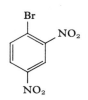

2,4-dinitrobromobenzene

Figure 61.3. Infrared spectrum of 2,4-dinitrobromobenzene; CHCl₃ solution.

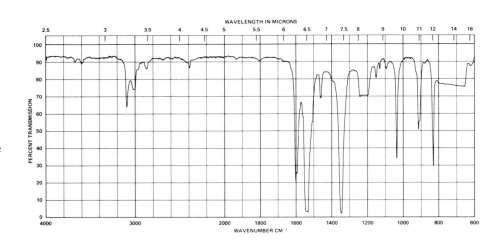

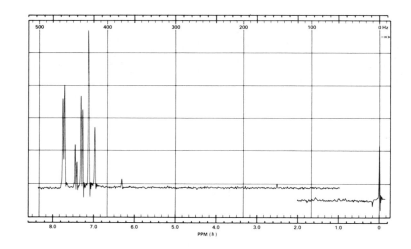

2,4-dinitrobromobenzene

Figure 61.4. NMR spectrum of 2,4-dinitrobromobenzene; CDCl₃ solution.

---

§62 NUCLEOPHILIC AROMATIC SUBSTITUTION REACTIONS OF 2,4-DINITROBROMOBENZENE

Certain benzene derivatives can undergo nucleophilic substitution. The more electron-withdrawing substituents there are on the ring, especially in the positions *ortho* and *para* to where the substitution takes place, the faster the reaction goes. This is interpreted in terms of a two-step addition–elimination mechanism in which the role of the electron-withdrawing substituent is to stabilize the negative charge on the benzene ring in the intermediate:

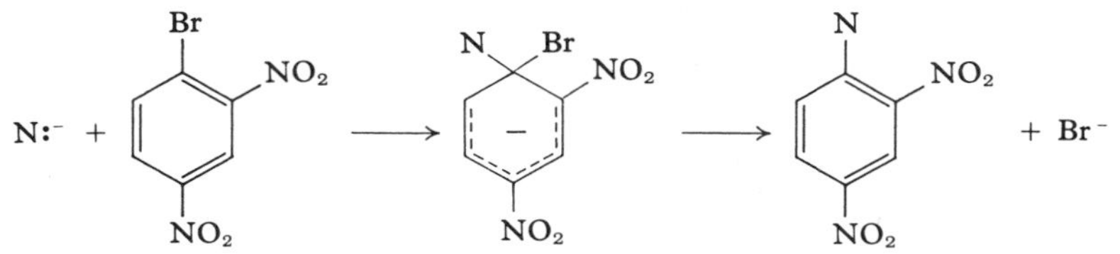

The same groups, for example the —NO$_2$ group, are activating for nucleophilic aromatic substitution and deactivating for electrophilic aromatic substitution; both effects are interpreted in terms of the electron-withdrawing ability of the groups.

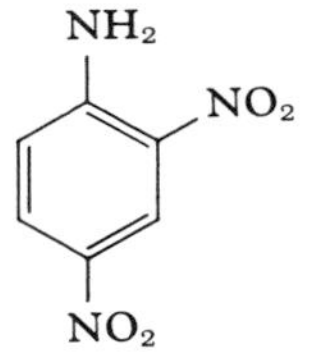

62.1 2,4-DINITROANILINE

Procedure. Place 25 ml of ethanol and 2.47 g (0.010 mole) of 2,4-dinitrobromobenzene in a 250-ml boiling flask. Heat the mixture gently to dissolve the solid. Add 20 ml (18 g; 0.32 moles) of conc. ammonium hydroxide. Relieve the resulting cloudiness of the solution by adding another 20 ml of ethanol and warming slightly. Allow the solution to stand for at least 96 hours (Note 1). ■ Then collect the yellow solid by suction filtration and recrystallize it from a mixture of 7.5 ml of water and 20 ml of ethanol. Yield: about 60%.

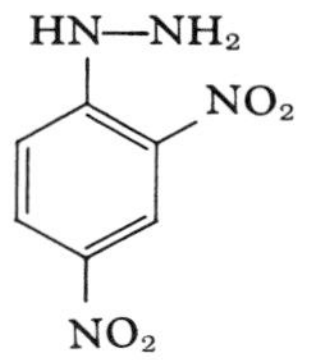

Note

1. A shorter period of standing will give a lower yield. Heating the mixture seems to speed loss of ammonia more than the desired reaction.

   Time: 1 hour plus four day reaction time.

2,4-dinitroaniline

2,4-dinitrophenylhydrazine

62.2 2,4-DINITROPHENYLHYDRAZINE

Procedure. Prepare a solution of 0.5 g (2 mmole) of 2,4-dinitrobromobenzene in 7.5 ml of 95% ethanol. Heat the solution almost to boiling and add to it a solution of 0.5 ml (10 mmole) of 64% hydrazine in 2.5 ml of 95% ethanol. Allow the light orange solution, which rapidly turns a deep red-purple, to cool undisturbed for 15–20 minutes. Collect the resulting crystals by suction filtration and wash them with a little ethanol. The product sometimes crystallizes as red-purple prisms alone, and sometimes as red-purple prisms and orange blades. Recrystallization of either form from boiling ethyl acetate (50 ml per gram) affords a product in the form of orange plates.

   Time: 2 hours.

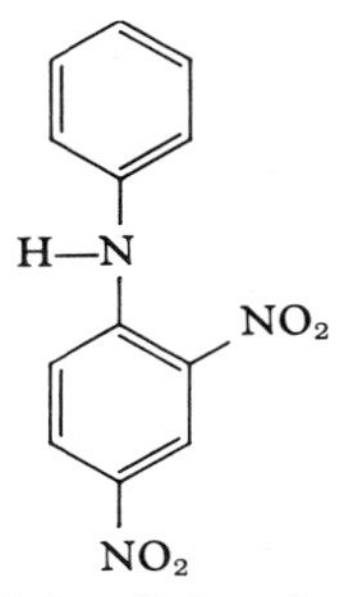

2,4-dinitrodiphenylamine

62.3 2,4-DINITRODIPHENYLAMINE

Procedure. Place 20 ml of ethanol, 1.84 g (7.5 mmole) of 2,4-dinitrobromobenzene, and 1.5 g (1.5 ml; 16 mmole) of aniline in a 250-ml boiling flask. Gently boil the mixture under reflux until all the solid has been dissolved. After allowing the mixture to cool slowly to room temperature (Note 1), collect the red needles by suction filtration. The product may be recrystallized from boiling ethanol (Note 2).

Notes

1. The solution should be allowed to stand for at least an hour before filtering.
2. The recrystallization will require 100–135 ml. The product dissolves slowly in boiling ethanol.

Time: 2 hours.

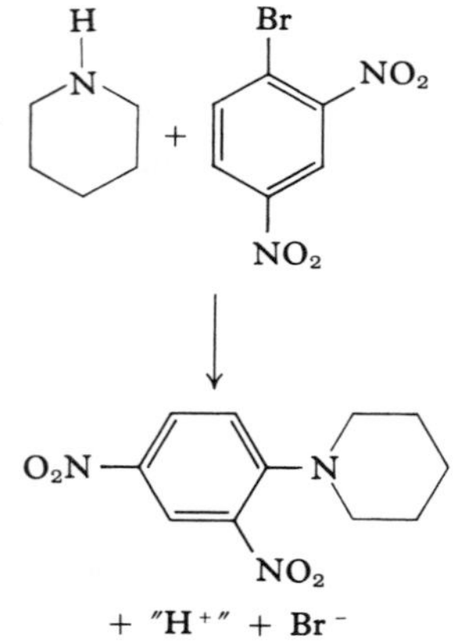

62.4 2,4-DINITROPHENYLPIPERIDINE

Procedure (Reference 1). Place 25 ml of ethanol and 1.84 g (7.5 mmole) of 2,4-dinitrobromobenzene in a 250-ml boiling flask. Heat the mixture gently to dissolve the solid. Then add 1.72 g (2.0 ml; 20 mmole) of piperidine and boil the mixture under reflux for 10 minutes. After allowing the solution to cool slowly to room temperature (Note 1) and finally cooling it in an ice bath, collect the orange needles by suction filtration. The product may be recrystallized from 20 ml of ethanol.

Note

1. The solution tends to become supersaturated. Crystallization may be induced by adding a seed crystal or by tapping the flask with a spatula.

Time: 2 hours.

Reference

1. J. W. McFarland, *Organic Laboratory Chemistry*, C. V. Mosby, St. Louis, 1969, p. 209.

§63 DIAZONIUM SALTS OF AROMATIC AMINES

Diazonium salts are prepared from the corresponding aromatic amine by treatment of an acidic solution of the amine with one equivalent of aqueous sodium nitrite solution:

$$\underset{X^-}{Ar\text{—}\overset{+}{N}H_3} + H\text{—}O\text{—}\ddot{N}\text{=}O \longrightarrow \{\underset{X^-}{Ar\text{—}\overset{+}{N}\equiv N\text{:} \longleftrightarrow Ar\text{—}\ddot{N}\text{=}\overset{+}{N}\text{:}}\} + 2\,H_2O$$

The mechanism appears to involve the electrophilic attack of the conjugate acid of nitrous acid upon the lone pair of electrons on the nitrogen of the free amine, which is in equilibrium with its conjugate acid:

$$Ar\text{—}\overset{+}{N}H_3 \rightleftarrows H^+ + Ar\text{—}\ddot{N}H_2 \xrightarrow[\text{slow}]{H\text{—}\overset{+}{O}H\text{—}\ddot{N}\text{=}O} Ar\text{—}\overset{+}{N}H_2\text{—}\ddot{N}\text{=}O + H_2O$$

The final steps of the reaction are fast and essentially involve the enolization of the N-nitrosoamine and ionization of the resulting diazohydroxide:

$$Ar-\overset{\overset{H}{|}}{\underset{\underset{H}{|}}{\overset{+}{N}}}-\ddot{N}=O \rightleftharpoons H^+ + Ar-\overset{\overset{}{|}}{\underset{\underset{H}{|}}{\ddot{N}}}-\ddot{N}=O \rightleftharpoons Ar-\ddot{N}=\ddot{N}-O-H$$

$$Ar-\ddot{N}=\ddot{N}-O-H + H^+ \rightleftharpoons \{Ar-\ddot{N}=\overset{+}{N}: \longleftrightarrow Ar-\overset{+}{N}\equiv N:\} + H_2O$$

The diazotization is carried out as near 0°C as possible in order to minimize the rate of the reaction of the diazonium salt with water to form the corresponding phenol:

$$\{Ar-\ddot{N}=\overset{+}{N}: \longleftrightarrow Ar-\overset{+}{N}\equiv N:\} + H_2O \longrightarrow Ar-OH + :N\equiv N: + H^+$$

Any phenol that forms will react with another molecule of the diazonium salt to form a highly colored azo compound (compare Section 65.2). If the aqueous solution of the diazonium salt is allowed to warm to room temperature, the formation of the phenol can become a significant reaction.

Since diazotization is an exothermic reaction, one might attempt to minimize the temperature rise and the subsequent formation of phenol by very slow addition of the sodium nitrite solution. However, the diazonium salt can react with undiazotized amine to form an azo compound (Section 65.4). Because of this, it is desirable to diazotize as quickly as possible in order to minimize the time during which the diazonium salt and undiazotized amine are present together. As a compromise, diazotization is usually carried out as fast as possible with good stirring and cooling, but slowly enough that the temperature of the solution does not rise above about 5°C. The lower the temperature is, the slower, also, will be the reaction between the diazonium salt and undiazotized amine.

An excess of sodium nitrite is usually to be avoided since in some cases the product of a subsequent reaction can react with excess nitrous acid.

63.1 BENZENEDIAZONIUM CHLORIDE FROM ANILINE

Procedure. Dissolve 9.1 ml (9.3 g; 0.10 mole) of aniline in a mixture of 25 ml (0.30 mole) of conc. hydrochloric acid and 25 ml of water. Cool the solution to 0°C in an ice bath, and add with good stirring and cooling (Note 1) a solution of 7.5 g (0.108 mole) of sodium nitrite in 25 ml of water at such a rate that the temperature of the solution does not exceed 5°C. Add only enough of this solution to give a slight excess of nitrous acid (Note 2). The resulting solution of benzenediazonium chloride must be stored in the ice bath and used fairly quickly.

Time: 1 hour.

63.2 p-TOLUENEDIAZONIUM CHLORIDE FROM p-TOLUIDINE

Procedure. Dissolve 10.7 g (0.10 mole) of p-toluidine in 15 ml of water by adding 10 ml (0.12 mole) of conc. hydrochloric acid and gently heating the mixture. Now add another 15 ml (0.18 mole) of conc. hydrochloric acid and cool the mixture to 0°C in an ice bath (Note 3). With good stirring and cooling (Note 1), add a solution of 7.5 g (0.108 mole) of sodium nitrite in 25 ml of water at such a

:NH₂

aniline

$H-O-N=O \quad \underset{0°C}{\overset{HCl/H_2O}{|}}$

$N_2^+Cl^-$

benzenediazonium chloride

rate that the temperature of the solution does not exceed 5°C. Add only enough of this solution to give a slight excess of nitrous acid (Note 2). The resulting solution of *p*-toluenediazonium chloride must be stored in an ice bath and used fairly soon.

Time: 1 hour.

63.3 *p*-NITROBENZENEDIAZONIUM SULFATE FROM *p*-NITROANILINE

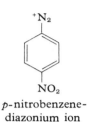

$^+N_2$

NO_2

p-nitrobenzene-
diazonium ion

Procedure. Make a solution of 20 ml (0.36 mole) of conc. sulfuric acid in 100 ml of water. Add to this 13.8 g (0.10 mole) of *p*-nitroaniline and heat the mixture gently to bring the amine into solution. Cool the mixture to about 10°C in an ice bath and add to the suspension of *p*-nitroaniline sulfate, with stirring and cooling, a solution of 6.9 g (0.10 mole) of sodium nitrite in 20 ml of water at such a rate that the temperature does not rise above about 10°C. Store the resulting solution of *p*-nitrobenzenediazonium sulfate in the ice bath.

Notes

1. The efficiency of cooling may be increased by adding pieces of ice to the reaction mixture.

2. Since nitrous acid oxidizes iodide to iodine, the presence of nitrous acid in the mixture may be detected by placing a drop of the solution on a piece of starch potassium iodide test paper. An immediate blue color will be produced if nitrous acid is present. If the quantities of reagents have been carefully measured, one need not start testing for excess nitrous acid until most of the solution of sodium nitrite has been added. The test for nitrous acid should be positive for at least 1–2 minutes after the last addition has been made in order to safely assume that slightly more than an equivalent amount of sodium nitrite has been added.

3. *p*-Toluidine hydrochloride will crystallize out of solution. As the diazotization of the dissolved salt proceeds, the precipitated salt goes into solution.

Time: 1 hour.

---

§64 REPLACEMENT REACTIONS OF DIAZONIUM SALTS

The two most generally useful types of reactions of aryl diazonium salts are the replacement reactions and the coupling reactions. The procedure of this section illustrates one of the possible replacement reactions, and Section 65 will present some examples of coupling reactions.

64.1 CHLOROBENZENE FROM BENZENEDIAZONIUM CHLORIDE

$$\{\phi-\overset{+}{N}\equiv N\colon \longleftrightarrow \phi-\overset{..}{N}=\overset{+}{N}\colon\} + CuCl_2^- \xrightarrow{\text{cold}} \phi-\overset{+}{N_2}\, CuCl_2^-$$

$$\phi-\overset{+}{N_2}\, CuCl_2^- \xrightarrow{\text{warm}} \phi-Cl + CuCl + N_2$$

The mechanism of the decomposition of the double salt is not completely agreed upon, but the reaction does appear to involve radical intermediates.

Figure 64.1 shows the infrared spectrum of chlorobenzene.

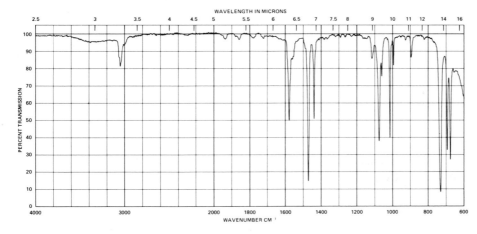

chlorobenzene

Figure 64.1. Infrared spectrum of chlorobenzene; thin film.

Procedure. Dissolve 30 g (0.12 mole) of copper sulfate pentahydrate in 100 ml of hot water in a 500-ml boiling flask. Then add and dissolve 10 g (0.17 mole) of sodium chloride. Prepare a solution of sodium sulfite by dissolving 7 g (0.067 mole) of sodium bisulfite and 4.5 g (0.11 mole) of sodium hydroxide in 50 ml of water. Reduce the copper by adding the solution of sodium sulfite in several portions over 2–3 minutes with good mixing. Allow the mixture containing the precipitated cuprous chloride to stand and settle in a pan of cold water while you prepare a solution of benzenediazonium chloride as described in Section 63.1.

When the diazotization is complete, decant the supernatant liquid from the precipitated cuprous chloride, and wash the precipitate by adding water, swirling, allowing to settle, and decanting. Then dissolve the cuprous chloride by adding 45 ml (0.54 mole) of conc. hydrochloric acid and swirling the mixture. Cool this solution in the ice bath and slowly add the solution of benzenediazonium chloride. Allow the mixture to stand at room temperature with occasional swirling for about 10 minutes. During this time, the initially formed double salt begins to decompose to give nitrogen and chlorobenzene. Cautiously heat the mixture to 50°C on the steam bath to complete the decomposition. Steam distill the mixture to isolate the chlorobenzene (Section 11). Separate the product from the distillate by extraction with ether, washing with dilute sodium hydroxide solution, drying, and distilling.

Time: 3 hours.

64.2 *p*-CHLOROTOLUENE FROM *p*-TOLUENEDIAZONIUM CHLORIDE

p-Chlorotoluene may be prepared by the preceding procedure, substituting *p*-toluenediazonium chloride (Section 63.2) for benzenediazonium chloride.

The infrared spectrum of *p*-chlorotoluene appears in Figure 64.2.

TRANSFORMATIONS

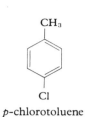

p-chlorotoluene

Figure 64.2. Infrared spectrum of p-chlorotoluene; thin film.

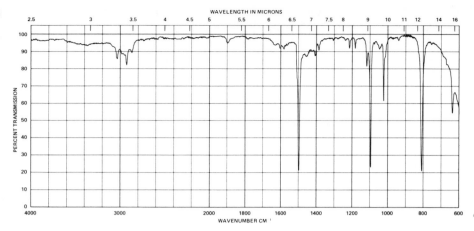

64.3 o-CHLOROTOLUENE FROM o-TOLUIDINE

o-Chlorotoluene may be prepared from o-toluidine by carrying out the diazotization as in Section 63.1 or Section 63.2. The replacement reaction may be carried out as described in Section 64.1.

Figure 64.3 shows the infrared spectrum of o-chlorotoluene.

o-chlorotoluene

Figure 64.3. Infrared spectrum of o-chlorotoluene; thin film.

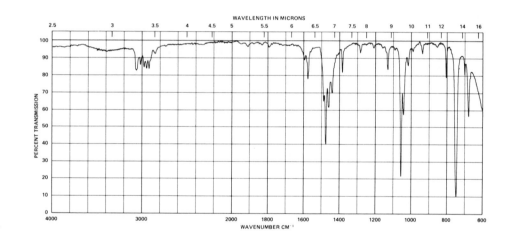

§65 ELECTROPHILIC AROMATIC SUBSTITUTION BY DIAZONIUM IONS; COUPLING REACTIONS OF DIAZONIUM SALTS

The positively charged diazonium ion can effect electrophilic substitution reactions on highly activated aromatic compounds such as amines and phenols. The molecules produced by such reactions have the —N=N— (azo) system linking two aromatic rings. This system of extended conjugation causes light to be absorbed in the visible region of the spectrum (Section 24.3). The resulting partial reflection or transmission of white light is the reason that these compounds and their solutions appear colored.

Procedure. Prepare a solution of benzenediazonium chloride according to the procedure of Section 63.1, using one-tenth of the quantities (0.010-mole scale).

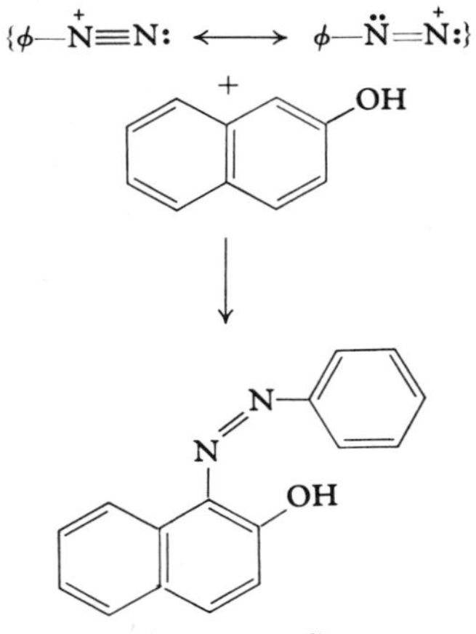

Prepare a solution of 1.44 g (0.010 mole) of β-naphthol in 5 ml of 3 M sodium hydroxide (0.015 mole). After cooling this solution to 5°C, add slowly and with good stirring the solution of benzenediazonium chloride. After allowing the reaction mixture to stand in the ice bath for 15 minutes, collect the product by suction filtration and wash it thoroughly with cold water. The product may be recrystallized from ethanol or from glacial acetic acid.

Time: 3 hours.

(orange-red)

65.2 p-NITROBENZENEDIAZONIUM SULFATE AND PHENOL: p-(4-NITROBENZENEAZO)-PHENOL

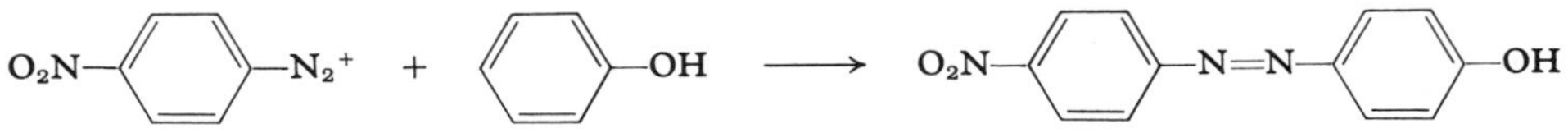

(orange-red)

Procedure. Prepare a solution of p-nitrobenzenediazonium sulfate according to the procedure of Section 63.3, using one-tenth of the quantities (0.010-mole scale).

Prepare a solution of 0.94 g (0.010 mole) of phenol in 5 ml of 1 M sodium hydroxide. Cool this mixture to 5°C and add to it the diazonium salt solution. Stir the mixture for 5 minutes and collect the product by suction filtration. The product may be recrystallized from toluene.

Time: 3 hours.

65.3 p-NITROBENZENEDIAZONIUM SULFATE AND β-NAPHTHOL: 1-(p-NITROPHENYLAZO)-2-NAPHTHOL (PARA RED; AMERICAN FLAG RED)

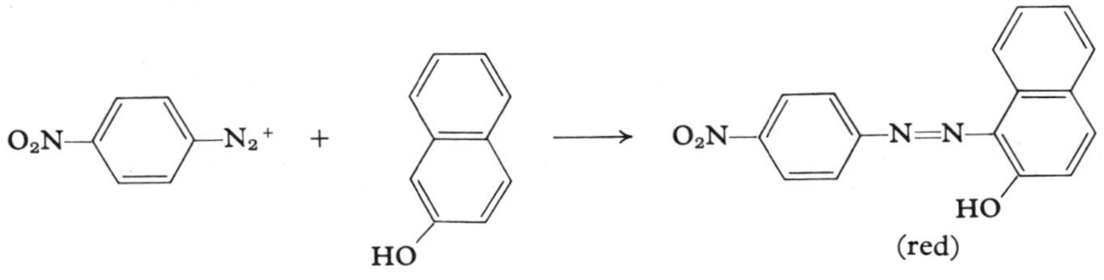

(red)

Procedure. Prepare a solution of p-nitrobenzenediazonium sulfate according to the procedure of Section 63.3, using one-tenth of the quantities (0.010-mole scale).

Prepare a solution of 1.44 g (0.010 mole) of β-naphthol in 25 ml of 10% sodium hydroxide solution (0.042 mole). When this solution has been cooled to 10°C, pour it into the diazonium salt solution. Acidify the mixture to litmus, and collect the red product by suction filtration. The product may be recrystallized from toluene or acetic acid.

Time: 3 hours.

65.4 *p*-NITROBENZENEDIAZONIUM SULFATE AND DIMETHYLANILINE: *p*-(4-NITROBENZENEAZO)-DIMETHYLANILINE

(reddish-brown)

Procedure. Prepare a solution of *p*-nitrobenzenediazonium sulfate according to the procedure of Section 63.3, using one-tenth of the quantities (0.010-mole scale).

Prepare a solution of 1.21 g (0.010 mole) of dimethylaniline in 1.5 ml of 1 *M* hydrochloric acid. Cool this solution in the ice bath and add the diazonium salt solution; mix well. Add 10 ml cold 1 *M* sodium hydroxide solution. Collect the product by suction filtration.

Time: 3 hours.

---

§66 THE DIELS-ALDER REACTION

The Diels-Alder reaction is one of the most interesting reactions in organic chemistry. It involves a reaction between an alkene and a conjugated diene in which a six-membered ring is formed:

diene   dienophile   adduct

The reaction appears to involve only a single step (one transition state).

Many different combinations of diene and dienophile will undergo this kind of reaction, and the experiments in this section illustrate a few of the great variety of structures that can result.

66.1 BUTADIENE (FROM 3-SULFOLENE) AND MALEIC ANHYDRIDE

Diels-Alder reactions involving butadiene must generally be carried out in a pressure vessel, since the temperature at which reaction takes place readily is usually far above the boiling point of butadiene, $-3°C$. However, Sample and Hatch (1) have described a convenient way to avoid the use of a pressure vessel in the preparation of the Diels-Alder adduct between butadiene and maleic anhydride. In their procedure, butadiene is generated by the thermal decomposition of 3-sulfolene, a sort of reverse Diels-Alder reaction itself:

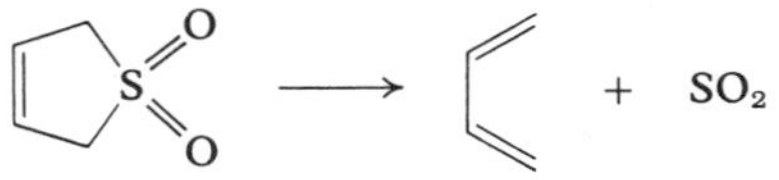

Apparently, at the temperature of the reaction, butadiene reacts with maleic anhydride more rapidly than it can be lost from the reaction mixture:

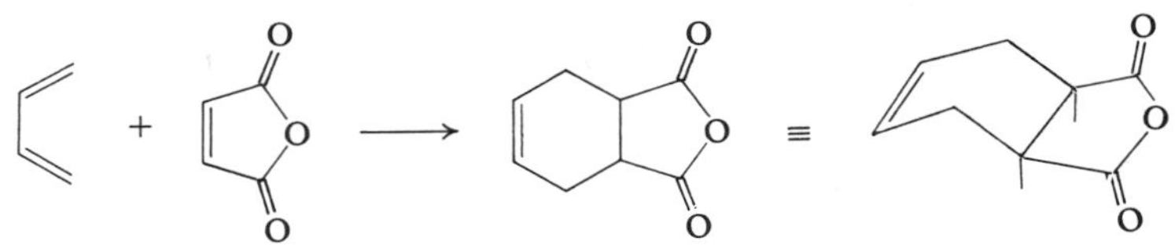

m.p. 104°C

Procedure (Reference 1).　Place 8.5 g (0.072 mole) of 3-sulfolene and 4.5 g **319** (0.050 mole) of powdered maleic anhydride in a 125-ml Erlenmeyer flask. Add 4 ml of xylene, fit the flask with a condenser, and stopper the condenser with a plug of absorbent cotton (Note 1). Swirl the flask gently to effect partial solution. Remove the cotton plug and cautiously heat the mixture to a slow boil (Notes 2 and 3). After boiling the mixture for about one-half hour, allow it to cool for about 5 minutes (Note 4) and then add 50 ml of toluene (Note 5) and 0.5 g of decolorizing carbon. Heat the suspension on the steam bath and filter it by gravity. Reheat the filtrate until the product redissolves, and then add with swirling 20–25 ml of petroleum ether (Note 5) until a slight cloudiness persists. Rewarm the mixture until it is substantially clear, and finally allow the solution to cool in an ice bath for crystallization.

Notes

1. The mixture will become quite cool. The absorbent cotton will prevent water vapor from entering the flask and causing hydrolysis of the maleic anhydride.
2. Since sulfur dioxide will be evolved, the reaction must be run in the hood or some provision must be made to absorb the fumes.
3. Because the reaction is exothermic, some care must be taken to avoid overheating at the beginning of the heating period.
4. If the cooling period is too long, the product will separate as a hard cake that will be difficult to redissolve.
5. The solvents should be dry in order to avoid hydrolyzing the product to the corresponding diacid. Since the diacid is much less soluble, it cannot be removed by crystallization.

Time: 3 hours.

Reference

1. T. E. Sample, Jr., and L. F. Hatch, *J. Chem. Educ.* **45**, 55 (1968).

66.2 CYCLOPENTADIENE AND MALEIC ANHYDRIDE

In all Diels-Alder reactions, the structure of the product indicates that the reaction takes place with *cis* addition to the double bond of the dienophile. In the case of cyclopentadiene, *cis* addition can take place in two different ways to give products with the same structure but with different configurations.

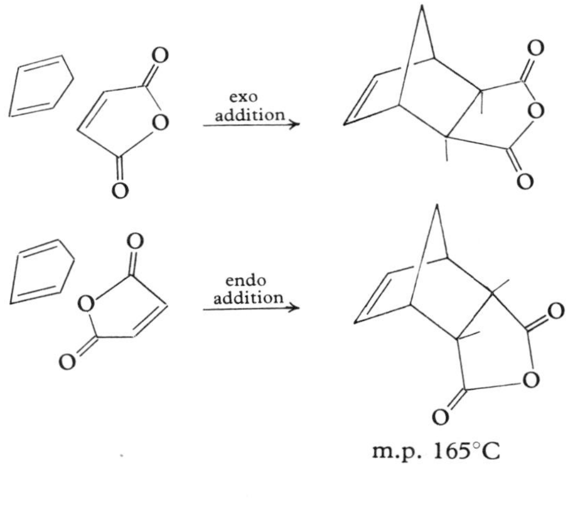

m.p. 165°C

With butadiene (Section 66.1) the two different modes of reaction would give two equivalent (but presumably rapidly interconverting) conformations of the same molecule, and thus there is no way of knowing whether one path was followed in preference to the other. With cyclopentadiene, under the conditions of this experiment, the isomer corresponding to endo addition may be isolated in high yield.

The predominance of the endo isomer in Diels-Alder reactions is generally observed, and, since the exo isomer is usually the more stable, the predominance of the endo isomer must be the result of kinetic control rather than thermodynamic control. The predominance of endo addition (Alder Rule) has been interpreted in terms of inductive effects, electrostatic effects due to an initial charge transfer between diene and dienophile, "maximum accumulation of unsaturation," and, most recently, orbital symmetry effects.

In those cases in which the exo isomer is more stable, the use of relatively high temperatures and long reaction times can result in formation of the exo isomer. In certain cases, the endo isomer may be converted to the exo isomer by heating. While it would be reasonable to propose that the exo isomer may be formed via dissociation of the endo isomer to diene and dienophile, followed by recombination, the isomerization of the endo adduct of cyclopentadiene and maleic anhydride takes place at 190°C in an open flask without observable loss in weight. One would expect free cyclopentadiene to be lost under these conditions, although the success of the procedure of Section 66.1 indicates that this loss may not necessarily take place.

Cyclopentadiene dimerizes on standing via a Diels-Alder reaction. It must be prepared by thermal decomposition of the dimer:

endo-dicyclopentadiene

The *endo-cis* diacid may be prepared as described in Exercise 1 at the end of Section 8.

Procedure.   Place 15 ml of dicyclopentadiene in a 50-ml boiling flask. Fit the flask with a fractionating column and connect an adapter with a condenser set for downward distillation on the top of the column. Heat the dicyclopentadiene with a flame until it boils, and then continue to heat at such a rate that the cyclopentadiene distills at about its boiling point of 43°C (Note 1). Cool the receiver in an ice bath.

During the distillation, dissolve 6 g (0.061 mole) of maleic anhydride in 20 ml of ethyl acetate by heating on the steam bath. Add 20 ml of ligroin, and cool the mixture thoroughly in an ice bath (Note 2). Add 6 ml (4.8 g; 0.072 mole) of dry cyclopentadiene (Note 3) and swirl the mixture in the ice bath until the product separates as a white solid. Heat the mixture on the steam bath until it is all dissolved (Note 4), and allow it to stand for crystallization; slow crystallization will result in a beautiful display of crystal formation.

Notes

1. A fair fractionating column and some patience are needed. It may take 30 minutes to collect a little more than the 6 ml needed.

2. Some maleic anhydride may crystallize.

3. Water in the distilled cyclopentadiene will cause it to be cloudy. The cyclopentadiene may be dried with a little anhydrous calcium chloride.

4. Any diacid formed by water that got into the reaction mixture will remain undissolved at this point; it should be removed by gravity filtration.

Time: 3 hours.

Question

1. Propose an alternative procedure for this reaction in which the cyclopentadiene is produced in situ, in analogy with the in situ production of butadiene in Section 66.1.

References

1. W. J. Sheppard, *J. Chem. Educ.* **40**, 40 (1963).
2. L. F. Fieser, *Organic Experiments*, Heath, Boston, 1964, p. 83.

66.3 FURAN AND MALEIC ANHYDRIDE

In the Diels-Alder reaction between furan and maleic anhydride, the only adduct that has been isolated is one whose m.p. has been variously reported as 116–117°C; 122°C; 125°C with foaming and decomposition. The stereochemistry of this product was shown to correspond to the exo mode of addition. It is suspected that this is the thermodynamically more stable isomer.

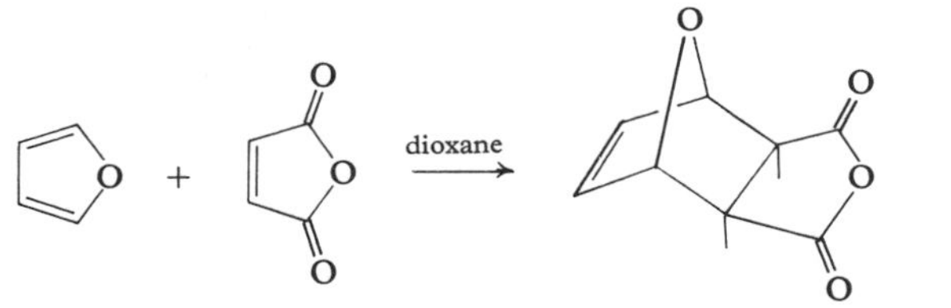

furan

product of exo addition:
m.p. 117°C

Procedure. Place 10.0 grams (0.1 mole) of maleic anhydride in a 125-ml ground-glass-stoppered Erlenmeyer flask. Add 25 ml of dioxane and swirl the mixture until the anhydride has all dissolved. Add to the flask 7.5 ml (7 grams; 0.1 mole) of furan, swirl to mix, and allow the solution to stand for at least 24 hours. Isolate the crystalline adduct by suction filtration, washing it with a little ether. Yield: about 11 grams (65%; Note 1).

Note

1. If 12.5 ml of dioxane is used the yield will be greater than 80% but it will separate from solution in hard lumps.

Time: 1 hour plus a reaction time of at least 24 hours.

TRANSFORMATIONS

Tetracyanoethylene is a very reactive dienophile in the Diels-Alder reaction. In its reaction with anthracene, a transient green color is produced that has been interpreted in terms of a rapidly and reversibly formed intermediate which more slowly goes on to form the observed Diels-Alder product. The transient color observed in some other Diels-Alder reactions has also been interpreted in this way. However, there is also the possibility that the species responsible for the color does not lie on the reaction path leading to the Diels-Alder product. Experiments remain to be done that can rule out one of these possibilities.

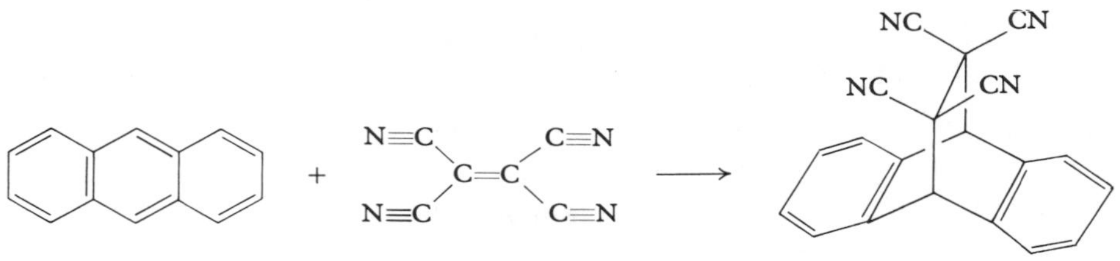

Procedure (References 1 and 2).   Prepare a solution of 0.45 g (2.5 mmole) of anthracene in 25 ml of benzene in a 125-ml Erlenmeyer flask. Prepare a solution of 0.33 g (2.6 mmole) of tetracyanoethylene (*caution:* Note 1) in 15 ml of benzene. If necessary, warm the mixtures slightly to effect solution, and then cool them to room temperature. Pour the tetracyanoethylene solution into the anthracene solution and allow the mixture to stand for 30 minutes. Collect the crystals by suction filtration and wash them with a little benzene. The product sublimes rapidly at the melting point. It may be recrystallized from acetone.

Note

1. Tetracyanoethylene hydrolyzes to form hydrogen cyanide. Replace the cap on the bottle tightly, and do not breathe near the top of the bottle.

   Time: 2 hours.

References

1. F. D. Popp and H. P. Schultz, *Organic Chemical Preparations*, Saunders, Philadelphia, 1964, p. 196.
2. J. C. Kellett, Jr., *J. Chem. Educ.* **40**, 543 (1963).

66.5 TETRAPHENYLCYCLOPENTADIENONE AND DIPHENYLACETYLENE: PREPARATION OF HEXAPHENYLBENZENE

The formation of the product isolated in this reaction can be rationalized in terms of the initial formation of a Diels-Alder adduct that loses carbon monoxide under the conditions of the reaction:

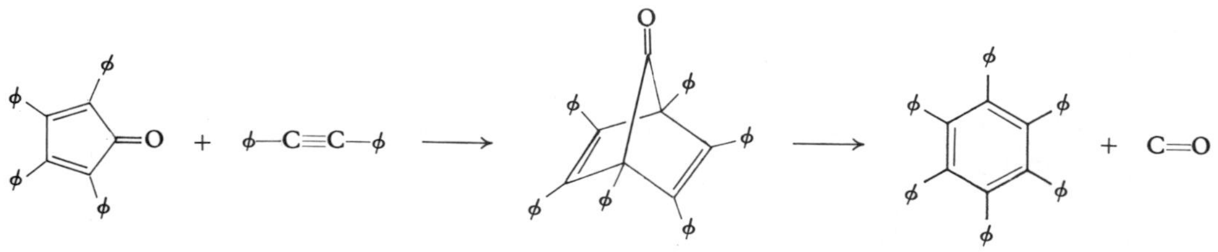

Procedure (Reference 1). Place 0.5 g (1.3 mmole) of tetraphenylcyclopenta-dienone (Section 74.4) and 0.5 g (2.8 mmole) of diphenylacetylene in a 25-mm × 150-mm test tube supported by a clamp, and heat the mixture with a flame. Soon after the reactants have melted together, a white solid appears. Continue heating to cause the excess diphenylacetylene to reflux (b.p., 300°C) on the sides of the tube, and then remove the excess hydrocarbon by sticking in a stirring rod and allowing the diphenylacetylene to condense and solidify on it (m.p., 62°C). Remove the stirring rod before the hydrocarbon melts, wipe it clean, cool it in cold water (wipe it off), and repeat this operation until it is possible to melt the entire reaction mixture by strong heating. Then allow the mixture to cool and solidify. Add 10 ml of diphenyl ether (b.p. 259°C) and heat the mixture to bring the product into solution. After allowing the solution to cool for crystallization, thin the mixture with 10 ml of toluene, and collect the product by suction filtration.

Time: 2 hours.

Reference

1. L. F. Fieser, *Organic Experiments*, Heath, Boston, 1964, p. 307.

§67 ALDOL CONDENSATIONS AND RELATED REACTIONS

A generally useful type of reaction in which a new carbon–carbon double bond is formed occurs when a molecule of water is eliminated between an acidic methylene and a carbonyl group.

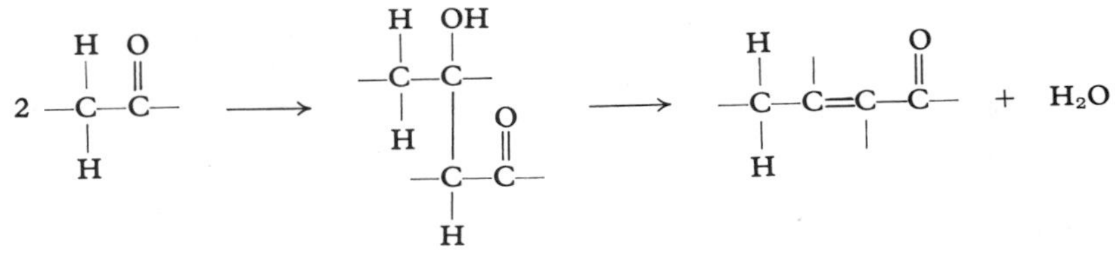

Although acid catalysis is sometimes possible, the reaction is usually carried out under base catalysis:

Formation of the Carbanion

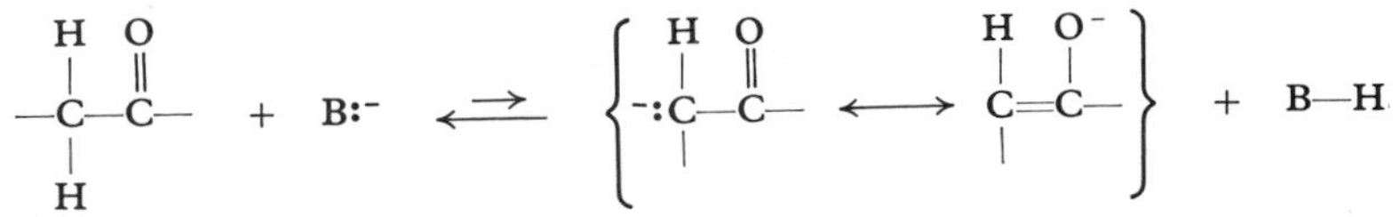

resonance-stabilized carbanion

Addition of the Carbanion to the Carbonyl Group

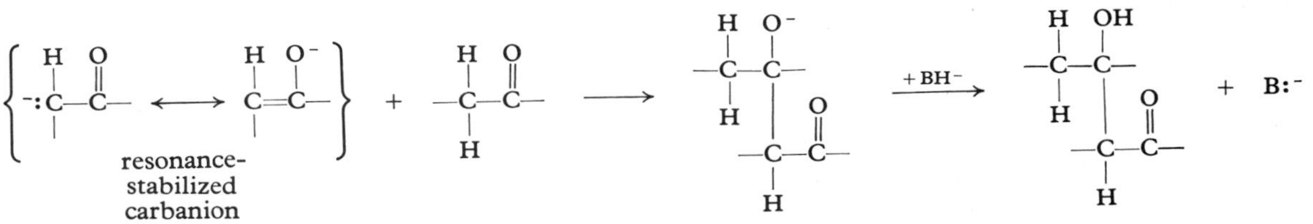

resonance-stabilized carbanion

It is occasionally possible to isolate the initially formed β-hydroxy aldehyde or ketone (Aldol condensation).

If two different carbonyl compounds are used, as many as four different products of this type of reaction can be formed. In order for the reaction between two different carbonyl compounds to be useful in synthesis, usually one must have no alpha hydrogen atoms. In such cases, there remain only two possibilities: the desired mixed condensation, and the self-condensation of the other compound.

67.1 SELF-CONDENSATION OF PROPIONALDEHYDE: 2-METHYL-2-PENTENAL

The infrared and NMR spectra of 2-methyl-2-pentenal are shown in Figures 67.1 and 67.2; Figures 67.3 and 67.4 show ultraviolet spectra.

Procedure (Reference 1). Place 5 ml of 10% sodium hydroxide in a 100-ml boiling flask, add a magnetic stirring bar, and fit the flask with a Claisen adapter. Fit a separatory funnel in the center neck of the Claisen adapter and a condenser in the other neck. Place 25 ml (20 g; 0.35 mole) of propionaldehyde (Note 1) in the separatory funnel, and add it slowly and with good stirring to the flask. After the addition is complete, stir the mixture until it has cooled nearly to room temperature. Remove as much of the aqueous layer as possible by means of a pipet or long medicine dropper and distill the residue, using a fractionating column, keeping the temperature at the still-head from rising above about

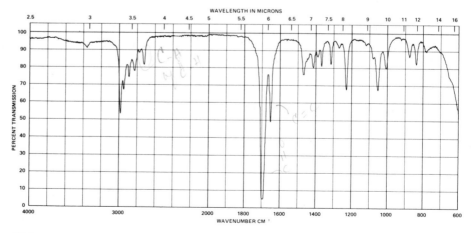

Figure 67.1. IR spectrum of 2-methyl-2-pentenal; thin film.

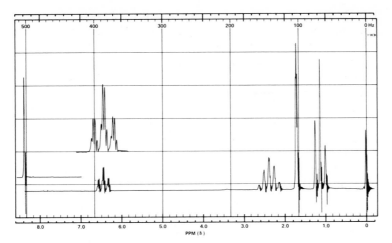

Figure 67.2. NMR spectrum of 2-methyl-2-pentenal; CCl₄ solution.

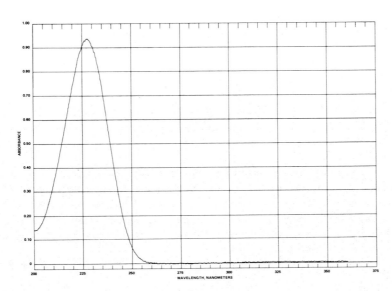

Figure 67.3. Ultraviolet spectrum of 2-methyl-2-pentenal; 5.3×10^{-5} molar in 95% ethanol.

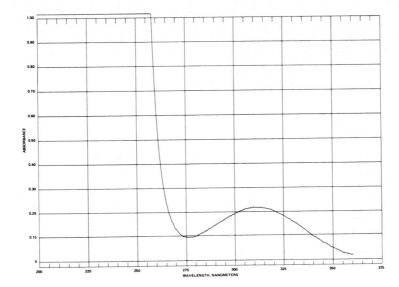

Figure 67.4. Ultraviolet spectrum of 2-methyl-2-pentenal; 5.3×10^{-3} molar in 95% ethanol.

140°C. When the temperature at the still-head begins to fall, discontinue the distillation and remove the water from the distillate with a pipet or medicine dropper. Add a little anhydrous calcium chloride to the distillate, and over a period of about 15 minutes, remove the aqueous phase as it forms; add more solid calcium chloride if necessary. ■ Redistill the product, collecting as 2-methyl-2-pentenal the material boiling between 133 and 137°C.

Note

1. The propionaldehyde must be free of propionic acid. It can be purified by distillation.

Time: 3 hours.

Reference

1. C. A. MacKenzie, *Experimental Organic Chemistry*, 3rd edition, Prentice-Hall, Englewood Cliffs, N.J., 1967, p. 161.

67.2 MESITYL OXIDE AND DIETHYL MALONATE: 5,5-DIMETHYL-1,3-CYCLOHEXANEDIONE; DIMEDON; METHONE

The initial base-catalyzed addition of diethyl malonate to the double bond of the α,β-unsaturated ketone is an example of the Michael condensation. This is followed by a type of Dieckmann, or cyclic Claisen, condensation. The final steps involve the hydrolysis and acid-catalyzed decarboxylation of a β-ketoacid.

The product Dimedon is useful for characterizing aldehydes (Section 29.4).

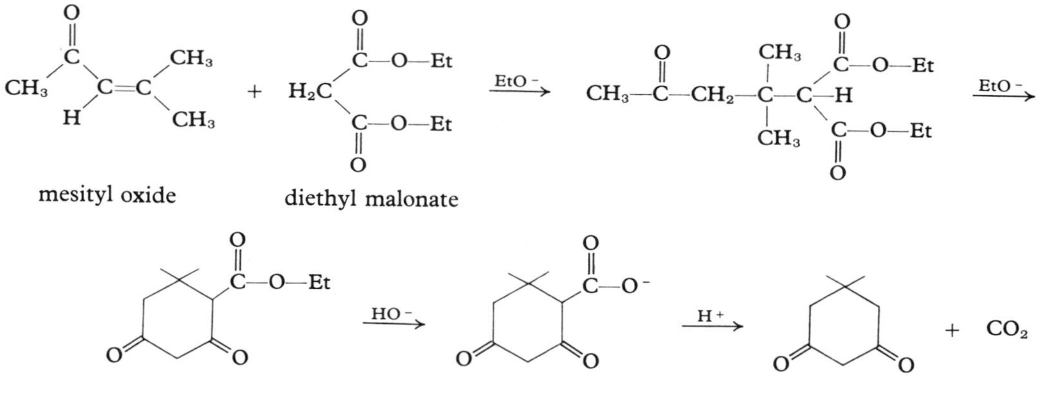

Procedure (Reference 1).   Place 2.5 ml of methanol in a 25-ml boiling flask, and add in portions and with cooling 0.15 g (6.5 mmole) of sodium (*caution:* Note 1). When all the sodium has dissolved, add 1.0 ml (1.05 g; 6.5 mmole) of diethyl malonate. Fit the flask with a condenser, and heat the mixture to boiling on the steam bath. By means of a pipet, cautiously add down the condenser 0.7 ml (0.60 g; 6.1 mmole) of mesityl oxide (Note 2). After a vigorous reaction has occurred, heat the mixture under reflux for one-half hour. At this time, add 5 ml of 2 *M* sodium hydroxide and boil the mixture for another hour and a half. At the end of the heating period, set the condenser for downward distillation and remove most of the alcohol by distillation on the steam bath. Prepare a mixture of 2.5 ml conc. hydrochloric acid and 5 ml of water, and when the removal of the alcohol is complete, heat the mixture to boiling with a very small flame and cautiously add the hydrochloric acid solution until the mixture is acid to methyl orange (Note 3). Cool the mixture in an ice bath, collect the product by suction filtration, and wash it with a little cold water. Dimedon may be recrystallized from acetone, using about 8 ml per gram; m.p., 147–148°C.

Notes

1. Sodium reacts violently with water, and the hydrogen formed usually ignites in air. See Section 1.7 for advice on handling sodium. An equivalent amount of sodium methoxide may be substituted for the sodium.
2. It appears that a recently redistilled sample of a good grade of mesityl oxide is required in this reaction.
3. The evolution of carbon dioxide will cause considerable foaming.

Time: 4 hours.

Question

1. Give the details of the mechanism of the reaction, taking care to specify the state of ionization of the intermediates.

Reference

1. J. H. Wilkinson, *Semi-micro Organic Preparations*, 2nd edition, Oliver and Boyd, Edinburgh (Scotland), 1958, p. 161.

§68 TWO THERMOCHROMIC COMPOUNDS: DIXANTHYLENE AND DIANTHRAQUINONE

Compounds whose color depends upon temperature are said to be *thermochromic*. Dixanthylene is a very pale yellow-green solid at room temperature, but it becomes

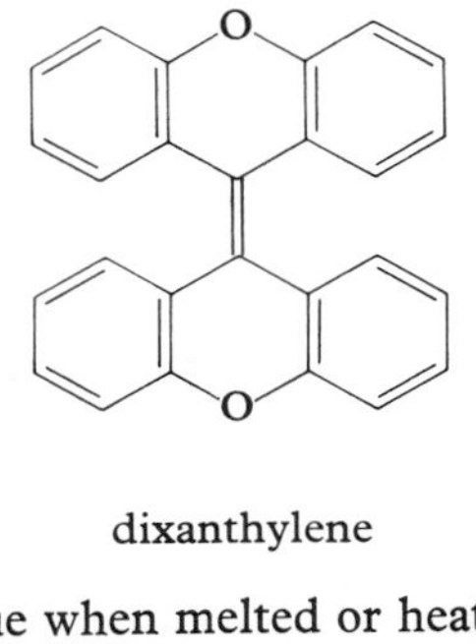

dixanthylene

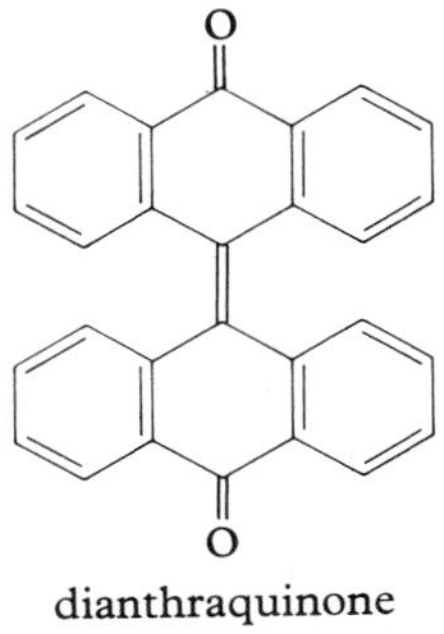

dianthraquinone

dark blue when melted or heated in solution; at liquid nitrogen temperature it is completely colorless. Dianthraquinone is a bright canary-yellow at room temperature, but it becomes a brilliant parrot-green when heated in solution. The thermochromism of each compound can be observed very nicely during recrystallization from mesitylene.

68.1 DIXANTHYLENE FROM XANTHENE AND XANTHONE VIA 9-HYDROXYDIXANTHYL

The preparation of dixanthylene may be carried out by the light-catalyzed addition of xanthene to xanthone to give 9-hydroxydixanthyl, followed by elimination of water.

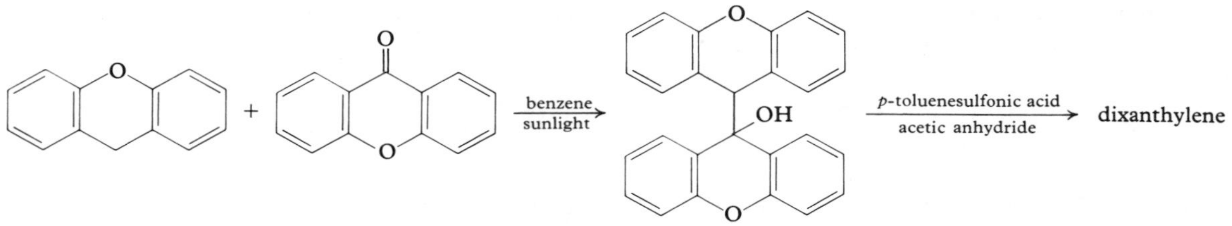

xanthene   xanthone   9-hydroxydixanthyl

Alternatively, dixanthylene may be prepared by the reductive dimerization of xanthone by means of zinc and acid:

$$2 \text{ xanthone} \xrightarrow[\text{hydrochloric acid; acetic acid}]{\text{zinc dust}} \text{dixanthylene}$$

9-Hydroxydixanthyl from Xanthone and Xanthene

Procedure. Place 3.23 g (16.5 mmole) xanthone and 3.00 g (16.5 mmole) xanthene in a 125-ml Pyrex Erlenmeyer flask. Add 75 ml of benzene and bring the solids into solution with swirling and gentle heating. Stopper the flask and allow it to stand in bright sunlight until it appears that the formation of the coarse, granular crystalline material has ceased (Note 1). ■ At this time, collect the crystalline 9-hydroxydixanthyl by suction filtration, and wash it with a little benzene. Yield: about 5.3 grams (85%); m.p.: 193.5–194.5°C; literature: 194°C (Reference 1).

Note

1. Three to fourteen days in the sun will be required; overexposure does no apparent harm.

Time: 1 hour plus the time required for irradiation.

Dixanthylene from 9-Hydroxydixanthyl

Procedure. Place 1.0 g (2.65 mmole) of 9-hydroxydixanthyl in a 25-ml Erlenmeyer flask. Add 14 ml of acetic anhydride and boil the mixture under reflux until all the solid has dissolved. Add 7.0 g (3.7 mmole) of p-toluenesulfonic acid monohydrate down the condenser and continue to boil the mixture until no more solid appears to form and no further color change occurs; about 3 minutes will be required; the color changes from red to dark brown to dark green. Allow the mixture to cool to room temperature for 1 hour and then collect the dixanthylene by suction filtration, washing it with a little acetic acid. Recrystallize the dixanthylene by heating it under reflux with about 10 ml of mesitylene (adding more down the condenser, if necessary, to bring all the solid into solution) and allowing the blue solution to cool to room temperature. Yield: about 400 mg (about 40%) of yellowish crystals; literature m.p.: 315°C (Reference 2).

Time: 2 hours.

Dixanthylene from Xanthone

Procedure.   Place 2.5 g (12.7 mmole) of xanthone in a 250-ml boiling flask, add 100 ml of glacial acetic acid, and swirl until most of the solid has dissolved. Add 0.8 g (12 mmole) of zinc dust and 2 drops of conc. hydrochloric acid, and boil the mixture under reflux for about 90 minutes. Every 5 minutes or so, add 2 drops of conc. hydrochloric acid down the condenser, taking care not to add the acid so often that the solution takes on a red color. Add three more 0.8-g portions of zinc dust, one at a time at intervals of about 20 minutes, adding the first about 20 minutes after the mixture has been brought to a boil. At the end of the period of boiling under reflux, allowing the solution to cool completely to room temperature, collect the solid by suction filtration, and wash it thoroughly with water. Separate any zinc from the product by boiling it with mesitylene to dissolve the dixanthylene (about 25 ml will be required), filtering the hot, blue solution by gravity, reheating the filtrate if necessary to redissolve any material that may have crystallized, and allowing the filtrate to cool to room temperature. Collect the very pale yellow-green crystals by suction filtration and wash them with a little benzene. Yield: about 0.8 g (35%).

The thermochromism of dixanthylene may be observed by heating a sample in a melting-point capillary over a small burner flame, or, better, by recrystallization from boiling mesitylene, using about 19 ml per gram (recovery: about 90%).

Time: 3 hours.

References

1. A. Schonberg and A. Mustafa, *J. Chem. Soc.* **1944**, 67.
2. G. Gurgenjanz and S. von Kostanecki, *Chem. Ber.* **28**, 2310 (1895).
3. A. Ault, R. Kopet, and A. Serianz, *J. Chem. Educ.* **48**, 410 (1971).

68.2 DIANTHRAQUINONE FROM ANTHRONE VIA 9-BROMOANTHRONE

Dianthraquinone may be prepared by bromination of anthrone to give 9-bromo-anthrone, followed by treatment of this product with diethylamine in chloroform:

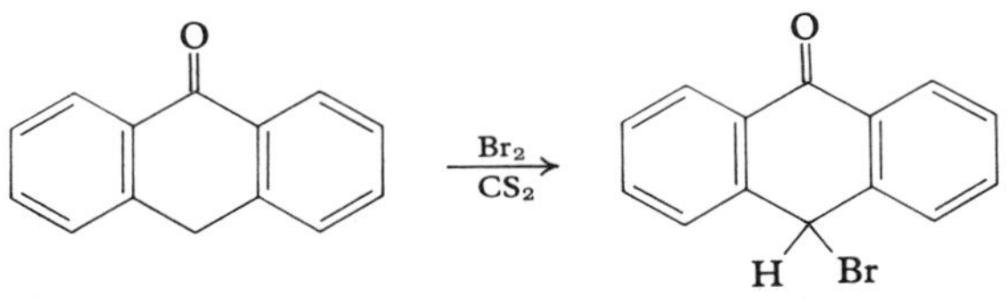

2 9-bromoanthrone $\xrightarrow[\text{CHCl}_3]{\text{Et}_2\text{NH}}$ dianthraquinone

9-Bromoanthrone from Anthrone

Procedure (Reference 1).   Suspend 5.0 grams (26 mmole) of anthrone in 15 ml of carbon disulfide. Add bromine dropwise with stirring over a period of about 15 minutes until the bromine color persists; about 4.1 grams (1.4 ml; 1 equivalent) will be required (Note 1). Collect the tan crystalline product by suction filtration and wash it with a little toluene. Yield: 6.0 grams (85%).

Time: 1 hour.

Dianthraquinone from 9-Bromoanthrone

Procedure (Reference 2).   Dissolve 2.0 grams (7.3 mmole) of 9-bromoanthrone (Note 2) in 30 ml of chloroform (Note 3). Add 2.0 ml (1.4 g; 19 mmole) of diethylamine, swirl to mix, and allow the resulting warm solution to stand for 2 hours. During this time, the color of the solution will change from yellow to a dark red-brown. At this time, add 75 ml of ether slowly with swirling. After 5 minutes, collect the yellow precipitate by suction filtration and wash it thoroughly with ether. Suspend the precipitate in about 10 ml of 95% ethanol and then collect the canary-yellow dianthraquinone by suction filtration, washing it with a little ethanol. Yield: about 0.8 g (55%).

The thermochromism of dianthraquinone can be observed very nicely upon recrystallization from boiling mesitylene using about 50 ml per gram.

Notes

1. Since hydrogen bromide gas is evolved in this reaction, it must be carried out in the hood or some provision must be made to absorb the gas produced (see Section 35).
2. The 9-bromoanthrone must be prepared just before use; otherwise the yield of dianthraquinone is greatly reduced.
3. Any insoluble material should be removed at this point by gravity filtration.

Time: 3 hours.

References

1. E. Barnett, J. W. Cook, and M. A. Matthews, *J. Chem. Soc.* **123**, 1994 (1923); and K. H. Meyer, *Annalen* **379**, 62 (1911).
2. E. Barnett, J. W. Cook, and H. H. Grainger, *J. Chem. Soc.* **121**, 2059 (1922).
3. A. Ault, R. Kopet, and A. Serianz, *J. Chem. Educ.* **48**, 410 (1971).

---

§69 A PHOTOCHROMIC COMPOUND: 2-(2,4-DINITROBENZYL)PYRIDINE

Crystals of the compound 2-(2,4-dinitrobenzyl)pyridine have the most unusual property of turning a very deep blue color when exposed to sunlight. During storage in the dark the crystals revert to their original sandy color. Color formation takes only a few minutes in bright sunlight, but the loss of color takes about a day.

The interconversion appears to be completely reversible any number of times. One explanation of the phenomenon proposes the formation of a tautomeric form by action of the light (1).

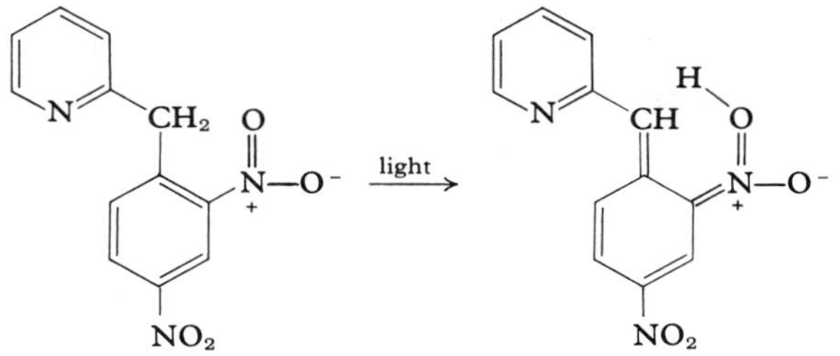

2-(2,4-dinitrobenzyl)pyridine

Procedure (Reference 2). Place 25 ml of conc. sulfuric acid (0.45 mole) in a 250-ml boiling flask. Cool the acid to 5°C or below by means of an ice bath. Arrange the flask so that it can be stirred or swirled in an ice bath and add gradually with good mixing 5.0 ml (5.3 g; 0.031 mole) of 2-benzylpyridine. To this cooled and well-stirred mixture add dropwise, over a period of about 3 minutes, 3.0 ml of red fuming nitric acid (density, 1.5 g/ml; 0.070 mole). The addition of the first few drops of the nitric acid will cause a color change to deep brown, but the mixture will become lighter in color as the remainder of the acid is added. After all the nitric acid has been added, heat the mixture for about 20 minutes on the steam bath.

At the end of the heating period, pour the mixture onto about 400 g of ice in a 2-l flask. Basify the solution to a pH of about 11 by adding almost all of a solution of 40 g (1 mole) of sodium hydroxide in about 500 ml of water. Toward the end of the addition of base, the product separates to give a milky yellow suspension. Add about 400 ml of ether and stir the mixture for 10–15 minutes to extract the product into the ether layer. Separate the ethereal solution, dry it over anhydrous magnesium sulfate for a few minutes, filter, and distill to reduce its volume to about 50 ml (Note 1). The product will sometimes crystallize during the concentration of the solution. Complete the crystallization by cooling the mixture in an ice bath. Collect the large sandy prisms by suction filtration, and wash them with a small amount of cold 95% ethanol. Yield: about 4 g (50%). The product may be recrystallized from 95% ethanol with 90% recovery using about 10 ml per gram; use of decolorizing carbon helps a little.

Note

1. Collect the ether and return it to the container provided so that it can be used by others.

Time: 3 hours.

References

1. J. A. Sousa and J. Weinstein, *J. Org. Chem.* **27**, 3155 (1962); and A. L. Bluhm, J. Weinstein, and J. A. Sousa, *J. Org. Chem.* **28**, 1989 (1963).
2. A. Ault and C. Kouba, *J. Chem. Educ.* **51**, 395 (1974).
3. K. Schofield, *J. Chem. Soc.* **1949**, 2411; and A. J. Nunn and K. Schofield, *J. Chem. Soc.* **1952**, 586
4. A. E. Tschitschibabin, B. M. Kuindshi, and S. W. Benewolenskaja, *Chem. Ber.* **58**, 1580 (1925).

§70 A CHEMILUMINESCENT COMPOUND: LUMINOL

The production of "cold light" by the American firefly is a familiar example of light emission that is nonthermal in origin; it is luminescence rather than incandescence. The light of the firefly is produced during oxidation of luciferin to oxyluciferin by oxygen in the presence of adenosine triphosphate (ATP) magnesium ion, and the enzyme luciferase. It appears that oxyluciferin is produced in

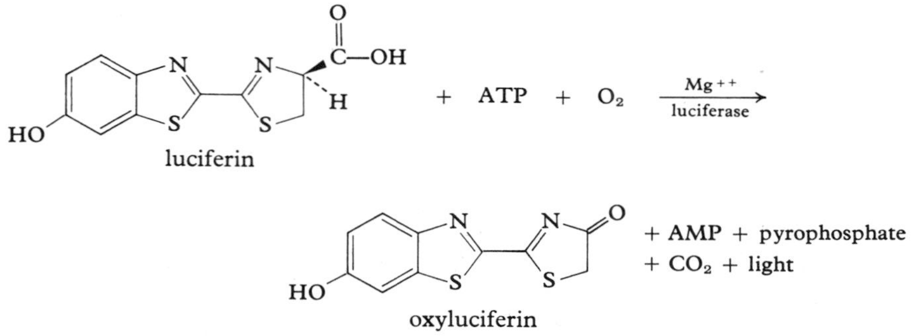

luciferin

$+$ ATP $+$ O_2 $\xrightarrow[\text{luciferase}]{\text{Mg}^{++}}$

oxyluciferin

$+$ AMP $+$ pyrophosphate $+$ CO_2 $+$ light

an electronically excited state, and that the energy released when the corresponding ground state is formed is given off as light.

A number of other substances have been shown to undergo chemical reaction with simultaneous production of light. A good example that is easy to observe is that of luminol reacting with oxygen and base in dimethylsulfoxide solution.

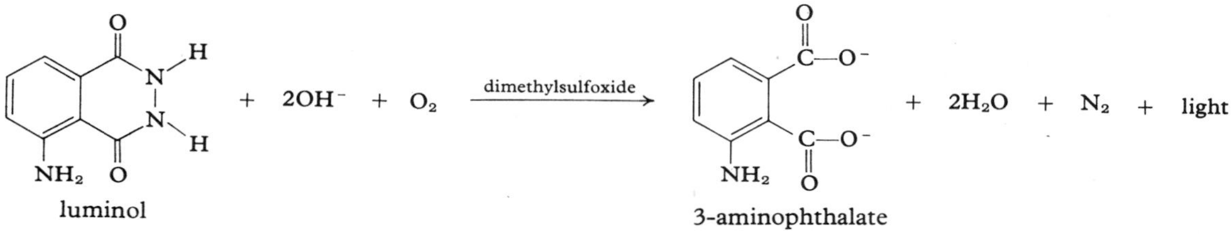

luminol

$+$ $2OH^-$ $+$ O_2 $\xrightarrow{\text{dimethylsulfoxide}}$

3-aminophthalate

$+$ $2H_2O$ $+$ N_2 $+$ light

Luminol may be prepared from 3-nitrophthalic anhydride by treatment with hydrazine followed by reduction of the first product by sodium dithionite.

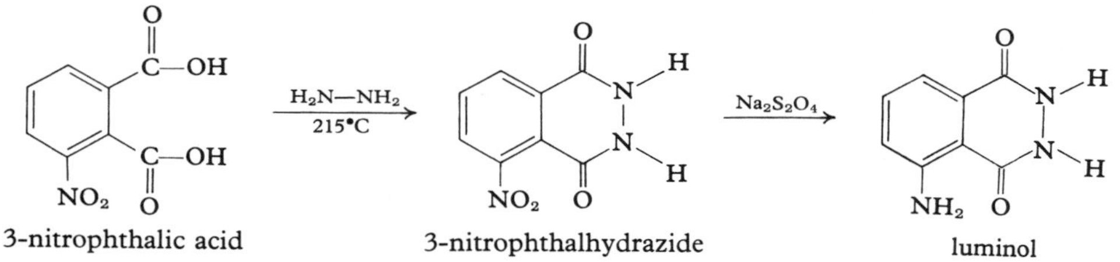

3-nitrophthalic acid

$\xrightarrow[215°C]{H_2N-NH_2}$

3-nitrophthalhydrazide

$\xrightarrow{Na_2S_2O_4}$

luminol

Preparation of Luminol from 3-Nitrophthalic Acid via 3-Nitrophthalhydrazide

Procedure.   First put a flask containing 15 ml of water on the steam bath to get hot. Then add to a 20 mm × 150 mm test tube 1.0 gram (4.7 mmole) of 3-nitrophthalic acid and 2 ml of an 8% aqueous solution of hydrazine (5 mmole; Note 1). Heat the mixture over a small burner flame until all the solid has gone into solution. Now add 3 ml of triethylene glycol, clamp the tube in a vertical position over the burner, and insert both a thermometer and a tube connected to the water aspirator. Boil the solution vigorously to distill the excess water, removing the vapors by means of the aspirator. Let the temperature rise rapidly until (in 3–4 minutes) it reaches 215°C. Remove the burner, notice the time, and by intermittent gentle heating maintain the temperature at 215–220°C for 2 minutes. At this time remove the burner, allow the temperature of the mixture to fall to 100°C (crystals of product may appear), add the 15 ml of hot water (Note 2), and collect the precipitated 3-nitrophthalhydrazide by suction filtration.

Transfer the damp nitro compound to the test tube in which it was prepared; do not clean the tube before making the addition. Add 5 ml of 10% sodium hydroxide solution (14 mmole), stir to dissolve, and to the resulting deep brown-red solution add 3.3 grams of "90% practical" sodium dithionite (17 mmole). Wash the solid down from the walls of the test tube with a little water. Heat the mixture to the boiling point, stir it, and keep it hot for 5 minutes; during this time some of the reduction product may separate from the yellow-brown solution. Now add 2 ml of acetic acid (35 mmole), cool the mixture thoroughly by swirling the test tube in a beaker of cold water, and collect the resulting precipitate of light yellow luminol by suction filtration. The damp crude product is suitable for demonstration of chemiluminescence.

Note

1. The 8% hydrazine solution may be prepared either by diluting 30.0 ml (31.2 grams) of 64% hydrazine to 250 ml or by diluting 22.3 ml (23.5 grams) of 85% hydrazine to 250 ml.

2. The reason for using hot water rather than cold is that the solid is then produced in a form that is more easily filtered.

Time: 2 hours.

Demonstration of the Chemiluminescence of Luminol

Procedure. Put 100 ml of dimethylsulfoxide in a 250-ml bottle or flask, add 1 ml of 50% aqueous sodium hydroxide, and swirl to mix. Then add 10–20 mg of luminol, stopper the container, and shake the contents vigorously. Within 60 seconds the contents of the bottle or flask will emit a blue-green light, which will be sufficiently bright to make it possible to easily read this book in the dark. When the light fades it can be restored by shaking the bottle or flask again. When all the luminol has reacted, further portions can be added, even after several weeks.

References

1. L. F. Fieser, *Organic Experiments*, 2nd edition, Heath, Boston, 1968, p. 239.
2. E. H. White and D. F. Roswell, *Accounts Chem. Res.* **3**, 54 (1970); and E. H. White, *J. Chem. Educ.* **34**, 275 (1957).
3. H. W. Schneider, *J. Chem. Educ.* **47**, 519 (1970); and M. T. Beck and F. Joo, *J. Chem. Educ.* **48**, A559 (1971).
4. W. D. McElroy and M. DeLuca, in *Chemiluminescence and Bioluminescence*, Cormier, Hercules, and Lee, editors, Plenum Press, New York, 1973, p. 285.

§71 THIAMINE-CATALYZED FORMATION OF BENZOIN FROM BENZALDEHYDE

$$2 \; \phi\text{—C(=O)—H} \longrightarrow \phi\text{—C(H)(OH)—C(=O)—}\phi$$

benzaldehyde   benzoin

$$R\text{—}\overset{\text{--}}{\underset{..}{C}}\text{=}\overset{..}{\underset{..}{O}}$$

acyl carbanion

Thiamine pyrophosphate (TPP) is a necessary coenzyme for a number of enzyme-catalyzed reactions, all of which involve, in effect, the transfer of an acyl group, R—C=O, as a carbanion. Organic chemists try to avoid proposing acyl

thiamine pyrophosphate (TPP)

carbanions as free intermediates since they lack the structural features (electron accepting groups) that are thought to stabilize such ions. The way in which acyl groups are transferred in biochemical reactions at a pH of 7 and at 37°C, then, remained a mystery until Professor Breslow at Columbia University proposed a role for thiamine (1). Breslow remembered that a famous old reaction of organic chemistry, the cyanide-catalyzed condensation of benzaldehyde to form benzoin (Section 74.1) appeared to involve the equivalent of an acyl carbanion. Cyanide

$$\text{"}\phi\text{—}\overset{O}{\overset{||}{C}}\text{:}^-\text{"} \; + \; H\text{—}\overset{O}{\overset{||}{C}}\text{—}\phi \longrightarrow \phi\text{—}\overset{O}{\overset{||}{C}}\text{—}\overset{O^-}{\underset{H}{C}}\text{—}\phi \xrightarrow{+H^+} \phi\text{—}\overset{O}{\overset{||}{C}}\text{—}\overset{OH}{\underset{H}{C}}\text{—}\phi$$

acyl carbanion   benzoin

ion had long been known as a specific catalyst for this reaction and cyanide was thought to act by adding to the carbonyl group of benzaldehyde to form the cyanohydrin, which could then lose the aldehyde hydrogen to form a resonance-stabilized anion (2).

$$\phi\text{—}\overset{OH}{\underset{H}{C}}\text{—C≡N:} \xrightarrow{-H^+} \left\{ \phi\text{—}\overset{OH}{C}\text{—C≡N:} \longleftrightarrow \phi\text{—}\overset{OH}{C}\text{=C=}\overset{..}{N}\text{:}^- \right\}$$

resonance stabilized anion

This anion could add to the carbonyl group of a second molecule of benzaldehyde to form the cyanohydrin of benzoin, which would then lose HCN to give benzoin and regenerate the catalyst.

$$\text{anion} + \phi\text{—}\overset{O}{\overset{||}{C}}\text{—H} \xrightarrow{+H^+} \phi\text{—}\overset{HO}{\underset{:N≡C}{C}}\text{—}\overset{OH}{\underset{H}{C}}\text{—}\phi \longrightarrow \phi\text{—}\overset{O}{\overset{||}{C}}\text{—}\overset{OH}{\underset{H}{C}}\text{—}\phi + HCN$$

benzoin

Breslow also knew that a number of thiazolium salts, including thiamine, had been reported to catalyze the conversion of benzaldehyde to benzoin, and not to give the product that was originally expected. In addition, he had determined that the C-2 proton of the thiazolium ring of thiamine was rapidly exchanged for deuterium in D_2O at room temperature. For these and other reasons he suggested that thiamine could catalyze the benzoin condensation in the following way.

Thiamine loses a proton to the solvent (or to the enzyme):

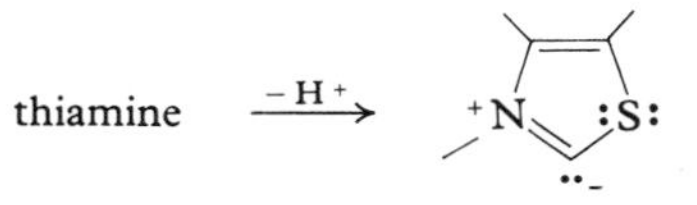

$$\text{thiamine} \xrightarrow{-H^+} \text{conjugate base of thiamine}$$

The conjugate base of thiamine adds to benzaldehyde:

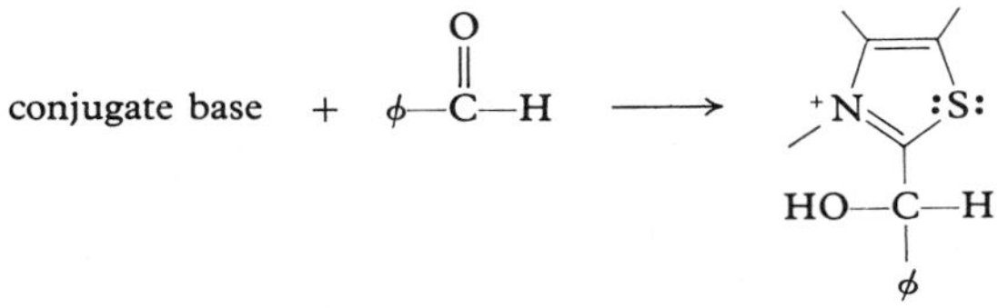

$$\text{conjugate base} \;+\; \phi\text{—C—H (O)} \longrightarrow \text{thiamine-benzaldehyde}$$

Thiamine-benzaldehyde loses a proton to give a resonance-stabilized equivalent of an acyl carbanion:

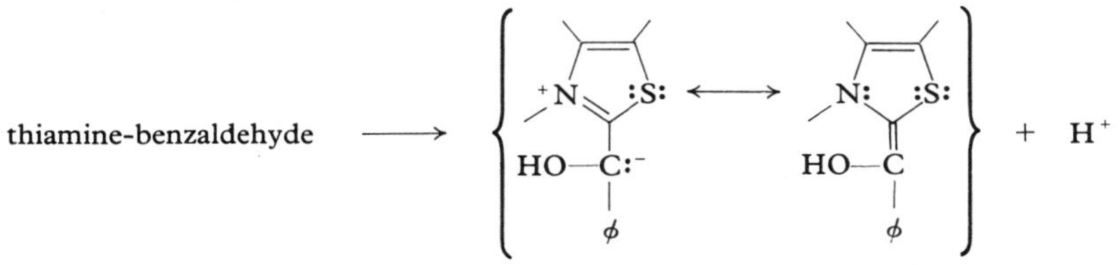

$$\text{thiamine-benzaldehyde} \longrightarrow \left\{ \cdots \right\} + H^+$$

benzaldehyde anion equivalent

The equivalent of the anion of benzaldehyde then adds to a second molecule of benzaldehyde to give a product that eliminates thiamine to give benzoin and to regenerate the catalyst:

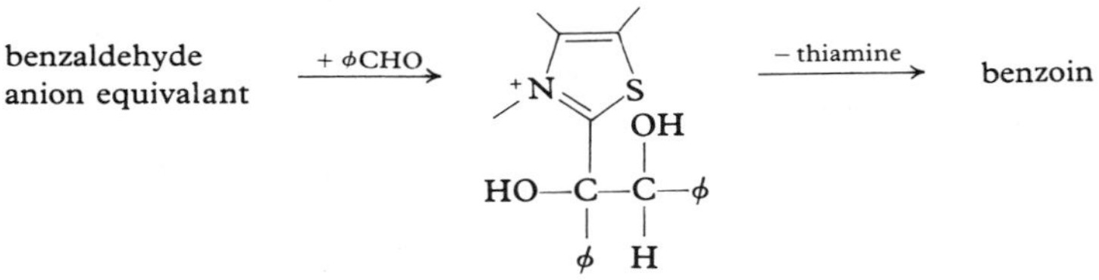

$$\text{benzaldehyde anion equivalant} \xrightarrow{+\,\phi CHO} \cdots \xrightarrow{-\text{thiamine}} \text{benzoin}$$

Reasoning from this model, Breslow proposed a similar role for thiamine pyrophosphate in the enzyme-catalyzed systems: resonance stabilization of the carbanion equivalent of an aldehyde.

That is, "$R\text{—}\overset{O}{\underset{}{\overset{\|}{C}}}\text{:}^-$" is

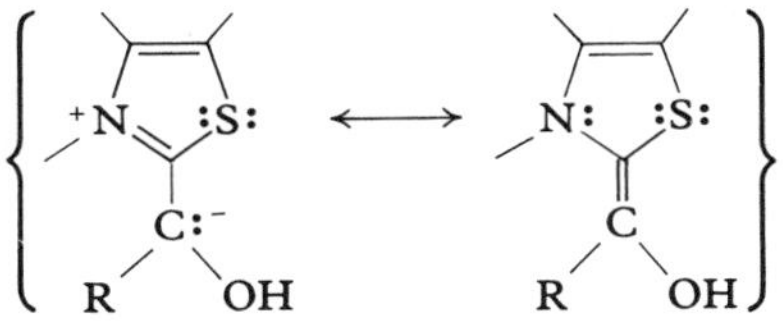

$$\left\{ \cdots \longleftrightarrow \cdots \right\}$$

Breslow was very careful not to draw conclusions that were not supported by the evidence. However, his proposal has now come to be accepted as representing the mechanism of thiamine action in enzymatic reactions involving thiamine pyrophosphate as a coenzyme. The role of the enzyme itself still remains to be determined.

The following procedure illustrates the use of thiamine as a catalyst in the benzoin condensation.

Procedure.   Place 3.5 grams (0.01 mole) of thiamine hydrochloride in a 125-ml Erlenmeyer flask. Add 10 ml of water and swirl the mixture until the thiamine hydrochloride has all dissolved. Then add 25 ml of 95% ethanol, 10 ml of 2 M NaOH (0.02 mole), and 10 ml (10.4 grams; 0.1 mole) of benzaldehyde (Note 1) swirling the flask during and after each addition in order to thoroughly mix the contents of the flask. Allow the resulting mixture to stand at room temperature for at least 6 hours (Note 2). Collect the resulting crystals of benzoin by suction filtration, washing them with about 20 ml of a cold mixture of ethanol and water (5:1). Yield: about 7 grams (about 70%).

Notes

1. We have always used a fresh bottle of benzaldehyde.
2. The mixture may be allowed to stand for 24 hours or for a week. It is helpful to add a seed crystal. By avoiding supersaturation crystallization starts sooner and the resulting crystals are larger and are more easily washed on the filter paper after collection by suction filtration.

Time: 2 hours plus time required for reaction to occur.

Questions

1. Draw the structure of "thiamine hydrochloride" (thiamine chloride hydrochloride; Vitamin B_1).
2. The free acyl carbonium ion, $R—\overset{+}{C}=\overset{..}{O}:$, is sometimes invoked as an intermediate in the aluminum chloride catalyzed acylation of aromatic compounds. Does it have any resonance stabilization? Explain.
3. Thiamine pyrophosphate is a required coenzyme in the enzyme-catalyzed decarboxylation of pyruvic acid to form acetaldehyde. Show in detail how thiamine could be involved in this reaction.

$$\underset{\text{pyruvic acid}}{CH_3—\overset{\overset{O}{\|}}{C}—\overset{\overset{O}{\|}}{C}—OH} \longrightarrow \underset{\text{acetaldehyde}}{CH_3—\overset{\overset{O}{\|}}{C}—H} + CO_2$$

4. Thiamine catalyzes both the self-condensation of acetaldehyde to acetoin and the reaction of pyruvic acid with acetaldehyde to give acetoin. Show in detail how this might occur.

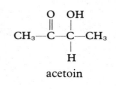

$$CH_3—\overset{\overset{O}{\|}}{C}—\overset{\overset{OH}{|}}{\underset{\underset{H}{|}}{C}}—CH_3$$
acetoin

References

1. R. Breslow, *J. Am. Chem. Soc.* **80**, 3719 (1958).
2. A. Lapworth, *J. Chem. Soc.* **83**, 995 (1903).
3. T. S. Bruice and S. Benkovic, *Bioorganic Mechanisms*, Volume 2, W. A. Benjamin, Inc., New York, 1966, Chapter 8.

---

§72 A MODEL FOR THE BIOCHEMICAL REDUCING AGENT NADH

Many chemical oxidation-reduction reactions are rather brutal (see Sections 28.27, 53, 73.3, and 74.2). In contrast, biochemical redox reactions take place rapidly at about room temperature and at a pH of about 7. Chemists continue to be interested in trying to understand and reconstruct these gentle biochemical reactions. One approach is to study the reactions of related but less complex compounds, to study model reactions. In this experiment* we will work with a model that has been used to try to understand the mechanism of the biochemical reducing agent NADH (Nicotinamide Adenine Dinucleotide, reduced form). The reducing agent, NADH, is represented by formula (1)

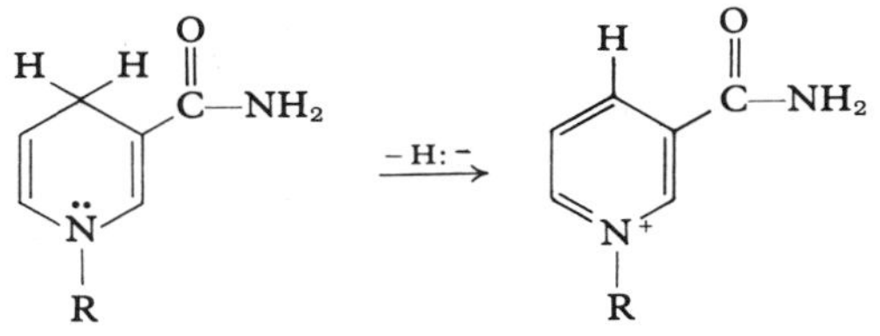

where R is

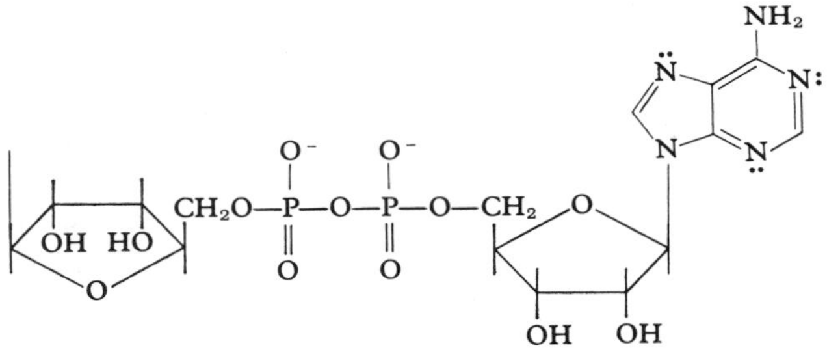

Formula (2) represents the oxidized form, NAD^+.

In our simple model, R will be benzyl: $\phi—CH_2—$. This model is one of several that were studied some years ago by Professor Westheimer at Harvard (1). In that research it was shown by deuterium labeling experiments that in the reduction of malachite green by 1-benzyldihydronicotinamide the hydrogen lost by the reducing agent is transferred directly to the hydrogen acceptor:

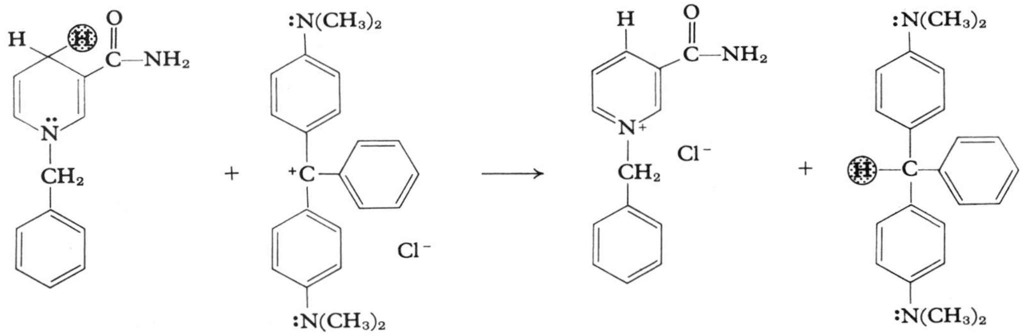

| yellow 1-benzyldihydronicotinamide | green malachite green | colorless benzylnicotinamide chloride | colorless leuco base |

This experiment involves the two-step preparation of 1-benzyldihydronicotinamide and a demonstration of its ability to reduce malachite green under mild conditions.

---

* This experiment was suggested by Dr. Bernard Golding, University of Warwick, Coventry, England.

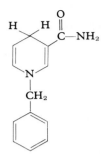

Figure 72.1. The infrared spectrum of 1-benzyldihydronicotin-amide; CHCl₃ solution.

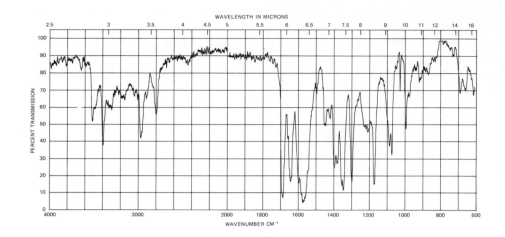

1-Benzylnicotinamide Chloride from Nicotinamide

nicotinamide   benzyl chloride

$$H_3C\!-\!\overset{\displaystyle O}{\underset{\displaystyle \|}{S}}\!-\!CH_3$$
dimethylsulfoxide
(DMSO)

Procedure. Clamp a 100 ml round bottom boiling flask into position over a wire gauze on an iron ring so that the flask can be heated with a burner flame. Add to the flask 5 ml of dimethylsulfoxide and 2 g (16.4 mmole) of nicotinamide. Heat the flask briefly with the burner and swirl the contents so as to dissolve the nicotinamide. When the solid has all dissolved, turn off the burner and add to the flask 10 ml (11 g; 88 mmole) of benzyl chloride (Note 1). Fit the flask with a Claisen adapter and by means of a thermometer adapter in the center neck of the Claisen adapter position a thermometer so that the bulb dips into the solution in the flask. Fit the side arm of the Claisen adapter with a reflux condenser. Using the burner again, heat the flask until crystals begin to separate from the solution (Note 2). At this time discontinue heating and allow the exothermic reaction to occur spontaneously (Note 3). After the reaction has taken place and the mixture has cooled to below 60° (Note 4) remove the condenser, adapters, and thermometer and add 10 ml of isopropyl alcohol. Stopper the flask and mix the contents of the flask by shaking it vigorously. Collect the product by suction filtration and wash it first with a little isopropyl alcohol and then with some hexane.

The crude product may be used to prepare 1-benzyldihydronicotinamide. It can also be recrystallized by heating it with 10 ml of 95% ethanol and adding just enough water (about 1 ml) to cause all the solid to dissolve.

Notes

1. Benzyl chloride is a lachrymator. It would be best to work in the hood while adding the benzyl chloride.
2. Crystals of the product start to appear when the temperature reaches about 135°C.

3. The temperature will rise to about 150–155°C before it starts to fall again.
4. When the temperature has fallen somewhat below 100°C the flask may be placed in a beaker of cold water in order to cool it faster.

Time: about 1 hour.

1-Benzyldihydronicotinamide from 1-Benzylnicotinamide Chloride

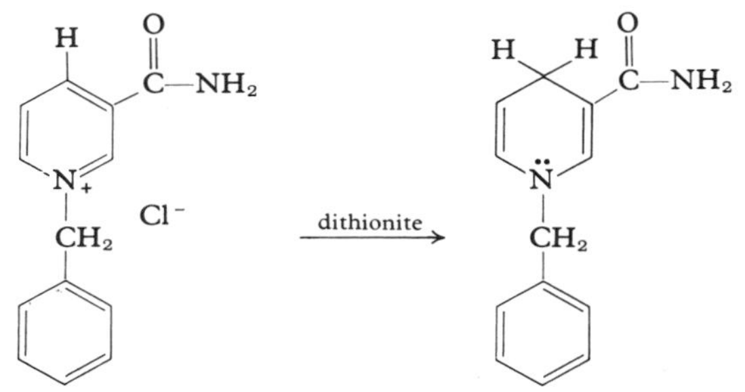

Procedure. Prepare a solution of 2.76 g of anhydrous sodium carbonate and 5.14 g (about 0.03 mole) of sodium dithionite in 40 ml of water in a 125-ml Erlenmeyer flask. Place a magnetic stirring bar in the flask and arrange for the solution to be stirred by a magnetic stirrer. Dissolve 2.0 g (8.0 mmoles) of 1-benzylnicotinamide chloride in 10 ml of water and add this solution all at once to the well-stirred solution of sodium carbonate and sodium dithionite. The resulting yellow-orange solution will suddenly produce a yellow precipitate in about 60 seconds. Continue to stir the mixture for another ten minutes and then collect the yellow solid by suction filtration using several small portions of water for rinsing and washing.

This product may be used directly for the reduction of malachite green, or it can be recrystallized to give beautiful yellow spars. To recrystallize, dissolve the crude product in 5 ml of hot 95% ethanol and filter the solution by gravity to remove a small amount of an insoluble impurity. Dilute the filtrate with 5 ml of warm water and, after swirling to mix, allow the solution to stand undisturbed for several hours (Note 1). Collect the crystals by suction filtration and wash them with a small amount of ice-cold 50% aqueous ethanol.

Note

1. It is helpful to seed the solution.

Time: less than 1 hour.

Reduction of Malachite Green by 1-Benzyldihydronicotinamide

Procedure. Dissolve about 10 milligrams (0.03 mmole) of malachite green in 1 ml of 95% ethanol in a small test tube. To this add 20–40 milligrams of 1-benzyldihydronicotinamide (0.1 to 0.2 mmoles). Swirl the mixture gently to dissolve the crystals. The intense green-blue color of the malachite green will give way to the yellow color of the excess reducing agent in 2 to 4 minutes.

Time: about 1/2 hour.

TRANSFORMATIONS

1. Malachite green can be thought of as a carbonium ion salt. What accounts for its great stability?

2. Ethyl benzoylformate is not reduced by 1-benzyldihydronicotinamide alone at room temperature in the dark in acetonitrile solution. In the presence of an equimolar amount of magnesium perchlorate, however, racemic ethyl mandelate is formed in quantitative yield (2). Explain.

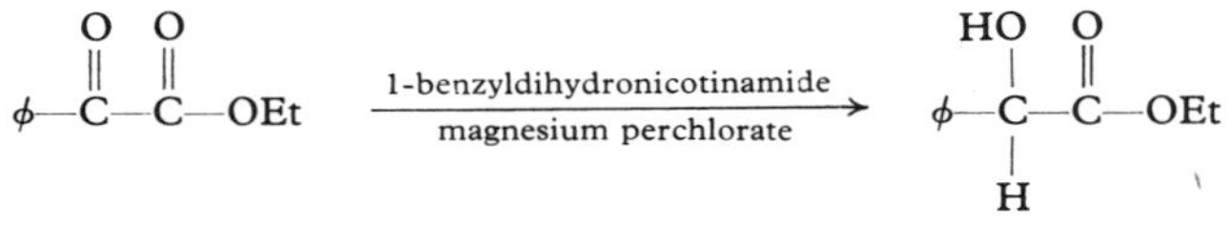

$$\phi-\overset{O}{\underset{\|}{C}}-\overset{O}{\underset{\|}{C}}-OEt \xrightarrow[\text{magnesium perchlorate}]{\text{1-benzyldihydronicotinamide}} \phi-\overset{HO}{\underset{\underset{H}{|}}{C}}-\overset{O}{\underset{\|}{C}}-OEt$$

ethyl benzoylformate   ethyl mandelate

3. When ethyl benzoylformate is treated with (R)-(−)-N-α-methylbenzyl-1-benzyl-dihydronicotinamide in the presence of an equimolar amount of magnesium perchlorate in acetonitrile solution the ethyl mandelate produced is 60% R and 40% S.

   a. Draw formulas indicating the structure and configuration of (R)-(−)-N-α-methylbenzyl-1-benzyldihydronicotinamide and of the two enantiomers of ethyl mandelate.

   b. Explain the results of the experiment.

References

1. D. Mauzerall and F. H. Westheimer, *J. Am. Chem. Soc.* **77**, 2261 (1954).
2. Y. Ohnishi, M. Kagami, and A. Ohno, *J. Am. Chem. Soc.* **97**, 4766 (1975).

Synthetic Sequences: Synthesis Experiments Involving a Sequence of Reactions

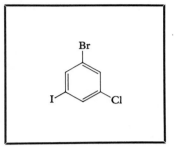

High yield is most important when each reaction uses as starting material the product of a previous reaction. For example, if three successive reactions can be carried out at 90 percent yield each, the overall yield at the end of the third reaction is $(0.9)^3 \times 100 = 73\%$. If each reaction is carried out at half that yield, the overall yield at the end of the third reaction is $(0.45)^3 \times 100 = 9\%$, or one-eighth as much. For this reason, synthetic sequences are a good test of one's laboratory skill.

Four synthetic sequences are outlined in this section. Many more can be made up from reactions in the preceding sections.

---

§73 STEROID TRANSFORMATIONS: Δ⁴-CHOLESTENE-3-ONE FROM CHOLESTEROL VIA CHOLESTEROL DIBROMIDE, 5α,6β-DIBROMOCHOLESTANE-3-ONE, AND Δ⁵-CHOLESTENE-3-ONE

As an initial step in exploring the chemistry of a rare and expensive substance, preliminary experiments may be done using a less expensive substance of similar structure as a *model compound*. Cholesterol, which can be isolated in quantity from beef spinal cord and brains (and from human gallstones; see Section 39), has often served as a model compound for exploring the chemistry of a class of polycyclic aliphatic compounds called *steroids*. Section 81 includes two projects which involve the transformation of a β,γ-unsaturated alcohol into the corresponding α,β-unsaturated ketone: androstenolone is converted into androstenedione (Section 81.1), and progesterone is prepared from pregnenolone (Section 81.2). In the sequence of reactions described in this section, exactly the same transformations are carried out, except that cholesterol is used as an inexpensive model compound. The reactions, which follow, consist of addition of bromine to the double bond (*anti* mechanism of addition) to protect it from the oxidizing agent, chromate oxidation of the secondary alcohol to the corresponding ketone, reductive elimination by zinc of bromine from the vicinal dibromide to regenerate the carbon–carbon double bond, and finally acid-catalyzed isomerization of the double bond from the unconjugated β,γ-position to the more stable conjugated α,β-position.

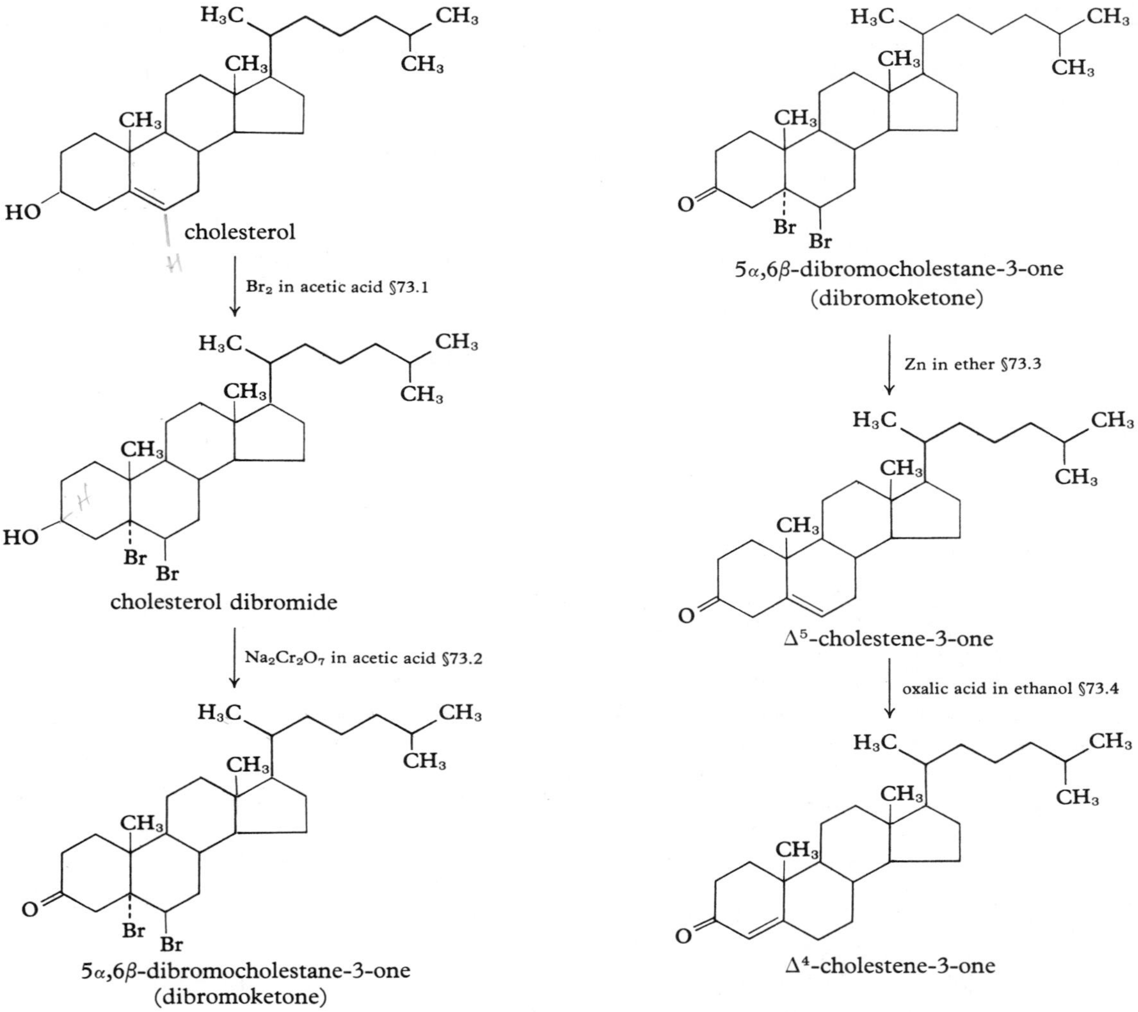

cholesterol

Br$_2$ in acetic acid §73.1

cholesterol dibromide

Na$_2$Cr$_2$O$_7$ in acetic acid §73.2

5α,6β-dibromocholestane-3-one
(dibromoketone)

5α,6β-dibromocholestane-3-one
(dibromoketone)

Zn in ether §73.3

Δ^5-cholestene-3-one

oxalic acid in ethanol §73.4

Δ^4-cholestene-3-one

There are no good stopping points (■) in this experiment until the Δ^5-cholestene-3-one solution has been set aside for crystallization (Section 73.3). If the bromine solution (Section 73.1) and the sodium dichromate solution (Section 73.2) are prepared ahead of time, it will take an inexperienced person about four hours to reach this point.

73.1 CHOLESTEROL DIBROMIDE FROM CHOLESTEROL

Procedure. Place 3.0 grams (7.8 mmole) of cholesterol in a 50-ml Erlenmeyer flask and add 20 ml of anhydrous ether. Dissolve the cholesterol by swirling the flask in a beaker of warm water until the ether just starts to boil. Cool the solution to about room temperature by swirling the flask in a beaker of cool water. Add 12 ml of a solution of bromine and sodium acetate in glacial acetic acid (Note 1; 8.5 mmole of bromine) and swirl to mix. The contents of the flask should promptly solidify to a stiff paste of cholesterol dibromide. Now place the flask in a beaker of cold water to cool. In the meantime, prepare a mixture of 9 ml of anhydrous ether and 21 ml of glacial acetic acid. Cool this mixture in an ice bath while you collect the cholesterol dibromide by suction filtration. Use the cooled mixture of ether and acetic acid to rinse the product from the Erlenmeyer flask into the funnel and to wash the product in the funnel free of the yellow mother liquor. Continue to apply suction to the product until the wash liquid has almost stopped dripping from the stem of the funnel. The damp dibromide is suitable for use in the next reaction.

Note

1. The solution may be prepared by dissolving 1.0 gram of anhydrous sodium acetate (0.012 mole) in 120 ml of glacial acetic acid and then adding 4.10 ml (13.6 grams; 0.085 mole) of bromine. This is enough reagent for ten 3-gram runs.

    Time: about 1 hour.

73.2 5α,6β-DIBROMOCHOLESTANE-3-ONE FROM CHOLESTEROL DIBROMIDE

Procedure. Add the moist dibromide obtained from the procedure of Section 73.1 to 40 ml of glacial acetic acid in a 125-ml Erlenmeyer flask. To this, add, all at once, 40 ml of a solution of sodium dichromate in acetic acid that has been preheated to 105°C. (Note 1; sufficient dichromate to oxidize 16 mmoles of secondary alcohol to the corresponding ketone). After the mixture has been swirled briefly, the temperature should rise to between 55 and 60°C. The temperature of the mixture must be maintained between 55 and 58°C for as long as it takes the solids to dissolve (from 3–5 minutes) and then for 2 minutes more (Note 2). After allowing the solution to stand at room temperature for about 20 minutes, add 8 ml of cold water, swirl to mix, and then cool the suspension of dibromoketone to about 15°C by means of a cold-water or ice bath. Collect the product by suction filtration, using 10–12 ml of cold methanol for rinsing and washing. Transfer the crude product to a 100-ml beaker containing 25 ml of ice-cold methanol. Stir the mixture so as to thoroughly wash the crystals and then collect the product again by suction filtration. The damp dibromoketone is suitable for use in the next reaction.

Notes

1. The solution may be prepared by dissolving 16 grams of sodium dichromate dihydrate (0.054 mole) in 400 ml of glacial acetic acid. This is enough reagent for ten runs of the scale described.
2. If the temperature fails to reach 55°C or shows signs of falling below 55°C, the mixture should be heated briefly in order to reach or maintain the specified temperature.

    Time: 1 hour.

73.3 Δ⁵-CHOLESTENE-3-ONE FROM 5α,6β-DIBROMOCHOLESTANE-3-ONE

If the reaction described in the procedure that follows is successful, the product will have an infrared spectrum like that of Figure 73.1. If the reaction does not go as expected, the dibromoketone will be recovered unchanged; see Note 4.

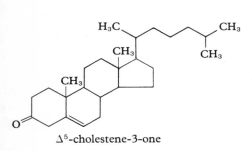

Δ⁵-cholestene-3-one

Figure 73.1. Infrared spectrum of Δ⁵-cholestene-3-one; CCl₄ solution.

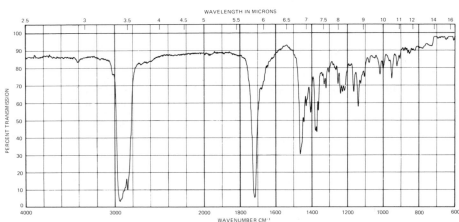

Procedure. Place the damp product obtained from the preceding oxidation in a 125-ml Erlenmeyer flask and add to the flask 40 ml of anhydrous ether and 0.5 ml of glacial acetic acid. Cool the contents of the flask to between 15 and 20°C by swirling it in a beaker of cold water. Then add in several portions over about 3–4 minutes 0.8 grams (12 mmole) of zinc dust. During this time the flask must be swirled vigorously to suspend the zinc; the temperature of the contents of the flask should show a definite tendency to rise but should be held below 20°C by very brief swirling of the flask in the beaker of cold water. Most but not all of the zinc will dissolve. After the zinc has been added and the mildly exothermic reaction is over, allow the mixture to stand at room temperature for 10 minutes. Then add 1.4 ml (17 mmole) of pyridine (to precipitate ionic zinc by complex formation), swirl well to mix, and remove the white precipitate by suction filtration (Note 1). Wash the filter cake well with several small portions of ether, collecting these washings along with the original filtrate. Transfer the filtrate to a separatory funnel and wash it with three 15-ml portions of water and one 15-ml portion of 5% sodium bicarbonate solution (Note 2). Dry the ethereal solution over anhydrous magnesium sulfate and then filter it by gravity into a 50-ml Erlenmeyer flask that has been marked to show levels at which it contains approximately 20 and 15 ml (Note 3). Add a boiling stone and evaporate or distill the ether until the volume has been reduced to 20 ml. The flask should be warmed either on the steam bath or in a beaker of warm water; no burners should be used because of the risk of setting ether vapors on fire. Add 10 ml of methanol to the flask and continue to concentrate the solution until the volume has been reduced to 15 ml. Stopper the flask and set it aside for crystallization. ■

After crystallization has proceeded at room temperature for at least one-half hour, cool the flask in an ice bath and then collect the unsaturated ketone by suction filtration (Note 4). A small amount of cold methanol may be used for rinsing and washing. Δ⁵-Cholestene-3-one is obtained as white granular crystals in a yield of about 1.5 grams (about 50% from cholesterol) with a melting point of 126–128°C. Reported literature values (Reference 1) include: m.p., 126–129°C; $[\alpha]_D$, −2.5° (2.03 grams/100 ml of chloroform).

Notes

SECTION 73

1. It is the *filtrate* you want; be sure your filter flask is clean.

2. The purpose of the sodium bicarbonate wash is to remove any trace of acid that could prematurely catalyze the isomerization of the double bond. Dip a piece of moist blue litmus paper into the ethereal solution to make sure it is not acidic.

3. This can be done by adding the appropriate volume of water to a second, similar, flask and then making a mark with a wax pencil on the flask you will use at about the same height as the level of liquid in the second flask.

4. If the debromination with zinc was unsuccessful, the material obtained at this point will be the unchanged $5\alpha,6\beta$-dibromocholestane-3-one. It will melt at about 70°C with decomposition to give a brown-orange liquid, and will turn pink to purple upon storage. If the debromination with zinc was unsuccessful *and* the solution was set aside for crystallization for a week, the product obtained may be 6β-bromocholest-4-ene-3-one. This substance is reported to have a melting point of 122–124°C, and can be prepared as follows. Add 1 gram of $5\alpha,6\beta$-dibromocholestane-3-one to 10 ml of pyridine at 5°C, stir the mixture at room temperature for 2 hours, pour the resulting solution into water, collect the precipitate by suction filtration, and recrystallize using 8 ml of hexane [see *Chemical Abstracts* **60**, 590d (1964)]. The infrared spectrum of 6β-bromocholest-4-ene-3-one is shown in Figure 73.2. This compound is recovered unchanged when treated according to the procedure of Section 73.4.

Time: 2 hours.

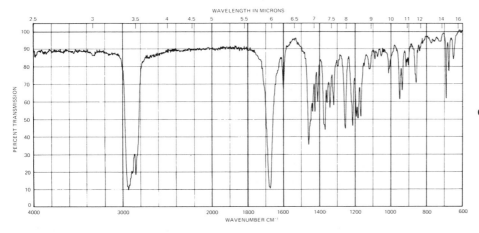

6β-bromocholest-4-ene-3-one

Figure 73.2. Infrared spectrum of 6β-bromocholest-4-ene-3-one; CCl_4 solution; see Note 4.

73.4 Δ⁴-CHOLESTENE-3-ONE FROM Δ⁵-CHOLESTENE-3-ONE

The infrared spectrum of Δ⁴-cholestene-3-one is shown in Figure 73.3.

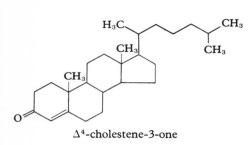

Δ^4-cholestene-3-one

Figure 73.3. Infrared spectrum of Δ^4-cholestene-3-one; CCl_4 solution.

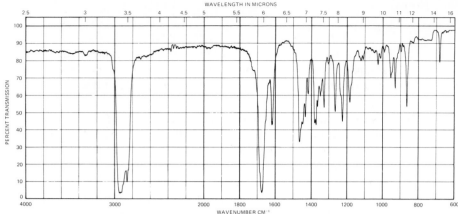

Procedure. Place 1.0 gram (2.6 mmole) of Δ^5-cholestene-3-one and 100 mg of anhydrous oxalic acid in a 25-ml Erlenmeyer flask. Add 8 ml of 95% ethanol, fit the flask with a reflux condenser, and boil the mixture gently until all the solids have gone into solution; about 15 minutes of boiling is required. Continue to heat the solution for 10 more minutes and then allow it to cool to room temperature. If necessary, induce crystallization by scratching the inner surface of the flask at the liquid level with a stirring rod or by adding seed crystals. After crystallization appears to be complete at room temperature, cool the flask in an ice bath for about 15 minutes and then collect the large colorless needles of Δ^4-cholestene-3-one by suction filtration. A small amount of cold methanol may be used for rinsing and washing. Yield: about 0.9 gram (about 90%) of crystals with a melting point of 79–81°C. Literature values (Reference 1) include: m.p., 81–82°C; $[\alpha]_D$, +92.0° (2.01 grams/100 ml of chloroform); λ_{max}, 242 nm; ϵ, 17,000 (ethanol).

Time: 2 hours.

Questions

1. How many stereoisomeric 5,6-dibromocholestane-3-ones can there be? How many of these stereoisomers could be formed by a *trans* mechanism of addition of bromine to cholesterol? Suggest an explanation for the fact that the 5α,6β-isomer is the only one formed. Which isomer would you expect to be the most stable?

2. Write a reasonable mechanism for the debromination of 5α,6β-dibromocholestane-3-one by zinc.

3. Comment on the mechanism and circumstances of formation of 6β-bromocholest-4-ene-3-one from 5α,6β-dibromocholestane-3-one as described in Note 4 of Section 73.3.

4. The acid-catalyzed isomerization of the β,γ-unsaturated ketone to the α,β-unsaturated isomer probably goes via the intermediate formation of the conjugated enol. Write out the details of the mechanism, being sure to indicate the possibilities for resonance stabilization of each intermediate cation.

References

1. L. F. Fieser, *J. Am. Chem. Soc.* **75**, 5421 (1953).
2. L. F. Fieser, *Organic Syntheses*, Coll. Vol. 4, Wiley, New York, 1963, p. 195.

Tetraphenylcyclopentadienone can be prepared from benzaldehyde and phenylacetic acid in four steps according to the following scheme.

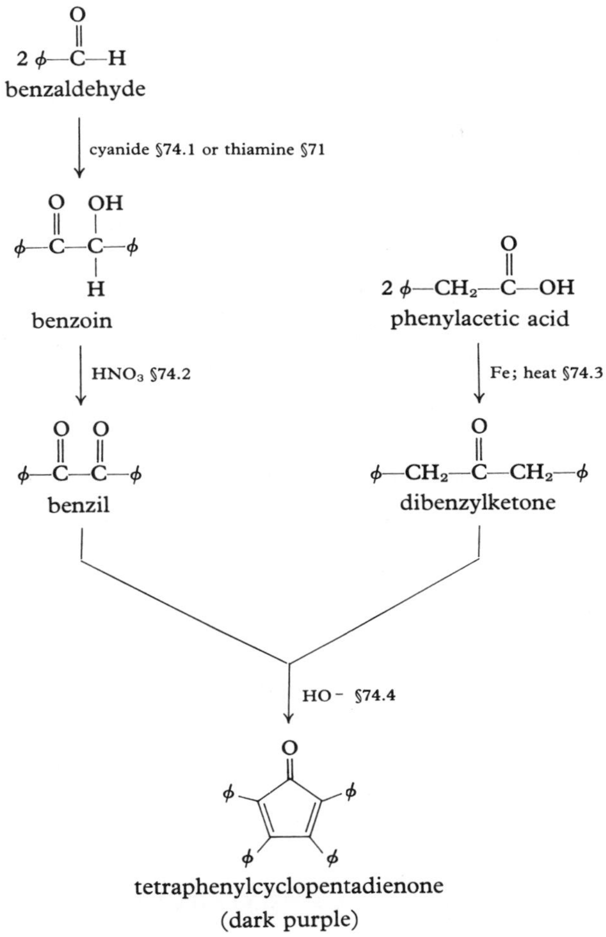

tetraphenylcyclopentadienone
(dark purple)

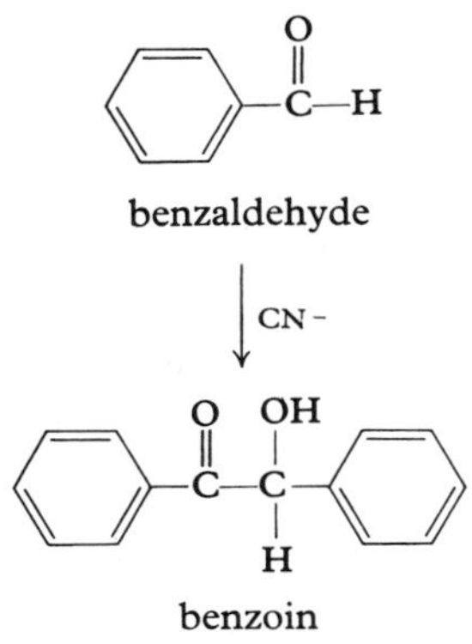

74.1 BENZOIN FROM BENZALDEHYDE

In this procedure, benzoin is prepared from benzaldehyde by use of cyanide as the catalyst. It can also be prepared by catalysis with thiamine (Section 71). The mechanisms of catalysis are outlined in Section 71.

Procedure. Dissolve 1.5 g (0.023 mole) of potassium cyanide (*poison*, see Section 1.7) in 15 ml of water in a 125-ml Erlenmeyer flask. Add 30 ml of 95% ethanol followed by 15 ml (15.6 g; 0.147 mole) of benzaldehyde (Note 1). Fit the flask with a condenser and boil the mixture on the steam bath for 30 minutes. Allow the mixture to cool to room temperature, swirling occasionally to speed crystallization. Cool the mixture in an ice bath, collect the benzoin by suction filtration, and wash it, first with two 15-ml portions of cold 50% ethanol and finally with water. Benzoin may be recrystallized from methanol using 12 ml per gram.

benzaldehyde

benzoin

Caution: cyanide

1. The benzaldehyde should be free of benzoic acid to avoid liberating hydrogen cyanide gas.

Time: less than 2 hours.

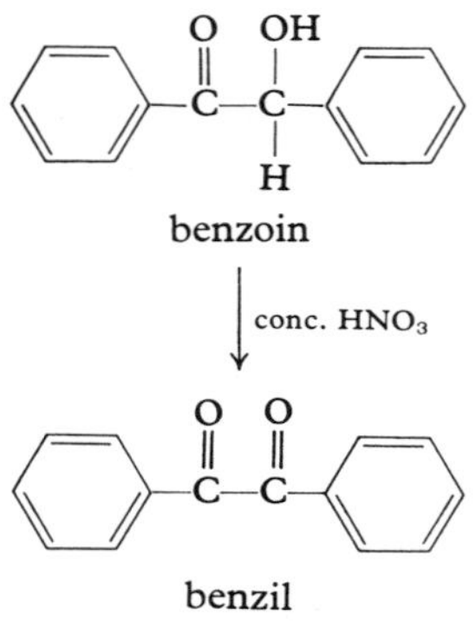

benzoin

conc. HNO₃

benzil

74.2 BENZIL FROM BENZOIN

Procedure. Place 4.2 g (0.02 mole) of benzoin in a 125 ml ground glass stoppered Erlenmeyer flask and then add 14 ml (20 g; 0.22 mole) of conc. nitric acid. Heat this mixture in the hood (Note 1) on a steam bath for 10 to 12 minutes. During the heating period, swirl the mixture occasionally (Note 2). At the end of the heating period add to the flask 75 ml of cold water and swirl the flask to mix the contents. Then add a seed crystal of benzil, stopper the flask, and shake it so as to cause the oily product to solidify in small lumps. Collect the bright yellow solidified benzil by suction filtration, washing it thoroughly with water to remove the nitric acid. Yield: about 4 g (about 95%). Benzil may be recrystallized from 95% ethanol, using 5–7 ml per gram (compare Exercise 5, Section 8.9).

Notes

1. Nitrous oxide fumes are evolved.
2. The solid will gradually dissolve and an oil (molten benzil) will form.

Time: 1 hour.

74.3 DIBENZYLKETONE FROM PHENYLACETIC ACID

$$Fe + 2\ \phi—CH_2—\overset{\overset{\displaystyle O}{\|}}{C}—OH \longrightarrow \phi—CH_2—\overset{\overset{\displaystyle O}{\|}}{C}—CH_2—\phi + H_2 + FeO + CO_2$$

This transformation takes place in two stages. The first is the reaction of the acid with elemental iron to form the ferrous salt of the acid:

$$2\ \phi—CH_2—\overset{\overset{\displaystyle O}{\|}}{C}—OH + Fe \longrightarrow (\phi—CH_2—\overset{\overset{\displaystyle O}{\|}}{C}—O^-)_2Fe^{2+} + H_2$$

The second stage is the loss of the elements of carbon dioxide and ferrous oxide by pyrolysis of the ferrous salt:

$$(\phi—CH_2—\overset{\overset{\displaystyle O}{\|}}{C}—O^-)_2Fe^{2+} \longrightarrow \phi—CH_2—\overset{\overset{\displaystyle O}{\|}}{C}—CH_2—\phi + CO_2 + FeO$$

The dibenzylketone produced in this experiment should have an infrared spectrum like that shown in Figure 74.1.

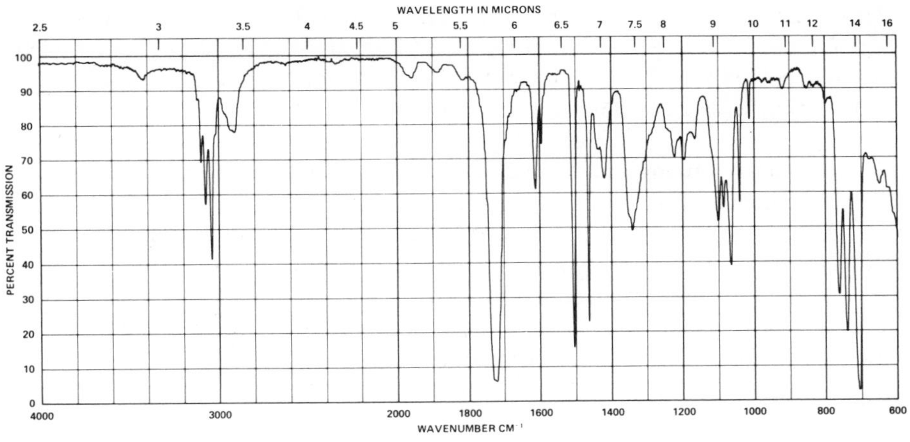

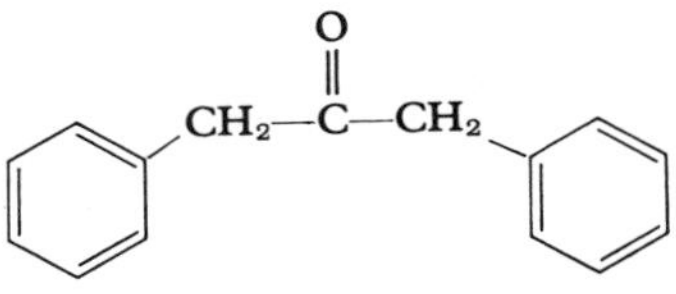

Figure 74.1. Infrared spectrum of dibenzylketone; thin film.

Procedure (Reference 1).   Place 13.6 g (0.10 mole) phenylacetic acid and 3.07 g (0.055 mole) iron powder in a 50-ml distilling flask (Note 1). Fit the flask with a side-arm test tube as a receiver and connect a length of rubber tubing to the side arm; lead the tubing to an area free of flames (Note 2). Arrange a thermometer in the top of the distilling flask so that the bulb almost touches the bottom of the flask. Heat the mixture of solids until it melts and gradually raise the temperature to 320°C (Note 3). When the temperature of the mixture reaches 320°, stop heating and raise the thermometer into position for a normal distillation. Then distill the contents of the flask, collecting the product at about 324°C (Note 4). Yield: about 60%.

Notes

1. Use an inexpensive flask as this reaction mixture etches the glass badly.
2. Hydrogen gas (flammable, explosive) is evolved.
3. During the heating period, which should be about 15 minutes, the contents of the flask will liquefy, gas will be evolved (H_2), and the contents of the flask will darken and solidify and then will remelt with further gas evolution (CO_2). Do not heat so quickly that anything distills over.
4. The product will be colored yellow to brown. Do not attempt to distill to dryness. To clean the distilling flask, use acetone and soapy water until they seem to have no more effect. Then cover the residue in the flask with conc. hydrochloric acid and allow it to stand in the hood.

   Time: 3 hours.

Reference

1. R. Davis and H. P. Schultz, *J. Org. Chem.* **27**, 854 (1962).

74.4 TETRAPHENYLCYCLOPENTADIENONE FROM BENZIL AND DIBENZYLKETONE

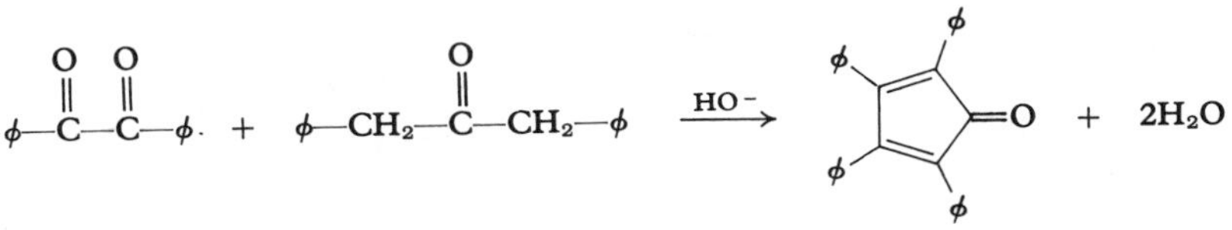

tetraphenylcyclopentadienone
(dark purple)

Procedure (Reference 1).   Dissolve 2.1 g (0.010 mole) of benzil (Section 74.2) and 2.1 g (0.010 mole) of dibenzylketone (Section 74.3) in 10 ml of triethylene glycol in a 25-mm × 150-mm test tube. While supporting the test tube in a clamp and stirring with a thermometer, heat the mixture with a small flame until the benzil is dissolved. Adjust the temperature to exactly 100°C, add 1 ml of 40% benzyltrimethylammonium hydroxide in methanol, and stir the mixture well with the thermometer.

When the temperature has dropped to 80°C, cool the mixture almost to room temperature, thin it with 10 ml of methanol, and collect the product by suction filtration. Wash it with methanol until the filtrate is purple-pink rather than brown. The very dark purple tetraphenylcyclopentadienone may be recrystallized from triethylene glycol, using 10 ml per gram.

Time: less than 2 hours.

Reference

1. L. F. Fieser, *Organic Experiments*, Heath, Boston, 1964, p. 303.

sulfanilamide

prontosil

§75 SULFANILAMIDE FROM BENZENE

Sulfanilamide was the first substance to be used systematically and effectively as a chemotherapeutic agent for the prevention and cure of bacterial infections in man. Although it was first prepared in 1908, more than 25 years had to pass before its therapeutic value was discovered. The discovery was, typically, indirect. In 1909, German dye chemists of the I. G. Farbenindustrie found that azo dyes containing sulfonamide groups were superior in colorfastness, especially toward the proteins of wool and silk. This led others to an investigation of the use of these dyes as selective staining agents for bacterial protoplasm. In the course of this work it was gradually recognized that certain dyes had the ability to kill some bacteria *in vitro* (in a bacterial culture outside a living host). This, reasonably enough, inspired the hope that a similar antibacterial action could occur *in vivo* (within a living host), but research along these lines was not pursued very vigorously. Then in the early 1930s, further work at the I. G. Farbenindustrie with a new sulfonamide-containing dye, Prontosil, showed that this new substance could protect mice with streptococcal and other bacterial infections, despite the fact that it had no *in vitro* antibacterial action. This discovery was recognized by the awarding of the Nobel prize in medicine for 1939 to Gerhard Domagk, one of the directors of research of the I. G. Farbenindustrie. In 1935, research workers at the Pasteur Institute in Paris who had been intrigued by the fact that Prontosil was effective *in vivo* but not *in vitro* reported a fascinating and far-reaching explanation: the reason was that, in a reaction that could take place only *in vivo*, Prontosil could be reduced by the host to sulfanilamide, which was the actual effective agent. Further research in England and the United States confirmed and extended this finding, with the result that within the next ten years over 5400 related substances had been prepared and tested for antibacterial activity. Fewer than 20 ever attained therapeutic importance.

As with all drugs, the next question was "How does it work?" How is it that sulfanilamide attacks bacteria specifically, with no apparent effect on the host? First, sulfanilamide does not usually kill the bacteria, but it greatly reduces their rate of growth and reproduction. On the molecular level, sulfanilamide competes with *p*-aminobenzoic acid in the bacterial synthesis of folic acid, a substance required for bacterial growth. Sulfanilamide rather than *p*-aminobenzoic acid

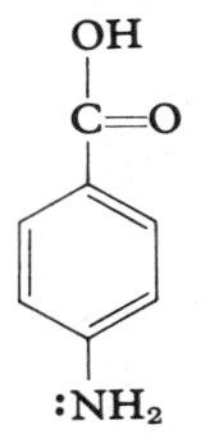

p-aminobenzoic acid

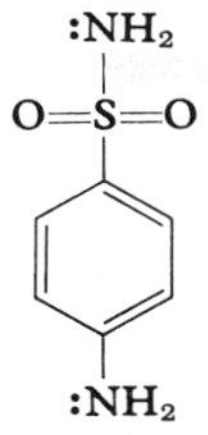

sulfanilamide

enters into the reaction that normally would produce folic acid, and the product, containing a $O=S=O$ group in place of the $C=O$ group of the amide function, is unable to carry out the function of folic acid. Since the bacteria are able under

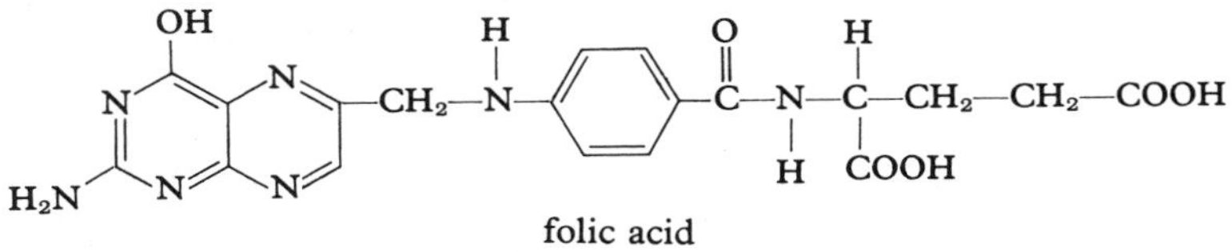

folic acid

normal conditions to synthesize their own folic acid, they have no mechanism for obtaining folic acid from their environment at the low concentrations at which it is present. Inhibition or diversion of folic acid synthesis, then, accounts for the bacteriostatic action of sulfanilamide. On the other hand, the host, a "higher organism," does not have the ability to synthesize folic acid but has instead the

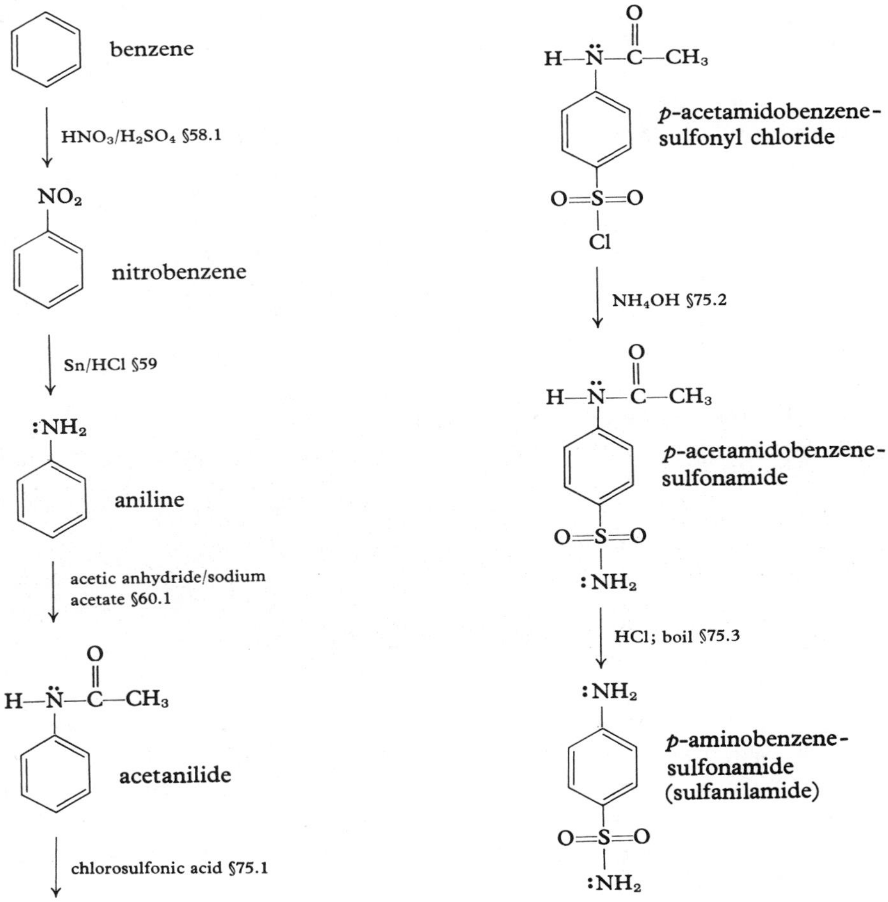

benzene

HNO₃/H₂SO₄ §58.1

nitrobenzene

Sn/HCl §59

aniline

acetic anhydride/sodium acetate §60.1

acetanilide

chlorosulfonic acid §75.1

p-acetamidobenzene-sulfonyl chloride

NH₄OH §75.2

p-acetamidobenzene-sulfonamide

HCl; boil §75.3

p-aminobenzene-sulfonamide (sulfanilamide)

capacity to make use of the folic acid present at low concentrations in its diet. Obviously, the presence or absence of sulfanilamide is irrelevant to this capacity to acquire folic acid from the environment, and this accounts for the lack of effect of sulfanilamide on the host organism.

Sulfanilamide can be prepared from benzene in six steps, as indicated by the scheme shown on p. 351. According to this approach, the aromatic amino group is first acetylated and then later deacetylated. There are at least two reasons for this. First, if one were to try to chlorosulfonate aniline itself, chlorosulfonic acid, an acid chloride, could be expected to react with the amino group of aniline. Then, even if it were possible to prepare *p*-aminobenzenesulfonyl chloride, the amino group of one molecule might be expected to react with the sulfonyl chloride group of another.

75.1 *p*-ACETAMIDOBENZENESULFONYL CHLORIDE FROM ACETANILIDE

Caution:
chlorosulfonic acid

p-acetamidobenzene-
sulfonyl chloride

Procedure. Place 6.75 g (0.050 mole) of completely dry (Note 1) acetanilide in a 125-ml Erlenmeyer flask. Gently heat the flask with a flame to melt the acetanilide, and swirl the flask containing the molten acetanilide as it cools, so that it will solidify in a thin layer on the bottom and lower walls of the flask. Prepare a trap to absorb hydrogen chloride and arrange for it to be connected to the Erlenmeyer flask by a length of rubber tubing (Section 35). Make certain that there is no possibility for water to be sucked back into the flask (*caution:* Note 1). Cool the flask in an ice bath. Then remove it and add all at once 17 ml (30 g; 0.26 mole) of chlorosulfonic acid (*caution:* Note 2) and connect the flask to the gas trap. Swirl the mixture until part of the solid has dissolved and hydrogen chloride is being rapidly evolved. Cool the flask in the ice bath if necessary to moderate the reaction. When all but a few lumps of acetanilide have dissolved (5–10 minutes), heat the mixture on the steam bath for 10 minutes and then cool the flask to room temperature. Slowly and cautiously (Note 1) pour the mixture onto 100 grams of ice in a beaker in the ice bath in the hood. Stir the precipitated product for a few minutes to obtain an even, granular suspension, and then collect it by suction filtration. The crude product should be used immediately.

Notes

1. Chlorosulfonic acid reacts violently with water.
2. Measure the chlorosulfonic acid in a *dry* graduate, as it will react violently with water.

Time: less than 2 hours.

Questions

1. Why is an excess of chlorosulfonic acid used?
2. Write out the reaction that takes place between chlorosulfonic acid and water.
3. What might be the product of reaction of chlorosulfonic acid and aniline?
4. What might be the product of reaction of *p*-aminobenzenesulfonyl chloride with another molecule of the same compound?
5. Why would we not expect a similar reaction between two molecules of *p*-acetamidobenzenesulfonyl chloride?

6. Although *p*-aminobenzenesulfonic acid is readily available, it can be acetylated only in the form of its sodium salt. Why should this be so? As a matter of fact, the sodium salt of *p*-aminobenzenesulfonic acid is not very soluble in acetic anhydride and the reaction proceeds so poorly that this is not a practical way to make *p*-acetamidobenzenesulfonic acid, which could then be converted to *p*-acetamidobenzenesulfonyl chloride.

75.2 *p*-ACETAMIDOBENZENESULFONAMIDE FROM *p*-ACETAMIDOBENZENESULFONYL CHLORIDE

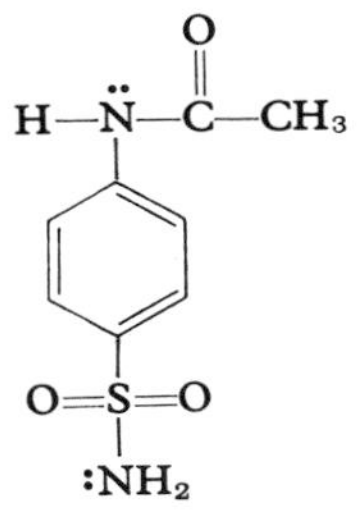

Procedure. Transfer the crude product of the procedure of Section 75.1 to a 125-ml Erlenmeyer flask, and add 20 ml (19 g; 0.32 mole) of conc. ammonium hydroxide and 20 ml of water. By means of a flame and in the hood, heat the mixture almost to boiling; in about 5 minutes, with occasional swirling, the granular sulfonyl chloride will be converted to the more pasty sulfonamide. Cool the mixture well in an ice bath, collect the product by suction filtration, and suck it as dry as possible. ■

Time: 1 hour.

p-acetamidobenzene-
sulfonamide

Question

1. Estimate the relative nucleophilicity of water, ammonia, and hydroxide ion toward the sulfonyl chloride group.

75.3 SULFANILAMIDE FROM *p*-ACETAMIDOBENZENESULFONAMIDE

Procedure. Transfer the moist product of the procedure of Section 75.2 to a 100-ml boiling flask, and add 7 ml (8.3 g; 0.09 mole) of conc. hydrochloric acid and 14 ml of water. Fit the flask with a condenser, and gently boil the mixture under reflux until all the solid has dissolved (about 10 minutes) and then for 10 minutes more. Cool the solution to room temperature (Note 1), add a little decolorizing carbon, and filter with suction. Transfer the filtrate to a beaker and add sodium bicarbonate as a saturated aqueous solution until the mixture is no longer acid to litmus (Note 2). Cool the suspension of precipitated sulfanilamide in the ice bath and then collect it by suction filtration. ■ Recrystallize the sulfanilamide from water, using 12 ml per gram (add decolorizing carbon, if necessary).

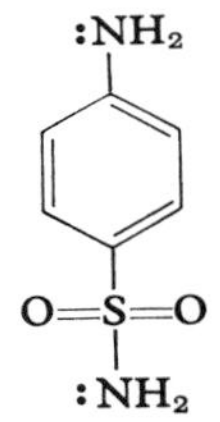

p-aminobenzene-
sulfonamide
(sulfanilamide)

Notes

1. If solid separates upon cooling, reheat and boil the mixture for a while longer.
2. About 7.5 g (0.09 mole) will be needed.

Time: less than 2 hours.

Question

1. Estimate the relative rates of hydrolysis of carboxylic acid amides and sulfonic acid amides.

§76 1-BROMO-3-CHLORO-5-IODOBENZENE FROM BENZENE

1-Bromo-3-chloro-5-iodobenzene may be prepared from benzene in eight steps according to the following scheme. Notice that procedures for the first four steps may be found in other sections of this book; only the last four steps are described in the present section.

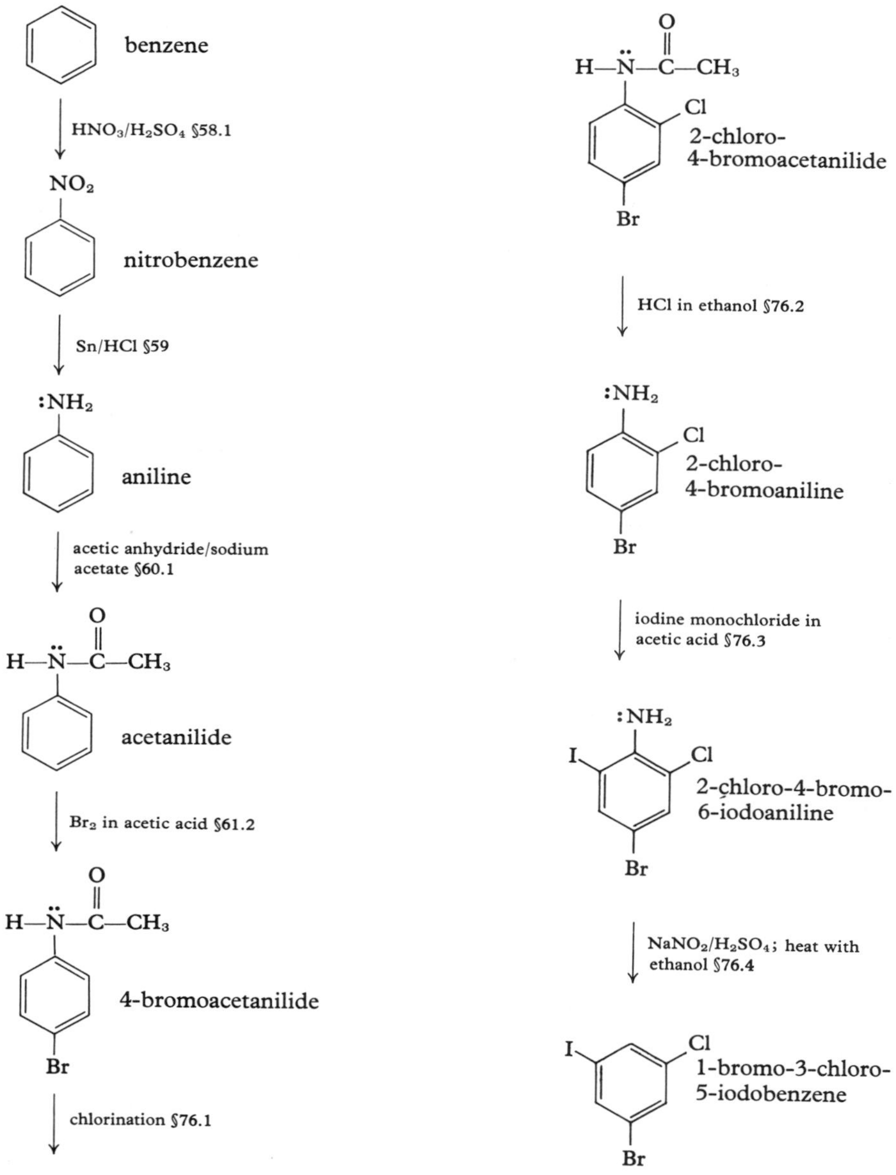

76.1 2-CHLORO-4-BROMOACETANILIDE FROM 4-BROMOACETANILIDE

Procedure a. Suspend 10.7 g (0.050 mole) 4-bromoacetanilide and 4.1 g (0.050 mole) of sodium acetate in 40 ml of glacial acetic acid in a 250-ml flask. The crude 4-bromoacetanilide may be wet with as much as an equal weight of water without affecting the reaction. While stirring the solution rapidly, pass in 0.055

mole chlorine gas (3.9 g; equivalent to 2.5 ml liquid chlorine) through a 7-mm tube which leads below the surface of the liquid in the flask (Note 1).

After the addition of chlorine is complete (in about 20 minutes), continue to stir the mixture for 2 or 3 minutes, and then slowly add 160 ml of cold water. During the addition of water, a homogeneous solution will be obtained, and then the product will begin to crystallize. Treat the resulting suspension with enough concentrated sodium bisulfite solution to remove the yellow-green color, and then collect the crude 2-chloro-4-bromoacetanilide by suction filtration, washing it with water and sucking it as dry as possible; yield: 11.8 g (95%). 2-Chloro-4-bromoacetanilide may be recrystallized from methanol, using 7 ml per gram, with 73% recovery; m.p., 154–156°C (literature value, 151°C).

Note

1. The chlorine is conveniently handled in the following way. A graduated receiver is fitted with a Claisen adapter, and a glass tube is positioned in the center neck of the Claisen adapter by means of a Teflon tubing adapter in such a way that the end of the tube extends to the top of the graduations of the receiver. A drying tube is placed in the outer neck of the Claisen adapter, and the glass tube is connected to a lecture bottle of chlorine gas by a length of Tygon tubing. While the receiver is cooled in a dry ice/acetone bath, chlorine gas is passed into it until the required volume of chlorine has been condensed. The end of the Tygon tubing attached to the lecture bottle is then connected to the 7-mm inlet tube of the reaction flask, the drying tube is replaced by a stopper, and the dry ice/acetone bath is removed. As the chlorine evaporates, it will pass into the reaction mixture at a convenient rate. If the chlorine starts to boil, the graduated receiver may be cooled momentarily with the cold bath.

Time: 2 hours.

Procedure b. (Reference 2.) Suspend 10.7 g (0.05 mole) of 4-bromoacetanilide in a mixture of 23 ml of conc. hydrochloric acid and 28 ml of glacial acetic acid in a 250-ml flask. Heat the mixture on the steam bath until all the solid has gone into solution. Cool the solution to 0°C in an ice bath. To the cold mixture gradually add a solution of 2.77 g (0.026 mole) of sodium chlorate dissolved in 7 ml of water. During the addition of the sodium chlorate solution some chlorine gas will be evolved (Note 1). As the addition is made, a yellow precipitate will form and the solution will turn yellow. After the addition is complete, allow the reaction mixture to stand at room temperature for an hour and then collect the solid by suction filtration.

2-chloro-
4-bromoacetanilide

Note

1. You must either work in the hood or make some other provision for removal of the chlorine gas produced (see Section 35).

Time: 2 hours.

Question

1. An attempt was made to prepare 2-chloro-4-bromoacetanilide from acetanilide without isolation of 4-bromoacetanilide, by treating 5 g of acetanilide and two

equivalents of sodium acetate dissolved in 40 ml of glacial acetic acid with one equivalent of bromine followed by one equivalent of chlorine. Instead of the anticipated product, 2,4-dibromoacetanilide was isolated in 92% yield. Consideration of this result led us to believe that it might be possible to prepare 2-chloro-4-bromoacetanilide by treating acetanilide, two equivalents of sodium acetate, and one equivalent of sodium bromide in 80% acetic acid with two equivalents of chlorine. The product, however, appears to be a mixture of 2,4-dihalogenated acetanilides. Interpret the observations.

76.2 2-CHLORO-4-BROMOANILINE FROM 2-CHLORO-4-BROMOACETANILIDE

The infrared and proton NMR spectra of 2-chloro-4-bromoaniline are shown in Figures 76.1 and 76.2.

Procedure. Place 12.4 g (0.050 mole) 2-chloro-4-bromoacetanilide in a 250-ml flask, and add 20 ml of 95% ethanol and 13 ml of conc. hydrochloric acid. Fit the flask with a reflux condenser and heat the mixture on the steam bath for 30 minutes. During this time a clear solution will be obtained, and then a white precipitate will form. At the end of the heating period, add 80 ml of hot water, and

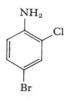

2-chloro-4-bromoaniline

Figure 76.1. Infrared spectrum of 2-chloro-4-bromoaniline; CCl$_4$ solution.

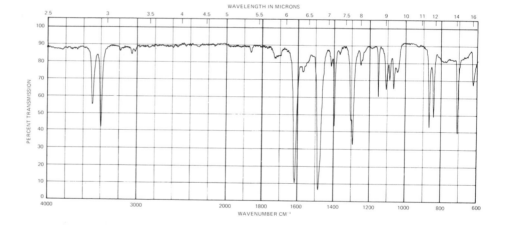

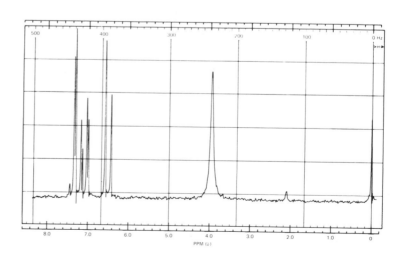

2-chloro-4-bromoaniline

Figure 76.2. NMR spectrum of 2-chloro-4-bromoaniline; CDCl$_3$ solution.

swirl the flask to dissolve the solid. Pour the solution onto about 150 g of ice, and then add to the well-stirred mixture 12 ml of 50% sodium hydroxide solution. Collect the crude product by suction filtration, washing it well with water and sucking it as dry as possible. Yield: 9.4 g (91%). 2-Chloro-4-bromoaniline may be recrystallized from hexane, using 2 ml per gram, with 91% recovery; m.p., 69–71°C (literature value, 70–71°C).

Time: less than 2 hours.

76.3 2-CHLORO-4-BROMO-6-IODOANILINE FROM 2-CHLORO-4-BROMOANILINE

Procedure. Dissolve 5.0 g (0.024 mole) 2-chloro-4-bromoaniline in 80 ml of glacial acetic acid, and add to the solution 20 ml of water. (If the crude 2-chloro-4-bromoaniline is wet with water, reduce accordingly the amount of water added to the reaction mixture.) Then add, over a period of 5 minutes, a solution of 5.0 g (1.6 ml; 0.03 mole) of technical iodine monochloride in 20 ml of glacial acetic acid. Heat the resulting mixture to 90°C on the steam bath, and add enough concentrated aqueous sodium bisulfite solution to lighten the color of the solution to bright yellow; about 20 ml will be needed. Then add an amount of water such that the volume of sodium bisulfite solution plus water added will equal 25 ml; about 5 ml will be required. Allow the solution to cool slowly to room temperature, ■ and finally cool it in an ice bath. Collect the 2-chloro-4-bromo-6-iodoaniline, which will have separated in a mass of long, almost colorless crystals, by suction filtration. Wash them with a little 33% acetic acid solution and then with water. 2-Chloro-4-bromo-6-iodoaniline may be recrystallized with 80% recovery by dissolving 1 g in 20 ml of glacial acetic acid, slowly adding 5 ml of water to the solution as it is heated on the steam bath, and allowing the resulting solution to cool slowly; m.p., 98–99°C (literature value, 97–97.5°C).

Time: less than 2 hours, plus the time required for the solution to cool.

Figure 76.3 shows the infrared spectrum of 2-bromo-4-chloro-6-iodoaniline.

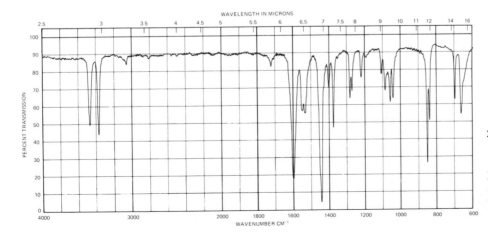

2-chloro-4-bromo-6-iodoaniline

Figure 76.3. Infrared spectrum of 2-chloro-4-bromo-6-iodoaniline; CCl_4 solution.

76.4 1-BROMO-3-CHLORO-5-IODOBENZENE FROM 2-CHLORO-4-BROMO-6-IODOANILINE

Procedure. Suspend 2.0 g (0.006 mole) 2-chloro-4-bromo-6-iodoaniline in 10 ml absolute ethanol in a 250-ml boiling flask, and add 4.0 ml of conc. sulfuric acid dropwise while stirring the mixture. Fit the flask with a reflux condenser. Add 0.70 g (0.01 mole) of powdered sodium nitrite down the condenser, in several portions, to the stirred solution. After the addition of sodium nitrite is complete, heat the mixture on the steam bath for 10 minutes, and then add 50 ml of hot water down the condenser. Steam distill the mixture, and collect about 100 ml of distillate. The product, which will have solidified in the condenser and receiver, is most conveniently isolated by dissolution and extraction with ether. Dry the extract with anhydrous magnesium sulfate, filter it, and distill off the ether. A residue of about 1.5 g (80%) crude 1-bromo-3-chloro-5-iodobenzene will remain. Recrystallization of the crude product from 10 ml of methanol will give about 0.9 g (47%) 1-bromo-3-chloro-5-iodobenzene in the form of long, almost colorless needles; m.p., 87.5–89°C (literature value, 85.5–86°C).

Time: about 3 hours.

The IR and NMR spectra of the product of this procedure are shown in Figures 76.4 and 76.5.

1-bromo-3-chloro-
5-iodobenzene

Figure 76.4. Infrared spectrum of 1-bromo-3-chloro-5-iodobenzene; CCl$_4$ solution.

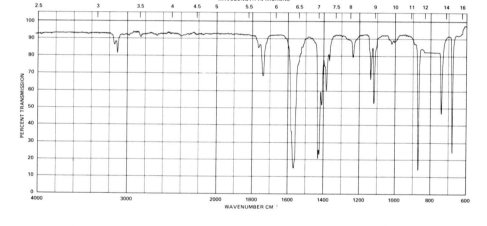

1-bromo-3-chloro-
5-iodobenzene

Figure 76.5. NMR spectrum of 1-bromo-3-chloro-5-iodobenzene; CCl$_4$ solution.

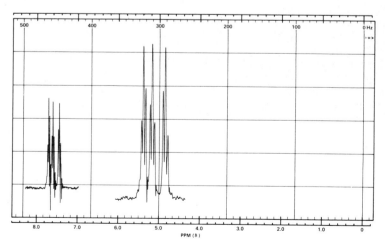

1. What do you suppose might happen if the "mixture of 2,4-dihalogenated acetanil-ides" described in the question in Section 76.1 were carried on through the rest of the procedures?

References

1. A. Ault and R. Kraig, *J. Chem. Educ.* **43**, 213 (1966).

2. R. M. Roberts, J. C. Gilbert, L. B. Rodewald, and A. S. Wingrove, *An Introduction to Modern Experimental Organic Chemistry*, 2nd edition, Holt, Rinehart and Winston, New York, 1974, p. 350.

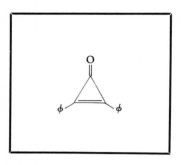

Projects

This final portion of the text presents a number of synthesis projects. The object is to prepare the compound of interest, using the given procedures as the starting point in your development of a synthesis. The procedures that are given are reproduced from the original literature and appear to be the best the literature has to offer. They range from excerpts from Chemical Abstracts *to procedures from* Organic Syntheses. *In some instances, you will have to supply most of the experimental details, and in others you will want to modify the procedure to fit your resources of equipment, glassware, and time. Still more modifications will usually come to mind after you have made a run or two.*

In some cases it may prove useful to look over the article from which an excerpt has been presented, since a general discussion or interpretation of the experimental approach and alternative methods is sometimes given.

§77 FLAVORS AND FRAGRANCES

An early and still continuing motivation for organic synthesis is the search for compounds useful as flavors and fragrances. Sometimes the aim is to duplicate at lower cost a substance that occurs in nature, sometimes it is to prepare a similar or related compound with the hope that it will have an interesting property, and sometimes it is to prepare a less costly substitute of unrelated structure. Certain substances used as flavors and fragrances, which appear to be moderately accessible by synthesis, are presented in the following sections. An interesting book about the sources, properties, and chemistry of flavors and fragrances is P. Z. Bedoukian, *Perfumery and Flavoring Synthetics* (New York: American Elsevier, 1967).

360

γ-Nonanolactone has been described as having "a strong coconut odor and flavor with a very faint anisaldehyde-like bynote" and "an odor indistinguishable from that of coconut." Up to this time, it has not been found to occur naturally.

Physical properties reported for the compound are: boiling point, 137–138°C/17 mm; $d_4^{19.5}$, 0.9672; $n_D^{19.5}$, 1.4462.

The abstracts presented in Figures 77.1 and 77.2 indicate that the synthesis of γ-nonanolactone may be carried out according to the following reactions, where R is *n*-amyl:

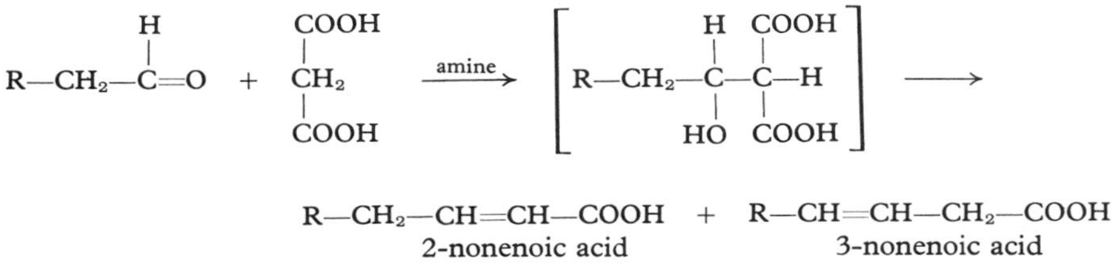

$$R-CH_2-CH=CH-COOH \quad + \quad R-CH=CH-CH_2-COOH$$

2-nonenoic acid   3-nonenoic acid

followed by:

γ-nonanolactone

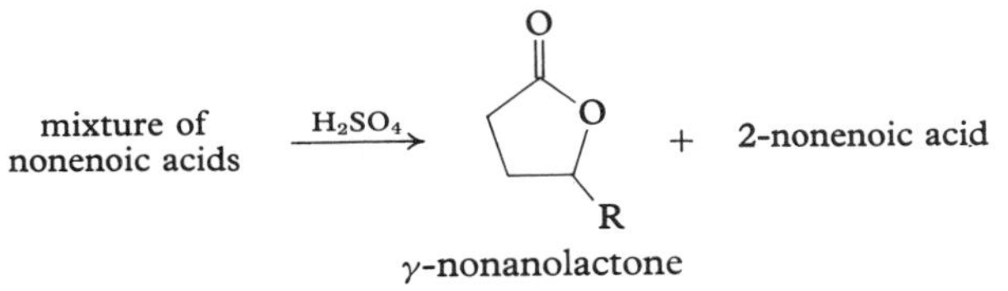

Figure 77.3 shows the infrared spectrum of the product, γ-nonanolactone.

Question

1. Consider the formation of both α,β- and β,γ-unsaturated isomers of nonenoic acid in this synthesis.
  a. Is the isomer ratio a result of kinetic control or equilibrium control?
  b. Attempt to explain the dependence of the isomer ratio on the structure of the base.

References

1. Preparation of α,β-nonenoic acid: *Organic Reactions* **1**, 252 (1942).
2. Knoevenagle condensation: *Organic Reactions* **15**, 204 (1967).

Preparation of "coconut aldehyde"

Synthesis and lactonization of olefinic acids. Kazuyoshi Fujiwara (Tokyo Inst. Technol.). *Nippon Kagaku Zasshi* **82**, 627–9(1961).—Aldehydes and $CH_2(CO_2H)_2$ (I) were condensed and the position of double bond detd. by lactone formation. $Me(CH_2)_5$-CHO (II) (11.4 g.), 10.4 g. I and 19 cc. PhNMe$_2$ in 45 cc. C$_6$H$_6$ were heated 1 hr. and the product acidified to give 35–49% nonenoic acid (III). Lactonization was carried out as follows. III (5.3 g.) and 5.8 g. 80% H$_2$SO$_4$ were heated 1 hr. at 80°, poured into 5% Na$_2$-CO$_3$, extd. with Et$_2$O to give γ-nonanolactone (IV) and the aq. layer acidified to give 2-nonenoic acid (V). Lactonization was carried out for 1, 2, 3, and 4 hrs. and it was found that heating 1 hr. was enough. III obtained with PhNMe$_2$ catalyst gave 66–9% IV and only 2–4% recovered V. Heating I and II with PhNMe$_2$ without C$_6$H$_6$ yielded 39–45% III from which 70–6% IV was isolated. The reaction was carried out using various amines (1.5 moles for 1 mole each of I and II) and the following results obtained (amine, % yield of III, % yield of IV from III, and % yield of V given): Et$_2$NPh, 53, 73, trace; N(CH$_2$CH$_2$-OH)$_3$, 63, 67, trace; Et$_3$N, 71, 79, trace; Bu$_3$N, 62, 67, trace; Bu$_2$NH, 27, 54, trace; pyridine, 59, 8.7, 46;

quinoline, 59, 20, 37. The results indicate that tertiary amines except pyridine and quinoline give 3-nonenoic acid selectively. Similarly PrCHO was condensed with **I** and the following results obtained (amine, yield of mixed hexenoic acid (**VI**), yield of

γ-caprolactone from **VI**, and % yield of 2-hexenoic acid given): PhNMe₂, 35, 75, trace; N(CH₂CH₂OH)₃, 54, 69, trace; Et₃N, 67, 74, trace; pyridine, 47, 4, 38; quinoline, 50, 9, 39.

M. Oki

Figure 77.1. From *Chemical Abstracts* **56**, 8549 (1962). Reproduced by permission. © American Chemical Society.

Preparation of "coconut aldehyde"

γ-**Nonalactone.** Kobayashi Perfume Co. (by Masaki Ota, Kazuyoshi Fujiwara, and Mitsutaka Takahashi). **Japan. 4060('63),** Apr. 22, Appl. Sept. 23, 1959; 2 pp. To 104 g. malonic acid and 151 g, Et₃N is added 114 g. enanthal, the mixt. heated 1 hr.,

cooled, dil. HCl added, the whole extd. with C₆H₆, and the ext. washed with H₂O, dried, and distd. *in vacuo* to give 110 g. mixt. of nonenic acids, b₂₂ 152–6°, which (110 g.) is heated 1 hr. at 80° with 120 g. 85% H₂SO₄, cooled, poured into 5% Na₂CO₃ soln., and extd. with Et₂O to give 87 g. γ-nonalactone, b₁₆ 143–4°.

Hiroshi Kataoka

Figure 77.2. From *Chemical Abstracts* **59**, 11265 (1963). Reproduced by permission. © American Chemical Society.

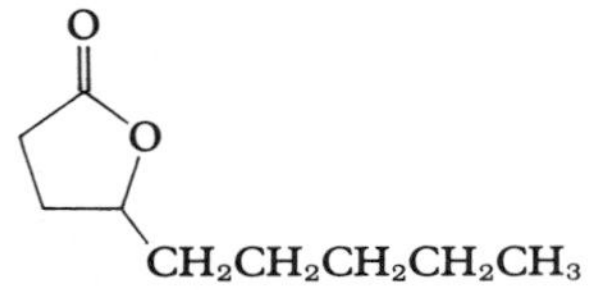

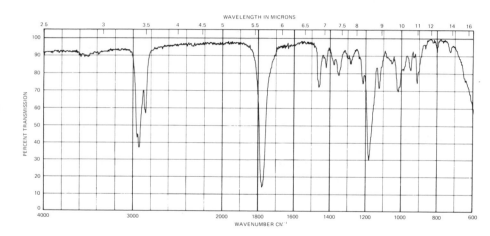

γ-nonanolactone

Figure 77.3. Infrared spectrum of γ-nonanolactone; thin film.

77.2 JASMINALDEHYDE; α-*n*-AMYLCINNAMALDEHYDE; 2-BENZYLIDENEHEPTANAL

This compound is reported to have a high-quality floral or jasmine odor. An obvious method of preparation is the mixed aldol condensation between benzaldehyde and heptanal (Figure 77.4).

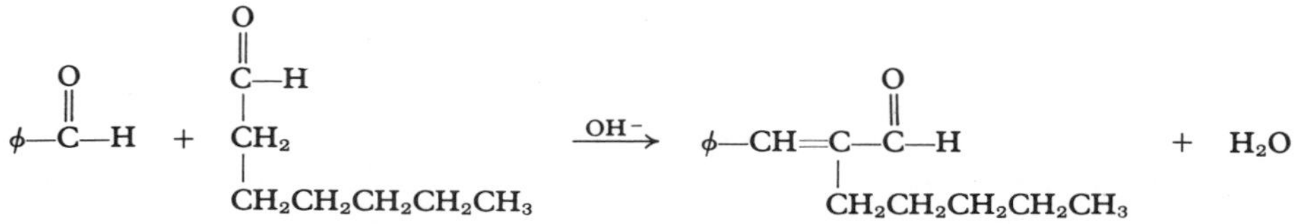

An important side product of this reaction is the compound that results from the self-condensation of heptanal (see Section 67). Some of the physical properties of the two products are:

362

	2-Benzylidene-heptanal	2-n-Heptanylidene-heptanal
d_0^{15}	0.9718	0.8520
n_D^{20}	1.5552	1.4596
b.p./5 mm	140°C	130°C

A second method for the preparation of jasminaldehyde, described in a U.S. Patent by W. C. Meuly (Figure 77.5), involves addition of the sodium bisulfite addition compound of heptanal to a basic solution of benzaldehyde. The high yield of the desired compound and the absence of the self-condensation product of heptanal are interpreted as being due to the slow release of heptanal from its bisulfite addition compound, possibly in a form that is particularly reactive toward benzaldehyde.

The infrared spectrum of α-n-amylcinnamaldehyde is shown in Figure 77.6.

Preparation of jasminaldehyde

Kondensation des Heptanals mit Benzaldehyd

Die Kondensation wurde in üblicher Weise ausgeführt, wobei als Medium wäßriger Alkohol diente. 10,6 g Benzaldehyd (0,1 Mol) und 11,4 g (0,1 Mol) Heptanal wurden in 100 ccm Alkohol gelöst und 500 ccm Wasser beigemischt. Als Kondensation bewirkendes Reagens wurde 10 prozent. Natronlauge angewandt, von der 7 ccm zur Reaktionsmasse beigefügt wurden. Nach zweitägigem Stehen bei Zimmertemperatur wurde die ölige Schicht abgetrennt. Der Rest wurde mit Essigsäure angesäuert und mit Äther extrahiert. Das abgetrennte Produkt, zusammen mit dem Ätherauszug, wurde getrocknet und der fraktionierten Destillation im Vakuum unterworfen. Bei den Versuchen konnte festgestellt werden, daß auch die Wasserdampfdestillation gute Dienste leistet, da das erhaltene Produkt viel schwerer mit dem Wasserdampf flüchtig ist, als die Ausgangsstoffe. Bei guter Mischung, z. B. auf einer kräftig arbeitenden Schüttelmaschine, verläuft die Kondensation auch ohne Alkoholzusatz.

Das erhaltene Produkt — α-n-Amyl-β-phenyl-acrolein — besaß folgende Eigenschaften. Die Substanz ist eine schwach gelbich gefärbte Flüssigkeit mit anhaftendem Geruch, der beim Verdünnen als angenehm bezeichnet werden kann. Sdp.$_{20}$ 174 bis 175°; spezifisches Gewicht d$_{20}^{20}$: 0,97108; Brechungsexponent n^{20} = 1,5381. Eine Bisulfitverbindung entsteht nur schwierig.

Figure 77.4. From B. N. Rutkowski and A. I. Korolew, in *Journal für Praktische Chemie* **119**, 273 (1928). Reproduced with permission.

Preparation of jasminaldehyde

A solution is prepared by dissolving 80 parts potassium hydroxide in 300 parts of water and 500 parts of methanol. 200 parts of this solution are added to 150 parts of benzaldehyde. Over a period of 3 hours there is gradually added at 25 to 30° C. the remaining 680 parts of the alkali solution and simultaneously the equivalent of 79 parts of n-heptylic aldehyde in the form of its sodium bisulfite compound. The reaction mass is diluted with 1500 parts water and the reaction product isolated according to known methods. After purification by distillation in a vacuum, there is obtained 120 to 124 parts of pure alpha-amyl-cinnamic-aldehyde and 72 parts of benzaldehyde.

Alpha-amyl-cinnamic aldehyde may be represented by the following formula

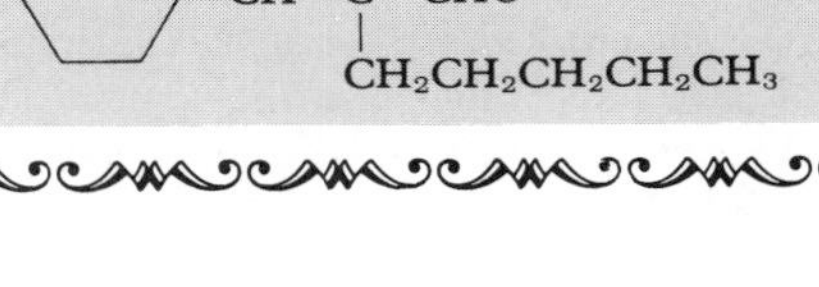

Figure 77.5. From W. C. Meuly, U.S. Patent 2,102,965.

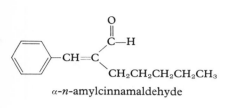

α-*n*-amylcinnamaldehyde

Figure 77.6. Infrared spectrum of α-*n*-amylcinnamaldehyde; thin film.

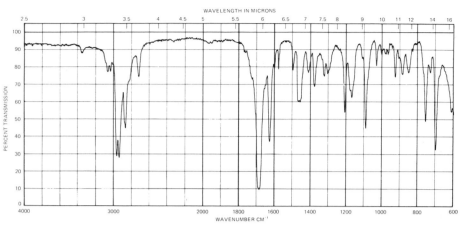

77.3 STRAWBERRY ALDEHYDE; ERDBEERALDEHYD

ethyl β-methyl-β-phenyl glycidate (one of four possible stereoisomeric forms)

This substance, which is not an aldehyde, is said to have a "penetrating strawberry-like odor." It may be prepared according to the procedure shown in Figure 77.7.

The infrared and proton NMR spectra of strawberry aldehyde, ethyl β-methyl-β-phenylglycidate, are shown in Figures 77.8 and 77.9. It is apparent from the NMR spectrum that comparable amounts of the two diastereomeric racemates are present in the sample.

Preparation of strawberry aldehyde

PHENYLMETHYLGLYCIDIC ESTER
(Hydrocinnamic acid, α,β-epoxy-β-methyl-, ethyl ester)

$$C_6H_5COCH_3 + ClCH_2CO_2C_2H_5 + NaNH_2 \rightarrow C_6H_5C(CH_3)\text{---}CHCO_2C_2H_5 + NaCl + NH_3$$

Submitted by C. F. H. Allen and J. VanAllan.
Checked by Nathan L. Drake and Carl Blumenstein.

1. Procedure

To a mixture of 120 g. (118.5 ml., 1 mole) of acetophenone (Note 1), 123 g. (109 ml., 1 mole) of ethyl chloroacetate, and 200 ml. of dry benzene in a 1-l. three-necked round-bottomed flask, fitted with a stirrer and low temperature thermometer, is added, over a period of 2 hours, 47.2 g. (1.2 moles) of finely pulverized sodium amide. The temperature is kept at 15–20° by external cooling (Note 2). After the addition has been completed, the mixture is stirred for 2 hours at room temperature, and the reddish mixture is poured upon 700 g. of cracked ice, with hand stirring. The organic layer is separated and the aqueous layer extracted once with 200 ml. of benzene. The combined benzene solutions are washed with three 300-ml. portions of water, the last one containing 10 ml. of

acetic acid. The benzene solution is dried over 25 g. of anhydrous sodium sulfate, filtered, the drying agent rinsed with a little dry benzene, and, after removal of the solvent, the residue is fractionated under reduced pressure, using a modified Claisen flask. The fraction boiling at 107–113°/3 mm. is collected separately and used for the preparation of hydratropaldehyde (p. 733). Redistillation yields a product which boils at 111–114°/3 mm. (Note 3). The yield is 128–132 g. (62–64%) (Note 4).

2. Notes

1. The practical grades of ketone and ester were used.
2. The reaction is strongly exothermic, and much ammonia is evolved.
3. Other boiling points are 272–275°/760 mm. with decomposition; 153–159°/20 mm.; and 147–149°/12 mm.
4. There are several side reactions which reduce the yield. There are always unchanged ketone and ester in

[1] Erlenmeyer, *Ann.*, **271**, 161 (1892).
[2] Darzens, *Compt. rend.*, **139**, 1215 (1904).
[3] Dutta, *J. Indian Chem. Soc.*, **18**, 235 (1941) [*C. A.*, **36**, 761 (1942)].
[4] Claisen and Feyerabend, *Ber.*, **38**, 702 (1905)
[5] I. G. Farbenind. A.-G., Ger. pat. 591,452 [*Frdl.*, **19**, 288 (1934)] [*C. A.*, **28**, 2367 (1934)].
[6] I. G. Farbenind. A.-G., Ger. pat. 602,816 [*Frdl.*, **20**, 217 (1935)] [*C. A.*, **29**, 1438 (1935)].

3. Methods of Preparation

This is an example of a general reaction by which haloesters are condensed with ketones by means of sodium,[1] sodium ethoxide,[2,3] or sodium amide.[4-6]

Figure 77.7. From *Organic Syntheses*, Coll. Vol. III, ed. by Horning, p. 727. Copyright © John Wiley & Sons, Inc., 1955. Reprinted by permission of John Wiley & Sons, Inc.

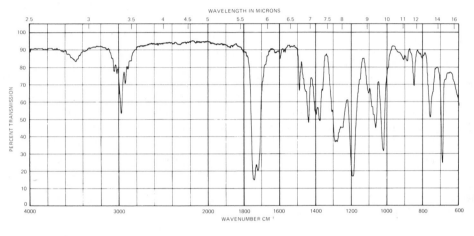

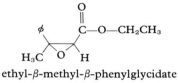

ethyl-β-methyl-β-phenylglycidate

Figure 77.8. Infrared spectrum of ethyl β-methyl-β-phenylglycidate; thin film.

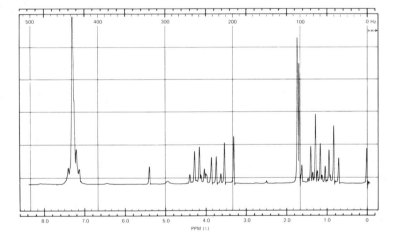

ethyl-β-methyl-β-phenylglycidate

Figure 77.9. NMR spectrum of ethyl β-methyl-β-phenylglycidate; CCl₄ solution.

77.4 HYDRATROPALDEHYDE; α-PHENYLPROPIONALDEHYDE

This substance is said to have a "green, honey-like" odor. It may be prepared from strawberry aldehyde (preceding section) by the procedure given in Figure 77.10.

The infrared and proton NMR spectra of α-phenylpropionaldehyde are shown in Figures 77.11 and 77.12. The sample used to obtain the spectra was described as technical grade, 90 percent pure; the sharp singlet in the proton NMR spectrum at δ = 2.5 was probably contributed by an impurity.

α-phenylpropionaldehyde

α-PHENYLPROPIONALDEHYDE
(Hydratropaldehyde)

$$C_6H_5C(CH_3)CHCO_2C_2H_5 \xrightarrow[\text{then HCl}]{C_2H_5ONa} C_6H_5CH(CH_3)CHO$$

Submitted by C. F. H. Allen and J. VanAllan.
Checked by Nathan L. Drake and Carl Blumenstein.

1. Procedure

An ethanolic solution of sodium ethoxide is prepared in a 1-l. round-bottomed flask from 15.5 g. (0.67 gram atom) of sodium and 300 ml. of absolute ethanol (Note 1). One hundred thirty-three grams (0.64 mole) of phenylmethylglycidic ester (p. 727) is added to this solution slowly and with shaking. The flask is then cooled externally to 15°, and 15 ml. of water is slowly added; considerable heat is evolved, and the sodium salt soon begins to separate. After the mixture has stood overnight, the salt is filtered by suction and rinsed with one 50-ml. portion of ethanol and a similar amount of ether. The dried salt weighs 102–108 g. (80–85%) and melts at 255–256° with decomposition.

The salt is added to a dilute solution of hydrochloric acid prepared by mixing 300 ml. of water and 56 ml. of concentrated acid (sp. gr. 1.18); the acid should be contained in a 1-l. flask under a reflux condenser. The mixture is warmed gently, whereupon carbon dioxide is evolved and an oil separates. The flask is heated on a steam bath for 1.5 hours, and the oil is then extracted from the cooled mixture with 150 ml. of benzene. The extract is washed once with 200 ml. of water and distilled under reduced pressure, using an ordinary 500-ml. Claisen flask. The hydratropaldehyde distils at 90–93°/ 10 mm. or 73–76°/4 mm. (oil bath at 120–130°), leaving only a slight residue (3–5 g.). The yield is 56–60 g. (65–70%) (Notes 2 and 3).

2. Notes

1. The metal is placed in the flask, and the ethanol is added through the condenser as fast as refluxing will allow (p. 215, Note 1).
2. The generality of this reaction is limited only by the availability of the glycidic esters.
3. The 2,4-dinitrophenylhydrazone forms yellow prisms that melt at 135°.

3. Methods of Preparation

Hydratropaldehyde has been prepared by hydrolysis of phenylmethylglycidic ester,[1–3] by chromyl chloride oxidation of cumene,[4] by the elimination of halogen acid or water from halohydrins or glycols,[5–7] and by the distillation at ordinary pressure of methylphenylethylene oxide.[8,9]

Hydratropaldehyde has also been prepared by the catalytic addition of hydrogen and carbon monoxide to styrene.[10]

[1] Claisen, *Ber.*, **38**, 703 (1905).
[2] I. G. Farbenind. A.-G., Ger. pat. 602,816 [*Frdl.*, **20**, 217 (1935)].
[3] Dutta, *J. Indian Chem. Soc.*, **18**, 235 (1941) [*C. A.*, **36**, 161 (1942)].
[4] v. Miller and Rohde, *Ber.*, **24**, 1359 (1889).
[5] Bougault, *Ann. chim. phys.*, (7) **25**, 548 (1902).
[6] Tiffeneau, *Bull. soc. chem. France*, (3) **27**, 643 (1902); *Compt. rend.*, **134**, 846 (1902); **137**, 1261 (1903); **142**, 1538 (1906); *Ann. chim. phys.*, (8) **10**, 353 (1907).
[7] Stoermer, *Ber.*, **39**, 2298 (1906).
[8] Klages, *Ber.*, **38**, 1971 (1905).
[9] Tiffeneau, *Compt. rend.*, **140**, 1459 (1905); *Ann. chim. phys.*, (8) **10**, 192 (1907).
[10] Adkins and Krsek, *J. Am. Chem. Soc.*, **70**, 383 (1948).

Figure 77.10. From *Organic Syntheses*, Coll. Vol. III, ed. by Horning, p. 733. Copyright © John Wiley & Sons, Inc. 1955. Reprinted by permission of John Wiley & Sons, Inc.

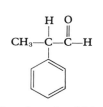

α-phenylpropionaldehyde

Figure 77.11. Infrared spectrum of α-phenyl-propionaldehyde; thin film.

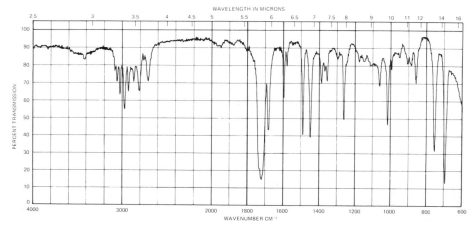

366

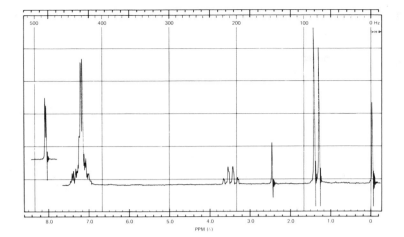

SECTION 77

H O

CH₃—C—C—H

α-phenylpropionaldehyde

Figure 77.12. NMR spectrum of α-phenyl-propionaldehyde; CCl₄ solution.

77.5 (R)-(−)-CARVONE FROM (R)-(+)-LIMONENE

Carvone can exist in two enantiomeric forms because of the presence of the single chiral carbon atom (∗).

By following the procedures given in Section 46, one enantiomer, the (R)-(−)-form, can be isolated from oil spearmint, and the other, the (S)-(+)-form, can be isolated from oil of caraway. These enantiomers are somewhat unusual in that most people perceive them to have quite different odors: the (R)-(−)-isomer smells like spearmint, and the (S)-(+)-isomer smells like caraway. The presence of the (R)-(−)-isomer in a well-known brand of toothpaste accounts for its spearmint flavor. However, since the quantity of carvone required for this use is rather large, it must be obtained by synthesis rather than by isolation from oil of spearmint. The commercial method of synthesis uses as the starting material (R)-(+)-limonene, which is obtained from orange and grapefruit rind (see Section 45), and involves the intermediate formation of limonene nitrosochloride and carvoxime.

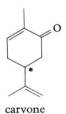

carvone

In this synthesis the chiral center originally present in limonene is undisturbed and ends up as the chiral center of carvone. This sequence of reactions is based on the procedures of Royals and Horne, and of Reitsema; these are shown in Figures 77.13 and 77.14. The procedure of Wallach referred to in Figure 77.13 is shown in Figure 77.15. A similar series of reactions is described in Reference 1.

At Cornell College we have tried to develop these procedures into a routine synthesis of (R)-(−)-carvone from (R)-(+)-limonene, which can be done by students in the introductory organic lab. At this time our yields of limonene nitrosochloride are rather modest and quite variable, and we are having a lot of difficulty with the hydrolysis of carvoxime. We would be pleased to learn about improved procedures for all three of these reactions.

(R)-(+)-limonene

↓ "NOCl"

limonene nitrosochloride

↓ heat; base

carvoxime

↓ H₂O; acid

(R)-(−)-carvone
spearmint

Preparation of *d*-Limonene Nitrosochloride.—The method of Wallach[4] was found to be quite suitable for the preparation of *d*-limonene nitrosochloride on a small scale. In our hands, the yields on 0.06 molar scale ranged from 32 to 60% depending upon the temperature of reaction. In general, the yield increased with lower reaction temperature. Quite small yields were obtained when attempt was made to scale up this procedure. Similarly, the procedures of Tilden[3] and of Rupe[5] gave quite low yields of *d*-limonene nitrosochloride on 0.5 molar scale (8–11 and 16%, respectively).

The following procedure was found quite suitable to the preparation of *d*-limonene nitrosochloride on 0.5 molar scale. A mixture of 68.1 g. (0.5 mole) of *d*-limonene and 85 ml. of ethyl alcohol was placed in a 500-ml., three-necked flask equipped with a mechanical stirrer, a thermometer and a gas inlet tube. The flask was surrounded by an ice–salt freezing mixture, and the contents were cooled to −10°. Gaseous ethyl nitrite was passed into the limonene solution while maintaining a temperature of −8 to −10°. The ethyl nitrite was generated by dropping a mixture of 32.2 g. (0.70 mole) of ethyl alcohol, 39 ml. of sulfuric acid, and sufficient water to make 390 ml. into a mixture of 95 g. (1.38 moles) of sodium nitrite, 32.2 g. (0.70 mole) of ethyl alcohol and sufficient water to make 390 ml. The generator flask was warmed slightly with a warm water-bath from time to time to ensure a steady evolution of ethyl nitrite. After all of the ethyl nitrite had been passed into the alcoholic limonene solution, 10 ml. of water was added, and moist hydrogen chloride was passed into the solution. The reaction was quite exothermic, but the introduction of hydrogen chloride could be so regulated as to maintain a reaction temperature of 0 to −10°, preferably −5°. The moist hydrogen chloride was generated by dropping 114 ml. of concentrated hydrochloric acid (1.5 moles of HCl) into sulfuric acid and passing the evolved hydrogen chloride through concentrated hydrochloric acid. After addition of the hydrogen chloride was complete, the reaction mixture was stirred for about thirty minutes to permit the reaction temperature to drop to −10°. The precipitated *d*-limonene nitrosochloride, 80 g. (80% yield), was filtered with suction and washed with cold ethyl alcohol. The addition of 10 ml. of water to the reaction mixture prior to introduction of hydrogen chloride was found to be absolutely necessary to the isolation of a good yield of product. When anhydrous conditions were maintained, yields of only 9–11% were realized.

A similar procedure using methyl nitrite gave yields of *d*-limonene nitrosochloride of the order of 70%.

Preparation of *l*-Carvoxime.—*d*-Limonene nitrosochloride was dehydrohalogenated by the pyridine procedure of Wallach[9] to afford *l*-carvoxime in 90–95% yield; m.p. 74° (reported[11] m.p. 72°).

Hydrolysis of Carvoxime.—The following procedures were found satisfactory for the hydrolysis of *l*-carvoxime to *l*-carvone:

Procedure A: A mixture of 10 g. (0.06 mole) of *l*-carvoxime and 100 ml. of 5% aqueous oxalic acid was heated to reflux for two hours. At the end of this time, the reaction mixture was steam distilled, and the distillate was extracted with ether. The ethereal solution was dried over sodium sulfate and fractionally distilled to give 7.1 g. (78%) of *l*-carvone.

Procedure B: A mixture of 37 g. (0.22 mole) of carvoxime and 500 ml. of 5% oxalic acid was steam distilled without previous reflux. The distillate was treated as described above to give 20.5 g. of *l*-carvone. Nine grams of carvoxime was recovered from the residue from steam distillation. The yield of *l*-carvone was 80% based on the unrecovered *l*-carvoxime.

The *l*-carvone prepared by either procedure showed the following properties: b.p. 88–90° (4 mm.); n^{20}D 1.4989; d^{25}_4 0.9673; $[\alpha]^{25}$D −54.2°. The properties reported[12] for natural *l*-carvone are: b.p. 97–98° (9 mm.); n^{20}D 1.4988; d^{15}_{15} 0.9652; $[\alpha]^{20}$D −62.46°. Our *l*-carvone gave a semicarbazone, m.p. 140–141.5°; reported[13] m.p. 141–142°.

A mixture of 20 g. (0.12 mole) of *l*-carvoxime and 100 ml. of 5 N hydrochloric acid was heated to reflux for 15 minutes. The mixture was steam distilled, the distillate was extracted with ether, and the ethereal solution was dried over anhydrous sodium sulfate. Removal of the ether left 13.3 g. of an oil which was completely soluble in 5% aqueous sodium hydroxide. This oil was characterized as carvacrol through the formation of 2-methyl-5-isopropylphenoxylacetic acid, m.p. 152.0–152.5°; reported[14] m.p. 150–151°. The yield of crude carvacrol was 74%. An attempt to prevent isomerization of carvone to carvacrol by steam distilling the hydrochloric acid solution without reflux led also to carvacrol in 77% yield.

An attempted oxime exchange between benzaldehyde and carvoxime in glacial acetic acid as solvent was unsuccessful; carvoxime was almost quantitatively recovered.

EMORY UNIVERSITY, GA.   RECEIVED MAY 28, 1951

(11) O. Wallach, *Ann.*, **305**, 324 (1899).

(12) J. L. Simonsen, "The Terpenes," Second Ed., Vol. I, The University Press, Cambridge, Mass., 1947, p. 396.

(13) H. Rupe and K. Dorschky, *Ber.*, **13**, 2112, 2372 (1906).

(14) C. F. Koelsch, THIS JOURNAL, **53**, 304 (1931).

Figure 77.13. From E. E. Royals and S. E. Horne, Jr., in *J. Am. Chem. Soc.* **73**, 5856 (1951).

Preparation of (R)-(−)-*carvone*

Nitrosochloride Syntheses and Preparation of Carvone

ROBERT H. REITSEMA[1]

Received July 21, 1958

A convenient preparation of limonene nitrosochloride has been developed. In place of the usual generation of nitrosyl chloride in a separate generator or the use of amyl nitrate it has been found that nitrosochlorides can be formed *in situ* by simultaneous addition of sodium nitrite and acid to the olefin. Yields by the procedure are high, and the quality of the product is satisfactory. Limonene nitrosochloride can be produced in yields of 80% with rotations of 226°.

The amount of acid used in the reaction is critical. The usual amount of concentrated aqueous hydrochloric acid used was 40–50 ml. per tenth mole of olefin. As low as 33.3 ml. per tenth mole of olefin gave nearly the same results. Use of 75 ml. of concentrated acid per tenth mole of olefin reduced the yield from the usual 75–80% range to less than 10%. In general, equimolar proportions of sodium nitrite were used although a 50% excess had no adverse effects. The temperature of the reaction had to be maintained below 10°. Operation at 20° resulted in a 26% yield of nitrosochloride. The rate of addition and the efficiency of stirring had to be adjusted to avoid localized heating and high acid concentrations even with optimum amounts of reagents.

Limonene nitrosochloride is relatively stable when purified by repeated washings with cold isopropyl alcohol. Some separation of forms of the derivative are observed and the less soluble material obtained upon excessive washing, in 63% yield, had a rotation of 340°. The nitrosochloride undergoes decomposition upon standing. A brown oil is formed and this then crystallizes after several days standing at room temperature. The latter material is carvoxime. The progress of the decomposition can be followed by the loss in weight as well as by the change from positive to negative rotation. Conversions of the nitrosochloride to carvoxime described in the literature using pyridine and acetone[2,3] or urea and isopropyl alcohol[3] were found to work well. It was observed, however, that the amounts of reagents specified are excessive and that the reaction occurs under much milder conditions with less decomposition. The progress of the reaction was followed by the change in rotation of the solution. The use of sodium hydroxide or McElvain's buffer[4] was not useful. The use of dimethyl formamide was found to work at least as well as pyridine or urea.

A significant improvement in the quality of the carvone produced by decomposition of carvoxime was obtained when the reaction was run on a continuous basis. The control of pH as suggested[3] was found to be desirable. In order to minimize the time of contact of the carboxime and carvone with the acidic media, carvoxime was added as a melt or as a slurry in water during the reaction. Carvone was steam distilled out of the reaction mixture as rapidly as it was formed. A yield of 83% of the theoretical amount of carvone from the oxime was obtained. The product had a rotation of −58.8°, which is significantly higher than from other methods of synthesis.

EXPERIMENTAL

Limonene nitrosochloride. A solution of 40.8 g. (0.3 mole) of (+)-limonene (α_D +95°) in 40 ml. of isopropyl alcohol was cooled below 10°. To this were added simultaneously through separate dropping funnels a solution of 120 ml. of concentrated hydrochloric acid in 80 ml. of isopropyl alcohol and a concentrated aqueous solution of 20.7 g. (0.3 mole) sodium nitrite. The addition was adjusted so as to maintain the temperature below 10°. The mixture was stirred for an additional 15 min. and was allowed to stand in the refrigerator for 1 hr. The solid, isolated by filtration, was washed with enough cold ethanol to make a thick slurry to provide a water white product. A small amount of additional solid could be isolated from the mother liquors from the washing by cooling and filtration. A total of 48.8 g. (80.7% of theory) of product with a rotation in ethanol of 226° was obtained.

Carvoxime. Eight grams of limonene nitrosochloride and 4 ml. of dimethylformamide were boiled 30 min. under reflux in 25 ml. of isopropyl alcohol. The product at the end of this time had a constant negative rotation. It was poured into 150 ml. of cracked ice and water, stirred vigorously, and was filtered after the ice was melted. The solid was washed three times with 10 ml. of cold water and once with 3 ml. of cold isopropyl alcohol. The dry product weighed 5.47 g. (83.5%) m.p. 66–69°, α_D −40.3°, and was pure enough for further work.

(2) C. Bordenca and R. K. Allison, *Ind. Eng. Chem.*, **43**, 1196 (1951).

(3) E. E. Royals and S. E. Horne, *J. Am. Chem. Soc.*, **73**, 5856 (1951).

(4) T. C. McElvain, *J. Biol. Chem.*, **49**, 183 (1921).

Carvone. A one liter three-necked flask was fitted with a Beckman high temperature glass electrode 8890-15, a reference electrode 8970-92, and outlets which allowed addition of sulfuric acid, addition of oxime, and distillation of water-product azeotrope. In the flask was placed 700 ml. of water which was brought to boiling, and the *p*H was adjusted to 0.8 with 3*N* sulfuric acid. About 50 ml. of acid was required. Then carvoxime was added dropwise from a separatory funnel in which solid oxime was melted in portions by means of an infrared lamp. Evolution of carvone was nearly instantaneous and the addition was so regulated that the carvone in the still pot could be voided within 5 min. About 1.25 hr. were required for the addition. Dilute sulfuric acid was added portionwise during the reaction to maintain the *p*H at 0.7–0.9. A total of 85 ml. of 3*N* sulfuric acid was used. The product was isolated in a continuous separator and the condensate water returned to the still pot. A total of 18.6 g. (83% yield) of white carvone with a rotation of −58.8° was obtained.

A similar reaction was run in which solid carvoxime was added through a condenser directly to the still pot without prior melting. The yield was 18.8 g. and the rotation again was −58.8°.

Figure 77.14. From R. H. Reitsema, in *J. Org. Chem.* **23**, 2038 (1958).

Preparation of carvoxime

Durch Zusammenbringen von Limonen, Äthylnitrit und konzentrierter Salzsäure in bekannter Weise dargestelltes Limonennitrosochlorid wird durch gutes Auswachen mit Aceton (oder auch durch Krystallisation daraus) gereinigt, wobei man nicht nötig hat, auf eine Trennung der bekanntlich gleichzeitig entstehenden α-und β-Modifikation Bedacht zu nehmen. 7 g Nitrosochlorid wurden mit 3 ccm Pyridin und 6 ccm Aceton übergossen und im Wasserbad 10—15 Minuten erwärmt. Es tritt vollkommene Lösung ein, und wenn man die ausgegossene noch warme Flüssigkeit mit etwas Wasser versetzt, scheidet sich das entstandene Carvoxim bald krystallinisch aus, oder wird nach dem Animpfen sofort fest. Die Ausbeute ist sehr befriedigend.

Figure 77.15. From O. Wallach, in *Liebig's Annalen* **414**, 257 (1918).

Another interesting possibility is that of preparing the (S)-(+)-isomer of carvone from (R)-(+)-limonene by leaving the ring double bond in its original position and introducing the carbonyl group at the position of the allylic methylene group.

Figures 77.16 and 77.17 show the infrared spectra of limonene nitrosochloride and of carvoxime.

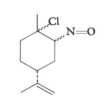

limonene nitrosochloride

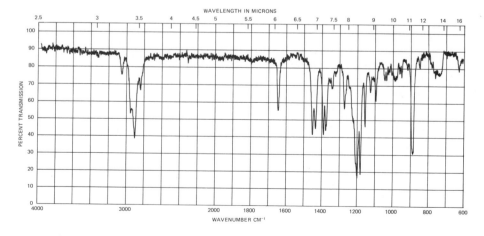

Figure 77.16. The infrared spectrum of limonene nitrosochloride; CCl_4 solution.

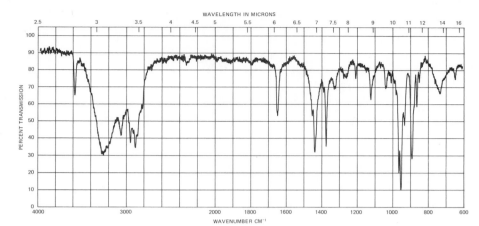

WAVELENGTH IN MICRONS

PERCENT TRANSMISSION

WAVENUMBER CM⁻¹

carvoxime

Figure 77.17. The infrared spectrum of carvoxime; CCl_4 solution.

Questions

1. How many stereoisomeric forms can there be of "limonene nitrosochloride"? How many of these could be formed in this experiment?

2. When (R)-(+)-limonene was oxidized to carvone with the pyridine complex of chromic oxide, racemic carvone was obtained. Explain.

References

1. C. Bordenca, R. K. Allison, and P. H. Durstine, *Ind. Eng. Chem.* **43**, 1196 (1951).

2. T. J. Leitereg et al. *J. Agr. Food Chem.* **19**, 785 (1971), using the procedure of Dauben, Lorber, and Fullerton, *J. Org. Chem.* **34**, 3587 (1969). Reference (2).

77.6 DIHYDROJASMONE AND JASMONE

Dihydrojasmone, found in bergamot oil and described as having an intensely fresh odor with fruity undertones, and jasmone, a constituent of the volatile oil of jasmine flowers, have been the objects of a great variety of syntheses (1 and 2).

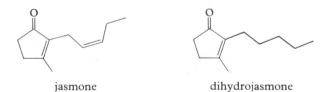

jasmone   dihydrojasmone

Their natural sources are quite limited, and the compounds are highly prized by the perfume industry. The projects in this section illustrate a few of the many possible syntheses of these compounds.

SYNTHESES INVOLVING 1,4-DIKETONE INTERMEDIATES

1,4-Diketones such as undecane-2,5-dione undergo a cyclic aldol-type condensation in the presence of base to give 2-alkyl-3-methylcyclopentenones.

The alternative cyclization products, 3-alkylcyclopentenones, are apparently formed much more slowly (3) and are less stable (4) than the observed product.

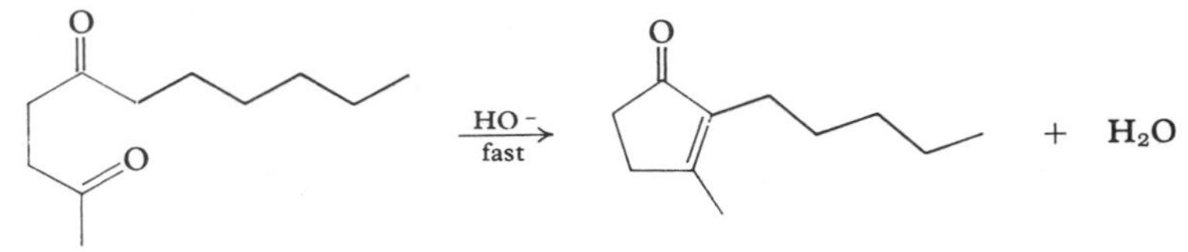

undecane-2,5-dione

2-pentyl-3-methylcyclopentenone
(dihydrojasmone)

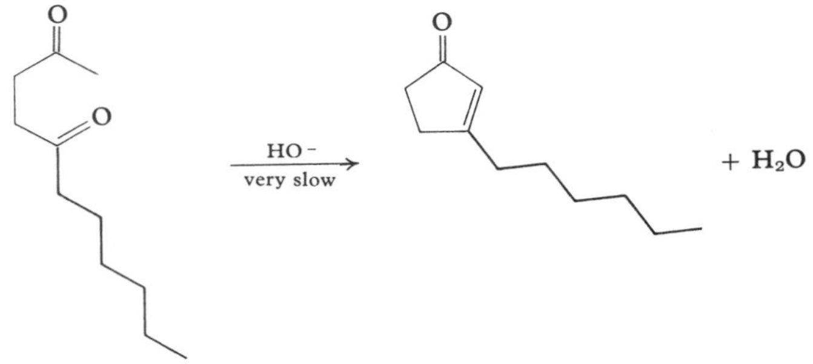

undecane-2,5-dione   3-hexylcyclopentenone

The diketone required for the preparation of dihydrojasmone, undecane-2,5-dione, can be prepared in many ways (1 and 2). For example, the mixed anhydride between levulinic acid and pivalic acid gives this compound in good yield upon treatment with *n*-hexyl magnesium bromide at −78° (Figure 77.18).

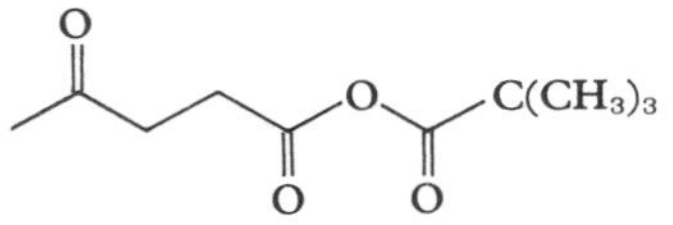

levulinic pivalic anhydride

↓ *n*-hexylmagnesium bromide

↓ water; ammonium chloride

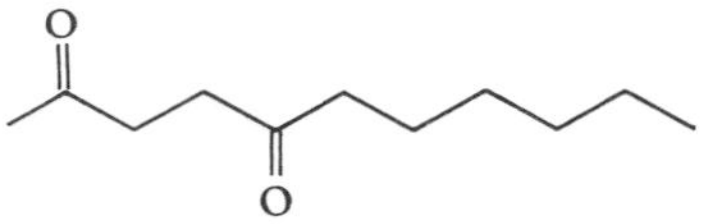

undecane-2,5-dione
2,5-dioxoundecane

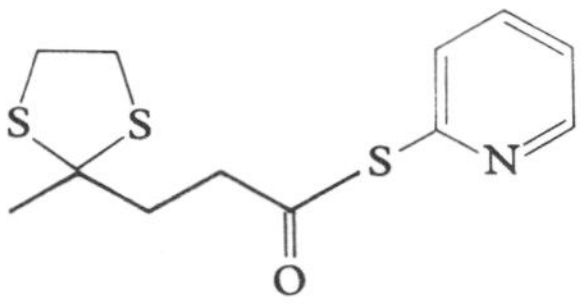

levulinic acid S-(2-pyridyl)thioate ethylenedithioacetal

↓ *n*-hexylmagnesium bromide

↓ water; ammonium chloride

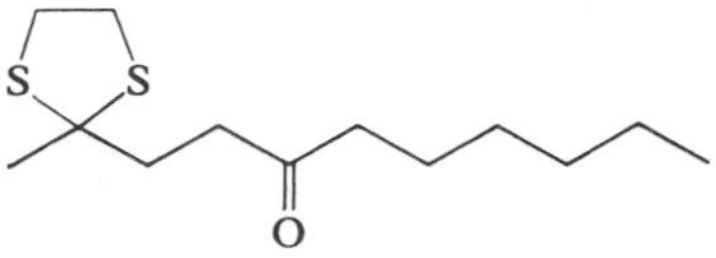

A similar procedure also starts with levulinic acid but involves protecting the ketonic carbonyl as the ethylenedithioketal and conversion of the carboxyl group to the S-(2-pyridyl)-thioester before treatment with the Grignard reagent (Figure 77.19).

The general procedure for the removal of the ethylenedithioacetal protecting group is shown in Figure 77.20.

Preparation of undecane-2,5-dione

For example, to a solution of levulinic acid (10 mmol) in 90 ml of dry THF was added triethylamine (10 mmol) and pivaloyl chloride (10 mmol) in this order at −30°C. After stirring for 20 minutes at −20 ~ −30°C, a THF solution containing 10 mmol of *n*-hexylmagnesium bromide was added at −78°C. The resulted reaction mixture was stirred for 30 minutes at −78°C and hydrolyzed with 10% aqueous ammonium chloride. After usual work-up, 2,5-dioxoundecane and 2,2-dimethyl-3-oxononane were isolated by silica gel column chromatography in 71 and 11% yields, respectively.

Figure 77.18. From M. Araki, S. Sakata, H. Takei, and T. Mukaiyama, in *Chemistry Letters* **1974**, 687.

Preparation of undecane-2,5-dione

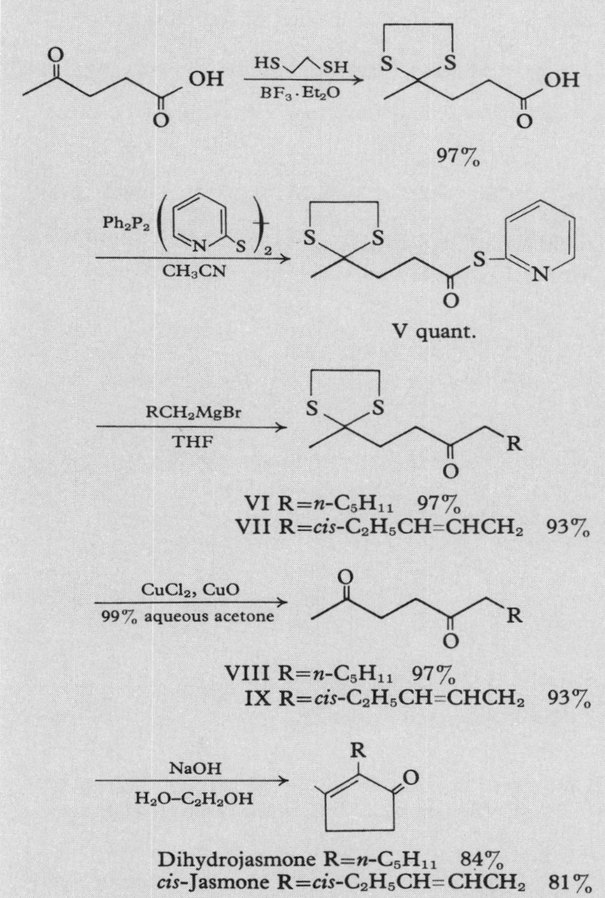
HS SH
BF$_3$·Et$_2$O
OH
97%
Ph$_2$P$_2$
CH$_3$CN
V quant.
RCH$_2$MgBr
THF
VI R=n-C$_5$H$_{11}$ 97%
VII R=*cis*-C$_2$H$_5$CH=CHCH$_2$ 93%
CuCl$_2$, CuO
99% aqueous acetone
VIII R=n-C$_5$H$_{11}$ 97%
IX R=*cis*-C$_2$H$_5$CH=CHCH$_2$ 93%
NaOH
H$_2$O-C$_2$H$_5$OH
Dihydrojasmone R=n-C$_5$H$_{11}$ 84%
cis-Jasmone R=*cis*-C$_2$H$_5$CH=CHCH$_2$ 81%

Scheme 1. Syntheses of *cis*-jasmone and dihydrojasmone

The ketone VI was obtained in 97% yield by the reaction of V with *n*-hexylmagnesium bromide, and the hydrolysis of VI by our method[7] afforded 1,4-diketone (VIII) in 98% yield. Dihydrojasmone was obtained from VIII according to the method of Hunsdiecker[8] in 84% yield. Analogously, olefinic 1,4-diketone IX was isolated in 82% yield by the reaction of V with 3-*cis*-hexenyl magnesium bromide, followed by the deprotection of the carbonyl group.[7] IX was converted to *cis*-jasmone in 81% yield by the ordinary procedure.[8]

Experimental

All the melting points and boiling points are uncorrected.

General Method for the Preparation of S-(-2 Pyridyl) Thioates. *S*-(2-Pyridyl) thioates were prepared by the following two methods. (A) To a solution of 2-pyridinethiol[9] (2.22 g, 20 mmol) and triethylamine (2.02 g, 20 mmol) in dry THF (30 ml) was added 20 mmol of an acid halide at 0 °C with stirring. The reaction mixture was stirred for 4 hr at room temperature. After removal of triethylamine hydrochloride by filtration, the solvent was removed under reduced pressure and the residue was dissolved in ether. The ether layer was washed with saturated sodium chloride solution and dried over sodium sulfate. Removal of the solvent gave a crude *S*-(2-pyridyl) thioate quantitatively. The crude thioate was purified by distillation or recrystallization. (B) To a mixture

7) K. Narasaka, T. Sakashita, and T. Mukaiyama, *ibid.*, **45**, 3724 (1972).
8) H. Hunsdiecker, *Chem. Ber.*, **75**, 460 (1942).
9) M. A. Phillips and H. Shapiro, *J. Chem. Soc.*, **1942**, 584.

of 2,2'-dipyridyl disulfide (2.20 g, 10 mmol), triphenylphosphine (2.62 g, 10 mmol), and a carboxylic acid (10 mmol) was added dry acetonitrile (30 ml) at once with stirring. The reaction mixture was stirred for 10 min and then the solvent was removed under reduced pressure. The residue was chromatographed over silica gel to give a S-(2-pyridyl) thioate.

S-(2-Pyridyl) thioates obtained by the method A or B were listed in Table 5.

TABLE 5. S-(2-PYRIDYL)THIOATES

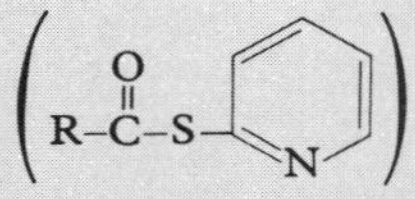

$$\left(\begin{array}{c} O \\ \| \\ R-C-S \end{array} \bigodot_N \right)$$

R	Method	Yield (%)	Mp °C (Bp °C/mmHg)
Ph	A	80	50.0—51.0[a]
CH$_3$	A	60	(98—99/4)
n-C$_5$H$_{11}$	A	84	(111—112/1)
PhCH$_2$CH$_2$	A	82	56.0—58.0[b]
(CH$_2$)$_4$	A	57	44.5—45.0[c]
CH$_3$CO(CH$_2$)$_4$	A	80[d]	oil
C$_2$H$_5$OCO(CH$_2$)$_4$	B	78[d]	oil

Recrystallized from a) benzene-hexane, b) cyclohexane, c) ether. d) Purified by silica gel column chromatography.

General Procedure for the Reaction of Thioates with Grignard Reagents under Argon. In a 50 ml two-necked flask equipped with dropping funnel and gas inlet tube were placed 20 ml of dry THF and 2 mmol of a thioate. The flask was cooled in ice-water bath and a THF solution of Grignard reagent (approximately 1 equivalent) was added from the dropping funnel until the thioates were consumed (checked by tlc). The reaction mixture was quenched with 10% ammonium chloride and the resulting solution was extracted with ether. The organic layer was washed with 1 M sodium hydroxide to remove the thiol formed, with saturated sodium chloride and dried over sodium sulfate. After removal of the solvent the residue, almost pure ketone, was further purified by silica gel chromatography or distillation.

Levulinic Acid Ethylene Dithioacetal. To a mixture of levulinic acid (0.231 g) and ethanedithiol (0.234 g) was added boron trifluoride etherate (0.2 ml) with stirring. An exothermic reaction occurred and after 2.5 hr, the resulting white solid was dissolved in ether. The ether layer was washed with water and dried over sodium sulfate. After removal of the ether, the obtained acid was purified by silica gel thin layer chromatography (0.370 g 97% mp 53.0—55.0 °C). In the case of larger scale experiment, recrystallization may be recommended instead of the chromatography.

Figure 77.19. Adapted from M. Araki, S. Sakata, H. Takei, and T. Mukaiyama, in *Bulletin of the Chemical Society of Japan* **47**, 1777 (1974).

A Convenient Method for Hydrolysis of 1,3-Dithiane Derivatives Using Cupric Chloride

Hydrolysis of dithioketals

Koichi NARASAKA, Takeshi SAKASHITA, and Teruaki MUKAIYAMA
Laboratory of Organic Chemistry, Tokyo Institute of Technology, Ookayama, Meguro-ku, Tokyo

(Received October 19, 1972)

A variety of methods for generation of carbonyl compounds from 1,3-dithiane derivatives has been developed, however, none of these methods is general.[1] Therefore, in connection with the studies on the activation of carbon-sulfur bond with cupric chloride[2], we investigated the hydrolysis of 1,3-dithiane derivatives with cupric chloride.

The mixture of 5 mmol of 2-methyl-2-phenethyl-1,3-dithiane (1), 10 mmol of cupric chloride and 20 mmol

of cupric oxide in 50 m*l* of aqueous 99% acetone[3] was refluxed for 1 hr. A precipitate was filtered off and the filtrate was condensed under reduced pressure. Ether

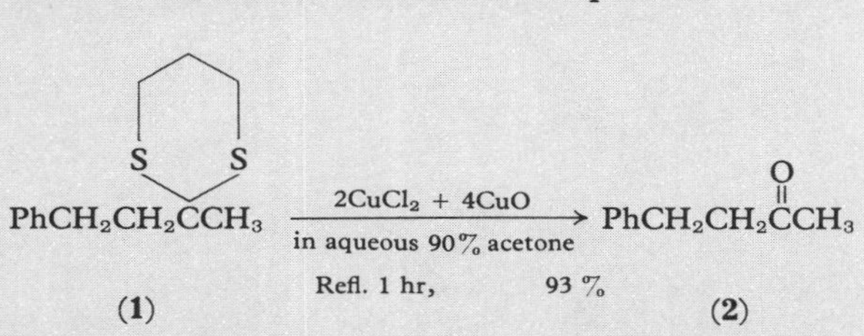

$$PhCH_2CH_2CCH_3 \xrightarrow[\text{in aqueous 90\% acetone}]{2CuCl_2 + 4CuO} PhCH_2CH_2CCH_3$$

Refl. 1 hr,   93%

(1)   (2)

1) a) K. Kuhn and F. A. Neugebauer, *Chem. Ber.*, **94**, 2629 (1961): b) D. Seebach, *Synthesis*, **1**, 17 (1969): c) M. L. Wolform, *J. Amer. Chem. Soc.*, **51**, 2188 (1929): d) E. J. Corey and D. Crouse, *J. Org. Chem.*, **33**, 298 (1968): e) E. J. Corey and B. W. Erickson, *ibid.*, **36**, 3553 (1971): f) E. Vedejs and P. L. Fuchs, *ibid.*, **36**, 366 (1971): g) W. Huurdeman and H. Wynberg, *Synthetic Commun.*, **2**, 7 (1972): h) T. Mukaiyama, S. Kobayashi, K. Kamino and H. Takei, *Chem. Lett.*, **1972**, 237.

2) T. Mukaiyama, K. Narasaka, and H, Hokonoki, *J. Amer. Chem. Soc.*, **91**, 4315 (1969); T. Mukaiyama, K. Mackawa and K. Narasaka, *Tetrahedron Lett.*, **1970**, 4669.

3) The reaction proceeds in various aqueous organic solvents (acetone, acetonitrile, methanol, and tetrahydrofuran), but it was noted that benzylacetone is obtained in the best yield in the case of aqueous 99% acetone.

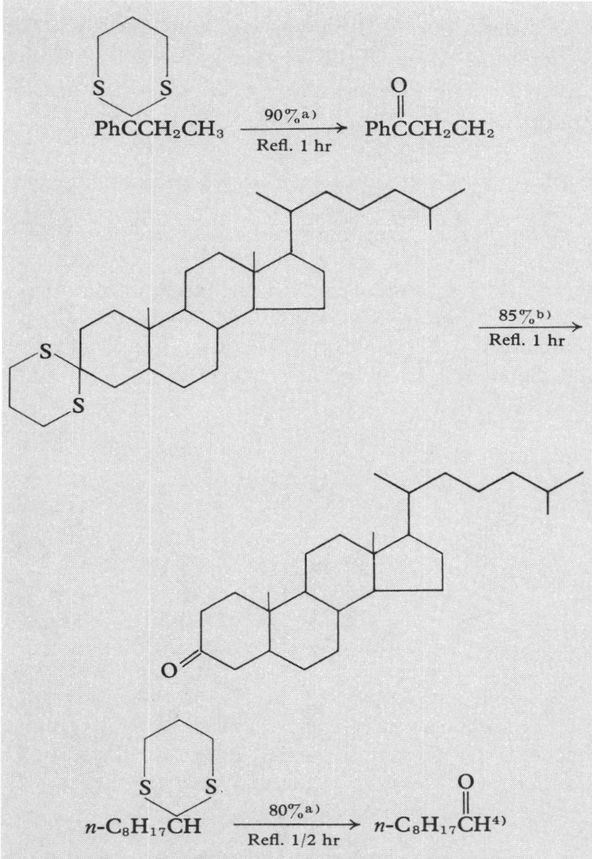

$$PhCCH_2CH_3 \xrightarrow[\text{Refl. 1 hr}]{90\%^{a)}} PhCCH_2CH_2$$

$$\xrightarrow[\text{Refl. 1 hr}]{85\%^{b)}}$$

$$n\text{-}C_8H_{17}CH \xrightarrow[\text{Refl. 1/2 hr}]{80\%^{a)}} n\text{-}C_8H_{17}CH^{4)}$$

a) Yield of pure carbonyl compound isolated
   by distillation.
b) Yield of recrystallized product.

was added to the residue and the resulting precipitate was filtered off. The filtrate, which was homogeneous by tlc and vpc, was distilled and benzylacetone (2) was obtained in 93% yield. When the reaction was carried out at room temperature for 2 hr, 2 was isolated in 83% yield.

The hydrolysis of several other 1,3-dithiane derivatives was examined by the same procedure to test the generality of this method.

From the results shown in the above equations, it was confirmed that this method is applicable to hydrolysis of a wide variety of 1,3-dithiane derivatives and the corresponding carbonyl compounds were isolated in good yields by the simple procedure.

In addition, it was found that 2-n-octyl-1,3-dithiane is readily hydrolyzed with cupric chloride to the corresponding aldehyde in higher yield as compared with the cases of the hydrolyses of 2-monoalkyl-1,3-dithianes, such as 2-benzyl-1,3-dithiane and 2-n-hexyl-1,3-dithiane with mercuric chloride[1e,f], n-hexyl-1,3-dithiane with boron trifluoride-mercuric oxide[1f] and 2-n-octyl-1,3-dithiane with silver perchlorate.[5]

In conclusion, it is noted that this method affords a convenient method for hydrolysis of 1,3-dithiane derivatives, a useful synthetic intermediates, to carbonyl compounds. The further application is now in progress.

4) In this case, the reaction was carried out under an argon atmosphere for 30 min.
5) T. Mukaiyama, S. Kobayashi, and K. Kamio, Unpublished work.

Figure 77.20. From K. Narasaka, T. Sakashita, and T. Mukaiyama, in *Bulletin of the Chemical Society of Japan* **45**, 3724 (1972).

Another procedure for the preparation of undecane-2,5-dione involves the oxidative coupling of the lithium enolates of acetone and 2-octanone by cupric chloride.

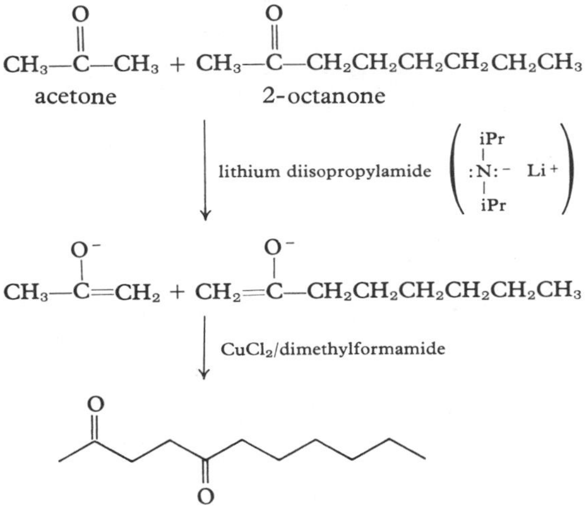

$$CH_3-\overset{O}{\underset{\|}{C}}-CH_3 \quad + \quad CH_3-\overset{O}{\underset{\|}{C}}-CH_2CH_2CH_2CH_2CH_2CH_3$$

acetone   2-octanone

lithium diisopropylamide $\left(\begin{array}{c} iPr \\ | \\ :N:^- \quad Li^+ \\ | \\ iPr \end{array} \right)$

$$CH_3-\overset{O^-}{\underset{\|}{C}}=CH_2 \quad + \quad CH_2=\overset{O^-}{\underset{\|}{C}}-CH_2CH_2CH_2CH_2CH_2CH_3$$

CuCl$_2$/dimethylformamide

375

Figure 77.21 gives a general description of this procedure as applied to the preparation of undecane-2,5-dione, the precursor of dihydrojasmone, and Figure 77.22 gives a more detailed description of the preparation of (Z)-8-undecene-2,5-dione, the precursor of jasmone.

Preparation of undecane-2,5-dione

A typical experimental procedure is as follows; a solution of diisopropylamine (5 mmol) in dry THF (5 ml) is treated with *n*-butyllithium (15% hexane solution, 5 mmol) at −78°, and after 15 min 3-methyl-butan-2-one (4.5 mmol) is added dropwise with stirring. After 15 min, anhydrous CuCl₂ (5 mmol) in DMF (5 ml) is added all at once at the same temperature. The dark green solution is stirred for an additional 30 min and then allowed to reach room temperature. The reaction mixture becomes dark brown and homogeneous. Acid work-up produces 2,7-dimethyloctan-3,6-dione (89%) with 3,3,6-trimethylheptan-2,5-dione (3%).

Next, the present method is successfully applicable to some cross couplings of two different methyl ketones and of methyl ketone with acetate, leading to unsymmetrical 1,4-diketones (RCOCH₂CH₂COR′) and γ-ketocarboesters (RCOCH₂CH₂CO₂R′), respectively (Table II). For instance, addition of CuCl₂ in DMF to a 1:3 mixture of lithium enolates of 2-octanone and

acetone at −78°, prepared by addition of a 1:3 mixture of 2-octanone and acetone into lithium diisopropylamide in THF according to the above procedure, produced undecan-2,5-dione in 73% yield (based upon the starting 2-octanone) together with 3-pentylhexan-2,5-dione (4%), hexadecane-7,11-dione (8%), and hexan-2,5-dione.

$$3CH_3COCH_3$$
$$+$$
$$1CH_3COC_6H_{13}$$
$$\xrightarrow[\text{2. } 4.5CuCl_2\text{-DMF}]{\text{1. } 4.5L\ iN(1\text{-Pr})_2}$$

$$CH_3COCH_2CH_2COC_6H_{13} \ (73\%)$$
$$CH_3COCH_2CH(C_5H_{11})COCH_3 \ (4\%)$$
$$C_6H_{13}COCH_2CH_2COC_6H_{13} \ (8\%)$$
$$CH_3COCH_2CH_2COCH_3$$

Besides the readily available starting materials and the manipulative simplicity, the high selectivity and good yield render this reaction most straightforward and useful in the preparation of undecan-2,5-dione, a precursor of dihydrojasmone.[7]

Figure 77.21. From Y. Ito, T. Konoike, and T. Saegusa, in *J. Am. Chem. Soc.* **97**, 2912 (1975).

Preparation of (Z)-8-undecene-2,5-dione

Oxidative Cross Coupling of Acetone and (Z)-5-Octen-2-one (14). Under nitrogen, a solution of diisopropylamine (9 mmol) in dry THF (10 ml) was treated with *n*-butyllithium (9 mmol, 15% hexane solution) at −78 °C, and after 15 min, a mixture of 0.26 g (2 mmol) of (Z)-5-octen-2-one and 0.35 g (6 mmol) of acetone was added dropwise. After 15 min, CuCl₂ (9 mmol) in 15 ml of DMF was added to the mixture of lithium enolates of acetone and **14** at the same temperature. The resulting green solution was stirred for an additional 30 min and then allowed to reach room temperature. The reaction mixture was treated with 3% aqueous HCl and extracted with ether. The ether solution was washed with 3% aqueous HCl and with water and dried over MgSO₄. After removal

of ether solvent, the residue was distilled in vacuo to afford 247 mg of (Z)-8-undecene-2,5-dione (**16**, 68% based on **14** used) and (Z)-4-acetyl-6-nonen-2-one (**17**, 1% based on **14** used) along with hexane-2,5-dione. Product **16** was separated and isolated by preparative GLC (1.5 m × 0.3 mm, 10% Silicone DC 550 on Shimalite W at 200 °C, H₂ carrier gas 1 kg/cm², retention time of 5.2 min). **16** was identified by comparison of its spectral data with those of authentic sample.[1]

Synthesis of *cis*-Jasmone (22). A mixture of 277 mg (1.50 mmol) of (Z)-8-undecene-2,5-dione (**16**) and 3 ml of 10% NaOH solution in methanol–water (1:1) was heated at 40 °C for 4 h. The reaction mixture was extracted with ether to afford 215 mg (87%) of jasmone (**22**). *cis*-Jasmone (**22**) was identified by comparison of its IR spectrum with that of an authentic sample.

Figure 77.22. From Y. Ito, T. Konoike, T. Harada, and T. Saegusa, in *J. Am. Chem. Soc.* **99**, 1487 (1977).

The final method that we will consider for the preparation of undecane-2,5-dione involves the addition of the equivalent of R—⁻C̈=O to the carbon-carbon double bond of methyl vinyl ketone.

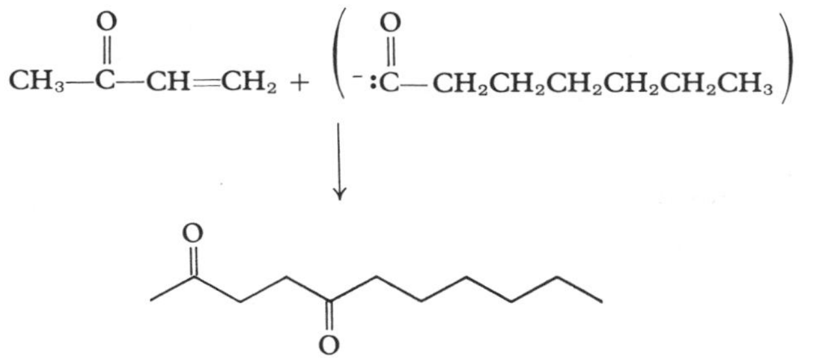

$$CH_3-\overset{\overset{\displaystyle O}{\|}}{C}-CH=CH_2 + \left({}^-:\overset{\overset{\displaystyle O}{\|}}{C}-CH_2CH_2CH_2CH_2CH_2CH_3 \right)$$

The desired carbanion equivalent is formed by treatment of heptanal with a thiazolium salt to form an intermediate that can lose the aldehyde proton to give a resonance stabilized carbanion. The process is exactly equivalent to that which is thought to account for the thiamine catalysis of the benzoin condensation (Section 71) and the decarboxylation of α-ketoacids in biological systems.

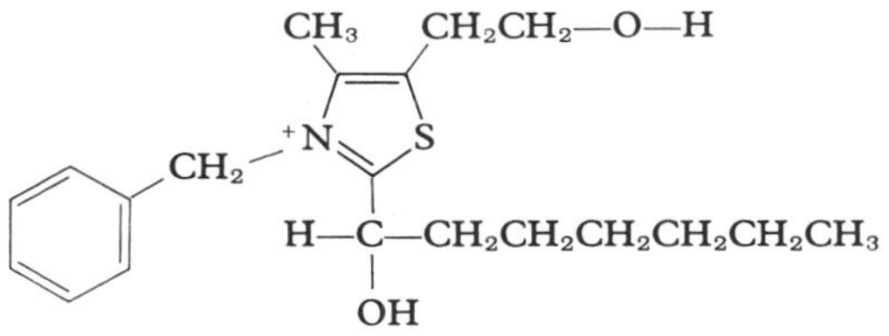

thiazolium-heptaldehyde intermediate

Figure 77.23 gives the procedures for this method of synthesis.

Procedures that are suitable for the cyclization of undecane-2,5-dione and (Z)-8-undecene-2,5-dione to dihydrojasmone and jasmone are included in Figures 77.22 and 77.23. Three additional methods are shown in Figures 77.24, 77.25, and 77.26.

Preparation of dihydrojasmone and jasmone

Addition von Aldehyden an aktivierte Doppelbindungen; VII[1]. Eine neue einfache Synthese von *cis*-Jasmon und Dihydrojasmon

Hermann STETTER, Heinrich KUHLMANN

Institut für Organische Chemie der Technischen Hochschule, D-51 Aachen, Prof.-Pirlet-Straße 1

In der letzten Mitteilung berichteten wir über Additionen aliphatischer Aldehyde an Butenon. Unter Katalyse mit Thiazoliumsalzen werden nach der Methode der katalytischen nucleophilen Acylierung in einfacher und glatter Reaktion in allen Fällen die erwarteten 1,4-Diketone erhalten. Durch alkalikatalysierte Reaktion lassen sich diese 1,4-Diketone zu 3-Methyl-2-alkyl-2-cyclopentenonen[2] cyclisieren. Die beiden Riechstoffe 3-Methyl-2-(*cis*-2-pentenyl)-2-cyclo-pentenon (*cis*-Jasmon, **3a**) bzw. 3-Methyl-2-pentyl-2-cyclo-pentenon (Dihydrojasmon, **3b**), denen zahlreiche Veröffentlichungen der letzten Zeit galten[3], sind aus den augehörigen 1,4-Diketonen, *cis*-8-Undecen-2,5-dion (**2a**) bzw. Undecan-2,5-dion (**2b**) zugänglich. Die 1,4-Diketone werden durch Addition

von *cis*-4-Heptenal (**1a**) bzw. Heptanal (Önanthaldehyd), (**1b**) an Butenon erhalten. Als Katalysator diente 3-Benzyl-5-(2-hydroxyäthyl)-4-methyl-1,3-thiazoliumchlorid (**7**).

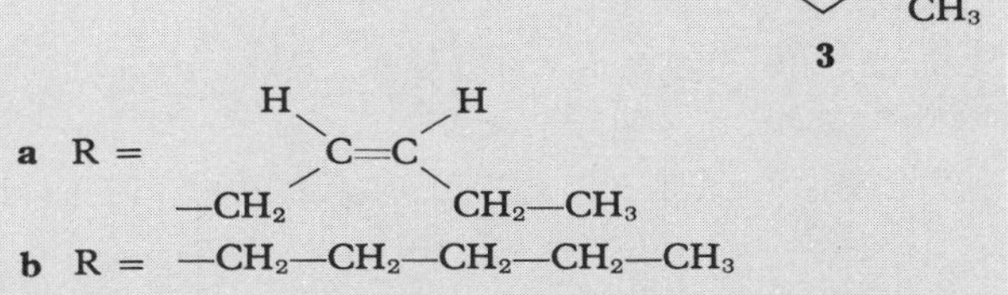

a R =

b R = $-CH_2-CH_2-CH_2-CH_2-CH_3$

Verwendet wurde käufliches Heptanal, das vor der Verwendung gereinigt wurde. Die Herstellung des *cis*-4-Heptenals erfolgte nach der Reaktionsfolge:

Aus *cis*-3-Hexenol (Blätteralkohol[4], **4**) wird 1-Bromo-*cis*-3-hexen (**5**) hergestellt[5]. Die Grignard-Verbindung von **5** wird mit Diäthyl-phenyl-orthoformiat (Diäthoxy-phenoxymethan[6]) umgesetzt. Nach der Hydrolyse erhält man 1,1-Diäthoxy-*cis*-4-hepten (**6**), das dann mit wasserfreier Ameisensäure[7] zu *cis*-4-Heptenal (**1a**) deacetalisiert wird.

1-Bromo-*cis*-3-hexen (5):
In Abänderung der Literaturvorschrift[5] konnten die Ausbeuten gesteigert werden. In einem 250-ml-Dreihalskolben mit Rührer, Trockenrohr und Tropftrichter mit Nadelventil wird ein Gemisch von *cis*-3-Hexenol (**4**; 100.1 g. 1 mol) und Pyridin (26.3 g, 0.33 mol) mit Eis–Kochsalz-Mischung gekühlt. Innerhalb von 3.5 h läßt man unter Rühren Phosphortribromid (108.3 g, 0.4 mol) hinzutropfen. Anschließend wird 2 h bei Zimmertemperatur gerührt. Das Gemisch wird auf ~190° Badtemperatur gebracht und über eine kurze Destillationsbrücke destilliert. Das Destillat wird in Äther aufgenommen, mit 10%iger Salzsäure, 10%iger Natronlauge und Wasser gewaschen, mit Magnesiumsulfat getrocknet und fraktionierend destilliert; Ausbeute: 119 g (73%) (Lit.[5], 50–60%); Kp: 149–151°/760 torr (Lit.[5], Kp: 149–151°/760 torr).

1,1-Diäthoxy-*cis*-4-hepten (6):
In üblicher Weise wird aus **5** (114 g, 0.7 mol) und Magnesium (17 g. 0.7 mol) in Äther (250 ml) die Grignard-Verbindung hergestellt. Wie in Lit.[6] läßt man Diäthoxy-phenoxy-methan (137.2 g 0.7 mol) in Äther (280 ml) hinzutropfen. Das Gemisch wird zur Aufarbeitung in Ammoniumchlorid-Lösung (30%; 750 ml) gegeben, gut durchgeschüttelt, abgetrennt, mit Wasser gewaschen, mit 10%iger Natronlauge und mit Wasser gewaschen. Nach dem Trocknen mit Kaliumcarbonat wird über eine Vigreux-Kolonne (30 cm) fraktionierend destilliert; Ausbeute: 92.5 g (70%); Kp: 82–84°/12 torr.

$C_{11}H_{22}O_2$ ber. C 70.92 H 11.90
(186.3) gef. 70.79 11.82

I.R. (kapillar) ν_{max} = 3000, 1650, 1125, 1060, 968 720 cm^{-1}.

^{1}H-N.M.R. (CDCl$_3$): δ = 5.33 (m, 2H), 4.47 (t, 1H, J = 5 Hz), 3.55 ppm (m, 4H).

cis-4-Heptenal (1a):
Wasserfreie Ameisensäure (98–100%; 415 g) und **6** (84.2 g, 0.45 mol) werden in einen Dreihalskolben mit Rührer, Rückflußkühler und Gaseinleitungsrohr gege-

ben. Unter Rühren wird 2 h bei 50° Außentemperatur im schwachen Stickstoff-Strom gehalten. Zur Aufarbeitung wird das Gemisch abgekühlt und in 1500 ml kaltes Wasser gegeben. Es wird mit Pentan extrahiert 4 × 100 ml). Danach wird mit verdünnter Natriumhydrogencarbonat-Lösung und mit Wasser gewaschen, mit Magnesiumsulfat getrocknet, das Pentan abdestilliert und das erhaltene Öl fraktionierend destilliert; Ausbeute: 33.6 g (67%); Kp: 42–44°/12 torr (Lit.[12], Kp: 28–30°/2 torr).

^{1}H-N.M.R.- und I.R.-Spektren wurden mit Lit.[12] verglichen.

3-benzyl-5-(2-hydroxyäthyl)-4-methyl-1,3-thiazolium-chlorid (7):
In einem 1000-ml Dreihalskolben mit Rührer, Rückflußkühler und Stopfen werden käufliches 5-(2-Hydroxyäthyl)-4-methyl-1,3-thiazol[8] (143.2 g, 1 mol), frisch destilliertes Benzylchlorid (126.6 g, 1 mol) und trockenes Acetonitril (500 ml) zusammengegeben. Das Gemisch wird 24 h unter Rückfluß erhitzt* und dann bei stetigem Rühren auf Zimmertemperatur gebracht. Das Produkt 7 wird durch Vakuum-Filtration isoliert, mit Acetonitril farblos gewaschen, vorgetrocknet und dann bei schwacher Drehung im Wasserstrahlvakuum (Badetemperatur 90°) getrocknet; Ausbeute: 220.8 g (82%); F: 140 140.5° (Lit.[9], F: 139–140.5°).

cis-8-Undecen-2,5-dion (2a):
In einen 250-ml-Dreihalskolben mit Rührer, Rückflußkühler (mit Trockenrohr), Tropfrichter und Gaseinleitungsrohr werden *cis*-4-Heptenal (**1a**; 22.4 g; 0.2 mol), Butenon (14 g, 0.2 mol) und das Salz 7 (5.4 g, 0.02 mol) unter Stickstoff auf 80° Badtemperatur erhitzt. Dann gibt man zügig Triäthylamin (12.2 g, 0.12 mol) zu und hält das Gemisch 12 h bei der gleichen Temperatur. Zur Aufarbeitung wird in 1%ige Schwefelsäure (500 ml) gegossen, gut durchgeschüttelt und mit Chloroform (4 × 100 ml) extrahiert. Die Chloroform-Phase wird mit Wasser, verdünnter Natriumhydrogencarbonat-Lösung und wieder mit Wasser gewaschen. Nach Abdestillieren des Lösungsmittels wird über eine Vigreux-Kolonne (15 cm) destilliert; Ausbeute: 27.6 g (76%); Kp: 90–91°/0.7 torr (Lit.[10], Kp: 135–137°/12 torr).

$C_{11}H_{18}O_2$ ber. C 72.49 H 9.96
(182.2) gef. 72.50 10.01

^{1}H-N.M.R.- und I.R.-Daten wurden mit Lit.[10] verglichen.

Undecan-2,5-dion (2b):
In gleicher Weise wie im vorstehenden Beispiel werden Heptanal (57.1 g, 0.5 mol), Butenon (35.1 g, 0.5 mol), das Salz 7 (13.5 g, 0.05 mol) und Triäthylamin (30.3 g,

---

* Bei Animpfen nach 12 h fällt das Produkt aus der siedenden Lösung aus. Kleinere Ansätze werden entsprechend ausgeführt.

0.3 mol) vereinigt, zur Reaktion gebracht und aufgearbeitet [Bei der Verwendung größerer Mengen wird die Umsetzung besser in Äthanol (300 ml für 1 mol Aldehyd) ausgeführt und bei Rückfluß gearbeit]; Ausbeute: 71.8 g (78%); Kp: 128°/8 torr; F: 33–34° (Lit.[2], Kp: 141°/14 torr, F: 33°).

cis-Jasmon (3a):

Nach der Vorschrift von Hunsdiecker[2] werden **2a** (25 g, 0.137 mol), 2%ige Natronlauge (220 g) und Äthanol (55 g) für 6 h bei Rückfluß unter Stickstoff erhitzt. Nach der üblichen Aufarbeitung wird fraktioniert; Ausbeute: 18.1 g (81%); Kp: 116–117°/9 torr (Lit.[10] Kp: 93–97°/0.8 torr).

---

[1] VI. Mitteilung: H. Stetter, H. Kuhlmann, *Tetrahedron Lett.* **1974**, 4505.
[2] H. Hunsdiecker, *Ber. dtsch. chem. Ges.* **75**, 447 (1942).
[3] R. A. Ellison, *Synthesis* **1973**, 397; Übersicht.
J. L. Herrmann, J. E. Richman, R. H. Schlesinger, *Tetrahedron Lett.* **1973**, 3275.
T. Cuvigny, M. Larchevêque, H. Normant, *Tetrahedron Lett.* **1974**, 1237.
P. A. Grieco, C. S. Pogonowski, *J. Org. Chem.* **39**, 732 (1974).
T. Wakamatsu, K. Akasaka, Y. Ban, *Tetrahedron Lett.* **1974**, 3883.
I. Kawamoto, S. Muramatsu, Y. Yura, *Tetrahedron Lett.* **1974**, 4223.
Einen vollständigen Überblick erhält man mit den in der angegebenen Literatur zitierten Stellen.

$C_{11}H_{16}O$ ber. C 80.44 H 9.83
(164.2) gef. 80.30 9.73

[1]H-N.M.R.- und I.R.-Daten wurden mit Lit.[10] verglichen.

Dihydrojasmon (3b):

In analoger Weise wie im vorstehenden Beispiel wird **2b** (70 g, 0.38 mol) mit 2%iger Natronlauge (610 g) und Äthanol (150 g) umgesetzt; Ausbeute: 56 g (89%); Kp: 113–115°/10 torr (Lit.[2], Kp: 122–124°/12 torr).

Die I.R.- und [1]H-N.M.R.-Daten wurden mit Lit.[11] verglichen.

Eingang: 5. Februar 1975

[4] F. Sondheimer, *J. Chem. Soc.* **1950**, 877.
[5] S. H. Harper, R. J. D. Smith, *J. Chem. Soc.* **1955**, 1512.
L. Crombie, S. H. Harper, R. E. Stedman, D. Thompson, *J. Chem. Soc.* **1951**, 2445.
[6] H. Stetter, E. Reske, *Chem. Ber.* **103**, 643 (1970).
[7] Nach der Methode von A. Gorgues, *Bull. Soc. Chim. France* **1974**, 529.
[8] 5-(2′-Hydroxyäthyl)-4-methyl-thiazol von Merck AG, Darmstadt, wurde ohne weitere Reinigung eingesetzt.
[9] R. F. Doerge, L. J. Ravin, H. C. Caldwell, *J. Pharm. Sci.* **54**, 1038 (1965).
[10] L. Crombie, P. Hemesley, G. Pattenden, *J. Chem. Soc.* [C] **1969**, 1024.
[11] E. Wenkert, et al., *J. Amer. Chem. Soc.* **92**, 7428 (1970).
[12] A. I. Meyers, et al., *J. Org. Chem.* **38**, 36 (1973).

Figure 77.23. From H. Stetter and H. Kuhlmann, in *Synthesis* **1975**, 379.

Preparation of dihydrojasmone

Beschreibung der Versuche.

1-Methyl-2-amyl-cyclopenten-(1)-on-(3) (Dihydrojasmon) aus Undecandion-(2.5) (Tafel 1, Nr. 4).

Man erhitzte 9.2 g Undecandion-(2.5) mit 80 g 2-proz. wäßr. Natronlauge und 20 g Alkohol 6 Stdn. am Rückflußkühler. Nach dem Abkühlen wurde ausgeäthert, der äther. Auszug mit Natriumsulfat getrocknet und der Äther abdestilliert. Der Rükstand siedete bei 122—124°/12 mm. Ausb. 7.6 g, d. s. 92% d. Th. an Dihydrojasmon

Mit Semicarbazidacetat in methylalkohol. Lösung fiel nach kurzem Sieden praktisch reines Semicarbazon vom Schmp. 176° aus. Aus diesem erhielt man nach der Zersetzung mit verd. wäßr. Säuren ein analysen- und geruchsreines Produkt vom Sdp.[12] 120–121°.

Figure 77.24. From H. Hunsdiecker, in *Berichte* **75**, 459 (1942).

Preparation of jasmone

Jasmon

1.2 g rohes Undecendion wurden mit 15 ccm 2-proz. Kalilauge 3 Stdn. gekocht. Nach der Aufarbeitung erhielt man 0.5 g rohes Jasmon vom Sdp.[14] 122—125°.

Das hieraus hergestellt Jasmon-semicarbazon schmolz bei 188—190°. Nach mehrfachem Umkrystallisieren stieg der Schmelzpunkt stetig bis auf 204°. Der Mischschmelzpunkt dieses gereinigten Semicarbazones mit natürlichem Jasmon-semicarbazon ergab keine Schmelzpunktserniedrigung.

Figure 77.25. From H. Hunsdiecker, in *Berichte* **75**, 464 (1942).

Preparation of jasmone

cis-**Jasmone** (**6**).—A mixture of undecenedione **5** (2.799 g), 7 ml of ethanol, and 25 ml of 0.5 *N* NaOH was allowed to reflux under nitrogen for 5 hr. The mixture was cooled, extracted with pentane, washed with water, dried over Na_2SO_4, and evaporated. The remaining oil (2.522 g) was distilled through a micro-spinning-band column at 80–100° (bath temperature) and 0.05-mm pressure, giving *cis*-jasmone (**6**) (2.087 g): $\nu_{max}^{CHCl_3}$ 2930, 1685, 1645, and 1385 cm^{-1}; nmr (CDCl$_2$), multiplet (2 H) at τ 4.53, doublet (2 H) at τ 6.97,

singlet (3 H) at τ 7.86, and triplet (3 H) at τ 8.98; λ_{max}^{EtOH} 234 mμ (ϵ 13.930). Thin layer chromatography (SiO$_2$, hexane + 20% ethyl acetate developed with phosphomolybdic acid) indicated two minor impurities with R_f 0.17 and 0.30 (*cis*-jasmone, R_f 0.25). Vapor phase chromatography using a Ucon Nonpolar 4-ft column at 165° revealed two components in a ratio of 8:92 with retention times of 26 and 31.5 min, respectively. The 2,4-dinitrophenylhydrazone was formed in 81% yield, mp 112–115°, and its melting point remained constant at 115–117° after one crystallization from ethanol.

Figure 77.26. From G. Büchi and H. Wüest, in *J. Org. Chem.* **31**, 977 (1966).

A SYNTHESIS FROM DIETHYL GLUTARATE, DIETHYL OXALATE, METHYL IODIDE, AND *n*-PENTYLMAGNESIUM BROMIDE

The last method that we will consider for the synthesis of jasmone is outlined in Figure 77.27. In this synthesis either the disodium salt of 1,4-dicarbethoxycyclopentane-2,3-dione (the disodium salt of 2 in Figure 77.27) or the dione itself is converted successively to 3, 4, 5, and then dihydrojasmone (1 in Figure 77.27) by a combination of alkylations, hydrolyses, and decarboxylations. The dione, compound 2, may be prepared by the procedure of Thaoker and Bagavant (5), which is to add dropwise to a refluxing solution of ethanolic sodium ethoxide (prepared from 5.0 g, or 0.22 mole of sodium and 20 ml of ethanol) a mixture of diethyl glutarate (18.8 g; 0.1 mole) and diethyl oxalate (14.6 g; 0.1 mole) and benzene (25 ml) over a period of about 30 minutes. After another 3 hours of reflux the

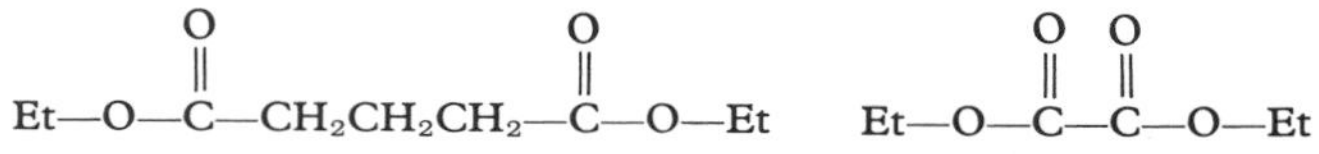

mixture is cooled, diluted with benzene, and the precipitated solid collected by suction filtration to give the disodium salt of 1,4-dicarbethoxycyclopentane-2,3-dione. Treatment of the salt with aqueous acid gives the dione of m.p. 115°C.

Preparation of dihydrojasmone

Oxocyclopentenes. Synthesis of 2-Methoxy-4-methyl-3-oxocyclopentene and its Conversion to Dihydrojasmone

A. Barco, S. Benetti, and G. P. Pollini

Instituto Chimico dell'Università, Via Scandiana 25, 1-44100 Ferrara, Italy.

Much attention has been made in recent years to the development of new synthetic routes to 1,2-disubstituted-3-oxocyclopentenes. Dihydrojasmone (**1**) represents a good test to demonstrate the feasibility of the synthetic procedures[1].

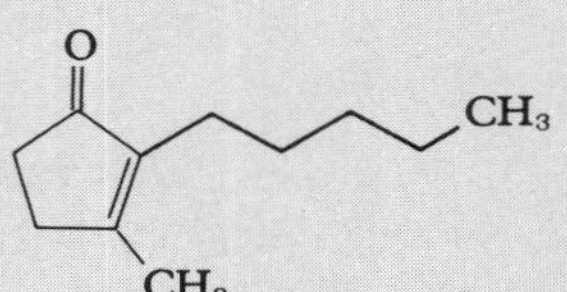

We wish to report a new synthesis of (**1**) starting from the disodium salt of 3,5-diethoxycarbonyl-1,2-dioxocyclopentane (**2**), readily available through the excellent procedure of Thaoker and Bagavant[2].

380

Alkylation of (2) with methyl iodide in acetonitrile afforded the doubly alkylated product (3) in 78% yield.

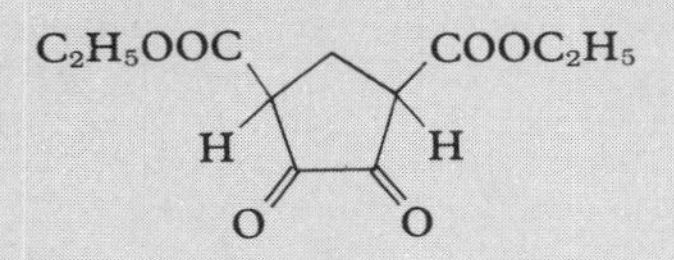

2

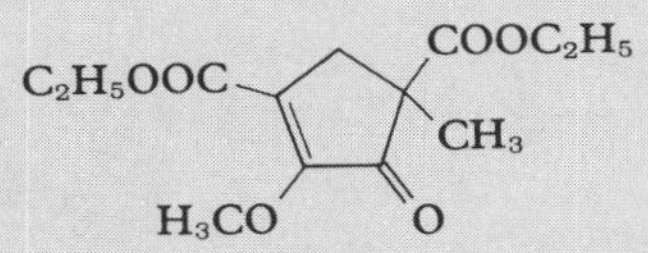

3

This concomitant C—O alkylation has been observed elsewhere [3,4] under different conditions as a by-product in respect to the predominant C-alkylation.

Saponification of (3) with aqueous 8% potassium hydroxide solution and careful acidification with 50% phosphoric acid at 0° occurred with spontaneous mono-decarboxylation, producing the acid (4) in 70% yield.

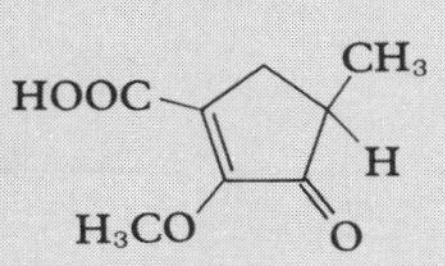

4

Distillation of 4 at 20 torr afforded almost quantitatively 2-methoxy-4-methyl-3-oxocyclopentene (5), reaction of which with n-pentylmagnesium bromide gave, after acid treatment, dihydrojasmone (1) in 89% yield.

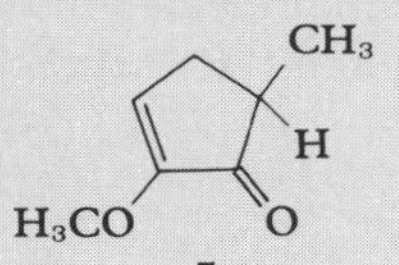

5

Compound (5) which could be a useful starting material for the preparation of 3-methyl-2-alkyloxocyclopentenes, is a new compound, since the relation product of 2-hydroxy-1-methyl-3-oxocyclopentene with diazomethane was incorrectly formulated as (5) by Hesse et al.[5] and has been proved to be the isomeric compound (6) by N.M.R. evidence[6].

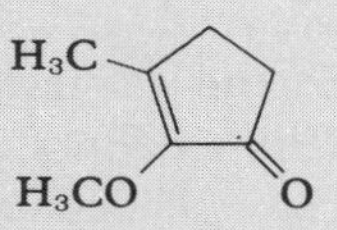

6

This method of synthesis of 2,3-disubstituted oxocyclopentenes appears to be quite general.

1,4-Diethoxycarbonyl-2-methoxy-4-methyl-3-oxocyclopentene (3):

A suspension of disodium salt of 2 (20 g), methyl iodide (40 g) in acetonitrile (200 ml) was heated under reflux for 22 hrs. Dilution with water and extraction with dichloromethane, gave, after the usual operations, an oily residue, which on distillation at 0.01 torr afforded 3; yield: 14.5 g (78%); b.p. = 100–102°. Reaction of 3,5-diethoxycarbonyl-1,2-dioxocyclopentane (2) with dimethyl sulphate in acetone solution in the presence of anhydrous potassium carbonate[7] also gave (3) in similar yield.

$C_{13}H_{18}O_6$ calc.   C 57.77   H 6.71
       found   57.63   6.69

2-Methoxy-4-methyl-3-oxocyclopentene-1-carboxylic Acid (4):

Compound 3 (20 g) was added to a solution of potassium hydroxide (30 g) in water (320 ml) and the mixture, made homogeneous with methanol, was left at room temperature for 24 hrs. The alkaline solution was acidified at 0°C with 50% phosphoric acid and the acid (4) extracted with dichloromethane. Yield 8.8 g. The acid (4), crystallized from diethyl ether, had m.p. 100–101.

$C_8H_{10}O_4$ calc.   C 56.46   H 5.92
       found   56.40   5.88

2-Methoxy-4-methyl-3-oxocyclopentene (5):

The acid 4 (5 g) was decarboxylated by distillation at 20 torr affording 3.6 g of (5) b.p. = 110–112°.

$C_7H_{10}O_2$ calc.   C 66.64   H 7.99
       found   66.58   7.94

I.R. (film): 1710 ($\nu_{C=O}$). 1625 cm^{-1} ($\nu_{C=O}$).

^{1}H-N.M.R.(CCl$_4$): δ = 1.08 (d, CH$_3$), 3.6 (s, OCH$_3$), 6.2 ppm (m, H vinylic).

The spectral properties of 2-methoxy-1-methyl-3-oxo-cyclopentene (6), prepared as described by Hesse et al.[3] are the following:

I.R. (film): 1700 ($\nu_{C=O}$), 1645 cm^{-1} ($\nu_{C=C}$).

^{1}H-N.M.R. (CCl$_4$): δ = 1.84 (s, CH$_3$), 2.20 (s, CH$_2$), 3.77 ppm (s, OCH$_3$).

Preparation of Dihydrojasmone:

Compound 5 (3 g) was added at 0° to an ethereal solution of n-pentylmagnesium bromide prepared from magnesium (0.6 g) and n-pentyl bromide (3.8 g). After an additional hour at room temperature, the mixture was treated with ice and diluted sulfuric acid (40%) and shaken for 4 hrs.

Ethereal layer was separated, dried and evaporated. The residue, on distillation, afforded 3.47 g of dihydrojasmone (1), b.p. = 60 at 0.01 torr identical with a sample prepared from 3-oxo-2-pentyl-cyclopentene following the method of Büchi[8].

2,4-Dinitrophenylhydrazone; m.p. 121–122° (from alcohol)[9].

$C_{17}H_{22}O_4N_4$ calc.   C 58.94   H 6.40   N 16.18
        found   58.99   6.35   16.02

Received: July 27, 1973; (Revised form: September 14, 1973)

[1] T. L. Ho, H. C. Ho, C. M. Wong, *Can. J. Chem.* **51**, 153 (1973) and refs. cited therein.
See also R. A. Ellison, *Synthesis* **1973**, 397.
[2] M. R. Thaoker, G. Bagavant, *J. Indian Chem. Soc.* **45**, 232 (1968).
[3] G. Komppa, *Liebigs Ann. Chem.* **370**, 218 (1910).
[4] M. A. Gianturco, P. Friedel, *Tetrahedron* **19**, 2039 (1963).

[5] G. Hesse, K. Breig, *Liebigs Ann. Chem.* **592**, 120 (1955).
[6] J. B. Son Bredenberg, *Acta Chem. Scand.* **13**, 1733 (1959).
[7] A. Barco, S. Benetti, G. P. Pollini, *Synthesis* **1973**, 316.
[8] G. Büchi, B. Egger, *J. Org. Chem.* **36**, 2021 (1971).
[9] R. A. Ellison, W. D. Woessner, *Chem. Commun.* **1972**, 529.

Figure 77.27. From A. Barco, S. Benetti, and G. P. Pollini, in *Synthesis* **1974**, 33.

References

1. R. A. Ellison, Methods for the Synthesis of 3-Oxocyclopentenes, *Synthesis* **1973**, 397.
2. T.-L. Ho, Synthesis of Jasmonoids, A Review, *Synthetic Communications* **4**, 265 (1974).
3. P. M. McCurry, Jr., and R. K. Singh, *J. Org. Chem.* **39**, 2316 (1974).
4. P. M. McCurry, Jr., and R. K. Singh, *J. Org. Chem.* **39**, 2317 (1974).
5. M. R. Thaoker and G. Bagavant, *Indian Journal of Chemistry* **1969**, 232. (The citation in footnote 2 of Figure 77.27 is incorrect.)

§78 HYDROCARBONS

Hydrocarbons of unusual structure have always been an object of synthesis. Sometimes the reason for synthesis is similar to that for first climbing Mt. Everest; sometimes it is hoped that the simplest possible version of an unusual structure will lead to an answer to an important theoretical question. Projects for the synthesis of several types of hydrocarbons are presented in the following sections. They include an allene, a cumulene, and an orange tetracyclic aromatic compound.

78.1 1,3,5-TRI*TERT*-BUTYLBENZENE

In Sections 58.3 and 58.4 procedures are given for the preparation of *tert*-butylbenzene and *p*-di*tert*-butylbenzene by treatment of benzene with *tert*-butyl chloride in the presence of aluminum chloride. Since there are reports in the literature that 1,3,5-tri*tert*-butylbenzene can be prepared by treatment of benzene (1), *p*-di*tert*-butylbenzene (1, 2, and 3) and *m*-di*tert*-butylbenzene (3) with *tert*-butyl chloride and aluminum chloride, it seems reasonable to hope that the procedures of Sections 58.3 and 58.4 could be modified so that 1,3,5-tri*tert*-butylbenzene could be isolated in a respectable yield. The changes would probably include the use of more *tert*-butyl chloride and a longer reaction time. Since *p*-di*tert*-butylbenzene is not very soluble in *tert*-butyl chloride it might be necessary or desirable to use a solvent; carbon disulfide and 1,2-dichloroethane have been used in similar reactions (see also 2).

References

1. L. R. Barclay and E. E. Betts, *Canadian Journal of Chemistry* **33**, 672 (1955).
2. P. D. Bartlett, M. Roha, and R. M. Stiles, *J. Am. Chem. Soc.* **76**, 2349 (1954).
3. S. Watarai, *Bull. Chem. Soc. Japan* **36**, 747 (1963). *Chemical Abstracts* **57**, 7393f (1963).

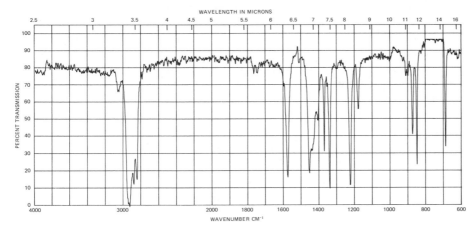

Figure 78.1. The infrared spectrum of 1,3,5-tri*tert*-butyl benzene.

78.2 *trans*-STILBENE AND *trans,trans*-1,4-DIPHENYLBUTADIENE; WITTIG SYNTHESES

One very important method of olefin synthesis is the Wittig reaction. In one version of this reaction a triphenyl alkyl phosphonium salt is treated with a strong base so as to form an inner salt, an *ylide*. Upon addition of a carbonyl compound the carbanion carbon of the ylide attacks the carbonyl carbon to form an intermediate that then eliminates triphenylphosphine oxide to give the olefin.

$$\phi\!-\!\overset{\displaystyle\phi}{\underset{\displaystyle\phi}{\overset{+}{P}}}\!-\!CH_2R \xrightarrow{\text{ base }} \phi_3\overset{+}{P}\!-\!\overset{..}{\overset{-}{C}}HR$$

ylide

$$\phi_3\overset{+}{P}\!-\!\overset{..}{\overset{-}{C}}HR + R\!-\!\overset{\displaystyle O}{\overset{\|}{C}}\!-\!H \longrightarrow \phi_3\overset{+}{P}\!-\!\overset{\displaystyle R\ \ H}{\underset{\displaystyle H\ \ R}{\overset{|\ \ \ |}{\underset{|\ \ \ |}{C\!-\!C}}}}\!-\!O^-$$

ylide   intermediate
(diastereomeric forms)

diastereomeric intermediates   ⟶   RHC=CHR   + $\phi_3\overset{+}{P}\!-\!O^-$
                             olefin   triphenylphosphine oxide
                          (Z and E forms)

It is believed that the first intermediate eliminates triphenylphosphine oxide via a second, cyclic, intermediate and that therefore the relative amounts of the Z and E forms of the olefin depend upon the relative amounts of the diastereomeric intermediates and their rates of elimination of triphenylphosphine oxide. At one

first intermediate   second intermediate   triphenylphosphine oxide

extreme the intermediates may be formed irreversibly, and so the relative amounts of Z and E olefin will depend upon the relative rates of formation of the diastereomeric intermediates since each must go on to product and each can give only one alkene. At the other extreme the intermediate may be formed reversibly. In this case it follows that the ratio of Z to E olefin then depends both upon the relative amounts of the two diastereomeric intermediates in the steady state and upon their relative rates of elimination to give olefin. If one diastereomeric intermediate eliminates rapidly compared to the other and they are present in the steady state in comparable amounts, the olefin formed from the faster reacting intermediate will predominate in the product. If the diastereomeric intermediates eliminate at comparable rates the product ratio will be determined by the ratio of the intermediates in the steady state.

In the paper by Seus and Wilson (Figure 78.2) a modification of the Wittig reaction is described in which the triphenyl phosphonium compound is replaced by a diethyl phosphonate ester. In this case the particular phosphonate ester is diethyl benzylphosphate, and it is converted to the corresponding ylide by treatment with sodium methoxide.

$$Et\text{—}O\text{—}\overset{\overset{\displaystyle :\ddot{O}:^-}{|}}{\underset{\underset{\displaystyle O\text{—}Et}{|}}{P^+}}\text{—}CH_2\phi \quad + \quad CH_3\text{—}O^-\,Na^+ \quad \longrightarrow \quad Et\text{—}O\text{—}\overset{\overset{\displaystyle :\ddot{O}:^-}{|}}{\underset{\underset{\displaystyle O\text{—}Et}{|}}{P^+}}\text{—}\ddot{C}H\phi$$

diethyl benzylphosphonate   ylide

Addition of benzaldehyde then leads to the formation of *trans*-stilbene.

$$ylide \quad + \quad \underset{H}{\overset{\phi}{{>}}}C{=}O \quad \longrightarrow \quad \underset{H}{\overset{\phi}{{>}}}C{=}C\underset{\phi}{\overset{H}{{<}}} \quad + \quad Et\text{—}O\text{—}\overset{\overset{\displaystyle :\ddot{O}:^-}{|}}{\underset{\underset{\displaystyle O\text{—}Et}{|}}{P^+}}\text{—}\ddot{O}:^-$$

benzaldehyde   *trans*-stilbene

Reactions with three other aldehydes are described in the paper, and presumably use of cinnamaldehyde would give *trans, trans*-1,4-diphenylbutadiene.

$$\underset{H}{\overset{\phi}{{>}}}C{=}C\underset{\underset{\displaystyle H}{\overset{\displaystyle |}{C{=}O}}}{\overset{H}{{<}}} \quad \longrightarrow \quad \underset{H}{\overset{\phi}{{>}}}C{=}C\overset{H}{\underset{\underset{\displaystyle \phi}{\overset{\displaystyle H}{C{=}C}}}{<}}\overset{H}{{>}}$$

cinnamaldehyde   *trans, trans*-1,4-diphenylbutadiene

Seus and Wilson also state that diethyl benzylphosphonate can be prepared from benzyl bromide and triethyl phosphite by the Michaelis-Arbuzov reaction. The less expensive benzyl chloride can be used just as well in this reaction, and the reaction can be carried out on a 0.05 mole scale by heating 5.8 ml (6.3 g; 0.05 mole) of benzyl chloride with 8.6 ml (8.3 g; 0.05 mole) of triethyl phosphite to a gentle boil under reflux for about one hour. The ethyl chloride that is formed in the reaction distills off and the boiling point of the liquid gradually rises to about 200°C. The crude diethyl benzylphosphonate can be used as soon as it has been cooled.

$$Et\text{—}O\text{—}\overset{\overset{\displaystyle O\text{—}Et}{|}}{\underset{\underset{\displaystyle O\text{—}Et}{|}}{P}}: \quad + \quad \phi\text{—}CH_2\text{—}Cl \quad \longrightarrow \quad Et\text{—}O\text{—}\overset{\overset{\displaystyle O\text{—}Et}{|}}{\underset{\underset{\displaystyle O\text{—}Et}{|}}{P^+}}\text{—}CH_2\text{—}\phi \;\; Cl^- \quad \longrightarrow \quad Et\text{—}O\text{—}\overset{\overset{\displaystyle :\ddot{O}:^-}{|}}{\underset{\underset{\displaystyle O\text{—}Et}{|}}{P^+}}\text{—}CH_2\text{—}\phi \quad + \quad Et\text{—}Cl$$

triethyl phosphite   diethyl benzylphosphonate

It also appears that *trans, trans*-1,4-diphenylbutadiene can be prepared by a further modification of the Wittig reaction, which employs the phenomenon of phase transfer catalysis (compare Section 52). According to the procedure described in the paper by Piechucki (Figure 78.3), a solution of diethyl benzyl-phosphonate and cinnamaldehyde in benzene will give *trans, trans*-1,4-diphenyl-butadiene when treated with 70% aqueous sodium hydroxide in the presence of tetra-*n*-butylammonium iodide (entry 5a, route A, in the table in Figure 78.3). Possibly 50% NaOH would be effective, and toluene could be used in place of benzene.

Figures 78.4 and 78.5 show the infrared spectra of *trans*-stilbene and *trans*-, *trans*-1,4-diphenylbutadiene.

Preparation of trans-*stilbene*

New Synthesis of Stilbene and Heterocyclic Stilbene Analogs

EDWARD J. SEUS AND CHARLES V. WILSON

Received July 14, 1961

In searching for a more convenient procedure for Wittig's olefin synthesis, Pommer[1] found that the phosphonates obtained from trialkyl phosphites *via* the Michaelis-Arbuzov reaction[2] underwent reaction with aldehydes or ketones in the presence of a strong base to give good yields of olefins. Extensive application of the reaction was made in the carotenoid field. We had applied this reaction to the synthesis of a number of known compounds when Wadsworth and Emmons[3] reported its versatility. Because of the latter publication, we should like to report some of our results in this field.

Stilbene and the heterocyclic analogs, 2-stilbazole, 2-styrylfuran, and 2-styrylthiophene, are obtained in greater than 75% yield by the reaction of diethyl benzylphosphonate with the corresponding aldehyde, using sodium methoxide in dimethylformamide. The

$$C_6H_6CH_2Br + (C_2H_5O)_3P \longrightarrow$$
$$C_6H_5CH_2PO(OC_2H_6)_2 + C_2H_5Br$$

$$C_6H_5CH_2PO(OC_2H_6)_2 + ArCHO \xrightarrow[\text{DMF}]{\text{NaOCH}_3}$$
$$C_6H_5CH{=}CHAr + (C_2H_5O)_2POONa + CH_3OH$$

Ar =

exothermic reaction gives the *trans*-olefin in a relatively pure state in 75–85% yields simply by treatment of the reaction mixture with water. The *trans* configuration is assigned to all these compounds on the basis of infrared analyses. A strong absorption at 10.2 to 10.4 μ in all these compounds is characteristic of *trans*-olefins.[4]

The best previous preparations of *trans*-stilbene are Emmons'[3] synthesis from diethyl benzylphosphonate and benzaldehyde with sodium hydride in dimethoxy-ethane and the Clemmensen reduction of benzoin.[5] *trans*-2-Stilbazole was obtained from 2-picoline and benzaldehyde[6] and from 2-picoline and toluene in acetic anhydride.[7] *trans*-2-Styrylthiophene[8] and *trans*-2-styrylfuran[9] were prepared by dehydration of the carbinol resulting from the action of benzylmagnesium chloride on the corresponding aldehydes.

The superiority of the present method is shown in Table I.

TABLE I

PREPARATION OF STYRYL DERIVATIVES

$$RCHO + (C_2H_5O)_2\overset{\text{O}}{\overset{\|}{P}}{-}CH_2C_6H_5 \xrightarrow[\text{DMF}]{\text{NaOCH}_3}$$
$$RCH{=}CHC_6H_5 + (C_2H_5O)_2POOH$$

R =	Yield, % Current	Yield, % Previous	M.P. Current[a]	M.P. Previous	Previous References
C₆H₅	85	53–57; 63	126–127[b]	123–124	5 3
(pyridine)	75	57; 51	92.5–	91	7
			93		6
(thiophene)	77	60	112–113[c]	111	8
(furan)	84	12	54–55[d]	49–50	9

[a] Melting points are corrected. [b] Recrystallized from ethanol-ethyl acetate. [c] Recrystallized from ethanol. [d] Re-crystallized from methanol.

(1) H. Pommer, *Angew. Chem.*, **72**, 911 (1960).

(2) G. M. Kosolapoff, *Organo Phosphorus Compounds*, 1st ed., J. Wiley and Sons, Inc., New York, N. Y. (1950), p. 121.

(3) W. S. Wadsworth, Jr., and W. D. Emmons, *J. Am. Chem. Soc.*, **83**, 1733 (1961).

(4) L. J. Bellamy, *The Infrared Spectra of Complex Molecules*, 2nd ed., J. Wiley and Sons, Inc., New York, N. Y. (1958), p. 49.

(5) R. L. Shriner and A. Berger, *Org. Syntheses*, Coll. Vol. **3**, 786 (1955).

(6) M. Ch. Chiang and W. H. Hartung, *J. Org. Chem.*, **10**, 21 (1945).

(7) J. Stanek and M. Horak, *Collection Czechoslov. Chem Communs.*, **15**, 1037 (1950); *Chem. Abstr.* **46**, 7100d (1952).

(8) Ng. Ph. Buu-Hoi, Ng. Hoan, and D. Lavit, *J. Am. Chem. Soc.*, **72**, 2130 (1950).

(9) R. B. Woodward, *J. Am. Chem. Soc.*, **62**, 1478 (1940).

EXPERIMENTAL

Diethyl benzylphosphonate. Diethyl benzylphosphonate was prepared in 85% yield by the Michaelis-Arbuzov reaction,[2] involving benzyl bromide and triethyl phosphite.

Stilbene and heterocyclic stilbene analogs. Diethyl benzylphosphonate (0.05M) and sodium methoxide (10% excess) were combined in dimethylformamide in a three necked flask fitted with a thermometer, drying tube, dropping funnel, and magnetic stirrer. The aromatic or heterocyclic aldehyde (0.05M) in dimethyl-

formamide (25–40 ml.) was added dropwise, with stirring and cooling in ice, at such a rate that the reaction temperature was maintained between 30° and 40°; a clear solution resulted. After standing for a short period of time, water was added to the solution, with cooling; the precipitated product was collected on a filter and washed with water. The compounds listed in the Table were prepared by this procedure.

RESEARCH LABORATORIES
EASTMAN KODAK CO.
ROCHESTER 4, N.Y.

Figure 78.2. From E. J. Seus and C. V. Wilson, in *J. Org. Chem.* **26**, 5243 (1961).

Preparation of trans,trans-*1,4-diphenylbutadiene*

Phase-Transfer Catalysed Wittig-Horner Reactions of Diethyl Phenyl- and Styrylmethanephosphonates; a Simple Preparation of 1-Aryl-4-phenylbuta-1,3-dienes

C. PIECHUCKI

Institute of General Chemistry, Technical University (Politechnika), 90–924 Łódź, Poland

The Wittig-Horner reaction is particularly useful for the synthesis of conjugated dienes and polyenes[1,2] owing to its known versatility. The advantages of this reaction over conventional Wittig olefinations result from the accessibility of starting phosphates, which are cheaper than the corresponding phosphonium salts, and from the much easier separation of the olefinic products from the reaction mixtures.

Wittig-Horner reactions of more active phosphonates were recently shown[3,4,5] to proceed under conditions of the phase-transfer catalysis[6,7] which generally simplifies numerous anionic reactions by eliminating expensive, anhydrous solvents and dangerous catalysts. It is now reported that the relatively less reactive diethyl phenylmethanephosphonate[8] (e.g. **1**) and its vinylogue **2** as well as other arylmethanephosphonates **1** can react similarly with cinnamaldehyde **3** and aromatic aldehydes **4** to give the expected 1-aryl-1-phenylbuta-1,3-dienes **5**.

The outlined reactions were carried out in an aqueous sodium hydroxide/benzene two-phase system using tetrabutylaminium iodide as the catalyst. Short heating at reflux temperature was necessary. The substrates **1** and **4** were selected so as to provide possibly representative data concerning the utility of these reactions for the preparation of dienes **5**.

An inspection of results (Table) indicates that *CH*—acidity of starting arylmethanephosphonates **1** does not influence decisively the yields of products **5a–c**. On the other hand, the pronounced carbonyl reactivity of 4-nitrobenzaldehyde and the hydrophilic properties of 3-formylpyridine are evidently responsible for considerable disproportionation of these substrates under the reaction conditions. The reactive and water-soluble 4-formylpyridine disproportionates to a much greater extent but even in this case, the desired diene **5f** was easily isolated. Besides polymeric products, only unchanged 2-pyridyl- and 4-bromophenylmethanephosphonates were obtained when these compounds and the standard 40% aqueous solution of glyoxal were used as starting materials in attempted phase-transfer catalyzed Horner syntheses of the corresponding 1,4-diaryl-buta-1,3-dienes. Addition of these alkali-labile aldehydes to a boiling benzene/aqueous sodium hydroxide system containing the phosphonate and a stoichiometric amount of tetrabutylaminium iodide gave no expected, appreciable effect.

The recently reported phase-transfer catalysed Wittig syntheses of 1,4-dipyridyl- and 1,4-dichinolylbuta-1,3-dienes are closely related to those described here but were only successful when the appropriate triphenyl-methylenephosphoranes were used as the substrates; alternative routes based on their vinylogues failed[9].

Possible extensions of the procedure given below to the synthesis of polyene analogues of the dienes **5** are exemplified by the preparation of 1,6-diphenylhexa-1,3,5-triene (**5h**).

Phase-Transfer Catalysed Wittig-Horner Reactions of Phosphonates 1 and 2; General Procedure:
A solution of phosphonate (25 mmol) and aldehyde (25 mmol) in benzene (5 ml) was added at room temperature to a stirred two-phase system consisting of benzene (20 ml), 50% aqueous sodium hydroxide (20 ml) and tetrabutylaminium iodide (1 mmol). The mixture was then refluxed for ~30 min to complete the reaction (I.R. analysis of the organic phase). The benzene layer was separated, washed with water (5 ml), and dried over magnesium sulphate. Evaporation of the solvent and recrystallisation of the solid residue from

methanol, ethanol, or petroleum ether afforded pure products **5**.

Reactions leading to dienes **5d–f** were accompanied by an exothermic effect observed on addition of substrates to a two-phase catalytic system. A water-soluble, white precipitate which simultaneously formed was probably the sodium salt of the acid resulting from disproportionation of aldehyde. In the case of **5f**, treatment of the neutralised aqueous layer with an acidified solution of copper sulphate and subsequent reaction of resultant copper isonicotinate (1.1 g, 60%) with hydrogen sulphide gave isonicotinic acid; m.p. 312–315°.

I.R. spectra of all compounds **5**, measured on Specord 7; IR instrument in $CHCl_3$ solution, showed in the region 943-1000 cm^{-1} only a single band indicating their all *trans* geometry[15]. U.V. spectral data concerning the all *trans* configuration of compounds **5a,c,e–h** are given in the corresponding references (Table). The structures of **5** were also confirmed by ^{1}H-N.M.R. spectra and by elemental analysis.

The assistance of Mr. P. Klosinski is acknowledged.

Received: October 16, 1975

[1] H. Pommer, *Angew. Chem.* **72**, 811, 911 (1960).
[2] J. Boutagy, R. Thomas, *Chem. Rev.* **74**, 87 (1974).
[3] C. Piechucki, *Synthesis* **1974**, 869.
[4] M. Mikolajczyk, S. Grzejszczak, W. Midura, A. Zatorski, *Synthesis* **1975**, 278.
[5] E. D'Incan, J. Seyden-Penne, *Synthesis* **1975**, 516.
[6] J. Dockx, *Synthesis* **1973**, 441.
[7] E. V. Dehmlow, *Angew. Chem.* **86**, 187 (1974); *Angew. Chem. Int. Ed. Engl.* **13**, 170 (1974).
[8] W. S. Wadsworth, W. D. Emmons, *J. Am. Chem. Soc.* **83**, 1733 (1961).
[9] S. Hünig, J. Stemmler, *Tetrahedron Lett.* **1974**, 3151.
[10] J. H. Pinckard, B. Wile, L. Zechmeister, *J. Am. Chem. Soc.* **70**, 1938 (1948).
[11] K. A. Huggins, O. E. Yokley, *J. Am. Chem. Soc.* **64**, 1160 (1942).
[12] H. J. Jahns, J. Müller, *Z. Chem.* **14**, 55 (1974).
[13] F. Bergmann, Z. Weinberg, *J. Org. Chem.* **6**, 134 (1941).
[14] K. Lunde, L. Zechmeister, *J. Am. Chem. Soc.* **76**, 2308 (1954).
[15] K. Lunde, L. Zechmeister, *Acta Chim. Scand.* **8**, 1430 (1954).

Table. 1-Aryl-4-phenylbuta-1,3-dienes(**5**) from Phase Transfer Catalysed Wittig-Horner Reactions of Phosphonates **1** and **2**

Diene	Ar	Yield (%)		m.p.	Brutto formula[a]
		Route A	Route B		
5a	(phenyl)	72[b]	70	147–149° lit.[10]	$C_{16}H_{14}$ (206.3)
5b	Br—(phenyl)	81	—	162–163° lit.[11]	$C_{16}H_{13}Br$ (285.2)
5c	(2-pyridyl)	75	—, 44[c]	121–123° lit.[12]	$C_{15}H_{13}N$ (207.3)
5d	O_2N—(phenyl)	—	57	170–172° lit.[13]	$C_{16}H_{13}NO_2$ (251.3)
5e	(3-pyridyl)	—	55, 59[c]	100–102° lit.[12]	$C_{15}H_{13}N$ (207.3)
5f	(4-pyridyl)	—	12, 52[c]	158–160° lit.[12]	$C_{15}H_{13}N$ (207.3)
5g	(2-furyl)	—	84, 38[c]	103–104° lit.[12]	$C_{14}H_{12}O$ (196.2)
5h	(phenyl)—CH=CH—	—	80	200–202° lit.[14]	$C_{18}H_{16}$ (232.3)

[a] All compounds gave satisfactory elemental analyses (C ± 0.3%, H ± 0.2%).
[b] 70% aqueous sodium hydroxide solution was used.
[c] From reactions performed in ethanolic solution of potassium *t*-butoxide[12].

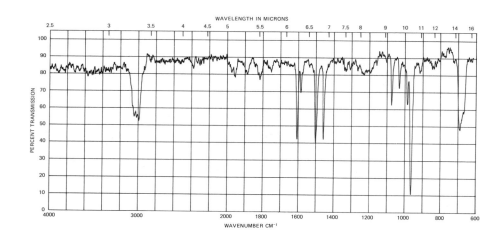

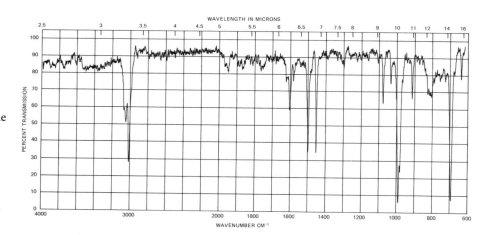

Figure 78.3. From C. Piechucki, in *Synthesis* **1976**, 187.

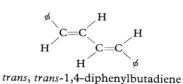

trans-stilbene

Figure 78.4. Infrared spectrum of *trans*-stilbene; CHCl$_3$ solution.

trans, trans-1,4-diphenylbutadiene

Figure 78.5. Infrared spectrum of *trans, trans*-1,4-diphenylbutadiene; CCl$_4$ solution.

Questions

1. Name *trans*-stilbene and *trans, trans*-1,4-diphenylbutadiene using the Z, E descriptors.

2. What is the total number of stereoisomeric forms possible for the first intermediate? In the case of the reaction to form stilbene, which stereoisomers will give the Z product and which will give the E product?

1. J. Reucroft and P. G. Sammes, Stereoselective and Stereospecific Olefin Syntheses, *Quarterly Reviews* **25**, 135 (1971).

2. F. A. Carey and R. J. Sundberg, *Advanced Organic Chemistry*, Part B, Plenum Press, New York (1977), pp. 53–59.

78.3 TETRAPHENYLALLENE

Tetraphenylallene is one of a series of compounds of the general type

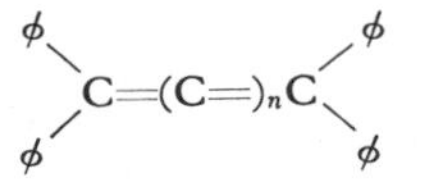

$$\phi_{\phi}\!\!>\!C\!=\!(C\!=\!)_n C\!<\!_{\phi}^{\phi}$$

The first member of the series is tetraphenylethylene ($n = 0$), and tetraphenylallene is the second ($n = 1$). A preparation for the next member of the series ($n = 2$) is presented in Section 78.5.

One synthetic route to the tetraphenylallene goes via 1,1-diphenylethylene and diphenylmethyl bromide, as described in Procedures a–c:

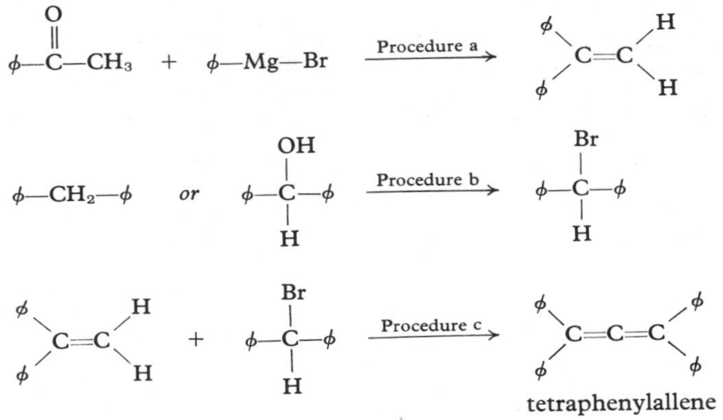

tetraphenylallene

Preparation of 1,1-Diphenylethylene (Procedure a)

Figure 78.6 presents a method for the preparation of 1,1-diphenylethylene. Some physical properties reported for this compound are: b.p., 92–93°C/0.8 mm, 113°C/2 mm; n_D^{20}, 1.6083, 1.6091.

The infrared spectrum of 1,1-diphenylethylene is shown in Figure 78.7.

Preparation of 1,1-diphenylethylene

Preparation of Some Arylalkenes

EDGAR W. GARBISCH, JR.

Received April 3, 1961

In connection with nitration studies of alkenes, the necessity arose for the preparation of a number of arylalkenes. The precursors for most of the arylalkenes were readily dehydrated tertiary arylcarbinols; however, the methods for the dehydration of many of these alcohols which have been described in the literature were found to be time-consuming.[1]

By using warm 20% sulfuric acid-acetic acid (by volume) it was observed that the crude alcohols could be converted within thirty seconds to the corresponding alkenes in over-all yields generally exceeding 70%

(based on the carbonyl compound used in the preparation).[2]

This method, as described, does not appear to be applicable to the dehydration of purely aliphatic tertiary alcohols or secondary arylcarbinols, since 1-methylcyclohexene and *trans*-stilbene were prepared in yields of only 20 to 30% from the corresponding tertiary and secondary alcohols respectively.

EXPERIMENTAL

General procedure. In each instance preparation of the carbinol was carried out by adding the carbonyl-containing reagent (Table I) to approximately 10% excess of appropriate Grignard reagent (generally 0.3–0.5 mole scale). The reaction mixture was then poured into cold ammonium chloride solution with stirring and the carbinol extracted with additional ether. The extract was washed with water, dried for a short period over sodium sulfate, and the ether was evaporated under reduced pressure on the steam bath. The resulting crude alcohol melt, while still warm, was treated with freshly prepared (warm) 20% (by volume) sulfuric acid—acetic acid (200 ml./mole of alcohol). The resulting mixture was swirled for 15–30 sec. (two phases separated immediately) and poured into ether-water (600 ml. of ether-1 l. of water per mole of alcohol). The ether was washed with water and dilute potassium bicarbonate, dried over calcium chloride, and evaporated under reduced pressure. The crude aryl-alkene residue was purified by vacuum distillation. On occasions where the crude alcohol did not melt on the steam bath, the solid was swirled with the sulfuric acid-acetic acid reagent until the solid phase had expired, and for several minutes thereafter. For those solid alkenes where purification by distillation was impracticable (see Table I), the acid dehydration mixture was poured into cold water containing excess sodium acetate. The solid was filtered, washed free of acetic acid, and recrystallized from methanol.

Acknowledgment. This work was supported by the National Science Foundation under Research Grant NSF-G7402.

DEPARTMENT OF CHEMISTRY
NORTHWESTERN UNIVERSITY
EVANSTON, ILL.

(1) Refluxing aqueous sulfuric acid has been widely employed to effect dehydration of tertiary arylcarbinols. See: (a) C. F. H. Allen and S. Converse, *Org. Syntheses*, **Coll. Vol. 1**, 266 (1941). (b) H. Adkins and W. Zartman, *Org. Syntheses*, **Coll. Vol. I**, 606 (1943).

(2) Acetic acid–sulfuric acid has been used by a number of previous investigators for dehydrations. The present method resembles most nearly that of R. Lagrave, *Ann. Chim.* (10), 8, 386 (1927).

Figure 78.6. Reprinted with permission from E. W. Garbisch, Jr., *J. Org. Chem.* **26**, 4165 (1961). Copyright by the American Chemical Society.

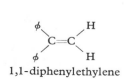

1,1-diphenylethylene

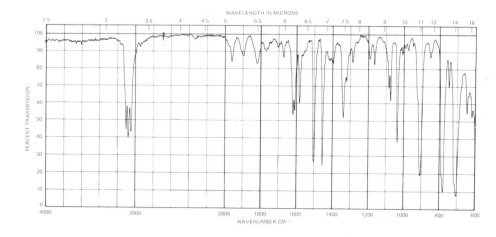

Figure 78.7. Infrared spectrum of 1,1-diphenylethylene; thin film.

Preparation of Diphenylmethyl Bromide (Benzhydryl Bromide) (Procedure b)

Figure 78.8 shows methods for the preparation of diphenylmethyl bromide from two different starting materials.

390

Preparation of diphenylmethyl bromide

DIPHENYLMETHYL BROMIDE

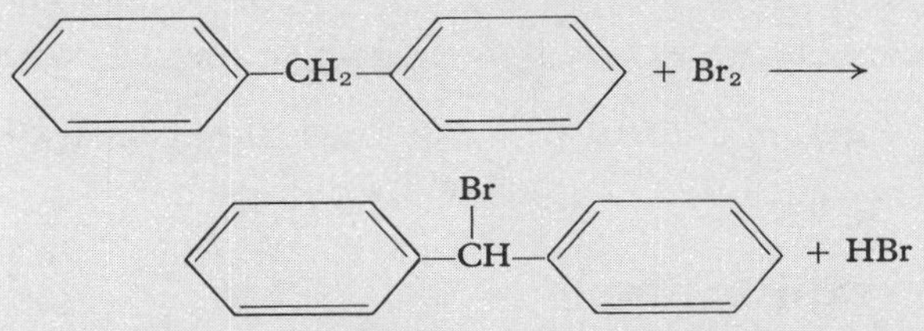

Norris, Thomas, and Brown, *Ber.*, **43** 2959 (1910).

Twenty grams (0.12 mole) of diphenylmethane is heated to 150–160° while 7 ml. (22 g., 0.14 mole) of bromine is added slowly. (*Hood.*) The product is distilled under vacuum, and diphenylmethyl bromide boiling at 183°/23 mm. is obtained in 86% yield.

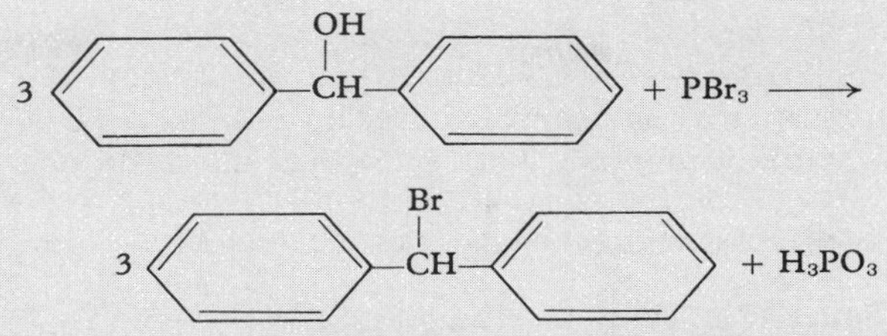

Claisen, Kremers, Roth, and Tietze, *Ann.*, **442**, 245 (1925).

To a solution of 42 g. (0.23 mole) of benzhydrol [*Org. Syntheses* Coll. Vol. **1,** 90 (1941)] in 125 g. of carbon tetrachloride is added slowly 27 g. (0.10 mole) of phosphorus tribromide. The mixture is allowed to stand for a day and then is heated to 60–70° for 6 hours. The reaction mixture is filtered, washed with ice water then with dilute sodium acetate solution, dried over calcium chloride, and distilled. The diphenylmethyl bromide is obtained in 65% yield (36 g.) boiling at 172–176°/14 mm. and melting at 43–45°.

Figure 78.8. From D. A. Shirley, in *Preparation of Organic Intermediates,* Wiley, New York, 1951, p. 136. Reproduced with permission.

Preparation of Tetraphenylallene (Procedure c)

Figure 78.9 gives a method for the preparation of tetraphenylallene from 1,2-diphenylethylene and diphenylmethyl bromide.

Preparation of tetraphenylallene

EXPERIMENTAL

1,1,3,3-*Tetra-arylpropenes.*—This is exemplified by the preparation of tetraphenylpropene. (*a*) A solution of diphenylmethyl bromide (2·47 g., 0·01 mole) or chloride (2·02 g), and 1,1-diphenylethylene (1·8 g., 0·01 mole) in dry benzene (30 c.c.) was refluxed on the water-bath for 6 hr., becoming yellow and then green with evolution of halogen acid. Benzene was distilled and the residue was digested for 30 min. with potassium hydroxide (1 g.) in alcohol (30 c.c.). The mixture was cooled, diluted with water (50 c.c.), and extracted with ether and the extract was washed with water. The ether was distilled off and the residue recrystallised from alcohol from which 1,1,3,3-tetraphenylpropene[1a] (2·1 g.) separated as colourless crystals, m. p. 127° (Found: C, 93·5; H, 6·4. Calc. for $C_{27}H_{22}$: C, 93·6; H, 6·4%). The treatment with alkali was merely precautionary because, as just indicated, the intermediate lost halogen acid spontaneously in the hot reaction mixture.

(*b*) Experiment (*a*) was repeated without use of a solvent, the mixture being heated in a boiling-water bath for 6 hr. and the product treated with alcoholic potassium hydroxide as above. 1,1,3,3-Tetraphenylpropene (2 g.) (from alcohol) had m. p. and mixed m. p. 127°.

(*c*) Hydrogen chloride or bromide gas was passed for 1 hr. through a boiling solution of diphenylmethanol (1·84 g., 0·01 mole) and 1,1-diphenylethylene (1·8 g., 0·01 mole) in dry benzene, then boiling was continued for further 5 hr., the same changes in colour as in (*a*) occurring; 1,1,3,3-tetraphenylpropene (2·1 g.) was obtained.

1,1,3,3-*Tetra-aryl-2-bromopropenes.*—To the tetra-arylpropene (0·01 mole) in ether (30 c.c.), bromine (1·6 g., 0·01 mole) in ether (10 c.c.) was added. After evaporation the residue was recrystallised from alcohol–acetone (2:1). Thus were obtained as colourless crystals in almost quantitative yield: 2-bromo-1,1,3,3-tetraphenylpropene,[1a,b] m. p. 124° (Found: C, 76·1; H, 4·8; Br, 18·4. Calc. for $C_{27}H_{21}Br$: C, 76·2; H,

Figure 78.9. From W. Tadros, A. B. Sakla, and A. A. A. Helmy, in *J. Chem. Soc.* **1961**, 2688. Reproduced with permission.

78.4 1,1,4,4-TETRAPHENYLBUTADIENE

1,1,4,4-Tetraphenylbutadiene has been prepared according to the following general scheme:

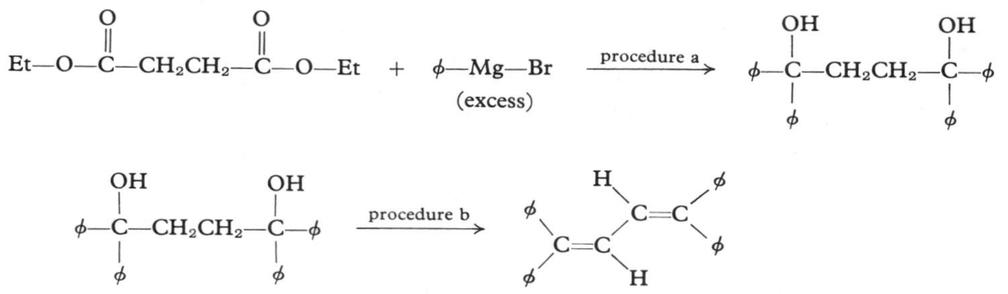

Preparation of 1,1,4,4-Tetraphenylbutane-1,4-Diol
(Procedure a)

Figures 78.10, 78.11 and 78.12, present procedures for the preparation of 1,1,4,4-tetraphenylbutane-1,4-diol.

Preparation of 1,1,4,4-tetraphenylbutane-1,4-diol

Oxalester.

39 g Bromobenzol und 6.5 g Magnesiumband werden mit 50 ccm absolutem Aether erwärmt bis alles Magnesium gelöst ist. Darauf lässt man die Lösung von 7 g Oxalsäureäthylester in der 4-fachen Menge Aether langsam zufliessen. Jeder Tropfen verursacht eine deutlich wahrnehmbare Reaction. Nach einigem Stehen wird in angesäuertes Eiswasser gegossen und der nach dem Verdunsten des Aethers hinterbleibende Rückstand aus Alkohol oder Eisessig umkrystallisirt. Nadeln (aus Eisessig) vom Schmp. 181–182°. Allen Eigenschaften zu Folge liegt β-Benzpinakolin vor. Ob Benzpinakon als Zwischenproduct der Reaction anzunehmen ist, konnte bisher nicht ermittelt werden. Bei der Darstellung darf nicht gekocht werden, da die sonst gute Ausbeute hierdurch verschlechtert wird.

Bernsteinsäureester.

Versuchsanordnung und Mengenverhaltnisse sind dieselben wie beim Oxalester. Schon beim Versetzen mit Wasser fällt der grösste Theil des Reactionsproductes als weisse Masse aus, den Rest gewinnt man durch Verdunsten der ätherischen Lösung. Ausbeute quantitativ. Nadeln aus Eisessig oder Alkohol vom Schmp. 202°. Die Lösung in concentrirter Schwefelsäure ist röthlich gefärbt.

0.1183 g Sbst.: 0.3680 g CO_2, 0.0731 g H_2O. —0.1708 g Sbst.: 0.5353 g CO_2, 0.1015 g H_2O.
$(C_6H_5)_2C(OH).CH_2.CH_2.C(OH)(C_6H_5)_2 = C_{28}H_{26}O_2$.

Ber. C 85.22,   H 6.66.
Gef. » 84.84, 85.48, » 6.87, 6.6

Zürich. Chemisches Institut der Universität.

Figure 78.10. From W. Dilthey and E. Last, in *Berichte* **37**, 2640 (1904).

Preparation of 1,1,4,4-tetraphenylbutane-1,4-diol

Action du bromure de phényle-magnésium sur l'oxalate de méthyle. — Dans une solution éthérée de bromure de phényle-magnésium dans l'éther, préparée au moyen de Mg 48 gr., C^6H^5Br 314 gr., on laisse tomber peu à peu une solution d'oxalate de méthyle dans l'éther anhydre. Il se produit une réaction très vive. Quand la totalité de l'oxalate de méthyle a été ajoutée, on termine la réaction en chauffant au bain-marie pendant 20 heures. On décompose ensuite par l'acide acétique étendu employé en quantité calculée; puis on chauffe le mélange au bain-marie pour déterminer la distillation de l'éther. On obtient ainsi une liqueur tenant en suspension un corps solide que l'on sépare et que l'on purifie par deux cristallisations dans l'acide acétique bouillant. On isole ainsi 90 gr. d'un corps blanc bien cristallisé dont le point de fusion varie avec la manière de le chauffeur. Si l'on chauffe assez rapidement, la fusion s'opère à 181°; au contraire si l'on élève lentement la température, la substance fond à 169° (Zagumeny indique 168° comme point de fusion de la benzopinacone). Dans les deux cas le liquide incolore ainsi obtenu ne reprend pas l'état solide par refroidissement.

Analyses. — (I) Subst., $0^{gr},2700$: CO^2, $0^{gr},8437$; H^2O, $0^{gr},1488$ — (II) Subst., $0^{hr},3033$: CO^2, $0^{gr},9476$; H^2O, $0^{gr},1650$ — soit en centièmes, trouvé: C, 85.22 et 85.20; H, 6.11 et 6.04 — calculé pour

$$(C^6H^5)^2C\!-\!\!-\!\!-\!\!-C(C^6H^5)^2$$
$$\quad\ |\qquad\quad\ |$$
$$\quad\ OH\quad\ OH$$

: C, 85.24; H, 6.01.

J'ai préparé de même cette benzopinacone par l'action de C^6H^5–Mg–Br sur le benzile C^6H^5-CO-CO-C^6H^5 d'après la réaction suivante:

I $\quad C^6H^5$-CO-CO-$C^6H^5 + 2C^6H^5MgBr$
$$= (C^6H^5)^2C\!-\!\!-\!\!-\!\!-C(C^6H^5)^2$$
$$\qquad\quad\ |\qquad\quad\ |$$
$$\qquad\quad\ OMgBr\quad\ OMgBr$$

II $\quad (C^6H^5)^2C(OMgBr)$-$C(OMgBr)(C^6H^5)^2 + 2H^2O$
$$= MgO + MgBr^2 + (C^6H^5)^2C(OH)\text{-}$$
$$C(OH)(C^6H^5)^2.$$

Le produit obtenu présentait les mêmes caractères que celui préparé à partir de l'oxalate et a donné à l'analyse. Trouvé: C, 85.32 et 85.15; H, 6.11 et 6.08 — calculé: C,85.24; H, 6.01.

Action de C^6H^5-Mg-Br sur le succinate d'éthyle. Tétraphénylbutanediol. — La réaction a été conduite comme dans le cas de l'oxalate en remplaçant toutefois la chauffe au bain-marie par un contact de 12 heures à froid. On obtient après décomposition par l'acide acétique un produit solide qui, après lavage, a été mis en suspension dans l'eau et soumis à l'action d'un courant de vapeur d'eau; celui-ci entraîne un produit qui se solidifie dans le réfrigérant en belles paillettes blanches nacrées. Ces paillettes ont été recueillies et purifiées par cristillisation dans l'alcool bouillant. Elles présentent tous les caractères du biphényle C^6H^5-C^6H^5 f. à 71°. Eb. 252°.

Analyses. — (I) Subs., $0^{gr},2767$; CO^2, $0^{gr},9434$; H^2O, $0^{gr},1653$ — (II) Subst., $0^{gr},3565$; CO^2, $1^{gr},2155$; H^2O, $0^{gr},2096$ — soit en centièmes, trouvé: C, 92.99 et 92.98; H, 6.63 et 6.53 — calculé pour C^6H^5-C^6H^5: C, 93.50; H, 6.49.

Le produit non entraîné par la vapeur d'eau est ensuite essoré, séché, puis soumis à la cristallisation dans le chloroforme ou l'acétone bouillant. On obtient ainsi, avec un rendement de 80 0/0, le tétraphénylbutanediol.

Le *tétraphénylbutanediol*

$$(C^6H^5)^2C\text{-}CH^2\text{-}CH^2\text{-}C(C^6H^5)^2$$
$$\qquad\quad |\qquad\quad\ |$$
$$\qquad\quad OH\qquad\ OH$$

est un corps blanc insoluble dans l'eau, assez peu soluble dans l'alcool même bouillant très soluble dans le chloroforme et dans l'acétone bouillants. Il cristallise dans ce dernier solvant en belles tables transparentes qui ne tardent pas à devenir opaques à l'air en perdant de l'acétone.

L'acétone a été dosée en mesurant la perte à 100°, $3^{gr},7469$ de substance ont ainsi perdu $0^{gr},5035$, soit 13,43 0/0; calculé pour $C^{28}H^{26}O^2 + C^3H^6O$; acétone 12,83 0/0. Le produit privé d'acétone fond à 208° et fournit à l'analyse les chiffres suivants: subst., $0^{gr},2411$: CO^2, $0^{gr},7552$ et H^2O, $0^{gr},1467$ — soit en centièmes trouvé: C, 85.42; H, 6.76 — calculé pour $C^{28}H^{26}O^2$: C, 85.27; H, 6.64.

Figure 78.11. From A. Valeur, in *Bull. Soc. Chim. Fr.* [3] **29**, 683 (1903).

Preparation of 1,1,4,4-tetraphenylbutane-1,4-diol

Action of Phenylmagnesium Bromide on Ethyl Oxalate.

Benzpinacone.—This substance was made to ascertain with what ease such glycols could be prepared from the esters of dibasic acids. As pointed out in the introduction, the work will be continued on a large number of aliphatic dibasic acids and the related hydroxy acids.

Dilthey and Last[2] state that only β-benzpinacolin, m. p. 182°, is formed by this reaction. I obtain pure benzpinacone.

To phenylmagnesium bromide (from 40 grams brombenzene and 10 grams magnesium turnings) ethyl oxalate (8 grams) was added slowly; both were in absolute ethereal solution. The mixture was boiled one hour in a reflux apparatus, then poured into cold, dilute sulphuric acid, well shaken, the ethereal layer dried with calcium chloride, the ether distilled and the residue crystallized from alcohol. Pure benzpinacone (12 grams), melting at 186°–187°, was obtained. When a specimen of this was mixed with one of benzpinacone, prepared from methyl benzsilicate and phenylmagnesium bromide,[1] the melting-point was unchanged.

As Dilthey and Last did not analyze their compound, it seems probable that it consisted of impure benzpinacone, because even a small amount of impurity lowers the melting-point of this substance considerably.

It was very important to determine if the normal compound (pinacone), or the rearranged substance (pinacolin), is the first product of the action of Grignard's reagent on the esters of these dibasic acids, for it is very probable, indeed nearly certain, from work that will be described below, that the substances obtained from succinic and glutaric esters and their homologues, eliminate water very readily and form inner anhydrides—derivatives of tetramethylene and pentamethylene oxides. Analysis:

0.1358 gram substance gave 0.4236 gram CO_2 and 0.0790 gram H_2O.

	Calculated for $C_{26}H_{22}O_2$.	Found.
C	85.20	85.07
H	6.07	6.46

Action of Phenylmagnesium Bromide on Ethyl Succinate.

Tetraphenyltetramethylene Glycol,
$(C_6H_5)_2CCH_2CH_2$—$C(C_6H_5)_2$.—This substance was
    OH   OH

prepared in the same manner as the benzpinacone. I have nothing to add to its properties, as described by Dilthey and Last,[2] except that my specimen melted at 206° instead of 202°.

Ten grams of ethyl succinate, in absolute ether, were added, gradually to a similar solution of phenylmagnesium bromide (from 40 grams of brombenzene and 7 grams of magnesium turnings), which was cooled in ice. The solution was allowed to stand several hours and was then poured slowly into a separatory-funnel containing finely divided ice. After thorough shaking, just enough dilute sulphuric acid was added to dissolve the magnesium hydroxide. The aqueous solution was removed and the cold ether layer filtered from the precipitated glycol. When dried and evaporated, the ethereal filtrate yields a residue containing more of the glycol. The latter is always accompanied by a substance, m. p. 163°–165°, formed by the loss of 1 molecule of water from a molecule of the glycol. When a mixture of the two is crystallized from acetone, the glycol is readily obtained pure, but two or three crystallizations of the mixture from boiling glacial acetic acid are sufficient to convert it completely into the anhydride. The same result is obtained by boiling the glycol for a short time with acetyl chloride. The pure glycol melts at 206°. Analysis:

0.1539 gram substance gave 0.4775 gram CO_2 and 0.0945 gram H_2O.

	Calculated for $C_{28}H_{26}O_2$.	Found.
C	85.22	84.62
H	6.66	6.82

[1] Acree: Ber. d. chem. Ges., **37**, 2761

[2] Ber. d. chem. Ges., **37**, 2639.   [2] *Loc. cit.*

Figure 78.12. From S. F. Acree, in *Am. Chem. J.* **33**, 190 (1905).

Preparation of 1,1,4,4-Tetraphenylbutadiene (Procedure b)

Figures 78.13 and 78.14 present procedures for the dehydration of 1,1,4,4-tetraphenylbutane-1,4-diol to 1,1,4,4,-tetraphenylbutadiene.

Preparation of 1,1,4,4-tetraphenylbutadiene

Beschreibung der Versuche.

1.1.4.4-Tetraphenyl-butadien-1.3.

1. Eine Lösung von 15 g 1.1.4.4-Tetraphenyl-butandiol-(1.4)[12] in 250 ccm Eisessig und 25 ccm konz. Salzsäure kocht man 10 Min., wobei sich das Butadien größtentiels ausscheidet. Nach dem Erkalten wird das gelbkrystalline Reaktionsprodukt abgesaugt und aus Benzol umkrystallisiert, aus dem es in farblosen, bläulich fluorescierenden Krystallen vom Schmp. 201° ausfällt. Beim Verreiben einer Probe der Substanz mit konz. Schwefelsäure verstärkt sich die

Fluorescenz, wobei eine rosa gefärbte Lösung entsteht, deren Farbe über hellgrün allmählich in schmutzigbraun übergeht.

II. Alle Versuche, im 1.1.4.4-Tetraphenyl-butandiol-(1.4) die beiden Hydroxylgruppen gegen Halogen auszutauschen, führten nicht zum 1.1.4.4-Tetraphenyl-1.4-dihalogen-butan, sondern zum 1.1.4.4-Tetraphenyl-butadien: 1. Sättigen einer Lösung des Glykols in Eisessig mit Bromwasserstoff, 2. Umsetzung des in Chloroform gelösten Glykols mit Thionylchlorid.

[12]) vergl. Valeur, Bull. Soc. chim. France [3] **29**, 683 [1903].

Figure 78.13. From G. Wittig and F. von Lupin, in *Berichte* **61**, 1630 (1928). Reproduced with permission of Verlag Chemie, GmbH, Weinheim.

Preparation of 1,1,4,4-tetraphenylbutadiene

Tétraphénylbutadiène $(C^6H^5)^2C{=}CH{-}CH{=}C(C_6{-}H^5)^2$. — Dans une solution bouillante de 15 gr. de tétraphénylbutanediol (ou de tétraphényltétrahydrofurfurane) dans 250 cc. d'acide acétique cristallisable, on laisse tomber 25 cc. d'acide chlorhydrique concentré et l'on maintient l'ébullition pendant une demi-heure. Le mélange se trouble de plus en plus et s'épaissit à mesure que l'opération s'avance; on laisse refroidir et l'on sépare la substance solide qui s'est déposée; on en recueille ainsi 14 gr. On peut, dans cette opération, remplacer HCl par 15 cc. de SO^4H^2; dans ce cas il suffit de chauffer quelques minutes seulement pour que le mélange se prenne en une masse feutrée d'aiguilles. Le tétraphénylbutadiène cristallise dans l'acide acétique bouillant en belles aiguilles feutrées à reflet violacé fusibles à 202°; il se dissout également dans le benzène bouillant et s'en dépose par refroidissement en beaux cristaux d'un blanc d'argent avec reflets verdâtres. Ces cristaux perdent lentement du benzène à la température ordinaire et rapidement dès qu'on les chauffe.

1^{gr},5658 maintenus à 100°, jusqu'à poids constant, ont perdu 0^{gr},2936, soit benzène 18,7 0/0; calculé pour $C^{28}H^{22} + C^6H^6$, benzène 17,8 0/0.

Le produit privé de benzène fournit à l'analyse les résultats suivants : subst., 0^{gr}.1940; CO^2, 0^{gr}6642; H^2O, 0^{gr},1103 — soit en centièmes, trouvé : C, 93.37; H, 6.31 — calculé pour $C^{28}H^{22}$: C, 93.85; H, 6.14.

Bien qu'il soit assez peu soluble dans le benzène froid, une détermination cryoscopique a pu néanmoins être effectuée dans ce solvant : subst., 0^{gr}6757; benzène, 60,36; abaissement, 0^0,16; M=342 — caculé pour $C^{28}H^{22}$: M=358.

Figure 78.14. From A. Valeur, in *Bull. Soc. Chim. Fr.* [3] **29**, 687 (1903).

78.5 1,1,4,4-TETRAPHENYLBUTATRIENE

The golden yellow hydrocarbon 1,1,4,4-tetraphenylbutatriene has been prepared by treatment of 1,1,4,4-tetraphenylbutyne-1,4-diol with various reducing agents.

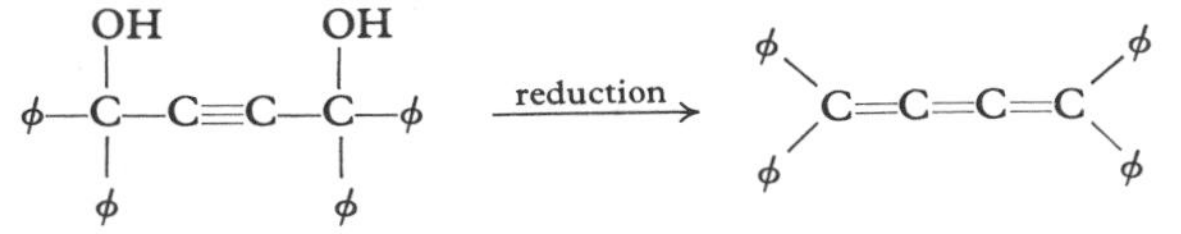

The diol has been prepared by several variations of a procedure which essentially involves the addition of one molecule of acetylene to two molecules of benzophenone.

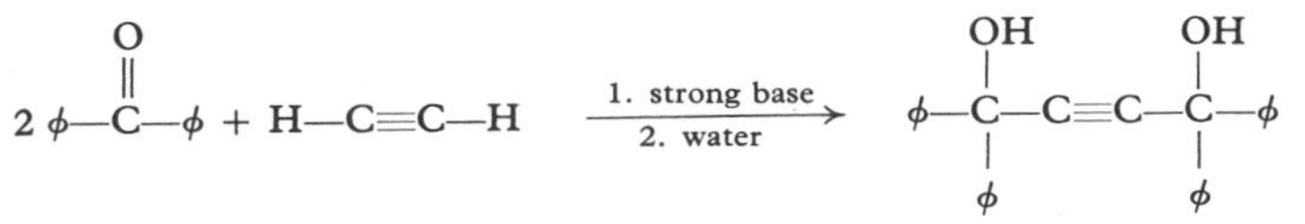

Figures 78.15 through 78.17 present three preparations of 1,1,4,4-tetraphenylbutatriene.

Preparation of 1,1,4,4-tetraphenylbutatriene

1,1,4,4-Tetraphenylbutyne-1,4-diol.—A solution of ethylmagnesium bromide was prepared from 24.3 g. (1.0 g.-atom) of magnesium turnings and 120 g. (1.1 moles) of ethyl bromide in 600 ml. of dry ether. Into this solution was passed purified acetylene[21] at a moderate rate, with stirring, at room temperature, for 18 hr. At this point, a solution of 200.2 g. of benzophenone in 600 ml. of dry benzene was added with rapid stirring, over 1 hr. The reaction mixture after remaining overnight was decomposed by pouring into excess acidified ice-water. The benzene–ether layer was separated and the remaining aqueous solution and solids were extracted with ether. These extracts were combined with the previously separated ether–benzene mixture and concentrated to give a slurry of fine white solid. Filtration gave 110 g. (28%) of the diol, m.p. 191–192° (reported[22] 193°). Recrystallization from toluene gave 102 g., m.p. 193.0–193.5°.

1,1,4,4-Tetraphenylbutatriene.—To a refluxing solution of 16.0 g. (0.041 mole) of 1,1,4,4-tetraphenyl-butyne-1,4-diol and 16.6 g. (0.10 mole) potassium iodide in 60 ml. of ethanol was added, dropwise, over 15 min., a solution of 4.0 g. of concd. sulfuric acid in 40 ml. of ethanol. Refluxing was continued for 1 hr., the mixture was allowed to cool to room temperature, and was filtered and washed with ethanol.

Golden yellow crystals, 11.5 g. (79%), m.p. 233–235° (reported[23] m.p. 235°) were obtained. Recrystallization from acetic acid, then xylene, brought the m.p. to 235–236°.

All melting points are uncorrected.
(21) Acetylene was purified by successively passing through a Dry Ice–acetone trap, concd. sulfuric acid, sodium hydroxide pellets and finally a column of Linde 4A molecular sieves.

Figure 78.15. Reprinted with permission from A. Zweig and A. K. Hoffmann, *J. Am. Chem. Soc.* **84**, 3282 (1962). Copyright by the American Chemical Society.

Preparation of 1,1,4,4-tetraphenylbutatriene

Alkali Metal Adduct Solutions.—All adducts were prepared in a three-neck flask equipped with a mechanical stirrer. The alkali metal acceptor, the solvent, and the metal were introduced in a stream of nitrogen.

1,1,4,4-Tetraphenylbutatriene (I).—Compound I was prepared from 1,1,4,4-tetraphenylbutyne-1,4-diol.[10] Dupont used ethylmagnesium bromide, acetylene, and benzophenone as his reactants. We found the following procedure to give better results. Sodium (47 g., 2 g.-atoms) was stirred for 12 hr. under nitrogen with 2.1 of tetrahydrofuran and 320 g. (2.06 moles) of biphenyl. Acetylene was then bubbled into the blue solution until decolorization was complete. To the white slurry of sodium acetylide, 365 g. (2 moles) of powdered benzophenone was added, and the mixture was refluxed, with stirring, for 1 hr. After cooling, the contents of the flask were poured into a large excess of water. Petroleum ether (b.p. 66–75°) (500 ml.) was added, and the butynediol was removed by filtration. The product was washed with water, petroleum ether, and dried, yield 86% m.p. 192° (lit.[10] m.p. 193°).

1,1,4,4-Tetraphenylbutyne-1,4-diol was converted to 1,1,4,4-tetraphenylbutatriene by the stannous chloride–hydrogen chloride method.[11] The product was recrystallized from xylene, yield 71%, m.p. 239° (lit.[11] m.p. 236°). The diol was also converted to the butatriene by the potassium iodide method,[12] yield 78%.

Figure 78.16. Reprinted with permission from R. Nahon and A. R. Day, *J. Org. Chem.* **30**, 1975 (1965). Copyright by the American Chemical Society.

Preparation of 1,1,4,4,-tetraphenylbutatriene

1,1,4,4-Tetraphenyl-2-butyne-1,4-diol (I).—Lithium acetylide–ethylenediamine complex[6] (100 g, *ca.* 1 mole) was refluxed, under nitrogen, with 1000 ml of tetrahydrofuran and 182 g (1 mole) of benzophenone for 2 hr. After cooling, the mixture containing the precipitated lithium glycolate was poured with stirring into 2000 ml of water containing 150 ml of concentrated hydrochloric acid. The diol was then extracted with ether. The ether layer was washed with 10% sodium bicarbonate, and dried (MgSO₄). The glycol, obtained by evaporating the ether, was washed with cyclohexane and recrystallized from toluene (yield 76%), mp 192–193° (lit.[7] mp 193°).

1,1,4,4-Tetraphenyl-1,4-dimethoxy-2-butyne (II).—1,1,4,4-Tetraphenyl-2-butyne-1,4-diol (10 g) was dissolved in 150 ml of warm methanol. A solution of 1 ml of concentrated sulfuric acid in 10 ml of methanol was added. After shaking for 30 min, the solution was allowed to stand for 24 hr. The solid product was washed with methanol, aqueous ammonia, and again with methanol. It was recrystallized from ethanol (yield 10.4 g), mp 111–112° (lit.[4] mp 112°).

1,1,4,4-Tetraphenylbutatriene (III). **Method A.**—This involved treating the acetylenic diol (I) with stannous chloride and hydrogen chloride in ether solution.[1,2]

Method B.—Potassium iodide (34 g, 0.204 mole) and 34.4 g (0.082 mole) of 1,1,4,4-tetraphenyl-1,4-dimethoxy-2-butyne were added to 200 ml of ethanol and the solution was heated to reflux. A solution of 8 g of concentrated sulfuric acid in 50 ml of ethanol was then added dropwise and the mixture was refluxed for 4 hr. After cooling, the solid was removed, washed with water, ethanol, and water, respectively, and recrystallized from xylene (yields 70–75%), mp 235–236°.

Figure 78.17. Reprinted with permission from S. F. Sisenwine and A. R. Day, *J. Org. Chem.* **32**, 1772 (1967). Copyright by the American Chemical Society.

78.6 TETRACENE

The tetracyclic aromatic compound tetracene absorbs light at a relatively long wavelength that gives it an orange color. The analogous substance with one less ring (anthracene) is colorless, while the substance with one more ring (pentacene) is blue. Figure 78.18 presents the procedures for the preparation of tetracene from phthalic anhydride and tetralin according to the following plan.

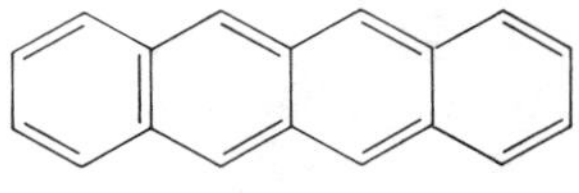

tetracene

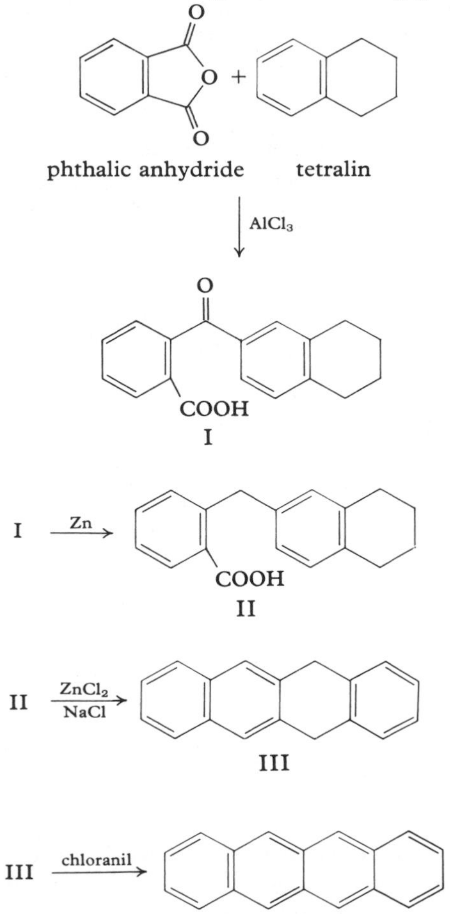

phthalic anhydride   tetralin

AlCl₃

I

I →Zn→ II

II →ZnCl₂/NaCl→ III

III →chloranil→ tetracene (orange)

The names of compounds **II** and **III** are given in Figure 78.18.

397

Beschreibung der Versuche.

2-[5.6.7.8-Tetrahydro-β-naphthylmethyl]-benzoesäure (II).

Zu 40 g gepulvertem Phthalsäureanhydrid und 40 g Tetralin in 120 ccm Tetrachloräthan fügt man ohne Kühlung allmählich 80 g gepulvertes Aluminiumchlorid[5]). Nach ½ Stde. Schütteln hat die Chlorwasserstoff-Entwicklung meist aufgehört. Man zerlegt mit Eis und verd. Salzsäure und treibt das Tetrachloräthan mit Wasserdampf ab. Der Rückstand wird von der wäßr. Lösung abgetrennt und nach dem Waschen unter weiterem Dampfeinleiten in 80 g Natronlauge und 2 l Wasser gelöst. Sodann wird filtriert und die klare Lösung unter Rückfluß und Rühren 5 Stdn. mit 80 g Zinkstaub zum Sieden erhitzt. Man filtriert dann und versetzt heiß mit Salzsäure. Die ölig ausgefallene Säure krystallisiert aus Eisessig in flachen Nadeln, die bei 145–147⁰ schmelzen und sich in konz. Schwefelsäure blaßgelb lösen.

21.31 mg Sbst.: 63.91 mg CO_2, 12.00 mg H_2O.
$C_{18}H_{18}O_2$ (266.33). Ber. C 81.17, H 6.81. Gef. C 81.84, H 6.30.

5.12-Dihydro-tetracen (III).

Das Rohprodukt von der obigen Reaktion wird mit 50 g Chlorzink und 10 g Natriumchlorid gemischt und so lange auf 300—310⁰ erhitzt, bis kein Wasserdampf mehr entweicht. Es haben sich dann zwei Schichten gebildet. Die untere besteht aus den beiden geschmolzenen Salzen und schützt die obere aus Dihydrotetracen bestehende vor örtlicher Überhitzung. Will man diese Vorsichtsmaßnahme nicht anwenden, so kann man auch mit geringeren Mengen der Salze, die nur katalytisch wirken, auskommen. Bei höherem Erhitzen destilliert dann das Dihydrotetracen ab; es erstarrt krystallin als gelbiches Destillat. Ausb. mindestens 30 g. Um die Gelbfärbung möglichst zu vermeiden, empfiehlt es sich, die Wasserabspaltung unter Kohlensäure und die Destillation im Vak. vorzunehmen.

Die Gelbfärbung vom nebenher entstandenen Tetracen läßt sich leicht durch Kochen des Rohprodukte in siedendem Xylol mit etwas Maleinsäureanhydrid entfernen. Reines 5.12-Dihydro-tetracen bildet farblose, glänzende. Nadeln aus Xylol, die in Übereinstimmung mit Dufraisse u. Horclois[4]) bei 212⁰. schmelzen und sich in konz. Schwefelsäure in der Kälte gelb, beim Erwärmen grün lösen. Beim Oxydieren in siedendem Eisessig mit Chromsäureanhydrid entsteht glatt Tetracenchinon-(5.12).

Tetracen (IV).

Außer der Dehydrierung durch Überleiten über auf 400⁰ erhitztes Kupfer in bekannter Weise kann man kleinere Mengen von Dihydrotetracen rasch wie folgt dehydrieren: 4 g Dihydrotetracen in 150 ccm siedendem Eisessig werden mit 7 g Chloranil versetzt. Die Lösung wird sofort orangerot und läßt schon in der Hitze 2.5 g schöne Blättchen von Tetracen ausfallen. Aus der Mutterlauge kann noch etwas weniger reines Tetracen gewonnen werden. Tetracen sublimiert sehr schön im Vak. im CO_2-Strom. Es schmilzt entsprechend den Angaben von Dufraisse u. Horclois[4]) im evakuierten Röhrchen bei 357⁰.

---

[4]) Bull. Soc. chim. France [5] **3**, 1887, 1892 [1936].
[5]) E. de Barry Barnett u. R. A. Lowry, B. **65**, 1650 [1932].

Figure 78.18. From E. Clar, in *Berichte* **75**, 1272 (1942). Reproduced with permission of Verlag Chemie, GmbH, Weinheim.

§79 CYCLOPROPENE DERIVATIVES

The synthesis of compounds with small rings has fascinated organic chemists for many years. The next two sections present easy syntheses of cyclopropene derivatives. Apparently, the resonance stabilization due to the two-electron π system present in these compounds more than compensates for the expected destabilization due to the unusual bond angles.

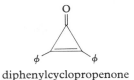

diphenylcyclopropenone

Diphenylcyclopropenone can be prepared by the procedure described in Figure 79.1 A preparation of dibenzylketone is given in Section 74.3.

It has been reported that some individuals have suffered severe allergic skin reactions to diphenylcyclopropenone (*J. Chem. Educ.* **53** 531 (1976)).

Preparation of diphenylcyclopropenone

Experimental

1. Synthesis of Diphenylcyclopropenone. A. From Dibenzyl Ketone. To a solution of 70 g. (⅓ mole) of commercial dibenzyl ketone in 250 ml. of glacial acetic acid, a solution of 100 g. (⅔ mole) of bromine in 500 ml. of acetic acid was added with stirring over 15 min. After the addition was complete, the mixture was stirred for an additional 5 min. and then poured into 1 l. of water. Solid Na_2SO_3 was added in small portions until the initial yellow color of the solution was discharged, and the mixture was allowed to stand for 1 hr. The slightly yellow mixture of *meso-* and *dl-α,α'*-dibromodibenzyl ketone was collected and air dried. Recrystallization from ligroin (*ca.* 1 l.) afforded 97 g. of white needles, m.p. 79–97°; an additional 11 g., m.p. 79–83°, was obtained by concentrating the mother liquors, and the two were combined.

This mixture of isomers (108 g.) was dissolved in 500 ml. of methylene chloride and the solution was added with stirring over 1 hr. to 100 ml. of triethylamine in 250 ml. of methylene chloride at room temperature. The mixture was stirred for an additional 30 min., extracted with two 150-ml. portions of 3 N HCl (discarded), and the organic phase was transferred to a flask and cooled in an ice bath. A cool solution of 50 ml. of H_2SO_4 in 25 ml. of water was slowly added. A slightly pink precipitate of diphenylcyclopropenone bisulfate separated and this was collected on a sintered glass funnel and washed with two 100-ml. portions of methylene chloride. The solid was then returned to the flask along with 250 ml. of methylene chloride and 500 ml. of water, and 5 g. of solid Na_2CO_3 was added in small portions. The organic layer was collected and the aqueous solution was extracted with two 150-ml. portions of methylene chloride. The combined organic layers were dried over $MgSO_4$ and evaporated to dryness. The impure diphenylcyclopropenone was recrystallized by repeated extractions with boiling cyclohexane (total 1.5 l.), the solution being decanted from a reddish, oily impurity. On cooling, the solution deposited white crystals, 29 g., and an additional 1 g. was obtained by concentrating the mother liquors to 150 ml. The combined 30 g. of diphenylcyclopropenone, m.p. 119–120°, represents an over-all yield of 45% based on dibenzyl ketone.

The mixture of stereoisomers of *α,α'*-dibromodibenzyl ketone could be resolved by fractional crystallization from cyclohexane into its components.

Isomer A, m.p. 96–96.5°, had n.m.r. signals at τ 2.8 (area 5) and 4.4 (area 1).

Anal. Calcd. for $C_{15}H_{12}Br_2O$: C, 48.96; H, 3.29; Br, 43.44. Found: C, 48.75; H, 3.33; Br, 43.71.

Isomer B, m.p. 115–116° (lit.[10] m.p. 114°), had n.m.r. signals at τ 2.8 (area 5) and 4.3 (area 1).

Anal. Calcd. for $C_{15}H_{12}Br_2O$: C, 48.96; H, 3.29; Br, 43.44. Found: C, 48.67; H, 3.19; Br, 43.50.

N.m.r. comparison of the τ 4.3 and 4.4 areas showed that the two isomers were present equally in the original mixture.

Treatment of the pure isomer B with triethylamine in the standard way afforded diphenylcyclopropenone in the same yield as was obtained from the 1:1 mixture of isomers A and B.

When the mixture of A and B (1.7 g.) was treated with 5 ml. of triethylamine in 35 ml. of dioxane and 5 ml. of H_2O, diphenylcyclopropenone was obtained in similar yield (0.34 g., 36%). No cyclopropenone could be detected (infrared band at 5.4 μ) when the mixture of A and B was treated with NaH in benzene, *t*-BuOK in benzene, or dry pyridine.

A mixture of *meso-* and *dl-α,α'*-dichlorodibenzyl ketones was prepared by reaction of dibenzyl ketone with SO_2Cl_2 in glacial acetic acid. The unfractionated product, m.p. 48–80°, n.m.r. τ 2.9 (area 5) and 4.45 and 4.75 (area 1), was used directly.

Anal. Calcd. for $C_{15}H_{12}Cl_2O$: C, 65.43; H, 4.33; Cl, 25.42. Found: C, 64.48; H, 4.32; Cl, 24.91.

Reaction of 10 g. of this dichloro ketone with triethylamine under the standard conditions afforded 0.9 g. (12% yield) of diphenylcyclopropenone.

Figure 79.1. Reprinted with permission from R. Breslow, T. Eicher, A. Krebs, R. A. Peterson, and J. Posner, *J. Am. Chem. Soc.* **87**, 1323 (1965). Copyright by the American Chemical Society.

A preparation of triphenylcyclopropenyl bromide is described in Figure 79.2. The procedure involves the generation of chlorophenyl carbene (ϕ—$\ddot{C}$—Cl) or its equivalent, and its reaction with diphenylacetylene (tolan).

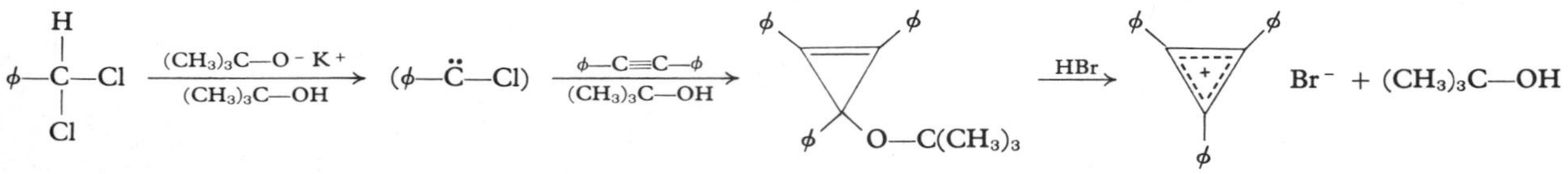

$$\phi-\underset{\underset{Cl}{|}}{\overset{\overset{H}{|}}{C}}-Cl \xrightarrow[\text{(CH}_3)_3\text{C—OH}]{\text{(CH}_3)_3\text{C—O}^- \text{K}^+} (\phi-\ddot{C}-Cl) \xrightarrow[\text{(CH}_3)_3\text{C—OH}]{\phi-C{\equiv}C-\phi} \quad \xrightarrow{\text{HBr}} \quad Br^- + (CH_3)_3C-OH$$

triphenylcyclopropenyl
bromide

The fact that the product exists in the ionic form is interpreted in terms of the aromaticity of the two-electron π system (compare tropylium bromide).

sym-**Triphenylcyclopropenyl Bromide (I).**—The conditions for best yield in the reaction between benzal chloride tolane and potassium *t*-butoxide were examined. In all cases the yield based on tolane was essentially quantitative, and the choice of high yield, based on the butoxide and benzal chloride, or of high conversion of tolane, which is easily recovered, dictates the proportions of reagents used. The following is the best procedure in which only a slight excess of carbene-forming reagents is employed, and in which consequently some tolane is recovered.

To a mixture of tolane (2.23 g., 0.0125 mole) and dry powdered potassium *t*-butoxide (3.5 g., 0.03 mole) in 50 ml. of dry benzene was added freshly distilled benzal chloride (2.42 g., 0.015 mole) with good stirring under nitrogen. The reaction mixture was refluxed for 3 hours, cooled, and water then was added to dissolve the inorganic salts. The layers were separated and the aqueous layer was extracted twice with ether, the ether extracts being pooled with the benzene layer. After drying (MgSO$_4$) the combined organic layers were saturated with anhydrous HBr, when crude *sym*-triphenylcyclopropenyl bromide[3] precipitated, m.p. 253–255° (2.6 g., 0.0075 mole). When the filtrate is evaporated and the residue percolated through alumina with hexane, tolane is recovered (0.8 g., 0.0045 mole) so the yield is 94%, the conversion 60%, based on tolane. The crude salt is suitable for most uses,

although it was recrystallized from acetonitrile to constant m.p. 269–271° dec., for the *pK* measurements. Similar percentage yields are obtained when the reaction is run on a molar scale, using one-half the proportionate amount of solvent.

When the reaction was carried out in air rather than under dry nitrogen the conversion was only 50%. Slightly lower conversions were realized consistently when the reflux period was only 1 hour; when the reflux was omitted no product was obtained. No cation was obtained when sodium ethoxide was substituted for the potassium *t*-butoxide, and the use of a *t*-butyl alcohol solution of potassium, rather than the dry powdered base, gave only half the usual conversion; sodium hydride in *t*-butyl alcohol gave only one-third the usual conversion. In all cases the yield was extremely high, however, since quantitative recovery of the unreacted tolane was possible. The use of a twofold excess of the benzal chloride, with a corresponding increase in the base, gave better than 90% conversion, as well as yield, based on the tolane.

When the usual preparative procedure was interrupted before addition of water and the inorganic materials were removed by filtration, concentration of the solution and addition of hexane caused the crystallization of **1,2,3-triphenylcyclopropenyl *t*-butyl ether** as white prisms, m.p. 143–144.5°.

Figure 79.2. Reprinted with permission from R. Breslow and H. W. Chang, *J. Am. Chem. Soc.* **83**, 2374 (1961). Copyright by the American Chemical Society.

It is now very clear that many organisms communicate with members of their own species and with other species by means of chemical signals (1, 2). For example, we all understand the message conveyed by the smell of the skunk, and bees are known to make use of quite an elaborate system of chemical signals. Substances that are used to carry messages between members of a particular species are often called *pheromones*, and several classes of pheromones have been recognized. These include sex attractants, alarm pheromones, trail marking pheromones, and aggregation pheromones. Some pheromones are quite ordinary compounds. For example, *n*-undecanal is a sex attractant for the female greater wax moth, valeric acid attracts the male sugar beet wireworm, and isoamyl acetate is one component of the alarm pheromone of the honeybee (see Section 56.1). Other pheromones are more complex, and the molecules may include one or more rings and several sites of unsaturation. Examples include (R)-(+)-limonene (Section 45), a substance that triggers a frenzy of biting and snapping among Australian harvester termites, and phenylacetic acid, one component of the stink of the stink pot turtle. Long-chain unsaturated, aliphatic alcohols, aldehydes,

undecanal

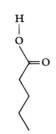

valeric acid

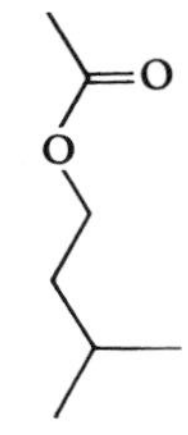

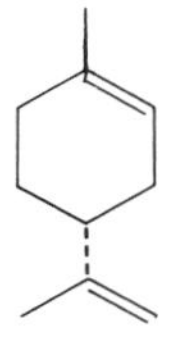

(R)-(+)-limonene

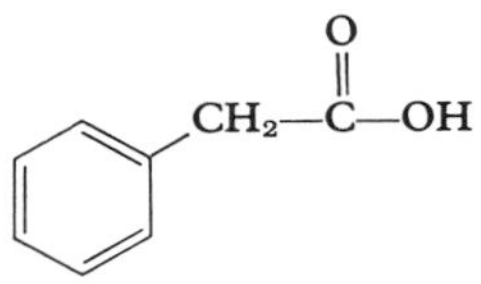

phenylacetic acid

isoamyl acetate

ketones, and esters seem to be the most common structures for most insect pheromones.

One of the most interesting aspects of pheromone communication is the structural and stereochemical specificity of the messenger molecules. For example, the sex attractant of the silkworm moth is 10,12-hexadecadien-1-ol. Of the four possible diasteromeric forms of this compound, the 10E,12Z isomer is at least 10^9 times more active than any of the other three isomers. Another example of

10E,12Z-hexadecadien-1-ol

specificity is that the New York version of the European corn borer is attracted to (E)-11-tetradecenyl acetate, while the Iowa version is attracted to the (Z) isomer.

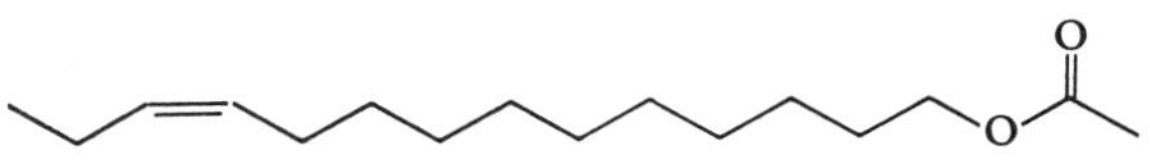

(Z)-11-tetradecenyl acetate

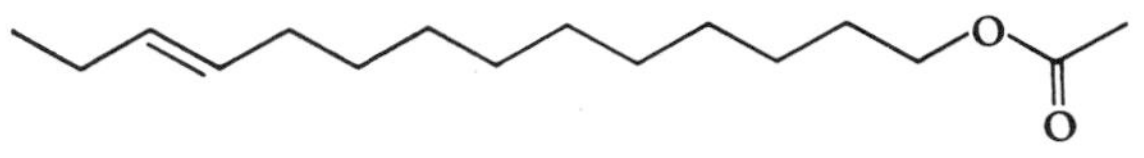

(E)-11-tetradecenyl acetate

Specificity also extends to enantiomeric forms of pheromones. Only the S-(+) isomer of 4-methyl-3-heptanone functions as an alarm pheromone for Texas leaf-cutting ants, and it appears that the (7R,8S)-(+) isomer of *cis*-7,8-epoxy-2-methyloctadecane (disparlure) is far more active as a sex attractant to the female gypsy moth than the enantiomeric (7S,8R)-(−) isomer. Enantiomeric specificity such as this should not be surprising, however, as there are known

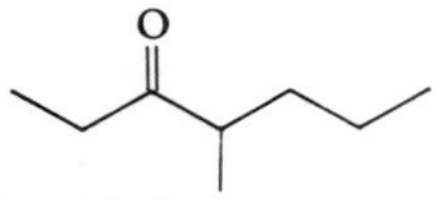

4-methyl-3-heptanone

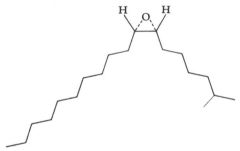

7R,8S-(+)-disparlure

many similar examples of this kind of biological specificity. Quite frequently it has been shown that only one stereoisomeric form of a drug is physiologically active, or that only one enantiomer of a substance, such as penicillamine, is toxic. Although S penicillamine is toxic, the R isomer is used as a chelating agent to accelerate the elimination from the body of normal heavy metals present in too high, nonphysiological concentrations (Wilson's disease) and of nonphysiological heavy metals (heavy metal poisoning). The R and S forms of amino acids generally have different tastes; only the S form of monosodium glutamate (MSG) has a meaty taste, the enantiomer of glucose is bitter, only the R enantiomer of luciferin (shown in Section 70) reacts enzymatically with the production of light, and, of course, the R and S isomers of carvone smell like spearmint and caraway (Section 46).

In this section we will describe some of the possible ways to synthesize four insect pheromones. These compounds are: 4-methyl-3-heptanol, one of the aggregation pheromones of the European elm beetle; 4-methyl-3-heptanone, the alarm pheromone for the harvester ant and the Texas leaf-cutting ant; (Z)-9-tricosene, or muscalure, the sex attractant of the common house fly, and 1,5-dimethyl-6,8-dioxobicyclo[3.2.1]octane, or frontalin, an aggregation pheromone of the Southern pine beetle.

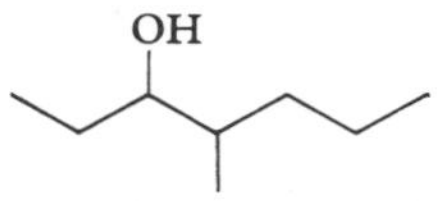

4-methyl-3-heptanol

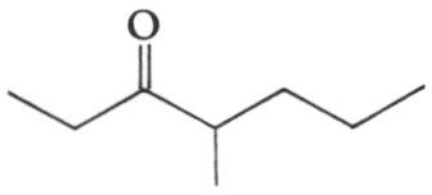

4-methyl-3-heptanone

(Z)-9-tricosene

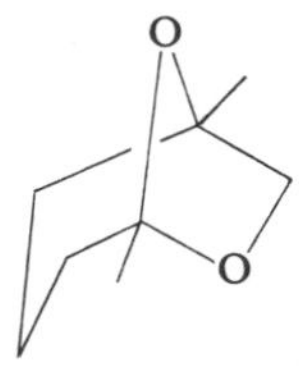

(1S,5R)-(−)-1,5-dimethyl-6,8-dioxobicyclo[3.2.1]octane
(−)-frontalin

Robert Einterz, Jay Ponder, and Ronald Lenox of Wabash College describe the preparation of these two compounds in an article in the *Journal of Chemical Education* (3). The alcohol is prepared by a Grignard reaction between the Grignard reagent made from 2-bromopentane and propionaldehyde, and the ketone by chromic acid oxidation of the alcohol. Perhaps these two compounds

could be prepared according to the procedures of Section 57 (The Grignard Reaction) and Section 53 (Cyclohexanone from Cyclohexanol). The ketone will be obtained as the racemic mixture, but only the S-(+) isomer is effective in the case of the Texas leaf-cutting ant. The biological specificity of the isomers of the alcohol is, apparently, unknown. The infrared spectra of 4-methyl-3-heptanol and 4-methyl-3-heptanone are shown in Figures 80.1 and 80.2.

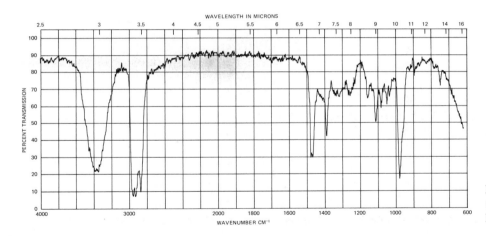

4-methyl-3-heptanol

Figure 80.1. Infrared spectrum of 4-methyl-3-heptanol; thin film.

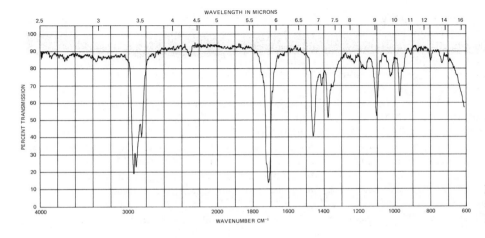

4-methyl-3-heptanone

Figure 80.2. Infrared spectrum of 4-methyl-3-heptanone; thin film.

PROJECTS

Two similar methods for the preparation of muscalure are shown in Figures 80.3 and 80.4. In the first, erucic acid is treated with methyl lithium to give an unsaturated ketone that is reduced under Wolff-Kishner conditions to the alkene.

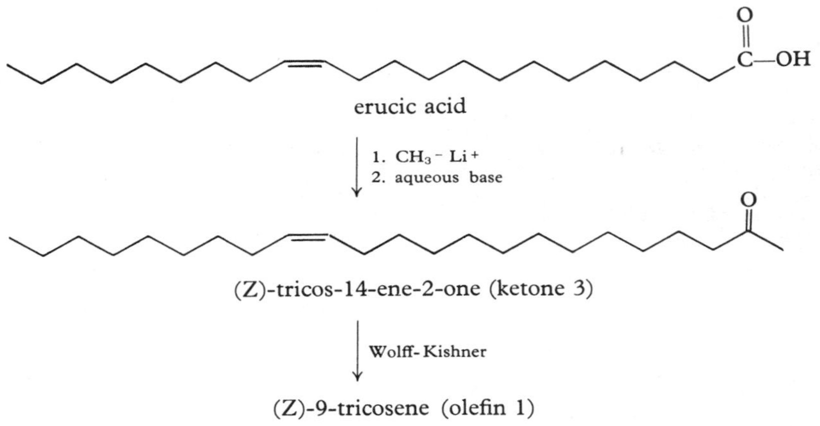

erucic acid

1. $CH_3{}^- Li^+$
2. aqueous base

(Z)-tricos-14-ene-2-one (ketone 3)

Wolff-Kishner

(Z)-9-tricosene (olefin 1)

In the second, an isomeric unsaturated ketone is formed either by treatment of oleic acid with pentyl lithium or of oleonitrile with pentylmagnesium bromide. This ketone can also be reduced under Wolff-Kishner conditions to muscalure.

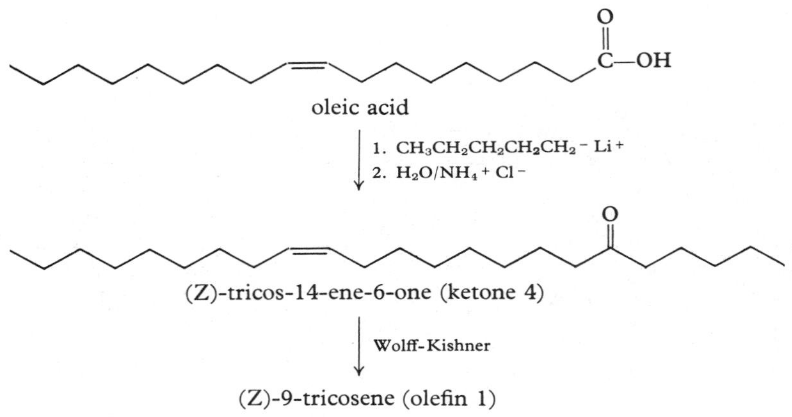

oleic acid

1. $CH_3CH_2CH_2CH_2CH_2{}^- Li^+$
2. $H_2O/NH_4{}^+ Cl^-$

(Z)-tricos-14-ene-6-one (ketone 4)

Wolff-Kishner

(Z)-9-tricosene (olefin 1)

Preparation of muscalure

(Z)-**Tricos-14-en-2-one** (**3**).—To a 10.7-g (31.3 mmol) quantity of erucic acid (**2**) (Columbia Organic Chemicals) in dry ether (200 ml) containing *o*-phenanthroline (5 mg) was added 31.5 ml (63.0 mmol) of 2.00 *M* methyllithium in hexane (Lithcoa), at a rate such that gas evolution was moderate. The reddish-brown reaction mixture was stirred for 30 min and quenched by cautiously adding 10% NaOH (100 ml), saturated $NaHCO_3$ (10 ml), and saturated $(NH_4)_2SO_4$ (10 ml). The phases were separated and the aqueous phase was extracted with ether (3 × 100 ml). The combined organic phases were dried ($MgSO_4$), concentrated, and distilled to give 9.77 g (93.3%) of ketone **3**: bp 140° (0.10 mm); $n^{23}{}_D$ 1.4572; ir (CCl_4) 3015 (olefinic CH), 1725 cm^{-1} (C=O); nmr (CCl_4) δ 0.88–

2.50 (m, 42 H, all protons except olefinic; CH_3 s at δ 2.00), 5.27 [t, *J* (apparent) ≅ 4.5 Hz, **2** H, olefinic]; mass spectrum (70 eV) showed M·$^+$ at *m/e* 336; vpc (3% SE-30, 8 ft × 0.125 in., 250°, 50 ml/min) showed one peak.

Anal. Calcd fro $C_{23}H_{44}O$: C, 82.07; H, 13.18. Found: C, 82.04; H, 13.30.

(Z)-**9-Tricosene** (**1**).—To a solution of 3.1 g (47 mmol) of 85% KOH in diethylene glycol (30 ml) was added 4.77 g (14.1 mmol) of ketone **3** and 2.0 g (40 mmol) of 85% hydrazine hydrate. The reaction mixture was heated at 140° until the water had been removed and then at 193° for 4 hr. The cooled reaction mixture was poured into ice–water (150 ml), neutralized with 6 *N* HCl, and extracted with pentane (5 × 100 ml). The combined extracts were dried ($MgSO_4$), concentrated, and distilled to give 4.55 g

(88.8%) of olefin **1**: bp 170–172° (0.5 mm); n^{23}D 1.4532 [lit.[1] bp 157–158° (0.1 mm); n^{26}D 1.4517]; ir (CCl$_4$) 3015 cm^{-1} (olefinic CH); nmr (CCl$_4$) δ 0.67–2.25 (m, 44 H, all protons except olefinic; CH$_2$CH═CHCH$_2$ m, at δ 1.98), 5.25 [t, J (apparent) ≅ 4.5 Hz, 2 H, vinyl]; mass spectrum (70 eV) showed M·$^+$ at m/e 322; vpc (3% SE-30, 8 ft × 0.125 in., 250°, 50 ml/min) showed one peak.

Anal. Calcd for C$_{23}$H$_{46}$: C, 85.63; H, 14.37. Found: C, 85.75; H, 14.30.

Registry No.—1, 27519–02–4; **3**, 36706–99–7.

(1) D. A. Carlson, M. S. Mayer, D. L. Silhacek, J. D. James, M. Beroza, and B. A. Bierl, *Science*, **174**, 76 (1971).
(2) Erucic acid is available commercially, or it can be isolated from rapeseed oil: C. R. Noller and R. H. Talbot, "Organic Syntheses," Collect. Vol. II, Wiley, New York, N. Y., 1943, p 258.
(3) M. J. Jorgenson, *Org. React.*, **18**, 1 (1970).
(4) Huang-Minlon, *J. Amer. Chem. Soc.*, **68**, 2487 (1946).

Figure 80.3. From R. L. Cargill and M. G. Rosenbloom, in *J. Org. Chem.* **37**, 3971 (1972).

Preparation of muscalure

(Z)-Tricos-14-en-6-one

(*a*) A solution of *n*-pentyl lithium (34 ml, 1.2 *M* in hexane) prepared by the halogen exchange method from lithium metal and 1-bromopentane was added dropwise to oleic acid (5.64 g, 20 mmol) in dry ether (75 ml). The mixture was stirred for 1 h at room temperature, then quenched with aqueous NH$_4$Cl. The layers were separated and the aqueous phase was extracted with ether (3 × 50 ml). The combined organic solutions were dried (MgSO$_4$), concentrated, and distilled to give ketone **4** (5.78 g, 86%), b.p. 170–174° (~0.6 Torr); i.r. (neat) ν_{max} 3010, 1720 cm^{-1}; p.m.r. (CCl$_4$) δ 0.90 (6H, t, J = 4.5 Hz, CH$_3$), 1.1–2.2 (32H, m, CH$_2$), 2.30 (4H, t, J = 6 Hz, COCH$_2$), 5.28 (2H, t, J = 4.5 Hz, CH═CH); m/e 336 (M$^+$).

Anal. Calcd. for C$_{23}$H$_{44}$O: C, 82.07; H, 13.18. Found: C, 82.30; H, 13.25.

(*b*) Oleonitrile (6.60 g, 25 mmol) was added to the Grignard reagent generated from magnesium (1.44 g) and 1-bromopentane (9.06 g, 60 mmol) in dry ether (100 ml). After stirring at room temperature for 5 h, the reaction mixture was poured into ice water containing hydrochloric acid. The layers were separated and the aqueous solution was extracted with ether (2 × 50 ml). The ethereal solutions were dried (MgSO$_4$), evaporated to dryness, and distilled to give ketone **4** (6.29 g, 75%), identical to the compound obtained above.

(Z)-9-Tricosene

A mixture of the ketone **4** (3.84 g, 11.4 mmol), 85% hydrazine hydrate (2.0 g, 40 mmol), potassium hydroxide (2.6 g), and diethylene glycol (25 ml) was heated to 140°. After most of the water had evaporated, the temperature was raised to 195–200° and maintained for 4 h. The cooled mixture was poured into water, and extracted with ether (2 × 40 ml). The extracts were dried, evaporated, and distilled to furnish the olefin **1** (3.30 g, 89%), b.p. 168–170° (~0.6 Torr); i.r. (neat) ν_{max} 3010, 1640 cm$^-$; p.m.r. (CCl$_4$) δ 0.89 (6H, t, J = 5 Hz, CH$_3$), 1.43 (34 H, br s, CH$_2$), 2.05 (4H, pseudo-*t*, allylic CH$_2$), 5.27 (2H, t, J = 4.5 Hz CH═CH).

Figure 80.4. From T.-.L. Ho and C. M. Wong, in *Canadian Journal of Chemistry* **52**, 1923 (1974).

80.3 FRONTALIN

Figure 80.5 presents a way to prepare one of the aggregation pheromones of the Southern pine beetle. The final steps of this synthesis, the conversion of 3 to 4, involve an oxymercuration-demercuration sequence. The first reaction is an electrophilic addition to a carbon-carbon double bond.

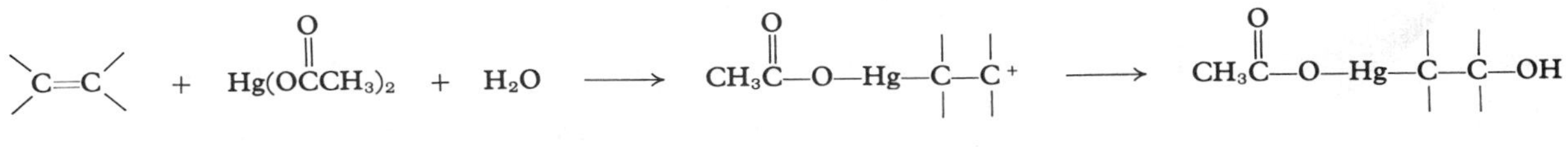

The second reaction is the removal of the mercury substituent by reduction.

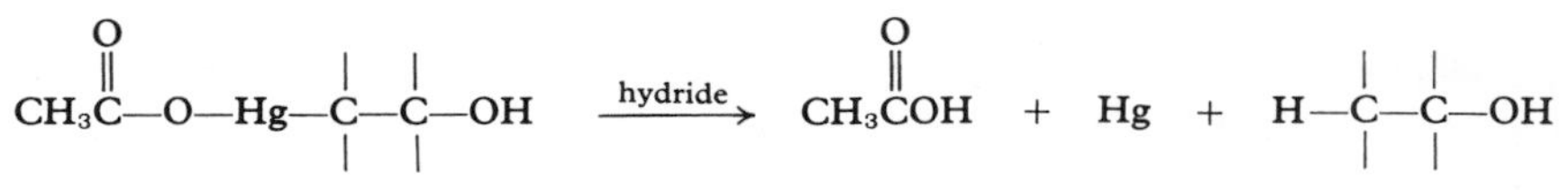

$$CH_3C(=O)-O-Hg-\overset{|}{\underset{|}{C}}-\overset{|}{\underset{|}{C}}-OH \xrightarrow{hydride} CH_3COH + Hg + H-\overset{|}{\underset{|}{C}}-\overset{|}{\underset{|}{C}}-OH$$

The alcohol formed in this particular instance spontaneously cyclizes to frontalin. This synthesis, of course, produces racemic frontalin. The 1S,5R-(−)-isomer, shown above, is the more active form.

Preparation of frontalin

Frontalin, 1,5-dimethyl-6,8-dioxabicyclo[3.2.1]octane is an aggregating pheromone of the southern pine beetle, *Dendroctonus frontalis*.[2] A Diels–Alder reaction of methyl vinyl ketone and methacrolein (**1a**) does not afford the correct adduct (**2a**) for further elaboration to frontalin (**4**). However, use of methyl methacrylate (**1b**) affords a mixture containing only the dimer of methyl vinyl ketone and **2b**. Lithium aluminum hydride reduction of **2b** gives **3** which is immediately cyclized to **4**. The yield of **4** from **2b** is 40%, and it seems reasonable that the overall conversion could be improved.[3]

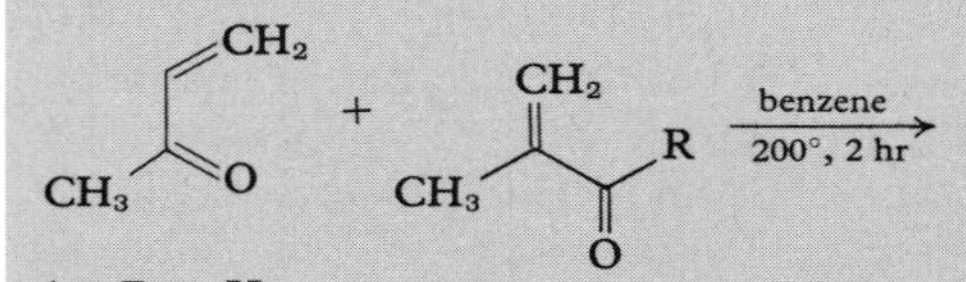

1a, R = H
  b, R = OCH₃

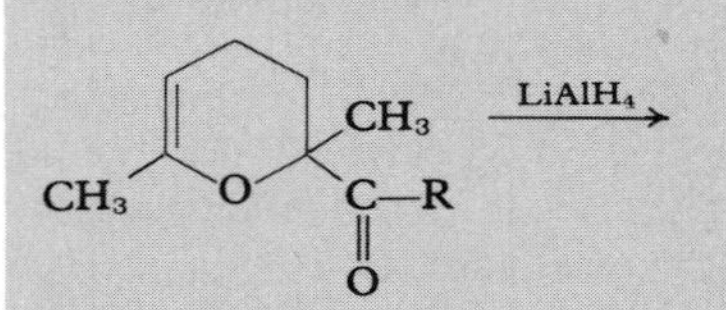

2a, R = H
  b, R = OCH₃

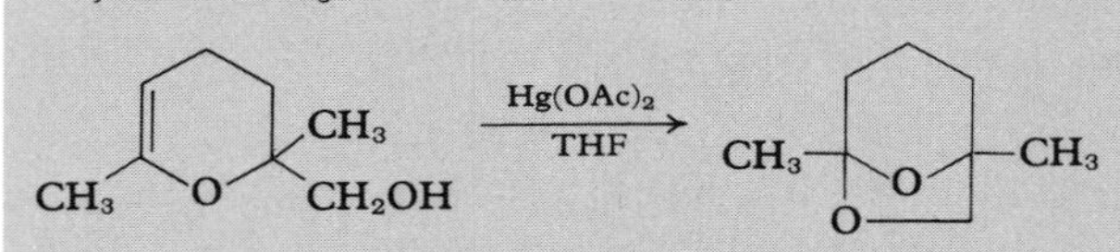

3   **4**

Methyl 2,3,4-Trihydro-2,6-dimethylpyran-2-carboxylate (2b).—A mixture of 14.0 g of methyl vinyl ketone, 20.0 g of methyl methacrylate, and 25 ml of benzene was heated for 2 hr at 200° in an autoclave. Distillation gave 13.0 g of a mixture composed of 67% **2b** and 33% methyl vinyl ketone dimer. Separation of the dimer from **2b** was achieved by way of a bisulfite addition complex. The ir spectrum of **2b** had major absorptions at 3020, 2910, 2830, 1750, 1730, 1680, 1455, 1435, 1380, 1315, 1295, 1220, 1190, 1168, 1113, 1105, 1070, 985, and 760 cm⁻¹. The nmr spectrum of **2b** exhibited a singlet at δ 1.49 (3 H), a singlet at 1.8 (3 H), a methylene envelope from 1.85 to 2.4 (4 H), a singlet at 3.72 (3 H), and a triplet at 4.50 (1 H).

Anal. Calcd for C₉H₁₄O₃: C, 63.49; H, 8.31. Found: C, 63.70; H, 8.61.

Preparation of Frontalin (4).—The 3.5 g of **2b** was reduced by 0.4 g of lithium aluminum hydride in anhydrous THF under dry nitrogen to give 2.4 g of a colorless liquid (**3**). This was treated with 6.0 g of mercuric acetate in 20 ml of dry THF. After it stirred at room temperature for 20 hr, 20 ml each of solutions containing 3 *M* potassium hydroxide, 0.5 *M* sodium borohydride in 3 *M* potassium hydroxide, saturated sodium chloride, and water were added in turn. Extraction with methylene chloride gave 1.8 g of colorless liquid which was 65% frontalin (**4**) by glc (20 ft × ³⁄₈ in. column packed with 30% SE-30 on Chromosorb W at 150° with a 150–200-ml/min flow rate). Ir and nmr[2a] spectra of a sample collected from preparative glc were identical with those of an authentic sample of frontalin obtained from the U. S. Forest Service.

(1) NDEA Predoctoral Fellow, 1968–1971.

(2) (a) G. W. Kinzer, A. F. Fentman, T. F. Page, R. L. Foltz, J. P. Vite, and G. B. Pitman [*Nature*, **221**, 477 (1969)] reported the isolation, identification, and synthesis of this pheromone; (b) W. D. Bedard, R. M. Silverstein, and D. L. Wood (*Science*, **167**, 1638 (1970)] discussed nomenclature problems associated with frontalin and questioned the importance of this compound as an active pheromone.

(3) The previously reported synthesis of frontalin[2a] gave no Experimental Section and was based on a Diels–Alder reaction of methallyl alcohol and acrolein which resulted in a direct 21% yield of 1-methyl-6,8-dioxabicyclo[3.2.1]octane [C. W. Smith, D. G. Morton, and S. A. Ballard, *J. Amer. Chem. Soc.*, **73**, 5270 (1951)]. We repeated the work of Kinzer, *et al.*,[2a] and have obtained a 6.7% yield of frontalin. Our reported synthesis with no attempt to maximize yields gave an equivalent overall yield and has the potential to be improved. For example, the separation of the methyl vinyl ketone dimer and **2b** *via* the bisulfite addition product was only attempted once giving a 37% recovery of **2b**.

Figure 80.5. From B. P. Mundy, R. D. Otzenberger, and A. R. DeBernardis, in *J. Org. Chem.* **36**, 2390 (1917).

1. a. How many stereoisomeric forms can there be of 4-methyl-3-hexanol?

   b. To what extent would it be possible to separate these isomers by distillation?

2. Draw out in detail the reactions involved in converting oleonitrile to (Z)-9-tricosane.

References

1. J. H. Law and F. E. Regnier, Pheromones, in *Annual Review of Biochemistry*, **40**, 533 (1971).

2. D. A. Evans and C. L. Green, Insect Attractants of Natural Origin, *Chemical Society Reviews* **2**, 75 (1973).

3. R. M. Einterz, J. W. Ponder, and R. S. Lenox, *J. Chem. Educ.* **54**, 382 (1977).

§81 SEX HORMONES

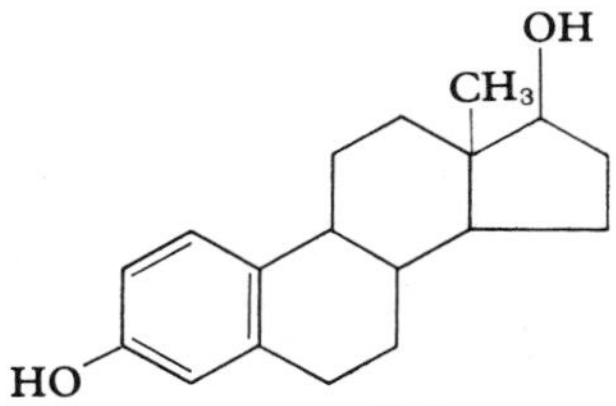

Sex hormones are the substances that are directly responsible for the development of sex characteristics and for the sexual functioning of the human organism. These substances are produced in the ovaries and the testes in response to stimulation by the gonadotropic hormones secreted in the anterior lobe of the pituitary gland. In contrast to the proteinaceous gonadotropic hormones, the sex hormones are basically polycyclic aliphatic substances, or *steroids*. The three primary human sex hormones are estradiol (the primary female sex hormone, or estrogen), testosterone (the primary male sex hormone, or androgen), and progesterone, a substance secreted by the corpus luteum of the female, which prepares the bed of the uterus for implantation of the fertilized ovum.

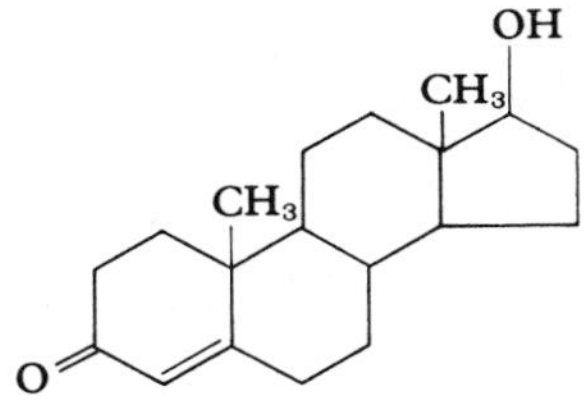

These substances were originally obtained by isolation from natural sources. Estradiol was first isolated in 1934 from sow ovaries at a yield of about 12 mg from some 3600 kg (four tons) of ovaries, testosterone in 1935 from steer testes at a yield of 10 mg from 100 kg of steer testis tissue, and progesterone in 1934 from sow ovaries at a yield of 20 mg from 625 kg of ovaries, from 50,000 sows. Further development of isolation methods later gave somewhat larger yields, but there was an obvious need for vastly more efficient procedures for the production of these hormones.

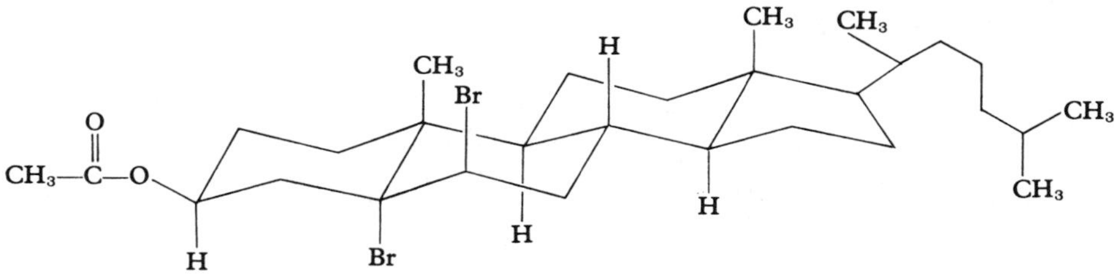

estradiol

testosterone

progesterone

cholesterol acetate dibromide

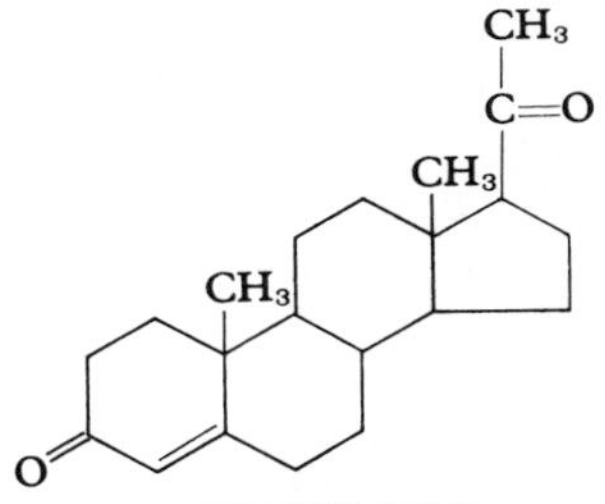

In 1935 several research groups determined that cholesterol (see Section 39) as the acetate dibromide could be degraded by vigorous chromic acid oxidation to give androstenolone as a minor product.

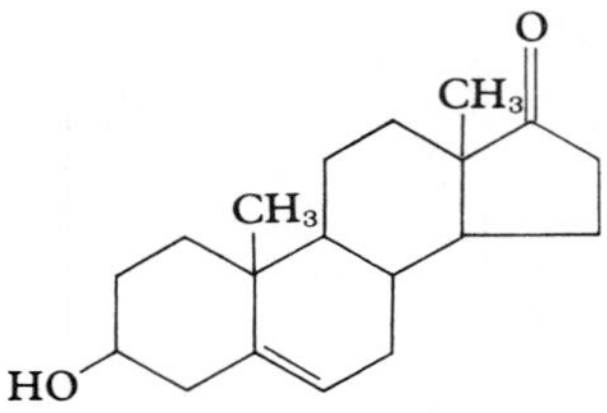

This substance, which had originally been isolated from male urine the year before, then served as a starting material for syntheses of the various sex hormones. The yield of androstenolone, isolated as the semicarbazone acetate, was originally less than 1 percent, but by various improvements in the method it was raised over the years to about 8 percent.

A far more satisfactory alternative for hormone syntheses was provided in 1939 by Russell Marker's discovery that certain substances of vegetable origin could be easily degraded in high yield to give suitable starting materials. For example, having established the spiroketal structure of the side chain of diosgenin,

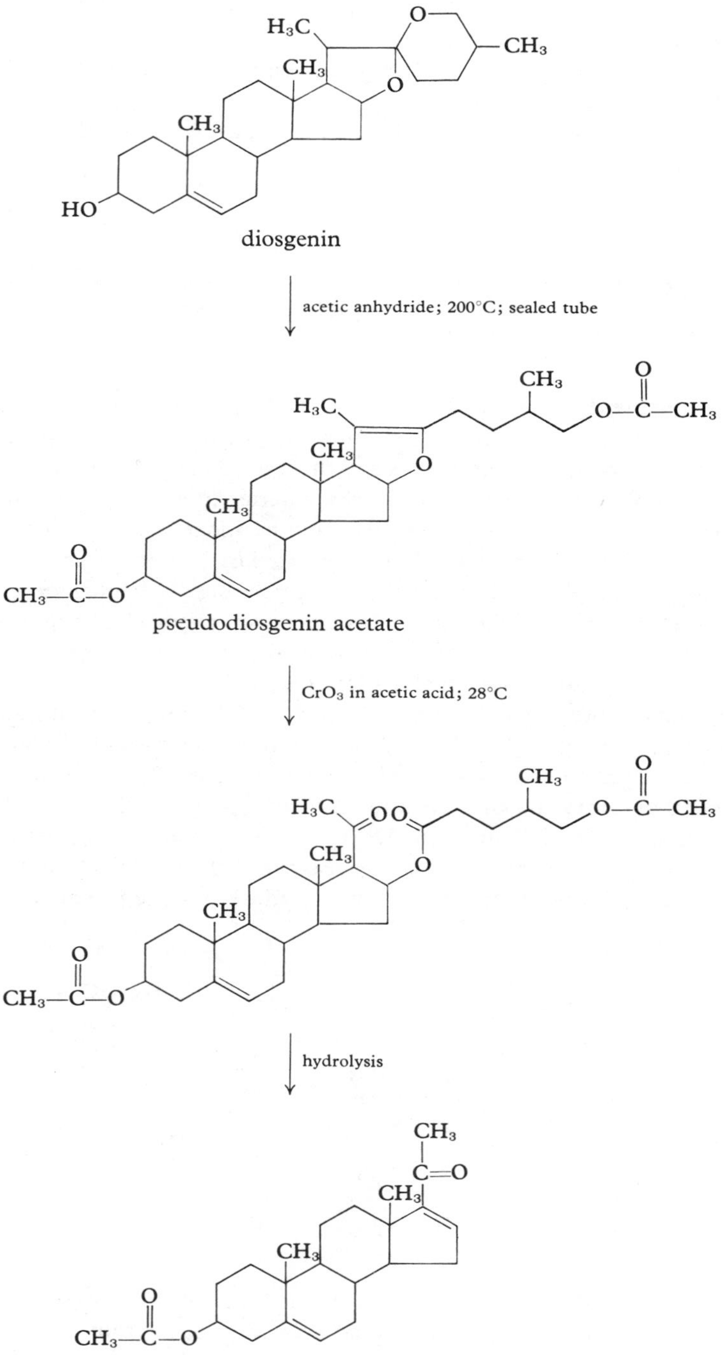

androstenolone

diosgenin

acetic anhydride; 200°C; sealed tube

pseudodiosgenin acetate

CrO₃ in acetic acid; 28°C

hydrolysis

16-dehydropregnenolone acetate

Marker showed that it could be converted in three steps to 16-dehydropregnenol-one acetate in over 60 percent yield.

The double bond in the five-membered ring could then be selectively hydrogenated to give pregnenolone acetate.

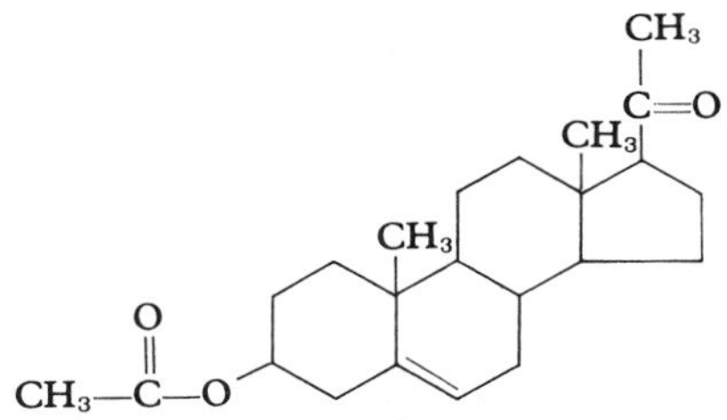

pregnenolone acetate

In 1943, Marker demonstrated the potential of his discovery by preparing 3 kg of progesterone from diosgenin. At that time this was worth a quarter of a million dollars and was equivalent to the yield from the corpora lutea of 30 million sows. The source of the diosgenin was the roots of the wild Mexican yam. Marker worked directly in Mexico, achieving his feat under incredibly primitive conditions with the help of untrained assistants. Marker's unusual and generally unappreci-ated professional life is reviewed in an article in the *Journal of Chemical Education* (1). The article ends with the following quotation:

> When I retired from Chemistry in 1949, after 5 years of production and research in Mexico, I felt I had accomplished what I had set out to do. I had found sources for the production of steroidal hormones in quantity at low prices, developed the process of manufacture, and put them into produc-tion. I assisted in establishing many competitive companies in order to insure a fair price to the public and without patent protection or royalties from the producers.
>
> Since retiring from the laboratory 20 years ago, I have never returned to chemistry or consulting, and have no shares of stocks in any hormone or related companies. My only appearance in public was recently on April 23, 1969 to accept an award by the Mexican Chemical Society showing their appreciation for the work I had accomplished.
>
> *Russell E. Marker, May 15, 1969**

The book *Steroids* by Fieser and Fieser gives many interesting details about the isolation, structure, determination, preparation, and total synthesis of sex hormones and other steroids (2).

The projects in this section have been divided into two groups. The first involves the preparation of testosterone from 16-dehydropregnenolone acetate via androstenolone and androstenedione.

The second group of projects is concerned with the preparation of progester-one from pregnenolone acetate via pregnenolone.

The starting material for the first project in each group is an intermediate that was obtained in the "Marker degradation" of diosgenin. The following outline summarizes all the projects and procedures in this section.

---

* Quoted with permission from the *Journal of Chemical Education* **50**, 199 (1973).

16-dehydropregnenolone acetate

pregnenolone acetate

androstenolone

pregnenolone

androstenedione

progesterone
(Group 2)

testosterone
(Group 1)

81.1 TESTOSTERONE FROM 16-DEHYDRO-PREGNENOLONE ACETATE VIA ANDROSTENOLONE AND ANDROSTENEDIONE

I. Androstenolone from 16-Dehydropregnenolone Acetate
   Alternative procedures:
   Figure 81.1 (Rosenkranz et al., 1956)
   Figure 81.2 (Testa and Fava, 1957)

II. Androstenedione from Androstenolone
   Alternative procedures:
   Figure 81.5 (Ruzicka and Wettstein, 1935)

Figure 81.6 (Butenandt and Kudszus, 1935)
Figure 81.7 (Oppenauer, 1937)

III. Testosterone from Androstenedione
Alternative procedures:
Figure 81.10 (Norymberski and Woods, 1955)
Figure 81.11 (Heyl and Herr, 1953)
Figure 81.12 (Dauben et al., 1954)
Figure 81.13 (Serini and Köster, 1938)
Figure 81.14 (Ercoli and de Ruggieri, 1953)

81.2 PROGESTERONE FROM PREGNENOLONE ACETATE VIA PREGNENOLONE

I. Pregnenolone from Pregnenolone Acetate
II. Progesterone from Pregnenolone
Alternative procedures:
Figure 81.18 (Fernholz, 1934)
Figure 81.19 (Butenandt and Westphal, 1934)
Figure 81.20 (Oppenauer, 1937)

In order to understand which compounds are referred to in the various excerpts from the literature, it is necessary to know something about steroid nomenclature. First, the name of the parent compound implies both a carbon skeleton *and an associated stereochemistry*. For example, androstane and pregnane are

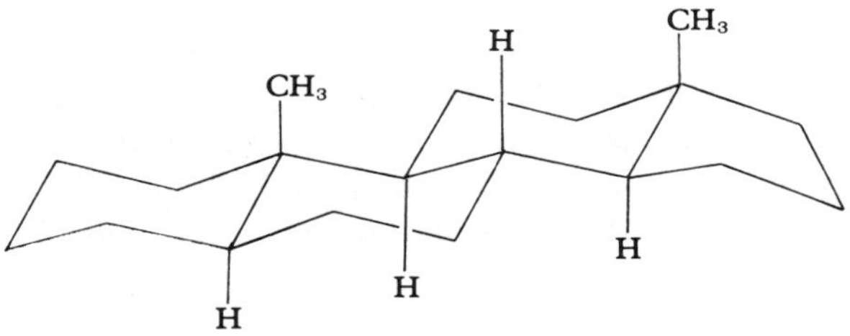

androstane

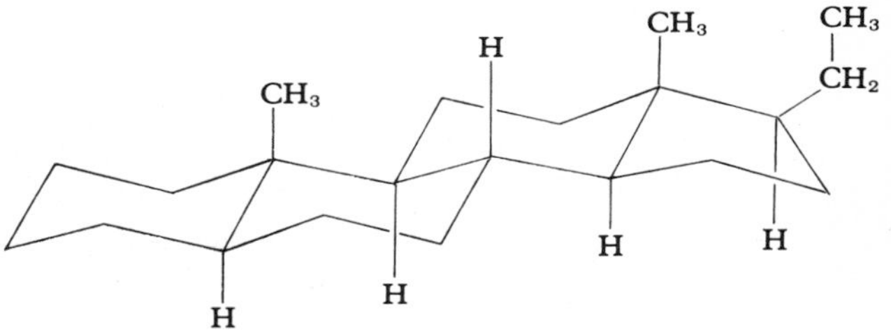

pregnane

and cholestane is

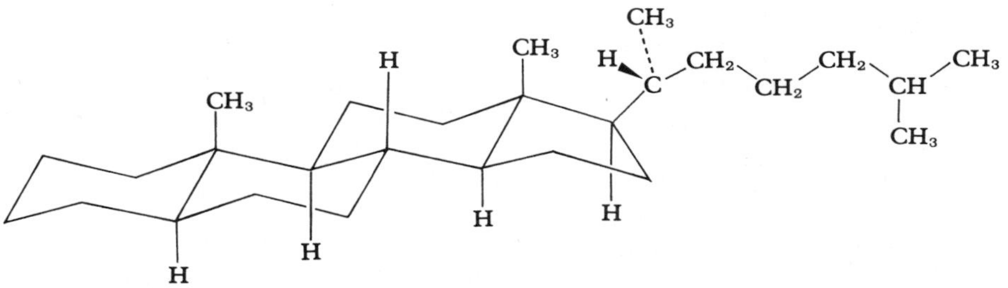

cholestane

In each case the stereochemistry at the six, seven, or eight chiral centers is as shown.* Those formulas, however, are usually written in the following way for the sake of convenience:

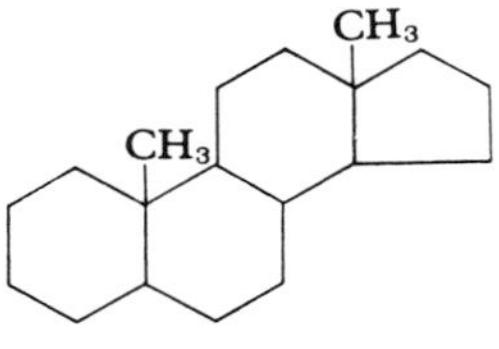

androstane 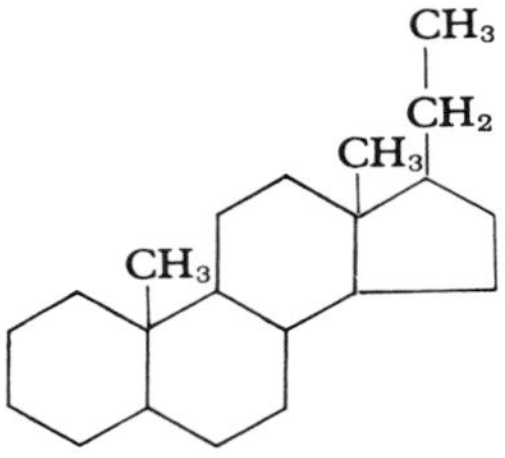   pregnane 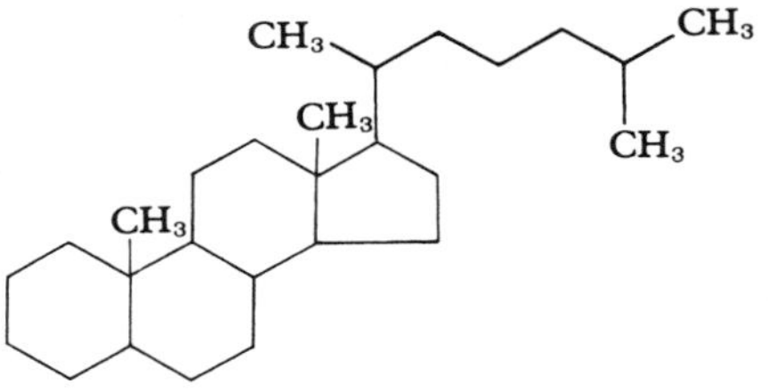   cholestane

Second, derivatives of these substances are named using the numbering system shown in the accompanying diagram to indicate the positions of double bonds and substituents. A double bond between the 5 and 6 carbon atoms, for

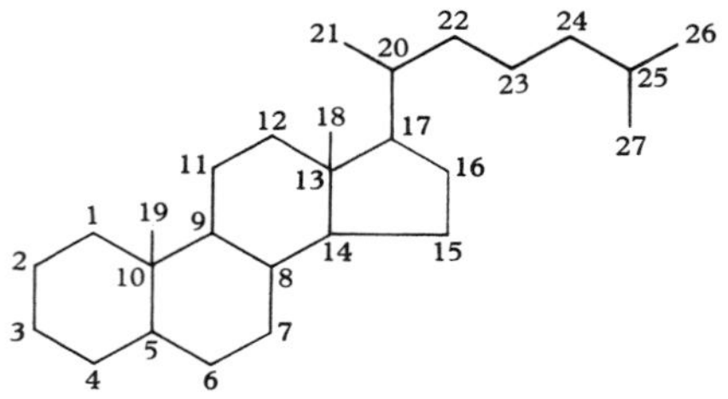

example, may be indicated by the prefix Δ^5; this prefix is pronounced "delta five." The stereochemical orientation of a substituent is given, when necessary, by the use of the modifier α or β to indicate that it is below or above the general plane of the molecule. For example, androstenedione would be named Δ^4-androstene-3,17-dione, and 16-dehydropregnenolone acetate would be named $\Delta^{5,16}$-pregnadiene-3β-ol-20-one acetate.

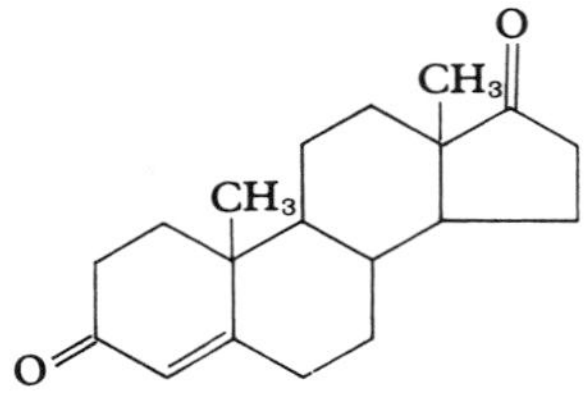

androstenedione;
Δ^4-androstene-3,17-dione

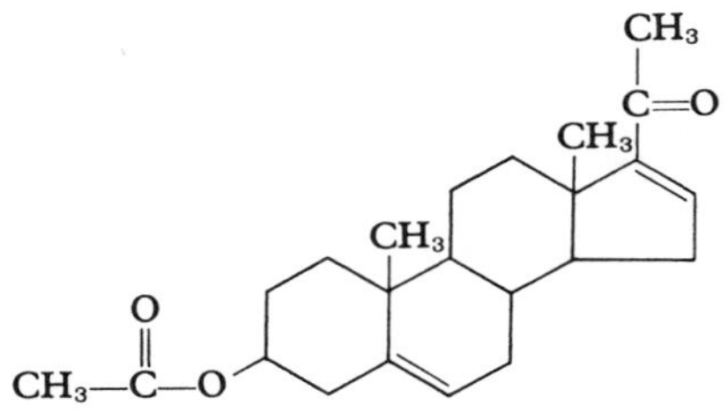

16-dehydropregnenolone;
$\Delta^{5,16}$-pregnadiene-3β-ol-20-one acetate

* The *Chemical Abstracts* system of naming differs slightly in that the configuration at C-5 is explicitly specified; the names of these compounds, according to the *Chemical Abstracts* system, are (5α)-androstane, (5α)-pregnane, and (5α)-cholestane.

Table 81.1.

Names	Crystalline Form	Recrystallization Solvent	m.p. (°C)	$[\alpha]_D$, ethanol	Merck Monograph[a]
16-Dehydropregnenolone acetate			166–169		
$\Delta^{5,16}$-Pregnadiene-3β-ol-20-one acetate					
Androstenolone	Needles	Benzene/petroleum ether	140–141	+10.9°	2846
Dehydroepiandrosterone					
Δ^5-Androstene-3β-ol-17-one	Leaflets		152–153		
$\Delta^{5,6}$-3-Oxy-ätio-cholenone-(17)					
Dehydroandrosteron (Fig. 81.6)					
Androstenedione	Needles	Acetone	142–144	+191°	675
Δ^4-Androstene-3,17-dione	Crystals	Hexane	173–174		
$\Delta^{4,5}$-Ätio-cholendione-(3,17)					
Androstendion (Fig. 81.6)					
Androst-4-ene-3:17-dione					
Testosterone	Needles	Dilute acetone	155	+109°	8890
Δ^4-Androstene-17β-ol-3-one					
Pregnenolone acetate	Needles	Alcohol	149–151	−22°	7533
Δ^5-Pregnene-3β-ol-20-one acetate					
Pregnenolone	Needles	Dilute alcohol	193	+28°	7533
Δ^5-Pregnene-3β-ol-20-one					
"Oxyketons I" (Figs. 81.17 and 81.18)					
Progesterone	Prisms	Dilute alcohol	127–131	+172–182°	7569
Δ^4-Pregnene-3,20-dione	Needles	Petroleum ether	121		

[a] *Merck Index*, 9th edition.

Table 81.1 gives common and systematic names for the key substances involved in these projects. The melting points and molecular rotations of these substances are also given.

81.1 TESTOSTERONE FROM 16-DEHYDROPREGNENOLONE ACETATE VIA ANDROSTENOLONE AND ANDROSTENEDIONE

The overall formation of testosterone from 16-dehydropregnenolone acetate has been divided into three parts. The first involves the preparation of androstenolone from 16-dehydropregnenolone acetate. The second is the conversion of androstenolone to androstenedione. The final part is the selective reduction of androstenedione to testosterone.

I: Androstenolone from 16-Dehydropregnenolone Acetate

This transformation, which involves the removal of the two-carbon side chain of 16-dehydropregnenolone acetate, is carried out by forming the oxime of the starting material, effecting a Beckmann rearrangement of this oxime, and then hydrolyzing the rearranged product.

Figures 81.1 and 81.2 present alternative procedures for these transformations. Figures 81.3 and 81.4 show infrared spectra of 16-dehydropregnenolone acetate and androstenolone.

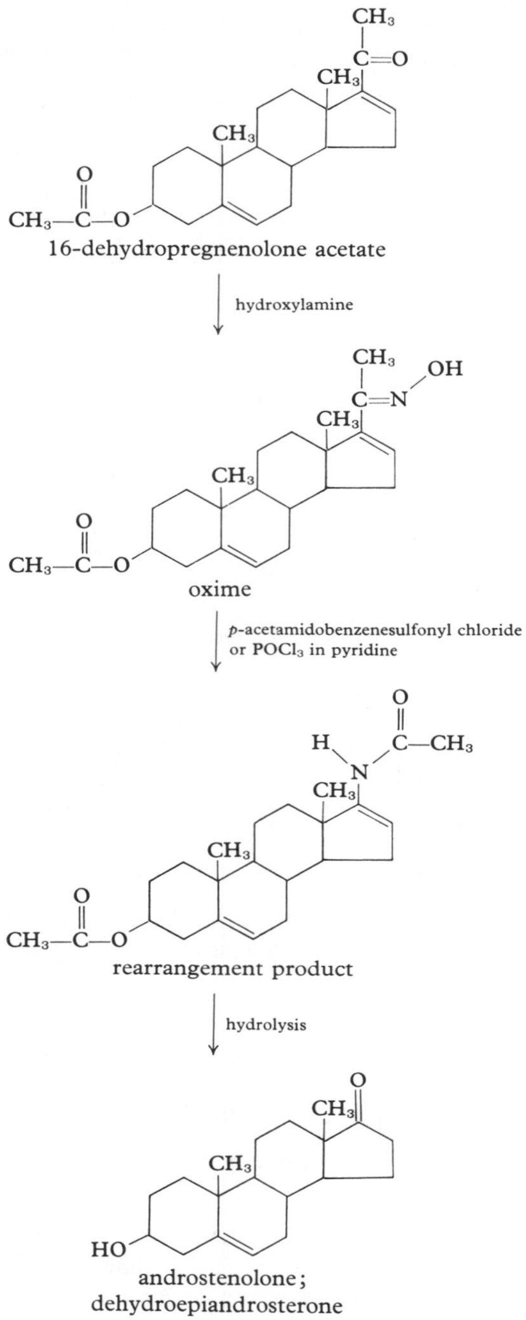

16-dehydropregnenolone acetate

↓ hydroxylamine

oxime

↓ *p*-acetamidobenzenesulfonyl chloride
or POCl₃ in pyridine

rearrangement product

↓ hydrolysis

androstenolone;
dehydroepiandrosterone

Preparation of androstenolone from 16-dehydropregnen-olone acetate

Conversion of $\Delta^{5,16}$-*pregnadien-3β-ol-20-one acetate* (XI) *to dehydroepiandrosterone* (XIVa). A mixture of 25 g. of XI, 30 cc. of pyridine, 130 cc. of 95% ethanol and 8.5 g. of hydroxylamine hydrochloride was refluxed for 30 minutes and cooled in ice. The precipitate was collected, washed with hot water and dried, thus giving 24.5 g., m.p. 228–230°. The oxime (XII), was dissolved in 70 cc. of anhydrous pyridine, cooled to 0° and a

solution of 30 g. of *p*-acetamidobenzenesulfonyl chloride in 70 cc. of pyridine was added with stirring, the temperature being maintained below 5°. After stirring for 2 hours at 10° and an additional 2 hours at room temperature, the mixture was poured into 500 g. of ice and 150 cc. of conc'd sulfuric acid and left in the refrigerator overnight. The product was collected, washed well with hot water and since infrared examination showed it to be a mixture of dehydroepiandrosterone (XIVa) and the corresponding 3-acetate (XIVb) it was directly saponified by refluxing with 150 cc. of 2.5% methanolic potassium hydroxide for 30 minutes. The usual work-up followed by recrystallization from methanol furnished 15.0 g. (74%) of dehydroepiandrosterone (XIVa), m.p. 151–153°, $[\alpha]_D$ +5.5°.

Beckmann rearrangement of $\Delta^{5,16}$-pregnadien-3β-ol-20-one 3-acetate 20-oxime (XII).[17] A solution of 5.0 g. of the oxime XII in 20 cc. of pyridine was treated with 5.0 g. of *p*-acetamidobenzenesulfonyl chloride and stirred for 2 hours at 10° and 2 hours at room temperature. The mixture was poured into ice-water, and the brown gummy precipitate was extracted with chloroform, washed with water, dried and evaporated. Crystallization of the residue from methylene chloride hexane (decanting first from a colored oil which separated) afforded 1.75 g. of yellowish crystals, m.p. 230–236° and an additional 0.3 g. (m.p. 233–236°) by chromatography of the mother liquors. The analytical sample of the amide (XIII) was obtained from the same solvent pair as colorless crystals, m.p. 237–240°, $[\alpha]_D$ −18°, $\lambda^{EtOH}_{max.}$ 240 mμ, log ϵ 3.82, $\lambda^{CHCl_3}_{max.}$ 2.90, 5.76, 5.90, and 6.58 μ (N—H deformation).

Anal. Calc'd for $C_{23}H_{33}NO_3$: C, 74.36; H, 8.95. Found: C, 74.81; H, 8.83.

The unsaturated amide (XIII) could be converted to the 17-ketone by acid hydrolysis as well as by alkaline treatment: 37.5 g. of amide was refluxed for 1 hour with 700 cc. of 5% ethanolic potassium hydroxide whereupon 22.5 g. of dehydroepiandrosterone (XIVa) was obtained.

Figure 81.1. Reprinted with permission from G. Rosenkranz, O. Mancera, F. Sondheimer, and C. Djerassi *J. Org. Chem.* **21**, 520 (1956). Copyright by the American Chemical Society.

Preparation of androstenolone from 16-dehydropregnenolone acetate

$\Delta^{5,16}$-*Pregnadien-3β-ol-20-one acetato ossima.* — In una beuta si sciolgono 30 g di $\Delta^{5,16}$-pregnadien-3β--ol-20-one acetato in 300 cm³ di piridina anidra. Alla soluzione si aggiungono g 6,750 di cloridrato di idrossilamina e si agita fino a soluzione totale. Si lascia riposare in recipiente ben chiuso a temperatura ambiente per quattro giorni, poi si versa in 1500 cm³ di acqua, si lava la beuta con acqua, si raccoglie il cristallo separatosi su Büchner, lo si lava con acqua e lo si secca fino a peso costante in stufa a 100°.

Resa g 30 (96,5% d.t.) p. f. 220-3°.

Δ^5-*Androsten-3β-ol-17-one acetato.* — *a*) Da $\Delta^{5,16}$--pregnadien-3β-ol-20-one acetato ossima: In un pallone munito di agitatore meccanico e termometro e posto in un bagno refrigerante con salamoia si sciolgono 10 g di $\Delta^{5,16}$-pregnadien-3β-ol-20-one acetato ossima in 50 cm³ di piridina anidra. Si raffredda la soluzione a —15° e mantenendo la temperatura fra —10° e —15° si fa gocciolare nella soluzione agitata una miscela raffreddata di 60 cm³ di piridina anidra c 20 cm³ di ossicloruro di fosforo. Terminata l'aggiunta si sostituisce il bagno refrigerante a salamoia con uno ad acqua e ghiaccio e si agita per 3 ore a 0° la sospensione gialloarancione. Il colore della sospensione incupisce e la massa cristallina tende ad aumentare. Trascorse le tre ore si toglie il bagno di acqua e ghiaccio e si lascia salire la temperatura fino a +8°/+10° provocando in tal modo una fluidificazione della massa. La miscela viene ora versata a filo un bicchiere contenente 140 cm³ di HCl conc. e 140 cm³ di ghiaccio, posto in un bagno di ghiaccio e agitato meccanicamente in modo molto efficace. La temperatura interna sale rapidamente a 40–45° e viente mantentuta durante tutta l'aggiunta entro questi limiti, compensado la reazione esotermica con un buon raffreddamento esteron. Al termine dell' aggiunta si agita per altri 10′, indi si diluisce con 750 cm³ di acqua e ghiaccio e si lascia 15′ in riposo. Sil filtra, si lava molto accuratamente il cristallo con acqua e si secca su b. m.

Polvere leggermente giallina a p. f. 166-7° (teorico 168-170°).

Resa g 8,4 (94% d.t.).

b) Dal $\Delta^{5,16}$-pregnadien-3β-ol-20-one acetato senza isolare l'ossima: In un pallone munito di agitatore meccanico e termometro si pongono g 20 di $\Delta^{5,16}$--pregnadien-3β-ol-20-one acetato, g 4,5 di idrossilamina cloridrato e 200 cm³ di piridina anidra. Si agita sino a soluzione completa, poi si abbandona a temperatura ambiente per quattro giorni. Trascorso questo tempo si raffredda la miscelain salamoia e, impedendo alla temperatura di superare i —10°, si aggiungono, poco a poco, 40 cm³ di ossicloruro di fosforo. Si procede poi esattamente come descritto all'esempio *a*) impiegando, come soluzione acida fredda, 560 cm³ di HCl 50%, e versando, dopo la reazione a 40-45°, in 1500 cm³ di acqua. Si ottengono così g 16,8 (94% d.t.) p. f. 167-8° (teorico 168-170°).

Δ^5-*Androsten-3β-ol-17-one.* — Grammi 16,8 di Δ^5-androsten-3β-ol-17-one acetato, sono scaldati a ricadere 30′ in 168 cm³ di una soluzione metanolica all'1% di KOH. La soluzione bruna e torbida viene trattata con nero animale, filtrata e diluita con 168 cm³ di acqua. Si separa un cristallo, che dopo 60′ di riposo a 0°, viene raccolto su Büchner e seccato sotto vuoto. Si ottengono g 12.4 (85% d.t.) di prodotto fondente a 145-9°. La resa complessiva in Δ^5-androsten-3β-ol-20-one dal $\Delta^{5,16}$-pregnadien-3β-ol-20-one acetato è di ca.

80% d.t. Il prodotto ottenuto può essere ulteriormente purificato sciogliendone 5 g in 150 cm³ di metanolo caldo, trattando la soluzione con poco nero, filtrando e concentrando a pressione ordinaria fino a volume di 100 cm³. Si raffredda e si raccoglie il prodotto su Büchner. Il prodotto viene ridisciolto in 10 cm³ di metanolo bollente e abbandonato, senza filtrare, a temperatura ambiente fino alla formazione di un abbondante cristallo. Si lascia la massa una notte a 0°, indi si filtra su Büchner ottenendo g 4,2 di prodotto purissimo (84% sul grezzo).

p. f. 148°-150° (teorico 149-150°).

[α]$_{20}^{D}$ + 12,05 (etanolo e = 1,2036)

(teorico da + 10 a + 12°).

Dalle acque madri metanoliche riunite della prima e seconda ricristallizzazione si ricupera dell'altro prodotto meno puro e leggermente colorato in giallo.

Figure 81.2. From E. Testa and F. Fava, in *Gazz. Chim. Ital.* **87**, 971 (1957). Reproduced with permission.

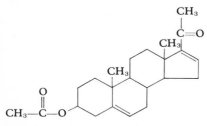

16-dehydropregnenolone acetate

Figure 81.3. Infrared spectrum of 16-dehydro-pregnenolone acetate; CCl₄ solution.

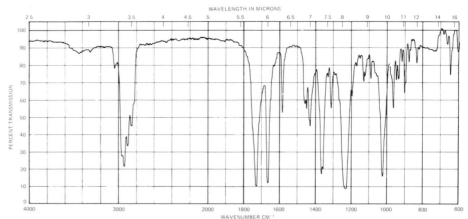

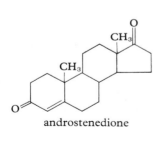

androstenedione

Figure 81.4. Infrared spectrum of androstenolone; CHCl₃ solution.

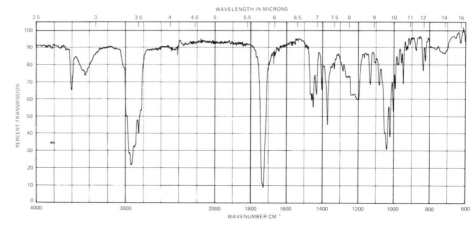

II: Androstendione from Androstenolone

One way to carry out this transformation is to protect the double bond by addition of bromine, to oxidize the secondary alcohol to the ketone with chromium trioxide, and then to restore the double bond by elimination with zinc in acetic acid. The acidic conditions of the elimination also cause the β,γ-unsaturated ketone to be converted to the more stable, conjugated α,β-unsaturated isomer. This sequence of reactions is analogous to that used in the preparation of Δ^4-cholestenone from cholesterol (see Section 73). Alternatively, these changes can be effected in one step by an Oppenauer oxidation. Again, the more stable α,β-unsaturated ketone accumulates under the highly basic reaction conditions.

416

Figures 81.5, 81.6, and 81.7 present the alternative procedures for these transformations. Figure 81.8 shows the infrared spectrum of androstenedione.

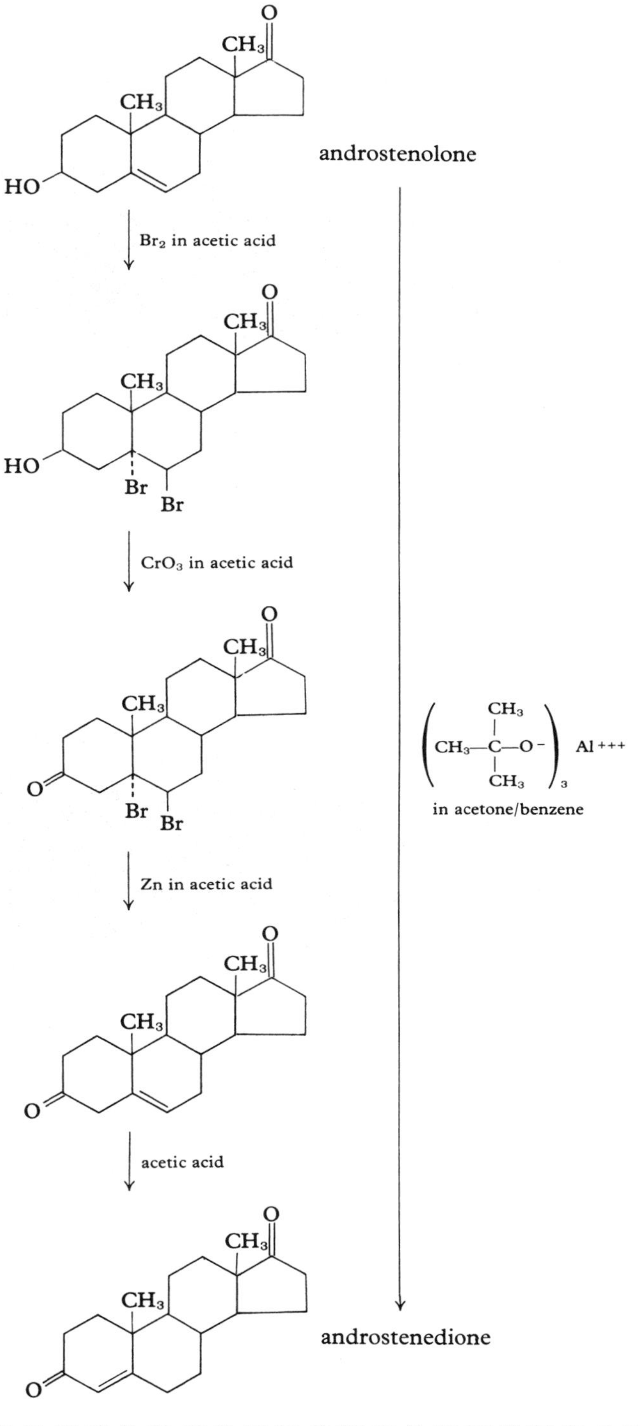

androstenolone

Br$_2$ in acetic acid

CrO$_3$ in acetic acid

$$\left(CH_3-\underset{\underset{CH_3}{|}}{\overset{\overset{CH_3}{|}}{C}}-O^- \right)_3 Al^{+++}$$

in acetone/benzene

Zn in acetic acid

acetic acid

androstenedione

Preparation of androstenedione from androstenolone
Δ^{4,5}-Ätio-cholendion-(3,17).

Zu einer Lösung von 200 mg Δ^{5,6}-3-Oxy-ätio-cholenon-(17) in 6 cm^3 Eisessig wird in der Kälte zuerst tropfenweise eine Lösung von 110 mg Brom in 2 cm^3 Eisessig und dann eine Lösung von 230 mg Chromtrioxyd in 9 cm^3 90-proz. Essigsäre hinzugefügt. Nach 14-stündigem Stehen giesst man in Wasser und saugt das ausfallende Diketon-dibromid ab. Es wird durch Erwärmen mit Zinkstaub und Eisessig auf dem Wasserbade entbromt. Aus der filtrierten Eisessiglösung fällt man das Diketon mit Wasser aus, saugt den flockigen Niederschlag ab, wäscht mit Wasser und trocknet über Phosphorpentoxyd. Durch Umkrystallisieren aus Hexan unter Zusatz von etwas Tierkohle erhält man das $\Delta^{4,5}$-Ätio-cholendion-(3,17) rein. Es schmilzt bei 173—174⁰.

4,120; 3,185 mg Subst. gaben 12,025; 9,28 mg CO_2 und 3,445; 2,63 mg H_2O

$C_{19}H_{26}O_2$ Ber. C 79,66   H 9,15%
Gef. ,, 79,60; 79,46   ,, 9,36; 9,24%

Das Diketon entfärbt eine verdünnt-alkoholische Permanganatlösung sofort. Mit Tetranitromethan hingegen zeigt es einen nur ganz schwachen Anflug von Gelbfärbung. Bekanntlich geben α,β-ungesättigte Ketone diese Reaktion nur schwach.

Dass es sich tatsächlich um eine solche Verbindung handelt, dass also mit der Oxydation der Hydroxylgruppe auch die Doppelbindung von der β,γ- in die α,β-Stellung verschoben ist, zeigte die spektrophotometrische Untersuchung[1]), die eine sehr starke Bande bei 2350 Å nachwies. Bei dieser Wellenlänge ergab sich der Maximalwert der molekularen Extinktion (d = 1 cm, c = 1 Mol pro Liter) zu: log ϵ = 4,25. Nach *Menschick*, *Page* und *Bossert*[2]) ist eine starke Absorptionsbande im Bereiche von 2400 Å charakteristisch für konjugierte Doppelbindungen, wie sie z. B. im Cholestenon und im Corpus luteum-Hormon vorkommen[3]).

Figure 81.5. From L. Ruzicka and A. Wettstein, in *Helv. Chim. Acta* **18**, 986 (1935). Reproduced with permission.

Preparation of androstenedione from androstenolone

Überführung des Dehydro-androsterons in Androstendion. 288,2 mg Dehydro-androsteron werden in 40 ccm Eisessig gelöst und bei Zimmertemperatur durch Eintropfen einer Lösung von 159,8 mg Brom in 3,5 ccm Eisessig bromiert. Die Reaktionslösung entfärbt sich rasch; sie wird anschließend tropfenweise mit 100 mg Chromsäure in 10 ccm Eisessig versetzt und 20 Stunden bei Zimmertemperatur sich selbst überlassen. Nach dem Verdünnen mit Wasser wird mit Äther ausgeschüttelt; die ätherische Lösung wird mit verdünnter Natronlauge und Wasser gewaschen, getrocknet und eingedampft. Der weiße krystalline Rückstand wird in 15 ccm Eisessig gelöst und ½ Stunde mit 1 g Zinkstaub geschüttelt. Zur vollständigen Entbromung wird unter gelindem Erwärmen auf dem Wasserbad nochmals 1 g Zinkstaub der Lösung zugefügt. Nach etwa 15 Minuten wird die Reaktion unterbrochen; die abgekühlte Lösung wird filtriert, mit Wasser verdünnt und mit Äther ausgeschüttelt. Man erhält 259 mg an ätherloslichen Neutralanteilen, die bei 150⁰ im Hochvakuum (0,001 mm) sublimiert und anschließend aus verdünntem Aceton umkrystallisiert werden. Das Androstendion scheidet sich in farblosen, mehrere Zentimeter langen Spießen vom Schmelzp. 169⁰ (unkorr.) ab. $[\alpha]_D^{18}$ = +185⁰. Ausbeute: 195 mg (68% d. Th.).

Figure 81.6. From A. Butenandt and H. Kudszus, in Hoppe-Seyler's *Z. Physiol. Chem.* **237**, 75 (1935). Reproduced with permission.

Preparation of androstenedione from androstenolone

Cholestenon aus Cholesterol[6]).   10 g reinstes, über das Dibromid gereinigtes Cholesterol[17]) werden in 120 g heissem Aceton gelöst, eine Lösung von 12 g kryst. tert. Al-Butylat in 300 cm³ Benzol zugefügt und 10 Stunden unter Rückfluss erwärmt (Siedeverzug!). Die gelbstichige Reaktionslösung wird hierauf gründlich mit verd. Schwefelsäure ausgeschüttelt, die nunmehr farblose benzolische Schicht mit Wasser gewaschen, mit Natriumsulfat getrocknet, das Benzol durch Eindampfen im Vakuum entfernt und der Rückstand noch ½ Stunde auf dem Wasserbad evakuiert. Es hinterbleibt ein schwach gelbstichiger Syrup, der nach dem Erkalten innerhalb weniger Minuten vollständig krystallisiert (10.6 g). Durch Umkrystallisieren aus Methanol und Äther können 8.9 g Cholestenon vom Schmp. 79—80° erhalten werden. Mischschmelzpunkt: 79—80°.

Androstendion aus Dehydroandrosteron[10]).   930 mg Dehydroandrosteron, in 20 g Aceton gelöst, werden mit einer Lösung von 2.5 g tert. Al-Butylat in 60 cm³ Benzol 14 Stunden unter Rückfluss erwärmt. Die Aufarbeitung (wie bei Cholestenon) ergibt einen gelblichen Rückstand von rohem Androstendion. Aus Äther umkrystallisiert, derbe Nadeln vom Schmp. 170.5—173.5°. $[\alpha]_D$ = +182° (Alkohol). Ausbeute 790 mg.

Figure 81.7. From R. V. Oppenauer, in *Rec. Trav. Chim.* **56**, 137 (1937). Reproduced with permission.

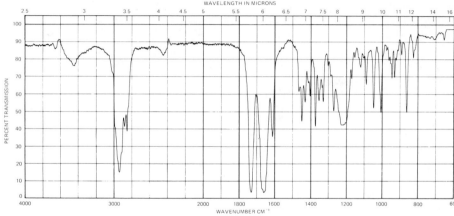

androstenolone

Figure 81.8. Infrared spectrum of androstenedione; CHCl₃ solution.

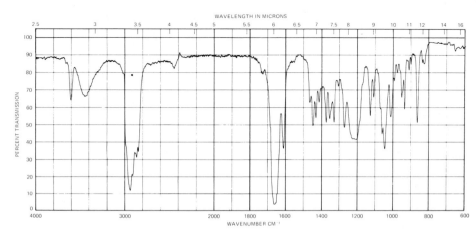

testosterone

Figure 81.9. Infrared spectrum of testosterone; CHCl₃ solution.

III: Testosterone from Androstenedione

Many methods have been used to selectively reduce the saturated 5-membered-ring ketone to the 17β alcohol without reducing the unsaturated 6-membered-ring ketone at the 3 position. A direct method makes use of sodium borohydride in methanol at 0°C. The testosterone, formed in 60–70 percent yield, is purified by chromatography. (For an infrared spectrum of this product, see Figure 81.9.) The procedure for this method is given in Figure 81.10).

Preparation of testosterone from androstenedione

Reductions with Sodium Borohydride. General Procedure.—A solution (0·4—0·6%) of the steroid in methanol (Burroughs' "A.R. quality") was treated with sodium borohydride (1·4—1·6 mole) for 1 hr. at 0°. Unless otherwise stated the mixture was worked up as follows. A few drops of acetic acid were added and the solution brought to dryness *in vacuo*. The residue was extracted with several portions of hot benzene or ethyl acetate. The crude product in benzene was absorbed on a column of alumina (100 parts) and subjected to gradient elution with benzene (50 c.c.)/benzene-ethyl acetate (1:1; 150 c.c.); 7-c.c. fractions were collected; finally the column was eluted with ethyl acetate (50—100 c.c.). In the following detailed descriptions the isolated products are recorded, under (i), (ii), &c., in the order of their elution from alumina.

Reduction of Androst-4-ene-3: 17-dione (IVa).—By the general procedure this dione gave: (i) Testosterone (Va) (170 mg.; m. p. 147—154°). A sample crystallised from acetone–hexane had m. p. and mixed m. p. 154—155°, [α]_D +110° (*c*, 0·7 in EtOH), λ_max. 240·5 mμ (ε 16,800) (Found: C, 79·3; H, 10·0. Calc. for C₁₉H₂₈O₂: C, 79·1; H, 9·8%); the infrared spectrum was identical with that of authentic material. (ii) and (iii) A mixture of diols (35 mg.) which was combined with similar material from replicated preparations: the aggregate (116 mg.) was separated on alumina into two main fractions. The fraction (50 mg.) more readily eluted crystallised from acetone-ethyl acetate in needles, m. p. 162—163°, $[\alpha]_D$ +48° [c, 1·14 in CHCl$_3$–EtOH (1:1)], no selective absorption above 210 mμ (Found: C, 78·8; H, 10.5. Calc. for $C_{19}H_{30}O_2$: C, 78·55; H, 10·4%), considered to be androst-4-ene-3β: 17β-diol (VI). Butenandt and Heusner (*loc. cit.*) recorded m. p. 153—154°, $[\alpha]_D$ +48·5° (in EtOH). Treatment with manganese dioxide in chloroform (cf. Sondheimer *et al.*, *loc. cit.*) afforded testosterone (Va) 90% by spectroscopic evidence), identified by m. p., mixed m. p., $[\alpha]_D$ +108° (c, 0·93), $\lambda_{max.}$ 240 mμ (ϵ16,500). The fraction (56 mg.) less readily eluted from alumina crystallised from acetone–hexane in prisms, m. p. 152—154°, $[\alpha]_D$ +30·5° [c, 0·83 in CHCl$_3$–EtOH (1:1)], no selective absorption above 20 μm [Found: C, 75·95; H, 11·0. $C_{19}H_{30}O_2,C_{19}H_{32}O_2,H_2O$ (?) requires C, 75·95; H, 10·7%]. Treatment of this material with manganese dioxide in chloroform gave testosterone (Va) (50% by spectroscopic evidence). Chromatography over alumina gave, in order of elution, (*a*) testosterone, m. p. and mixed m. p. 152—154°,

$[\alpha]_D$ +108° (c, 0·88 in EtOH), $\lambda_{max.}$ 240·5 mμ (ϵ 16,100), and (*b*) androstane-3β: 17β-diol (VII), m. p. 166—167 undepressed on admixture with authentic material (kindly provided by Dr. W. Klyne from the M.R.C. Steroid Reference Collection), $[\alpha]_D$ +8° (c, 0·36 in EtOH); the infrared spectrum was identical with that of authentic material. (iv) Sparingly soluble material (10 mg.) which from methanol afforded fine needles, m. p. 217—233°, $[\alpha]_D$ +70° (c, 0·3 in dioxan). Further characterisation of this material was not attempted. Repeated reductions of the dione (IVa) with sodium borohydride regularly furnished 60—70% of pure testosterone, 15—25% of the diol fraction, and 5—10% of the sparingly soluble material.

Reduction of the dione (IVa) (200 mg.) with potassium borohydride (27 mg., 0·7 mol.) at 22° under otherwise unchanged conditions afforded testosterone (135 mg.), identified by m. p., mixed m. p., and $[\alpha]_D$ +110° (c, 0·95 in EtOH).

Reduction of the dione (IVa) (200 mg.) with sodium trimethoxyborohydride (360 mg., 4·0 mol.) under otherwise unchanged conditions gave: (i) Androst-4-ene-3: 17-dione (45 mg.), identified by m. p., mixed m. p., and $[\alpha]_D$ +191° (c, 0·61 in EtOH). (ii) Testosterone (130 mg.), identified by m. p., mixed m. p., and $[\alpha]_D$ +109° (c, 0·75 in EtOH).

Figure 81.10. From J. K. Norymberski and G. F. Woods in *J. Chem. Soc.* **1955**, 3426. Reproduced with permission.

Preparation of testosterone from androstenedione

General Procedure for the Preparation of C$_3$-(N-Pyrrolidyl) Enamines.—The steroidal C$_3$-ketone was dissolved or suspended in 50 ml. of thiophene-free benzene per 0.01 mole of steroid and 4 mole equivalents of pyrrolidine was then added. Stirring was accomplished by means of a magnetic stirrer and the mixture was heated at reflux on a Glas-col mantle. The course of the reaction was followed by noting the amount of water collected in a Bidwell–Sterling moisture trap placed between the reaction flask and the condenser. The reaction was allowed to proceed until one mole equivalent of water was collected or until the amount was constant, at which point the mixture was concentrated to dryness *in vacuo*, using ordinary precautions to preclude moisture. The product was then triturated with methanol or acetone, cooled and recovered by filtration. This sufficed in most instances to give analytically pure material; however, for analyses the product was recrystallized from methylene chloride–methanol mixture, ethyl acetate, ether or benzene. The results obtained by this general procedure are shown in Table I.

TABLE I

C$_3$-(N-Pyrrolidyl) Enamines Prepared from Steroidal C$_3$-Ketones

Compd. No.	Parent C$_3$-ketone	Reflux time, hr.	Yield, %	M.p., °C. (dec.)	$[\alpha]_D$	λ^{ether}_{max}, mμ
I	Progesterone	1.5^a	97	170–175	− 22	281
II	4-Cholestene-3-one	4	79	138–140	− 110	280
III	Stigmastadienone^b	1	94	153–155	− 121	280
IV	Cholestan-3-one	3	94	105–110	+ 45	229
V	11-Ketoprogesterone^c	4	94	180–185	+ 34	282
VI	11α-Hydroxyprogesterone	2	86	145–152	− 126	281
VII	Methyl dehydrocholate	3	86	160–165		233
VIII	17α-Methyltestosterone	3	91	160–170	− 93	281
IX	4-Androstene-3,17-dione	2	90	200–205	− 135	281
X	Testosterone	5	86	133–137	− 116	280

Reduction of 3-(N-Pyrrolidyl)-3,5-androstadien-17-one (IX) to Form 3-(N-Pyrrolidyl)-3,5-androstadien-17β-ol (X) and Hydrolysis to Testosterone.—A solution of 3.4 g. (0.01 mole) of the ketoenamine (IX) in 70 ml. of tetrahydrofuran was added over 4 minutes to a stirred mixture of 1.9 g. of lithium aluminum hydride in 800 ml. of anhydrous ether and the reaction mixture brought to reflux. After 10 minutes

Figure 81.11. Reprinted with permission from F. W. Heyl and M. E. Herr, *J. Am. Chem. Soc.* **75**, 1918 (1953). Copyright by the American Chemical Society.

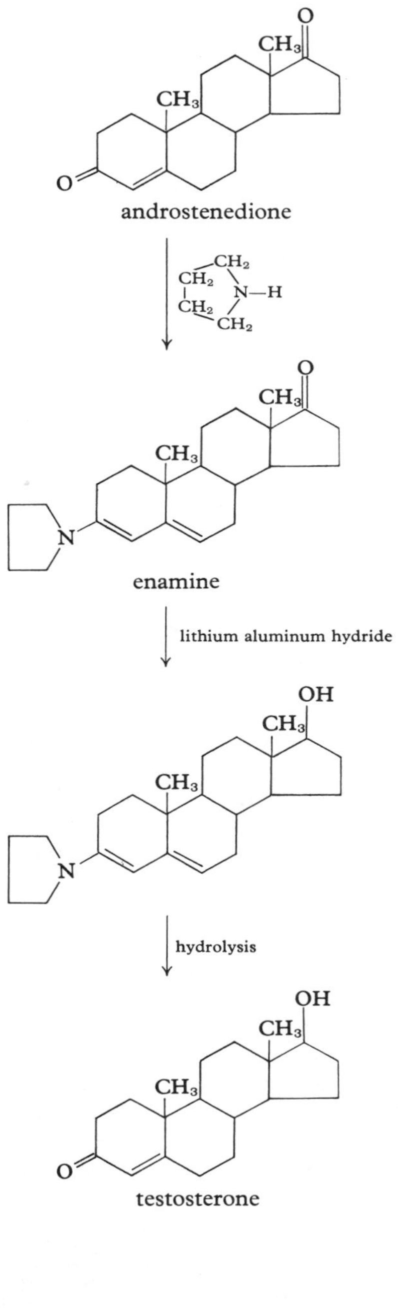

androstenedione

enamine

lithium aluminum hydride

hydrolysis

testosterone

Three indirect methods for the production of testosterone from androstenedione involve protection of the carbonyl group in the 3 position, as an enamine, as the ethylene ketal, or as an enol ether. The 17-keto group is then reduced to the 17β alcohol, and testosterone is obtained upon hydrolysis. These three methods are summarized on pages 421, 422, and 423. The first scheme, shown on p. 421, involves the intermediate formation of an enamine. The procedure is given in Figure 81.11.

The second scheme, shown below, in which the ethylene ketal is the intermediate, is given by the procedure of Figure 81.12.

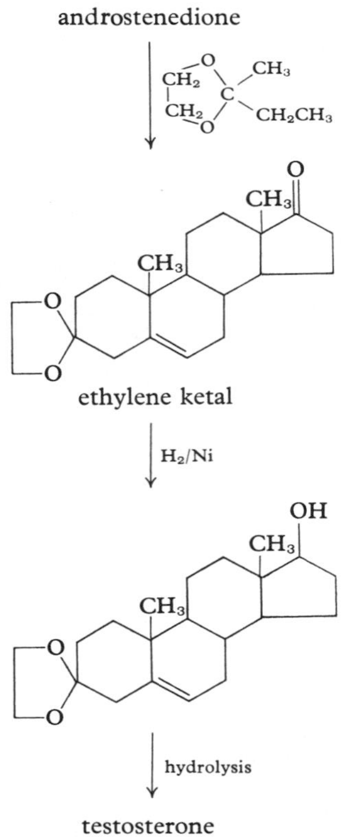

androstenedione

ethylene ketal

H₂/Ni

OH

hydrolysis

testosterone

Preparation of 1,3-Dioxolanes. 2-Methyl-2-ethyl-1,3-dioxolane.—A mixture of butanone (72.1 g., 89.6 ml., 1.00 mole; dried over Drierite and decanted), ethylene glycol (62.1 g., 55.8 ml., 1.00 mole), p-toluenesulfonic acid monohydrate (0.5 g.) and benzene (50 ml.) was heated under reflux in a modified Dean–Stark phase separator[20] until no more aqueous phase separated (*ca.* 1.1 moles, 20 ml., in about 30 hours). After neutralization by addition of excess anhydrous sodium carbonate, the unfiltered reaction mixture was subjected to fractional distillation in a packed column (38 × 1 cm. glass helices) with a total condensation, partial take-off head. After removal of the benzene, an intermediate fraction of slightly impure product (12 g., $n^{25.0}$D 1.4162) and a main fraction of 2-methyl-2-ethyl-1,3-dioxolane (90 g., 78%, b.p. 116.5–117°, $n^{24.8}$D (1.4087) were

collected.[21] Physical constants previously reported are: b.p. 118–118.5°, n^{20}D 1.4110[22]; b.p. 115.4–116.2° (763 mm.), n^{20}D 1.4096.[4b]

General Procedures for Exchange Dioxolanations. (A) By Refluxing with Undiluted 2-Methyl-2-ethyl-1,3-dioxolane[25] (suitable for saturated 3- and 20-ketones). **3-Ethylenedioxycoprostane.**—A solution of coprostanone (1.0 g.) and p-toluenesulfonic acid monohydrate (15 mg.) in 2-methyl-2-ethyl-1,3-dioxolane (20 ml.) was boiled under reflux and anhydrous conditions for 4 hours. The cooled reaction mixture was diluted with benzene, washed successively with 5% aqueous sodium bicarbonate and with water, dried over sodium sulfate and concentrated to dryness under reduced pressure. Crystallization of the residue from methanol containing a drop of pyridine gave 3-ethylenedioxycoprostanone (1.06 g., 96%, m.p. 51–52°, [α]²⁴D +27.6°).

Conversion of Δ^4-Androstene-3,17-dione into Testosterone. 3-Ethylenedioxy-Δ^5-androsten-17-one (VI).—A solution of Δ^4-androstene-3,17-dione (1.0 g.) and p-toluenesulfonic acid monohydrate (15 mg.) in pure, carefully fractionated[8,21] 2-methyl-2-ethyl-1,3-dioxolane (16 ml.) was distilled slowly through a glass helices-packed column for 5.5 hours (10 ml. of distillate collected). Work-up as in A followed by crystallization from methanol (1 drop of pyridine) gave the 3-monodioxolane product, 3-ethylenedioxy-Δ^5-androsten-17-one (840 mg., 74%, m.p. 197–198°, $[\alpha]^{24}$D + 15.4°; reported m.p. 199°[5b] and m.p. of 194° and 202°[5c]). No attempt was made to recover androstenedione by acid hydrolysis of mother liquors.[18]

3-Ethylenedioxy-Δ^5-androsten-17β-ol (VII).—3-Ethylenedioxy-Δ^5-androsten-17-one (535 mg.) in absolute ethanol (40 ml.) was added to prehydrogenated W-4 Raney nickel catalyst[29] (0.5 g.) in absolute ethanol (25 ml.) and hydrogenation at atmospheric pressure ceased in 1 hour after the absorption of one equivalent (39.8 ml. at 27°, 760 mm.; 1.00 equiv.). Concentration of the filtered solution, after addition of 2-methyl-2-ethyl-1,3-dioxolane (5 ml.), under re-duced pressure to remove the ethanol followed by refrigeration overnight yielded 3-ethylenedioxy-Δ^5-androsten-17β-ol (472 mg., 90%, m.p. 181–182°).[30] The twice recrystallized (from methanol, 1 drop of pyridine) product (m.p. 183–184°, $[\alpha]^{24}$D − 41.7°) was identical with authentic testosterone dioxolane (m.p. and m.m.p. 182–183°, $[\alpha]^{24}$D − 43.1°) prepared by exchange dioxolanation.

Testosterone (VIII).—3-Ethylenedioxy-Δ^5-androsten-17β-ol (1.0 g.) was dissolved in anhydrous acetone (50 ml.), p-toluenesulfonic acid monohydrate (50 mg.) added and the mixture boiled under reflux for 14 hours. Concentration of the resultant solution to a small volume (10 ml.) and precipitation with water gave a quantitative yield of slightly impure testosterone (0.87 g., 100%, m.p. 147–151°). Recrystallization from ether furnished the pure product (m.p. 152–154°, $[\alpha]^{24}$D +109°), identical in all respects with authentic testosterone.

(21) Complete purification may be attained alternately by redistillation from lithium aluminum hydride of material once distilled through an ordinary Claisen distillation apparatus.

Figure 81.12. Reprinted with permission from H. J. Dauben, Jr., B. Löken, and H. J. Ringold, *J. Am. Chem. Soc.* **76** 1359 (1954). Copyright by the American Chemical Society.

The third scheme, summarized below, involves an enol ether as an intermediate. The procedure is given in Figure 81.13.

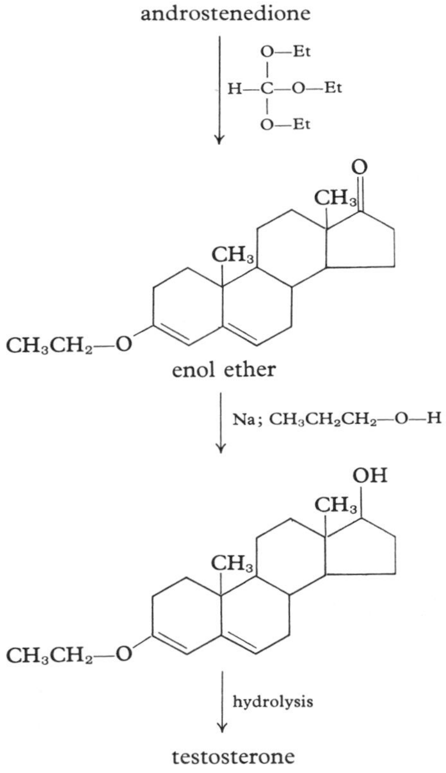

Preparation of testosterone from androstenedione
Cholestanon-diäthylacetal.

Eine Lösung von 7.7 g Cholestanon in 20 ccm Benzol wurde mit 3.4 g Orthoameisensäure-äthylester, 2.8 g absol. Alkohol und 20 Tropfen einer 8-proz. Lösung von Chlorwasserstoff in absol. Alkohol versetzt und 2 Stdn. auf 75° erhitzt. Nach dem Abkühlen wurde mit alkohol. Natronlauge alkalisch gemacht, in Wasser gegossen und mit Äther extrahiert; die ätherische Lösung wurde mit Wasser neutral gewaschen, mit Magnesiumsulfat getrocknet und verdampft. Es hinterblieb ein Öl, das beim Verreiben mit wenig Alkohol krystallisierte: 7.1 g Krystalle vom Schmp. 67—68.5° [6]). (Wir verwendeten in allen Fällan Alkohol mit geringem Pyridin-Zusatz, um die Anwesenheit von Säure auszuschließen.) Durch weitere Krystallisation aus pyridinhaltigem Alkohol wurde das Cholestanon-diäthylacetal rein erhalten. Schmp. 68—69.5°; $[\alpha]_D$: $+26°$ (Dioxan).

Androsten-dion-(3.17)-enoläthyläther-(3).

2.86 g Androsten-dion-(3.17) wurden in 10 ccm Benzol gelöst, mit 1.7 g Orthoameisensäure-äthylester, 1.4 g absol. Alkohol und 10 Tropfen etwa 8-proz. alkohol. HCl versetzt. Nach 2-stdg. Erhitzen auf 75° wurde wie in den anderen Acetalisierungsansätzen aufgearbeitet und das anfallende Krystallisat durch Umlösen aus pyridinhaltigem Alkohol gereinigt. Der Androsten-dion-(3.17)-enoläthyläther-(3) krystallisierte in Blättchen vom Schmp. 152°; $[\alpha]_D^{20}$: $—89°$ (Dioxan).

Testosteron aus
Androsten-dion-(3.17)-enoläthyläther-(3).

Eine Lösung von 2.0 g Androsten-dion-(3.17)-enoläther-(3) in 40 ccm n-Propylalkohol wurde auf 100° erhitzt. Dann wurden innerhalb von ¾ Std. 2.0 g Natrium in kleinen Portionen eingetragen.

Nach der Auflösung des Natriums wurde die Reaktionslösung mit Wasser versetzt, worauf sich der Testosteron-enoläthyläther als schön krystallisierter Körper abschied. Die Krystalle wurden mit Wasser gewaschen und getrocknet; sie schmolzen bei 118—122°.

Das Rohkrystallisat wurde in Äthanol gelöst und die Lösung nach Zusatz von etwas verd. Salzsäure 20 Min. auf dem Wasserblade erhitzt. Danach wurde mit Wasser versetzt, in Äther aufgenommen und die ätherische Lösung nach dem Waschen und Trocknen verdampft. Der krystallinische Rückstand zeigte den Schmp. 149° und die Drehung $[\alpha]_D^{20}$: $+107.5°$ ($C_2H_5 \cdot OH$).

Durch Krustallisation aus Essigester wurde Testosteron vom Schmp. 153—154° und der Drehung $[\alpha]_D^{20}$: $+109°$ ($C_2H_5 \cdot OH$) erhalten.

Figure 81.13. From A. Serini and H. Köster, in *Berichte* **71**, 1766 (1938). Reproduced with permission of Verlag Chemie, GmbH, Weinheim.

A variation of this last method involves the prior formation of a cyanohydrin at the 17 position and then conversion to the enol ether at the 3 position. Treatment with sodium in *n*-propyl alcohol gives reduction only at the 17 position, since the cyanohydrin is reversibly formed in basic media whereas the enol ether can be hydrolyzed only in acidic media. This scheme is summarized on p. 425, and the procedure for this variation is presented in Figure 81.14.

Preparation of testosterone from androstenedione

17-Cyanohydrin of Δ⁴-Androstene-3,17-dione (Ib).

(b) From Δ⁴-Androstene-3,17-dione (Ia).—Androstenedione (20 g.) was dissolved by gentle warming and swirling in 30 cc. of freshly prepared crude acetone cyanohydrin.[8] Crystallization began in a few minutes and was complete within 2 hr. at room temperature. The petroleum ether-washed crystals (19.90 g.) melted at 178° dec. The water-diluted mother liquor furnished another 1.20 g., m.p., 176–178° dec. The total yield was 21.1 g. (96%). The ultraviolet absorption spectrum was identical with that of the product obtained in (a). Also the analytical figures (Found: C, 76.57; H, 8.67; N, 4.42) agreed with the theoretical values.

3-Enol Ethyl Ether of Δ⁴-Androstene-3,17-dione-17-cyanohydrin (IIa).—From a suspension of 1.56 g. of the above androstenedione cyanohydrin (Ib) in 20 cc. of benzene, 10 cc. of the solvent was distilled. After the temperature was lowered to 65°, 1.76 g. of ethyl orthoformate, 0.9 cc. of absolute ethanol and 0.08 cc. of an ethanolic solution of hydrochloric acid (5.6 mg. of HCl) were added. After 15 min. all the product went into solution and then crystals of the enol ether began to separate. After a further 45 min. the mixture was cooled, and the crystals were filtered and then dried *in vacuo*; 880 mg., m.p. 207° dec. After addition of 0.014 cc. of pyridine the mother liquor was evaporated *in vacuo*. The crystalline solid, rapidly washed with a small quantity of ethanol containing a trace of pyridine, weighed 650 mg., m.p. 203–205° dec., total yield 1.53 g. (90%).

Anal. Calcd. for $C_{23}C_{31}O_2N$: C, 77.37; H, 9.15; N, 4.10. Found: C, 77.42; H, 9.11; N, 4.13.

The above product (100 mg.), dissolved in 2 cc. of ethanol containing two drops of pyridine, was warmed on the steam-bath for 20 minutes. Dilution with hot water afforded 80 mg. of the 3-enol ethyl ether of androstenedione (IIb), m.p. 152°, undepressed in mixture with an authentic specimen.

3-Enol Ethyl Ether of Testosterone (IIc).—To a boiling solution of 500 mg. of the above 3-enol ethyl ether of Δ⁴-androstene-3,17–dione-17-cyanohydrin (IIa) in 13 cc. of absolute *n*-propanol in which a small quantity of sodium had been previously dissolved, 500 mg. of sodium was added portionwise during half an hour. Dilution with hot water induced crystallization. The product obtained after cooling, filtering and washing with water weighed 440 mg., m.p. 119–121°, and did not depress the melting point of the 3-enol ethyl ether of testosterone.

A hot solution of the product (400 mg.) in 5 cc. of ethanol was diluted with warm 1 N hydrochloric acid solution. After cooling, 370 mg. of testosterone was obtained, m.p. (mixed) 151–153°.

Testosterone from Δ⁴-Androstene-3,17–dione-17-cyanohydrin (Ib) without Isolation of Intermediates.—The reaction of 1.56 g. of androstenedione cyanohydrin with ethyl orthoformate was carried out exactly as described above for the preparation of IIa. Instead of filtering the crystallized product, the whole mixture, after addition of 0.014 cc. of pyridine, was evaporated *in vacuo*. The reduction of the crystalline solid was performed as described above, with 40 cc. of *n*-propanol and 1.5 g. of sodium added during 40 minutes. After diluting with hot water, acidifying the mixture to congo red with 2 N hydrochloric acid, removing the main part of the propanol *in vacuo* and cooling, the crystalline product which separated was filtered, washed with water and dried; yield 1.31 g. (92%), m.p. 149–151°, undepressed on admixture with an authentic specimen of testosterone.

Figure 81.14. Reprinted with permission from A. Ercoli and P. de Ruggieri, *J. Am. Chem. Soc.* **75**, 650 (1953). Copyright by the American Chemical Society.

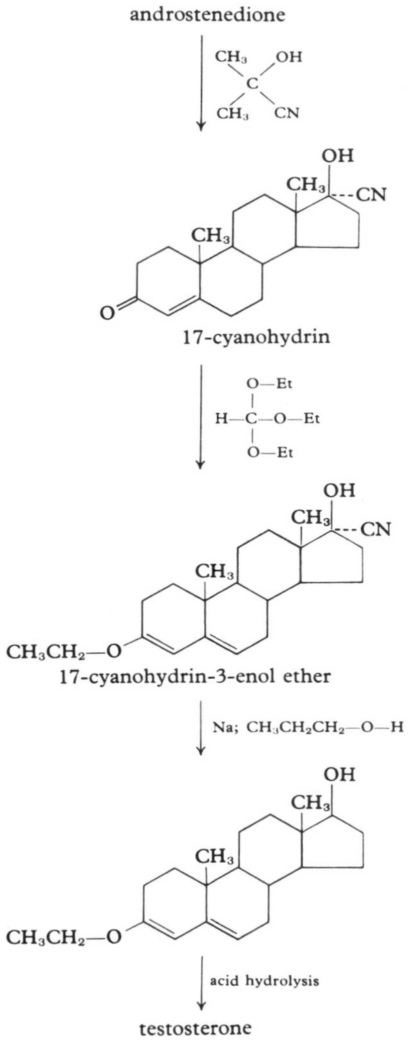

81.2 PROGESTERONE FROM PREGNENOLONE ACETATE VIA PREGNENOLONE

The preparation of progesterone from pregnenolone acetate has been divided into two parts. The first is the hydrolysis of the ester, pregnenolone acetate, to give pregnenolone. The second is the conversion of a β,γ-unsaturated alcohol to the corresponding α,β-unsaturated ketone. This latter transformation is analogous to the formation of androstenedione from androstenolone (see Figures 81.5, 81.6, and 81.7) and the conversion of cholesterol to Δ^4-cholestenone (see Section 73).

I: Pregnenolone from Pregnenolone Acetate

Butenandt, Westphal, and Cobler state that pregnenolone acetate can be hydrolyzed to pregnenolone on boiling for an hour in 3 *M* KOH in methanol [*Berichte* **67,** 1611 (1934)].

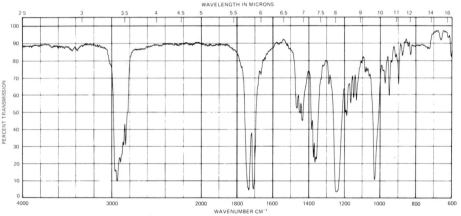

pregnenolone acetate → (3 *M* KOH in methanol; boil for 1 hour) → pregnenolone

The infrared spectra of the reactant pregnenolone acetate and the product pregnenolone are shown in Figures 81.15 and 81.16.

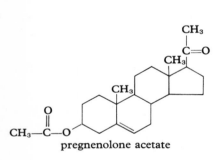

pregnenolone acetate

Figure 81.15. Infrared spectrum of pregnenolone acetate; CCl_4 solution.

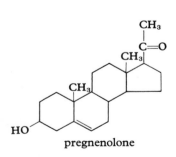

pregnenolone

Figure 81.16. Infrared spectrum of pregnenolone; $CHCl_3$ solution.

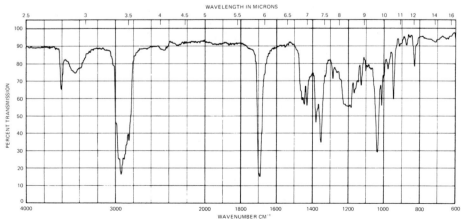

The transformation of pregnenolone to progesterone (see Figure 81.17 for the infrared spectrum of this product) can be effected either in several steps by

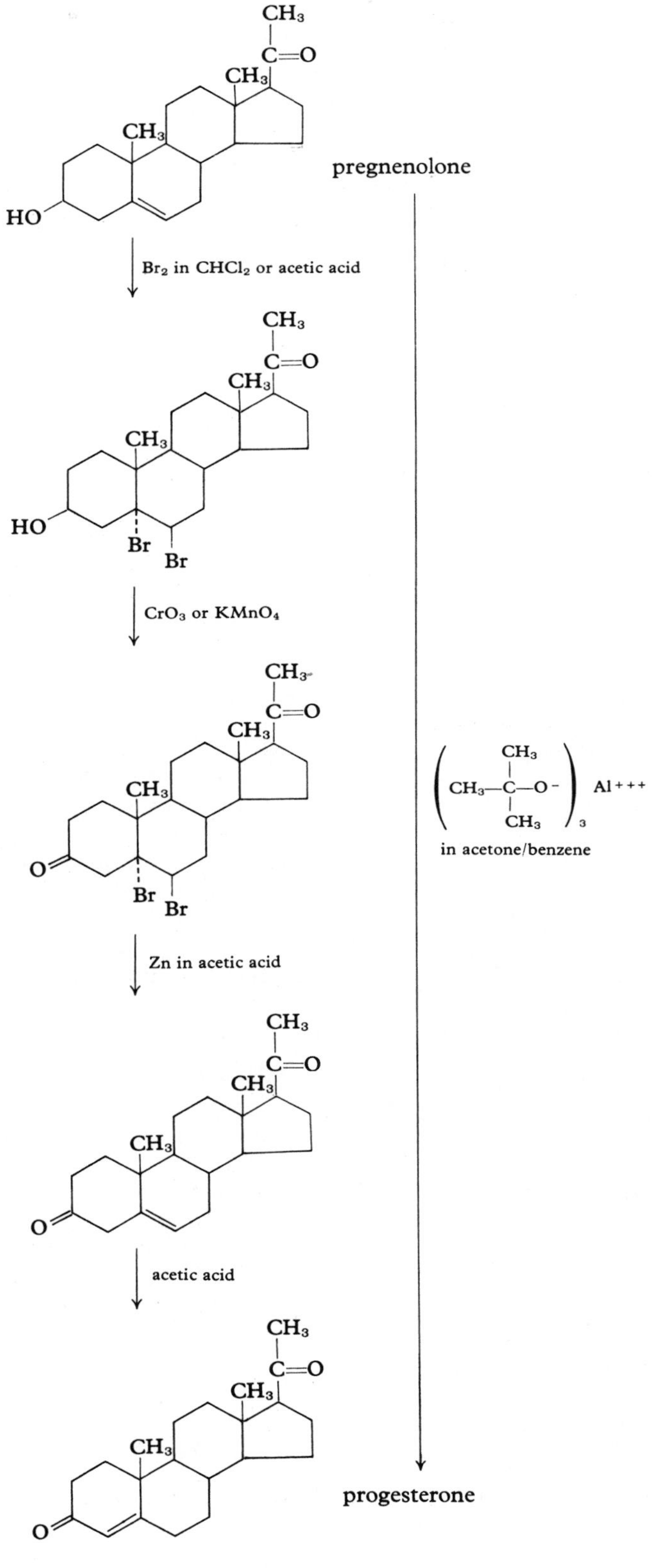

pregnenolone

Br₂ in CHCl₂ or acetic acid

CrO₃ or KMnO₄

$$\left(CH_3-\overset{\overset{\displaystyle CH_3}{|}}{\underset{\underset{\displaystyle CH_3}{|}}{C}}-O^- \right)_3 Al^{+++}$$

in acetone/benzene

Zn in acetic acid

acetic acid

progesterone

addition of bromine, oxidation, and debromination with isomerization, or in one step by an Oppenauer oxidation. The isomerization of the less stable β,γ-unsaturated ketone to the more stable α,β-unsaturated ketone takes place without isolation of the less stable isomer.

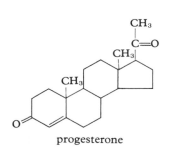

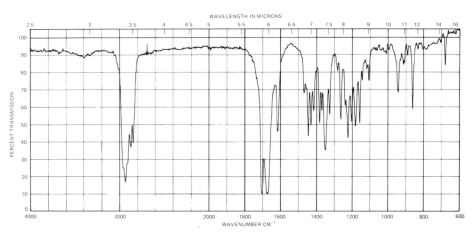

Figure 81.17. Infrared spectrum of progesterone; CCl_4 solution.

The alternative procedures for these reactions are shown in Figures 81.18, 81.19, and 81.20.

Preparation of progesterone from pregnenolone

Oxydation als Dibromid: 1.34 g des Oxyketons I wurden in 50 ccm gelöst und 0.68 g Brom in 30 ccm Chloroform aus einer Bürette in dünnem Strahle zulaufen gelassen. Das Lösungsmittel wurde dann bei 30° im Vak. abgedampft und das zurückbleibende Harz in 40 ccm Benzol gelöst. Diese Lösung wurde mit 30 ccm 5-proz. Kaliumpermanganat und 10 ccm 20-proz. Schwefelsäure 8 Stdn. geschüttelt. Der entstandene Braunstein wurde durch vorsichtigen Zusatz von schwefliger Säure gelöst und die Benzol-Schicht abgetrennt. Der wäßrige Teil wurde noch einmal mit Äther ausgeschüttelt. Äther- und Benzol-Lösung wurden vereinigt und nach Zusatz von 1 g Zinkstaub und 20 ccm Eisessig am absteigenden Kühler erhitzt, bis kein Destillat mehr überging. Zur Vervollständigung der Entbromung wurde der Rückstand nach Zugabe von 1 ccm Wasser noch ½ Stde. unter Rückfluß gekocht. Das organische Material wurde in Äther aufgenommen und durch Schütteln mit verd. Natronlauge von der geringen Menge saurer Oxydationsprodukte befreit. Der Äther hinterließ beim Verdampfen ein mit Krystallen durchsetztes Harz.

Dies wurde 10 Min. mit 10 ccm Essigsäure-anhydrid gekocht, um zu verhindern, daß gleichzeitig mit dem Diketon auch unverändertes Ausgangsmaterial auskrystallisierte. Nach Zerstörung des Anhydrids mit Wasser wurde das Harz in wenig Methanol gelöst, woraus sich bald Krystalle abzuscheiden begannen: 0.6 g vom Schmp. 115° (unscharf). Es wurde bis zur Konstanz von Schmp. und Drehung aus Methanol umkrystallisiert. Charakteristische kurze Prismen vom Schmp. 129°; leicht löslich in den meisten Lösungsmitteln, weniger leicht in Petroläther.

1.55 mg Sbst., in 2 ccm Chloroform: $\alpha_D = +1.55°$, $[\alpha]_D^{20} = +200°$.

Überführung in die Modifikation vom Schmp. 121°: 50 mg des Diketons vom Schmp. 129° wurden in wenig Alkohol gelöst und die bei Zusatz etwa des halben Volumens Wasser eingetretene Fällung mit Äther wieder in Lösung gebracht. Durch Übersaugen von Luft wurde die Lösung dann eingeengt, wobei sich dünne Nadeln abschieden. Schmp. 121°.

6.6 mg Sbst., in 2 ccm Chloroform: $\alpha_D = +0.66°$, $[\alpha]_D^{20} = +200°$.

Figure 81.18. From E. Fernholz, in *Berichte* **67**, 2027 (1934). Reproduced with permission of Verlag Chemie, GmbH, Weinheim.

Preparation of progesterone from pregnenolone

Bromierung des Δ^5-Pregnenolons-(20) (I).

70 mg des Oxy-ketons (I) wurden in 10 ccm Eisessig gelöst und tropfenweise mit einer Lösung von 35.3 mg Brom in 3 ccm Eisessig versetzt. Es trat schnelle Entfärbung des Broms ein. Die Lösung wurde direkt der Oxydation unterworfen.

Oxydation des Dibromides (II): Zu der Bromierungs-Lüsung wurde in der Kälte tropfenweise eine 5 Atomen Sauerstoff entsprechende Menge Chromtrioxyd in 13 ccm Eisessig hinzugegeben. Nach 20-stdg. Aufbewahren bei 20° wurde die Reaktions-Lösung in Wasser gegossen und mit Äther ausgeschüttelt. Nach dem Abdampfen der ätherischen Lüsung hinterblieb ein farbloses Öl, das mit 2 ccm Eisessig und etwa 1 g Zinkstaub 10′ auf dem Wasserbade erwärmt wurde; das Reaktionsprodukt wurde mit Wasser gefällt, in Äther aufgenommen und im Hochvakuum (130°, 10^{-3} mm) sublimiert. Das Rohkrystallisat zeigte einen Schmp. von 90—107°, es wurde aus verd. Alkohol bis zum konstanten Schmp. 128.5° umgelöst. Das Krystallisat zeigte alle Eigenschaften des natürlichen bzw. des aus Pregnandiol bereiteten Corpus-luteum-Hormons, mit dem es bei der Mischprobe keine Depression gab. $[\alpha]_{20}^{D} = +193.5°$,

Figure 81.19. From A. Butenandt and U. Westphal, in *Berichte* **67**, 2085 (1934). Reproduced with permission of Verlag Chemie, GmbH, Weinheim.

Preparation of progesterone from pregnenolone

Cholestenon aus Cholesterol[6]). 10 g reinstes, über das Dibromid gereinigtes Cholesterol[17]) werden in 120 g heissem Aceton gelöst, eine Lösung von 12 g kryst. tert. Al-Butylat in 300 cm³ Benzol zugefügt und 10 Stunden unter Rückfluss erwärmt (Siedeverzug!). Die gelbstichige Reaktionslösung wird hierauf gründlich mit verd. Schwefelsäure ausgeschüttelt, die nunmehr farblose benzolische Schicht mit Wasser gewaschen, mit Natriumsulfat getrocknet, das Benzol durch Eindampfen im Vakuum entfernt und der Rückstand noch ½ Stunde auf dem Wasserbad evakuiert. Es hinterbleibt ein schwach gelbstichiger Syrup, der nach dem Erkalten innerhalb weniger Minuten vollständig krystallisiert (10.6 g). Durch Umkrystallisieren aus Methanol und Äther können 8.9 g Cholestenon vom Schmp. 79—80° erhalten werden. Mischschmelzpunkt: 79—80°.

Progesteron aus Pregnenolon[11]). 212 mg Pregnenolon, 750 mg Al-Butylat (tert.) in 4 g Aceton und 10 cm³ Benzol werden 11 Stunden unter Rückfluss erwärmt. Die Aufarbeitung (wie bei Cholestenon) liefert 231 mg schwach gelb gefärbten Syrup, der in der Kälte bald erstarrt. Das Rohprodukt wird aus Äther umkrystallisiert. Dicke, stark lichtbrechende Prismen vom Schmp. 129.5—130.5°. $[\alpha]_D = +192°$ (Alkohol). Ausbeute 160 mg.

Figure 81.20. From R. V. Oppenauer, in *R. V. Trav. Chim.* **56**, 137 (1937). Reproduced with permission.

References

1. P. A. Lehmann F., A. Bolivar G., and R. Quintero R., *J. Chem. Educ.* **50**, 195 (1973).
2. L. F. Fieser and M. Fieser, *Steroids*, Reinhold, New York. 1959.

Chemical Substance Index

432

General Subject Index